MONSOONS OVER CHINA

ATMOSPHERIC SCIENCES LIBRARY

VOLUME 16

The titles published in this series are listed at the end of this volume.

Monsoons over China

by

DING YIHUI
Chinese Academy of Meteorological Sciences,
Beijing, China

KLUWER ACADEMIC PUBLISHERS
DORDRECHT / BOSTON / LONDON

Library of Congress Cataloging-in-Publication Data

Ding Yihui.
Monsoons over China / Y.H. Ding.
p. cm. -- (Atmospheric sciences library ; 16)
"December 1991."
ISBN 0-7923-1757-2
1. Monsoon--China. 2. Monsoons--East Asia. 3. Monsoons--China--Tibet, Plateau of. I. Title. II. Series.
QC939.M7T56 1991
551.5'184'0951--dc20 92-12017

ISBN 0-7923-1757-2

Published by Kluwer Academic Publishers,
P.O. Box 17, 3300 AA Dordrecht, The Netherlands.

Kluwer Academic Publishers incorporates
the publishing programmes of
D. Reidel, Martinus Nijhoff, Dr W. Junk and MTP Press.

Sold and distributed in the U.S.A. and Canada
by Kluwer Academic Publishers,
101 Philip Drive, Norwell, MA 02061, U.S.A.

In all other countries, sold and distributed
by Kluwer Academic Publishers Group,
P.O. Box 322, 3300 AH Dordrecht, The Netherlands.

Printed on acid-free paper

Printed in the Netherlands

TABLE OF CONTENTS

PREFACE

The monsoon over China is one of the major components of the general circulation on a global basis. Its activity bears a significant regional implication in East Asia and Southeast Asia. Recently, the remarkable relationship and teleconnection between this part of the monsoon and other regions over the world have been revealed. However, little of the overall picture of monsoons over China is known by meteorologists in English-speaking countries. This monograph provides the first opportunity to extensively introduce this subject and give a comprehensive and systematic description of the major aspects of monsoons over China, with a special emphasis on the fluctuations of the monsoon on various scales and the effects of the Tibetan Plateau on the monsoon. Much highly original material and achievements Chinese and Western meteorologists have made from the past 20 years have been incorporated with a unifying approach. In each chapter, the observational and theoretical (including modelling) treatment will be closely combined in order to fully illustrate the relevent problems. The unique thermal and dynamical effects of the Tibetan Plateau on the monsoon circulation features which are one of the central problems of the Asian monsoon are highlights of this monograph.

Researchers in meteorology and weather forecasters should find this book a very useful introduction to monsoons over China, not only for its systematic treatment of the subject, but also because of its considerable historical information. This monograph is equally suitable for graduates or more advanced students in meteorology, hydrology, and oceanography.

I would like to earnestly express my gratitude to Professors T. N. Krishnamurti, the State Florida University, Yeh Tucheng and Tao Shiyan, Institute of Atmospheric Physics, Academia Sinica, for their continued encouragement and valuable comments. Thanks also go to Mrs Liu Pin who typed the manuscript with great patience and enthusiasm, and Mrs Zhang Wanpei for carefully drawing all the illustrations. Without their help, I could not have completed this monograph.

DING YIHUI

Professor
Chinese Academy of Meteorological Science

INTRODUCTION

East Asia is situated in the eastern part of the Eurasian continent where the huge Qinghai-Xizang Plateau (the Tibetan Plateau) exists. Facing east and southeast, it adjoins the extensive North Pacific Ocean. This land-sea contrast and large-scale topography produces the salient atmospheric general circulation in East Asia and the noted monsoon climate which, in turn, exerts an important effect on the global circulation and the world climate.

Ramage (1971) defined the monsoon region using the following four criteria:

(1) the prevailing wind direction shifts by at least 120 degrees between January and July,
(2) the average frequency of prevailing directions in January and July exceeds 40%,
(3) the mean resultant winds in at least one of the months exceeds 3 m s^{-1}, and
(4) less than one cyclone-anticyclone alteration occurs on an average every two years in any one month in a 5 latitude-longitude rectangle.

According to this monsoon definition, the monsoon region mainly encompasses the regions between 35° N and 25° S and between 30° W and 170° E, but recent studies have indicated that its southern boundary may extend further southward down to most of the Australian area (Davidson, McBride and McAvaney, 1983). Among monsoon regions, however, the most important are the South and the East Asian monsoon regions. China is a major part of the East Asian monsoon region.

Due to China being greatly affected by monsoon activity, Chinese meteorologists have always paid great attention to the study of the East Asian monsoon. More than a half century ago, Prof. Zhu Kezhen (Zhu, 1934), the founder of modern meteorology in China, published his classical paper entitled "Monsoons in Southeast Asia and rainfall amount in China" in the first issue of the Journal of Geography (Acta Geographyca Sinica). The paper includes the following seven aspects:

(1) records concerning monsoons in ancient Chinese books;
(2) cause of monsoons;
(3) Indian monsoons;
(4) similarity and difference between the Indian monsoon and the monsoon over China;
(5) primary driving force of the precipitation in eastern China;
(6) an inquiry into whether or not the poem "Bo-Zhao wind" by Su Tongpo, a noted poem during the Soong Dynasty, is a true perspective description of monsoons;
(7) the reason why the Bo-Zhao wind can cause droughts.

During the last half century, great progress in the study of the above-mentioned aspects has been made (Zhu, Ding and Luo, 1990). Tu and Huang (1944) studied the advance and retreat of the summer monsoon in China. Later, a number of studies were undertaken of the monsoon climate in China and the circulation changes associated with the onset and withdrawal of the summer monsoon (Tao and Chen, 1957; Yeh *et al.*, 1959; Gao and Xu, 1962). Recent studies have been devoted to the following aspects:

(1) heat sources and sinks over Asia and their thermal effect on the monsoon circulation,
(2) thermal and dynamical effects of the Tibetan Plateau on the development of the Asian monsoon,
(3) intraseasonal low-frequency oscillations in the East Asian region,
(4) interannual variability of the monsoon and its relationship with droughts and floods in China,
(5) relationship between ENSO events and monsoon activity and associated teleconnection,
(6) summer monsoon rainfalls, including the Meiyu rainfall,
(7) winter monsoon and the cold surge,
(8) role of the monsoon in the seasonal change in the general circulation in East Asia.

The major achievements concerning the above-described problems will be incorporated in this book.

Based on studies by Chinese meteorologists over many years, it has been found that many differences exist between the monsoon circulation over India and that over East Asia. Sometimes, the monsoon activity over East Asia is opposite to that over India. This fact suggests that the structure and main components of the monsoon system over East Asia is likely to be independent of the Indian monsoon system, even though there exist some significant interactions. In other words, the huge Asian monsoon system can be divided into two subsystems, the South Asian (or Indian) and the East Asian monsoon systems, which are independent of each other and, at the same time, interact with each other (Chen and Jin, 1984). For example, during the episode of the intense ITCZ, when the summer monsoon over the South China Sea is active, the Indian summer monsoon often assumes a break period, and vice-versa. The year 1979 is such a case. Liang *et al.* (1981) indicated that in the transitional season from winter to spring, the summer monsoon over the South China Sea originates upstream in the Indian summer monsoon, whereas in mid-summer it has its origin in the cross-equatorial flow at about 105° E. Furthermore, there are some considerable differences between the summer monsoon over the South China Sea and the summer monsoon over the China Main. Thus, the summer monsoon over the South China Sea cannot be just thought of as the eastward extension of the Indian monsoon, on the one hand, and, on the other hand, the summer monsoon over the mainland of China cannot fully be taken to be the northward extension of the Indian monsoon. One must take into account their own regional characters. But Zhu *et al.* (1986) emphasized the interaction among them. They pointed out that this interaction may be accomplished through energy exchange, the propagation of low-frequency oscillation, and moisture transport. Within this context, the major problem on the East Asian monsoon will be discussed.

References

Chen Longxun, Jin Zuhui, 1984: 'The medium-range variations of the summer monsoon circulation system over East Asia'. *Adv. Atmos. Sci.*, **1**, 224–233.

Davidson, N. E., McBride, J. L., McAvaney, B. J., 1983: 'The onset of the Australian monsoon during Winter MONEX: Synoptic aspects'. *Mon. Wea. Rev.*, **111**, 496–516.

Gao Youxi, Xu Suyin, 1962: 'Some problems on monsoons over East Asia'. In *Collected Papers of the Institute of Geophysics and Meteorology*, Academia Sinica, No. 5, Science Press, Beijing. pp. 1–106.

Liang Biqi, Liang Mengxuan, Xu Xiaoying, 1981: 'The cross-equatorial current in the lower troposphere and the summer monsoon in the South China Sea'. *In Proc. of Symposium on the Summer Monsoon in South and East Asia*, Yunnan People's Press, pp. 39–48.

Ramage, C. S., 1971: *Monsoon Meteorology*. Academic Press, New York.

Tao Shiyan, Chen Longxun, 1957: 'The structure of the summer circulation over Asia'. *Acta Meteorol. Sinica*, **28**, 237–247.

Tu, Z. W., Hung, S. S., 1944: 'The advance and retreat of the summer monsoon'. *Meteorol. Magazine*, **18**, 1–20.

Zhu Baozhen, Ding Yihui, Luo Huibang, 1990: 'A review of the atmospheric general corculation and monsoon in East Asia'. *Acta Meteorol. Sinica*, **48**, 4–16.

Zhu Kezhen, 1934: 'Monsoons in Southeast Asia and rainfall amount in China'. *Acta Geographica Sinica*, **1**, 1–27.

Zhu Qiangen, He Jinhai, Wang Panxing, 1986: 'A study of circulation differences between East Asian and Indian summer monsoons with their interaction'. *Adv. Atmos. Sci.*, **3**, 466–477.

Yeh Tucheng, Tao Shiyan, Li Maicun, 1959: 'The abrupt change of circulation over the Northern Hemisphere during June and October', in Bolin, B. ed., *The Atmosphere and the Sea in Motion*, Rockefeller Inst. Press., New York, pp. 249–267.

CHAPTER 1

THE SUMMER MONSOON IN EAST ASIA

1.1. The Mean Conditions of the Summer Monsoon in East Asia

The Asian summer monsoon constitutes a major circulation feature in the Northern Hemisphere. Its development, maintenances and decay should be viewed as a large-scale phenomenon having thermal and dynamic causes. Therefore, it is quite necessary to first examine the large-scale mean conditions under which the Asian summer monsoon evolves. Figure 1.1 is the mean summer flow pattern for 1 June-30 August of 1979–1984 (Wu, 1987), showing that the cross-equatorial airflows coming from the South Indian Ocean turn into southwest monsoonal airflow over the Arabian Sea and, further, flow eastward, down to the South China Sea where the monsoonal aircurrent deflects northward. Southerly and southwesterly winds prevail over the eastern China equatorward of 40° N. One may also note another branch of the cross-equatorial aircurrent over the Malaysian-Borneo region, which seems to be much weaker than that off the East African coast (Somali jet). Therefore, the summer monsoonal aircurrents affecting China seem to originate in two regions: the Indian

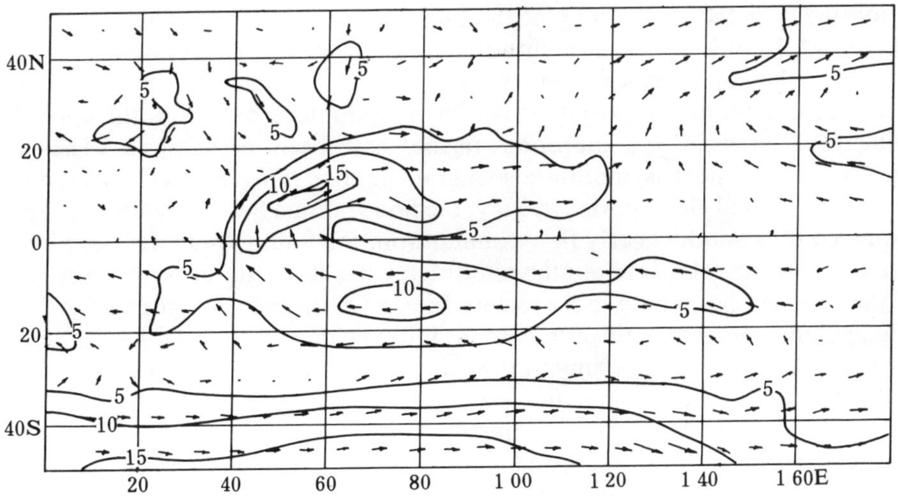

Figure 1.1. The mean wind field at 850 hPa averaged for June-August of 1979–1984. The full lines denote the isotachs. Units: m s^{-1} (Wu, 1977).

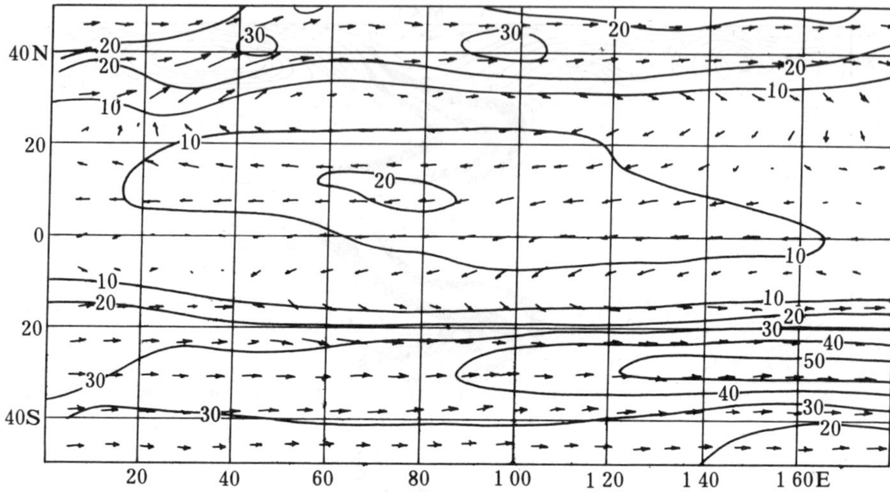

Figure 1.2. Same as Fig. 1.1, but for the summer mean 200 hPa wind field.

monsoon region and northern Australia and its neighboring sea region. As a matter of fact, the monsoonal airflow in East Asia has the third origin region, the tropical western Pacific, where the southeasterly monsoon airflow often develop at the western flank of the subtropical high. At 200 hPa (Fig. 1.2), a huge anticyclone dominates the tropical Africa-Asian region to the south of 40° N. The northeasterly flow prevails over East Asia, forming a prominent return flow lying over the low-level southwesterly and southerly monsoonal flows. To the south of the South Asian anticyclone an upper-level easterly jet exists, with a maximum wind speed of 20 m s^{-1} located over the southern tip of the Indian Peninsula.

Figures 1.3 and 1.4 show the month-to month variations of the flow patterns at 850 and 200 hPa for May-October, respectively (Krishnamurti, 1985). In these illustrations, the lower-tropospheric confluent asymptotic as well as the upper-level diffluent patterns in the monsoon region had been emphasized. The lower tropospheric confluent asymptotic also marks the climatological regions of monsoon rain and associated cloud cover. In May, the South China Sea and southern China is dominated by a southwesterly flow coming from the South Asian monsoon region. Note that one branch of the southwesterly flow has its origin in northern Australia and then crosses the equator in the eastern Indian Ocean and southern part of the Bay of Bengal. In June, the southwesterly monsoon flow prevails over most of the eastern China and Japan, corresponding to the rainy season over the Yangtze River Valley and Japan. In July, the monsoon flow has a more southerly component and may reach deep into northern China and even northeastern China. From August, the summer monsoon flow begins to weaken, and is completely replaced in October on the China Main by the winter northeasterly flow.

In the upper troposphere (Fig. 1.4), the climatological upper anticyclone makes a northward motion from northern Malaysia towards the Tibetan Plateau during July.

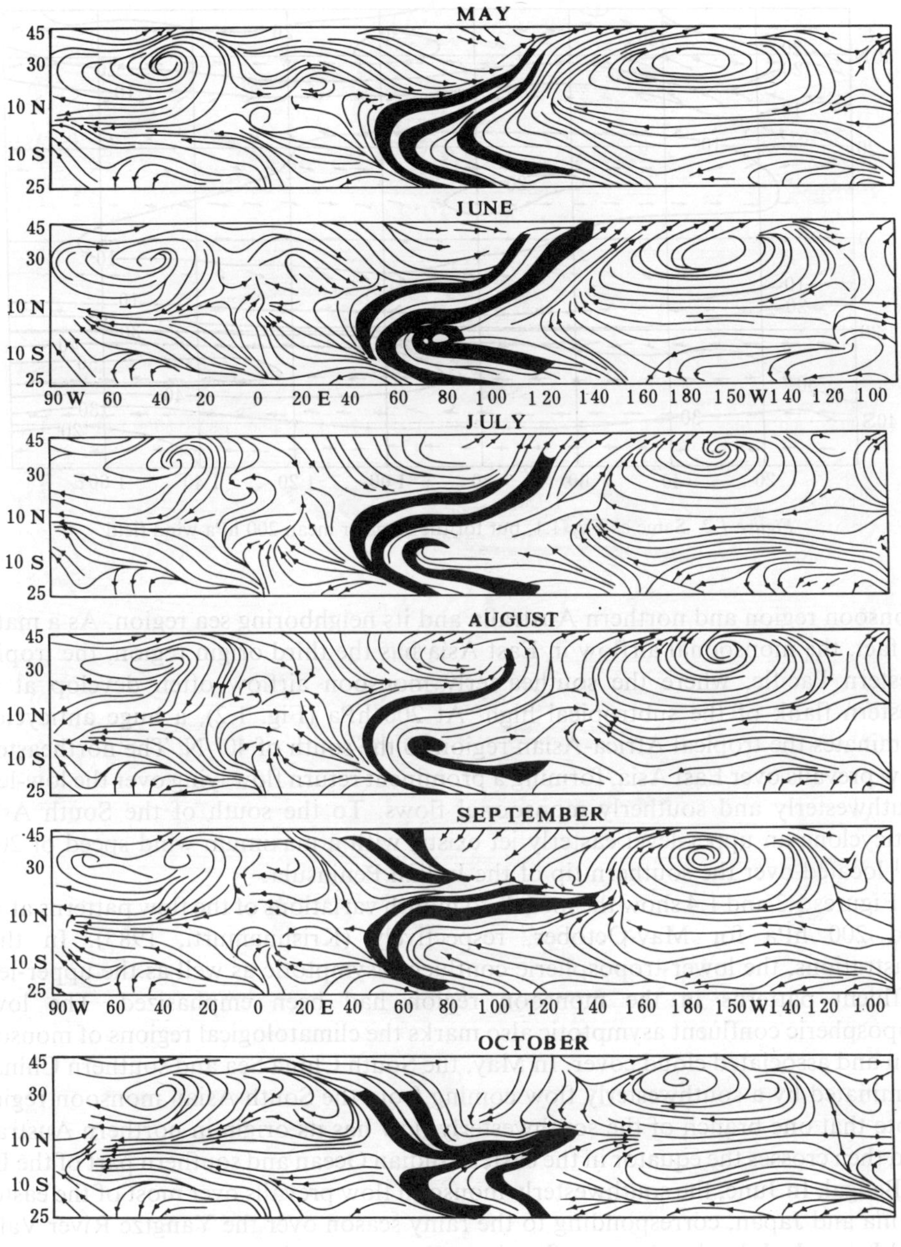

Figure 1.3. Monthly mean streamlines at 850 hPa from May to October (Krishnamurti, 1985).

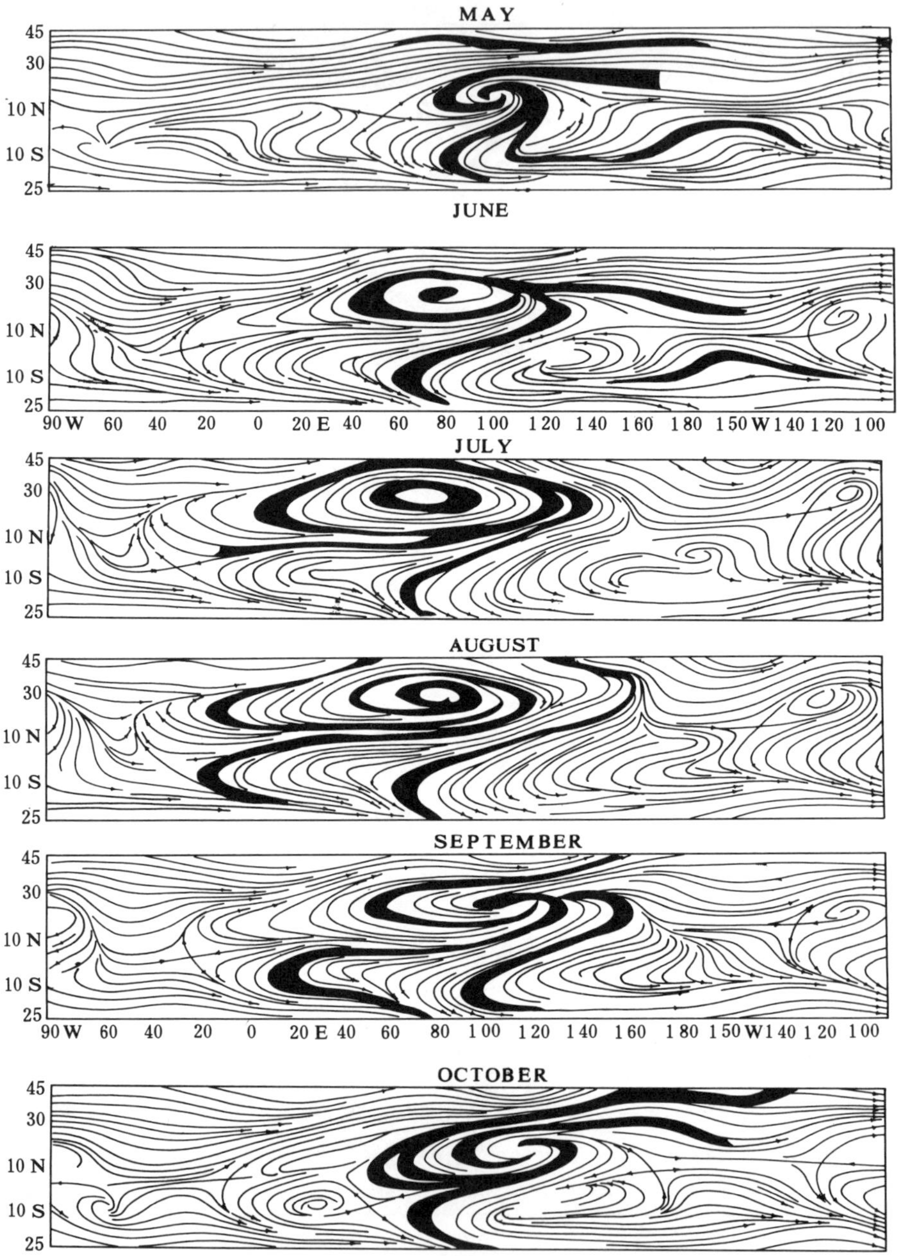

Figure 1.4. The same as Fig. 1.3, but for 200 hPa.

The anticyclone has its largest amplitude in the month of August. Diffluent upper-level flows are a characteristic feature above the regions of monsoon rainfall. As the withdrawal of the summer monsoon commences in September, the northern anticyclone moves towards the equator. There are, in fact, two upper anticyclones nearly all year round in each hemisphere. Note the pronounced cross-equatorial upper-level return flow, coming from the western Pacific, across the South China Sea. Therefore, the low-level summer monsoon flow is accompanied by a process of interhemispheric mass exchange.

Utilizing the data from 13 recent years, Ding *et al.* (1984a, b) analysed the long-term monthly mean wind field at 200 hPa in the global tropics and subtropics during the Northern summer. It was found that there are two upper-tropospheric anticyclonic centers in the Asian area, one located over the Tibetan Plateau, the other over the Iran-Afghanistan region. Their evolution is different, with an earlier northward advance and southward withdrawal of the Iranian high, from the Tibetan high. If one inspects the charts of monthly variations of the 200 hPa flow fields in East Asia (Fig. 1.5), it may be seen that the most drastic changes from May to June and from June to July occur over the region from the western part of the Tibetan Plateau to Iran. Namely, the most significant development of the anticyclonic circulation is observed in this region. But, in August, the changes in the intensity of the Iranian and the Tibetan highs are just the opposite, with the former rapidly weakening and retreating southward and the latter continually strengthening and advancing northward. This change clearly shows up in Figure 1.5c in which is observed a cyclonic circulation change over the Iran-Afghanistan region and an anticyclonic circulation change over the Tibetan Plateau and East Asia. On the month-to-month variation chart from August to September, the disappearance of the Iranian high and the southward retreat of the Tibetan high are reflected in an increase in the westerly

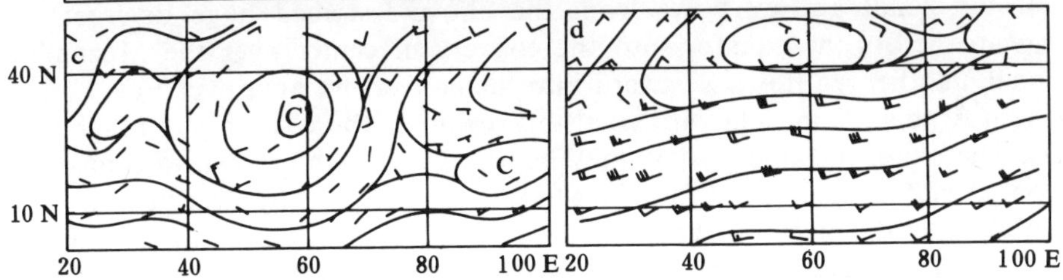

Figure 1.5. The month-to-month variations in the long-term mean 200 hPa flow fields over Asia, (a) from May to June, (b) from June to July, (c) from July to August, and (d) from August to September. A denotes the anticyclonic variation; C the cyclonic variation. Other symbols are conventional (Ding *et al.*, 1984).

components of the winds. The monthly changes in the flow field from September to October (not shown) indicate the occurrence of a strong westerly wind over the southern part of the Tibetan Plateau and northern India, which implies a weakening of the upper-level easterly jet and the re-establishment of the upper westerly jet over the region south of the Tibetan Plateau.

In addition, four interhemispheric mass exchange regions have also been indicated in the global tropics. Among them, the strongest is along 100° E and the Indian Ocean which is closely related to the activities of the Asian monsoon, as previously indicated (Ding *et al.*, 1984).

An important aspect of planetary-scale circulation is divergent circulation (Krishnamurti, 1971). Usually, the velocity potential at 200 hPa is used to display this field. The region with the strongest divergent wind at 200 hPa in summer is found over the Philippines and its adjoining sea area (15° N, 122° E). Note that the maxima of the divergent winds are located over China and the region from the Bay of Bengal to Indonesia, where there are maximum horizontal gradients of velocity potential. The strong divergent wind over China implies that the meridional circulation there is very vigorous.

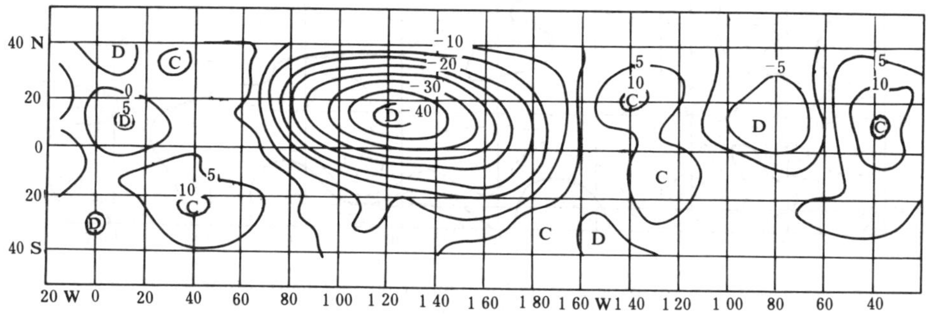

Figure 1.6. Mean field of the 200 hPa velocity potential in summer (June-August) for the time period 1968–1980. A full line denotes the isopleths of velocity potential in unit of $5 \times 10^5\ m^2\ s^{-1}$. D denotes the upper-level divergence and C denotes the upper-level convergence (Ding, Zhao and He, 1984).

The zonal circulations implied by the divergent wind field are very obvious. One leg of the zonal airflow flows from the above-described huge divergent area southward across the equator into the convergent center over the Mozambique Channel; another leg flows westward into the convergent area over Egypt and the northern Arabian Peninsula. Note that this region is where the western high system exists. Therefore, this high system may be associated with a downward motion, as compared with the eastern high system which is characterized by a deep rising motion. Eastward from the huge divergent center over the western Pacific, the airflow moves into a convergent area over the eastern Pacific Ocean, namely, the upper-level branch of the so-called Walker circulation.

A number of investigators have estimated the diabatic heating field over the summer monsoon region (Chen and Li, 1981; Johnson, 1985; Wu, 1987), indicating that, after the onset of the Indian summer monsoon, the most intense heat source is located around the Bay of Bengal and its neighboring areas (Fig. 1.7). This condition obviously results from the release of condensation latent heat caused by monsoon rainfalls. The other heat source is centered on the South China Sea. In contrast, there are heat sinks over the Madagascar and Australian regions, respectively. It is very natural to believe that the two pairs of the heat source around the Bay of Bengal and the heat sink over Madagascar, and the heat source over the South China Sea and the heat sink over Australia, may drive different monsoon systems, in the context of large-scale thermal causes of the formation of the monsoon circulation. These are the Indian summer monsoon and East Asian monsoon systems which will be discussed later.

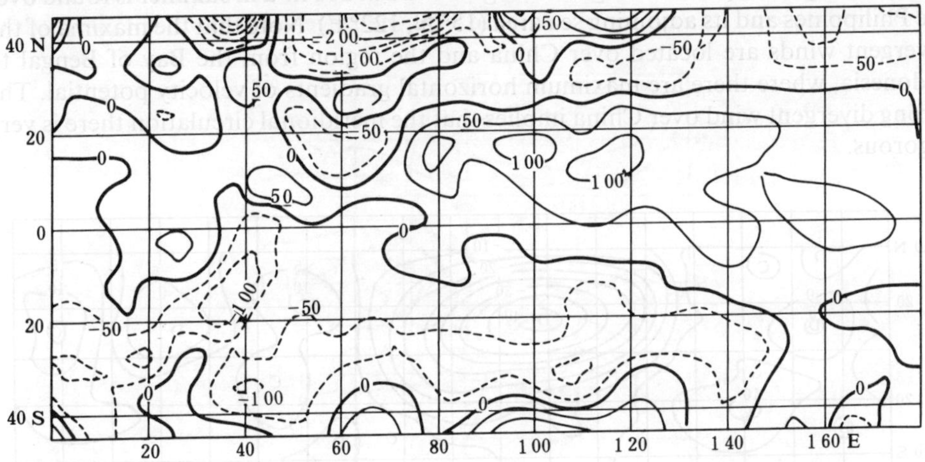

Figure 1.7. Net tropospheric diabatic heating for summer (June-August) of 1979–1984. Units: w m^{-2} (Wu, 1987).

In Figure 1.6, one may clearly see the effect of the above two pairs of heat source-sinks. The upper-level airflow flows southwestward from the strong divergent center over southern Asia and western Pacific down to the Madagascar area, converges and descends there, and flows back to the South Asian monsoon region at a low level as a cross-equatorial return flow (the Somali jet and southwest monsoon). On the other hand, over the western Pacific-Australian region, the upper-level airflow flows into the Southern Hemisphere as an upper-level branch of the Hadley cell and then back to the South China Sea and the western Pacific as cross-equatorial and south-westerly airflows.

From the month-to-month variations of the divergent circulation, one may see more clearly the effects of the above two heat source regions (Fig. 1.8). The most significant change, between May and June, is found over the Bay of Bengal (Fig. 1.8a), which is closely related to the onset and rapid spreading over extensive regions of the South Asian monsoon. Based on the estimates of heat sources (Chen and Li, 1981), the most intense heat source has been moved over this region. From June to

July, the most significant change in the divergent circulation shifts to the region of the western Pacific and the western part of the central Pacific, which may be associated with frequent activities of tropical disturbances and convection. Only a minor change may be observed, from July to August (not shown). Starting from September, the mean divergent circulation over Asia and the western Pacific tends to weaken, especially over the Indo-China Peninsula, the South China Sea and the western Pacific. This weakening is characterized by the convergent circulations in Fig. 1.8c, which reflect a weakening of intensity of heat sources from the Tibetan Plateau to the western Pacific. The retreat of the summer monsoon generally begins in the last part of September or the early part of October. From Fig. 1.8d, it can be seen that the convergent circulation is located over the Bay of Bengal, and the South China Sea and the western Pacific. The above discussion has indicated that the Asian summer monsoon may exert a very significant effect on the large-scale divergent circulation.

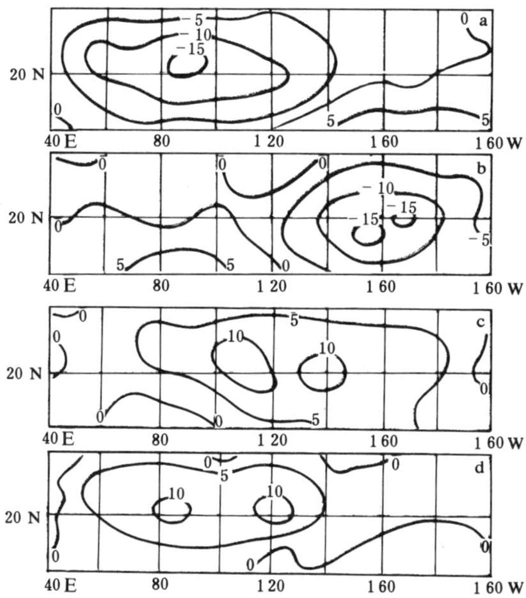

Figure 1.8. The month-to-month variations of long-term mean 200 hPa velocity potential for summer half year (May-October), (a) from May to June, (b) from June to July, (c) from August to September, and (d) from September to October. The symbols of the velocity potential are same as Fig. 1.6. The unit of velocity potential is 5×10^5 m^2 s^{-1} (Ding, Zhao and Zhang, 1987).

Figure 1.9 shows the mean summer moisture field (specific humidity) at 850 hPa that is characterized by a high moisture content in the Asian monsoon, especially in the East Asian monsoon region. This problem will be addressed in detail in Chapter 6.

Figures 1.10–1.12 indicate the meridional cross-sections of zonal and meridional wind components and the vertical velocity (ω = dp/dt) averaged for June-August 1979–1984 (Wu, 1987). It can be seen that the westerly monsoon may extend up to 500

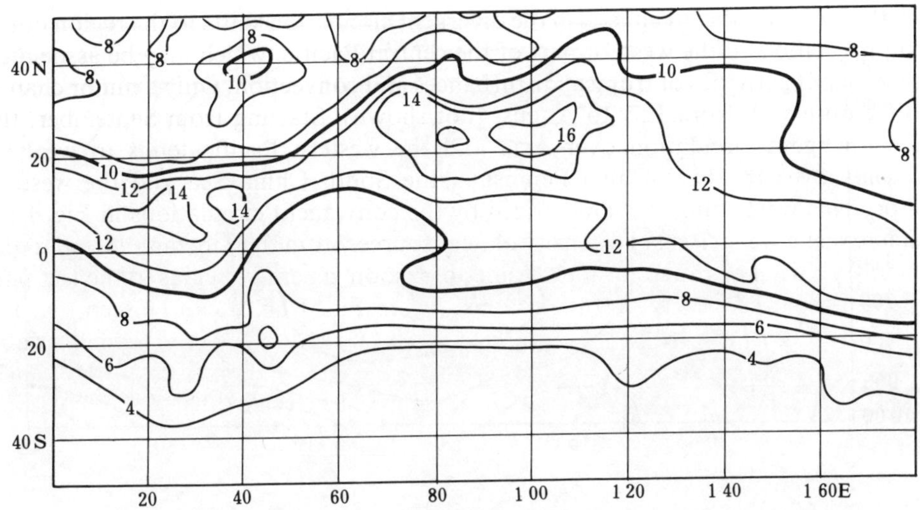

Figure 1.9. The summer (June-August) specific humidity field at 850 hPa averaged for 1979–1984. Units: g kg^{-1} (Wu, 1987).

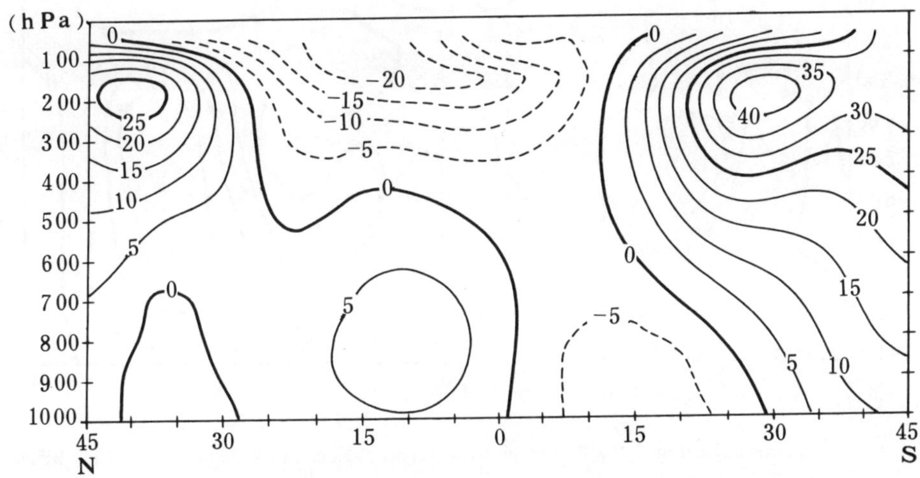

Figure 1.10. The meridional cross-section of the mean zonal wind component for summer (June-August) of 1979–1984 averaged for the longitudinal range of 55–140° E. Units: m s^{-1} (Wu, 1987).

hPa at the latitudinal range 0–30° N and further upward, the easterly wind is lying over it. Note that the southerly wind component corresponds to that of the westerly wind, while the northerly wind component corresponds to that of the easterly wind, implying that in the lower and middle troposphere, the southwesterly wind prevails while in the upper troposphere, the northeasterly wind prevails. The strongest upward motion is found at 20–30° N, in contrast to the downward motion in this latitudinal range outside the monsoon regions. Therefore, the normal-type Hadley cell should be

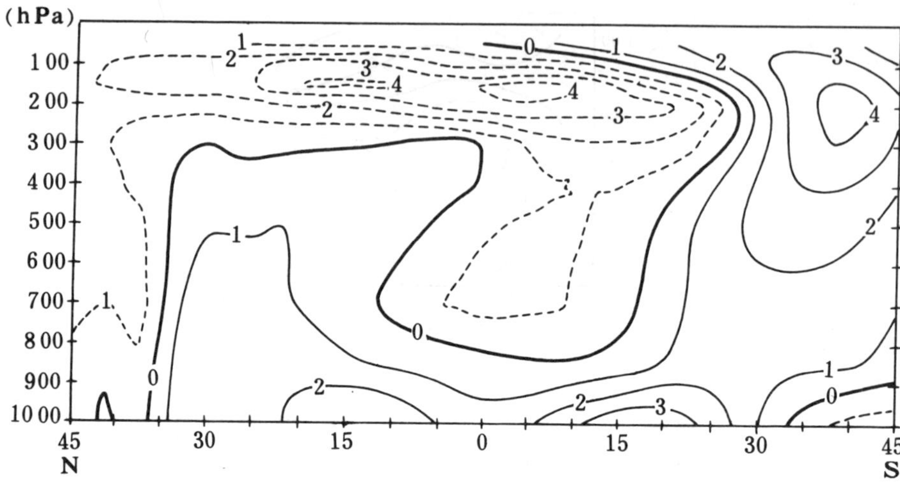

Figure 1.11. Same as Fig. 1.10, but for the mean meridional wind component.

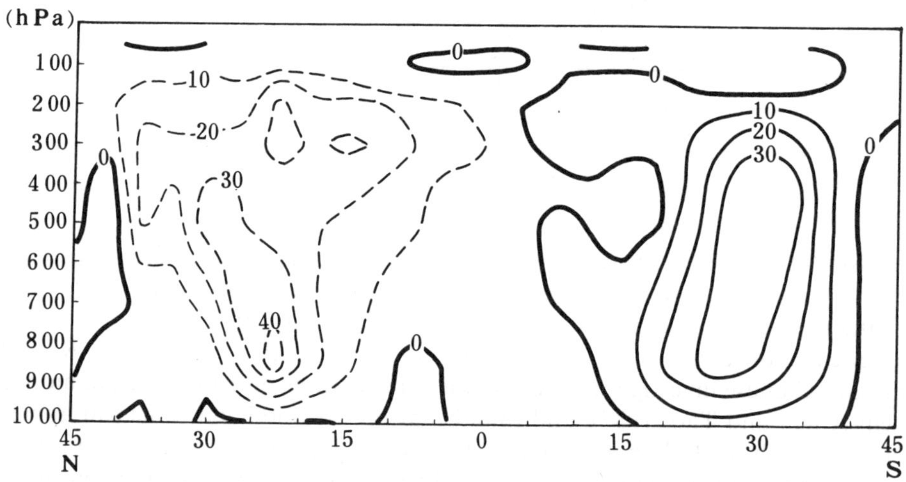

Figure 1.12. The same as for Fig. 1.10, but for the mean vertical velocity ($\omega = dp/dt$). Units: 10^{-3} hPa s^{-1}.

replaced by a meridional cell of the opposite sense. Chen *et al.* (1964) estimated this meridional cell associated with the East Asian monsoon region by using of observed wind data and first revealed the existence of this cell. It is now known as the monsoon meridional cell. Figures 1.13a, b are the mean meridional cells for the Asian monsoon region (55–140° E) and entire zonal circle for July (Yeh *et al.*, 1979; Oort and Rasmusson, 1971). The sense of the monsoon cell is just the opposite to that of the Hadley cell, with the ascending flow extending northward up to 40° N and a descending motion in the Southern Hemisphere.

From the observational evidences discussed above, one immediately gets the

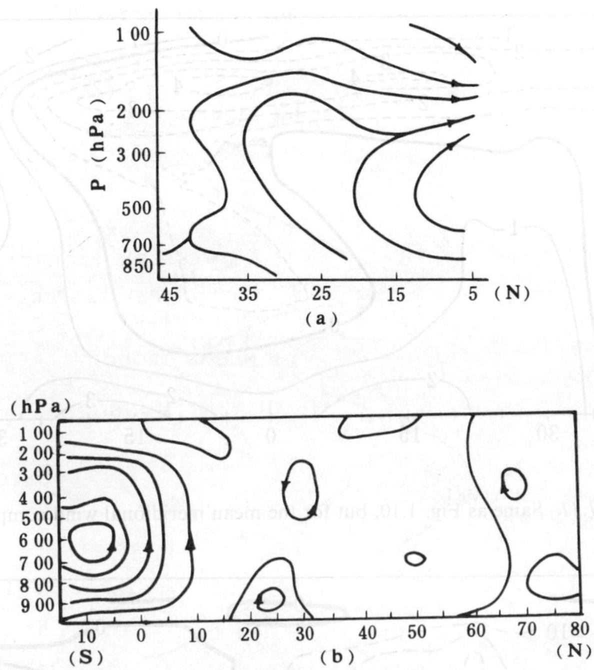

Figure 1.13. (a) The mean monsoon meridional cell for 55–140° E and (b) the zonal mean meridional cell for July (Yeh *et al.*, 1979).

impression that the East Asian summer monsoon cannot be simply viewed as northward and eastward extensions of the Indian summer monsoonal flow because it is characterized by several unique features, including a different wind regime and components, different heat sources and different weather conditions. Recently, numerous Chinese investigators have studied the features of the East Asian monsoon system (Chen and Tao, 1987), and found that on the one hand, it has an apparent interaction with the Indian monsoon system, and on the other hand, it is to a greater extent independent of the latter. Figure 1.14 demonstrates the comparison of these two summer monsoon systems. As Krishnamurti and Bhalme (1976) suggested, the Indian monsoon system includes the following principal components: the monsoon trough over Northern India, the southwest monsoon flow, the cross-equatorial, level-level jet (Somali jet) off East Africa, the Mascarene high, the anticyclone in the upper troposphere over the Tibetan Plateau, the tropical easterly jet stream and monsoon rainfall and cloud cover over the Indian Peninsula. The East Asian monsoon system includes the monsoon trough (or intertropical convergence zone (ITCZ)) in the South China Sea and the western Pacific, the cross-equatorial flow to the east of 100° E, the surface cold high in Australia, the subtropical high over the western Pacific, the upper-level northeasterly return flow, the convection along the monsoon trough, the Meiyu (or Baiu) frontal zones, and the mid-latitude disturbances. Therefore, the Asian monsoon system indeed comprises two systems: the Indian monsoon and

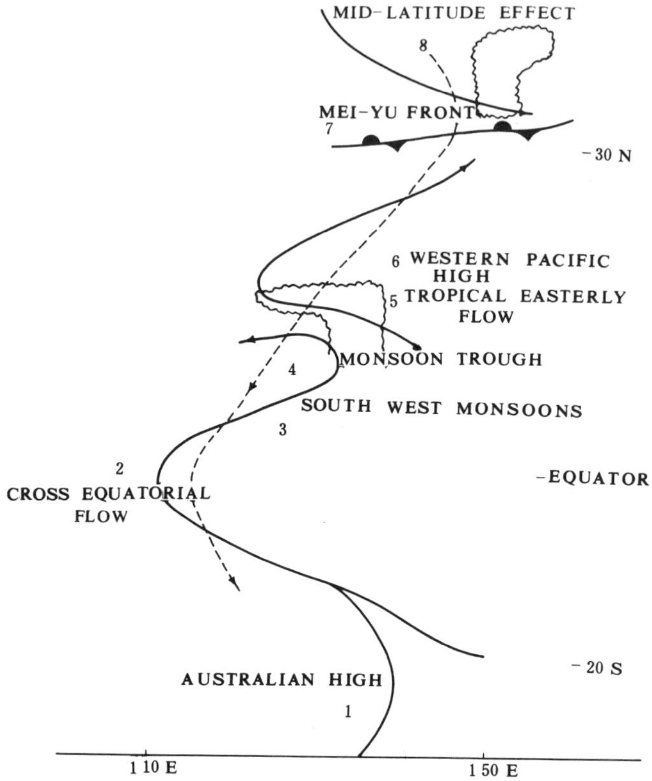

Figure 1.14. A schematic diagram of the summer monsoon systems in East Asia (Tao and Chen, 1987).

the East Asian monsoon which are driven by different heat sources and sinks. Below, we shall discuss in detail the major components of the East Asian monsoon system.

1.2. Advance and Retreat of the Summer Monsoon in China

China is greatly affected by monsoons. Every year, the summer monsoon brings about abundant precipitations. Like India, the activity and variability of the summer monsoon in China is intimately related to crop yields. Therefore, for a long time,the Chinese meteorologists have been engaged in numerous studies on the impact of monsoons on the weather and climate in China, especially those which regard to droughts and floods, and have obtained many important results. This section will only focus on the advance and retreat of the summer monsoon in China from the climatological viewpoint, which is a necessary basis and background for further understanding the behavior of the summer monsoon in China.

The summer monsoon affecting China originates in three airflows (Fig. 1.15): (1) the Indian summer monsoonal airflow. When it moves northward, mainly during

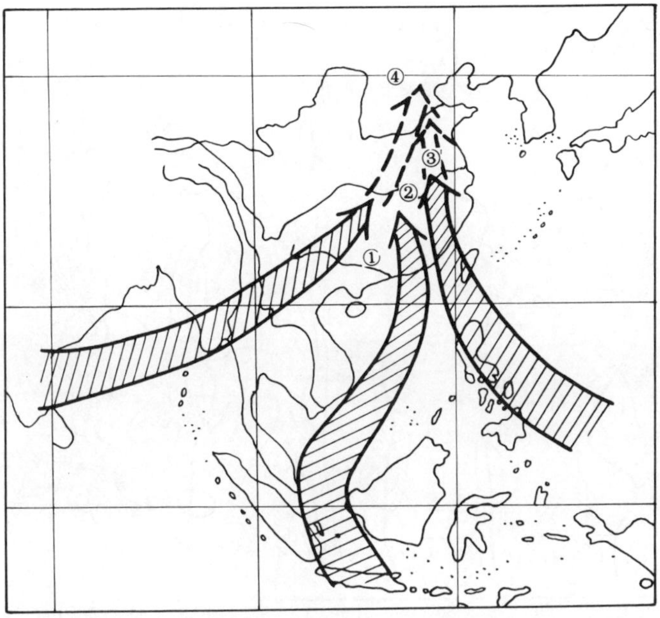

Figure 1.15. The main summer monsoonal airflows in China and their regions of origin. The subregions in the eastern China are indicated.

break in the monsoon period, the southwest monsoon may often advance deep into the China Main. (2) The cross-equatorial airflow flowing over Southeast Asia and the South China Sea, which is a kind of low-level southwest airflow. This airflow originates in the Australian region in the Southern Hemisphere and then streams across the equatorial region, mainly in the Singapore and North Borneo region. This cross-equatorial aircurrent is considered to be a significant mass exchange channel between the two hemispheres. Figure 1.16 shows the variation of the v-component with time at 850 hPa in the equatorial region from 40° E to 170° E in 1979 (Tan and Huang, 1980). A strip of stable and strong southerly wind is located between 40 and 50° E, which is the so-called Somali jet. Besides, near 105° E, another marked area of the southerly wind may be observed. This can flow northward over southern China, and even further over the Yangtze River Valley. At 700 hPa (Fig. 1.17), in contrast to the Somali aircurrents, it apparently cannot be detected. This fact also indicates that the Somali jet is the most stable and deepest cross-equatorial airflow in the Asian monsoon region, but at 105° E, there is a second cross-equatorial airflow. (3) The southeast monsoon coming from the southern flank of the subtropical high over the western Pacific. Sometimes, this airflow may turn into the southwest aircurrent while flowing along the western side of the high. In certain areas of China, it is often observed that one branch of the above three monsoon airflows or combinations thereof may affect the area. Therefore, the activity of the summer monsoon is much more complicated than in other monsoon regions.

The advance and retreat of the summer monsoon in China behave in stepwise way,

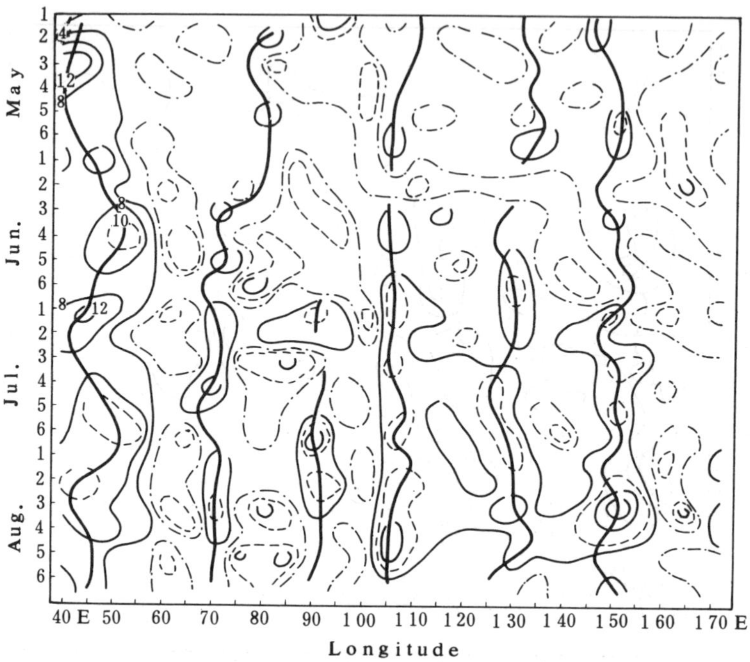

Figure 1.16. Temporal variations of the wind v-component (solid line: south wind) at 1000 hPa from 40 to 170° E in the equatorial zone in 1979. Units: m s^{-1} (Tan and Huang, 1980).

not continuous way. Generally, the isopleth of θ_{se} = 340 K may be used to characterize the thermodynamic property of the air mass of the summer monsoon. Figure 1.18 clearly shows the alternative changes in three abrupt northward shifts of the θ_{se} = 340 K isoline and four relatively stationary stays for a certain spell in the following areas (Lin, 1982): the northern part of southern China (from 4th pentad May to 2nd pentad June), the Yangtze River Valley (from 4th pentad June to 1st pentad July), the Yellow River Valley (from 2nd pentad to 4th July) and northern China (from 5th pentad July to 2nd August), respectively. Each stable stage corresponds to a main rainfall belt or belt of heavy rainfall, i.e., presummer rainy period in southern China, Meiyu or "Plum Rains" in the Yangtze River Valley, the rainy seasons in the Yellow River Valley, and the rainy seasons in the Yellow River Valley and northern China.

The three abrupt northward shifts of the summer monsoon and precipitation belts are closely related to seasonal change in the general circulation in East Asia, mainly the evolution of the upper-level planetary frontal zone, the westerly jet stream and subtropical high. Figure 1.19 is an example showing the sudden northward jump of the upper-level westerly jet stream starting from the 2nd pentad of June, with the wind speed maximum increasing at the same time. This shift of the jet position corresponds to the advance of the summer monsoon along the Yangtze River Valley where it becomes stationary. Meanwhile, the Meiyu begins to take place in this region. A similar change in the ridgeline position of the subtropical high is also observed (Fig.

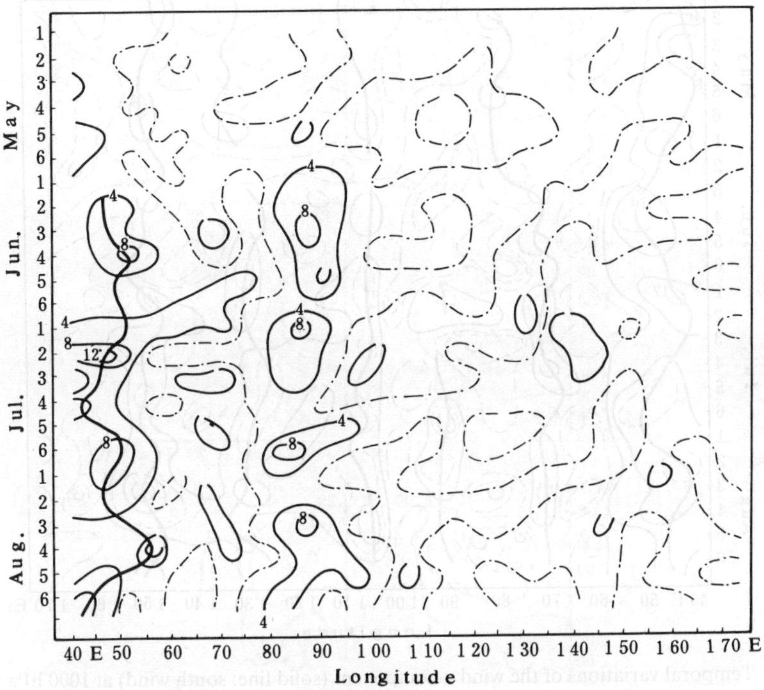

Figure 1.17. The same as for Fig. 1.16, but for 700 hPa.

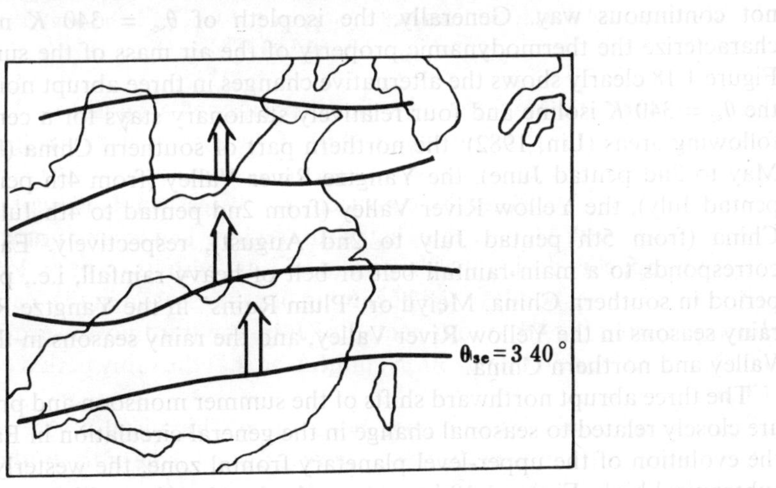

Figure 1.18. Temporal variation in the position of the average θ_{se} = 340 K isoline at 850 hPa. This mean position is obtained for 1971–1980 (see the text).

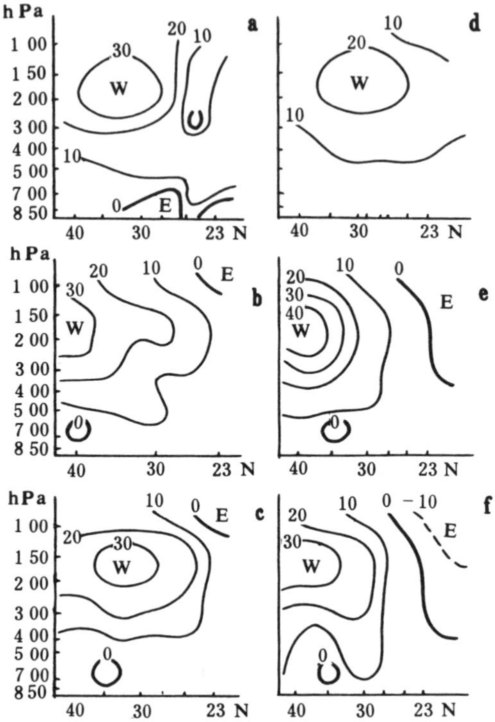

Figure 1.19. Mean pentad-to-pentad variation of the cross-sections of the u-component along 110° E from the 4th pentad of May (left top) to the 3rd pentad of June (right bottom), 1980. Units: m s^{-1} (Lu and Lin, 1982).

1.20). In April and May, the ridge-line is positioned around 15–18° N. Until the 2nd pentad of June, the first northward shift of the high brings its ridge line up to about 20° N, or even further. At the same time, the isoline of $\theta_{se} = 340\ K$ moves northward beyond 25° N. Therefore, the consistent southwest winds prevail over the Yangtze River Valley, with Meiyu occurring in this region. The second northward jump of the ridge line appears during the 1st pentad of July, with its position located near 25° N. This period corresponds to the beginning of rainy season in the Yellow River Valley. Finally, during the 4th or 5th pentad of July, the third northward jump of the ridge line brings its position up to 30° N, indicating that the rainy period comes to northern China.

The withdrawal of the summer monsoon in China is also very rapid, taking only about one month or even shorter to retreat from northern to southern China, i.e., in northern China, the summer monsoon starts withdrawing at the end of August or at the beginning of September and during the last 10 days of September or the first 10 days of October, the winter monsoon may set in over most of southern China.

Likewise, reverse changes in the general circulation in East Asia to the summer case can be observed. For example, the retreat of the summer monsoon corresponds to the three abrupt southward shifts of the ridge line of the subtropical high (Fig. 1.20). The

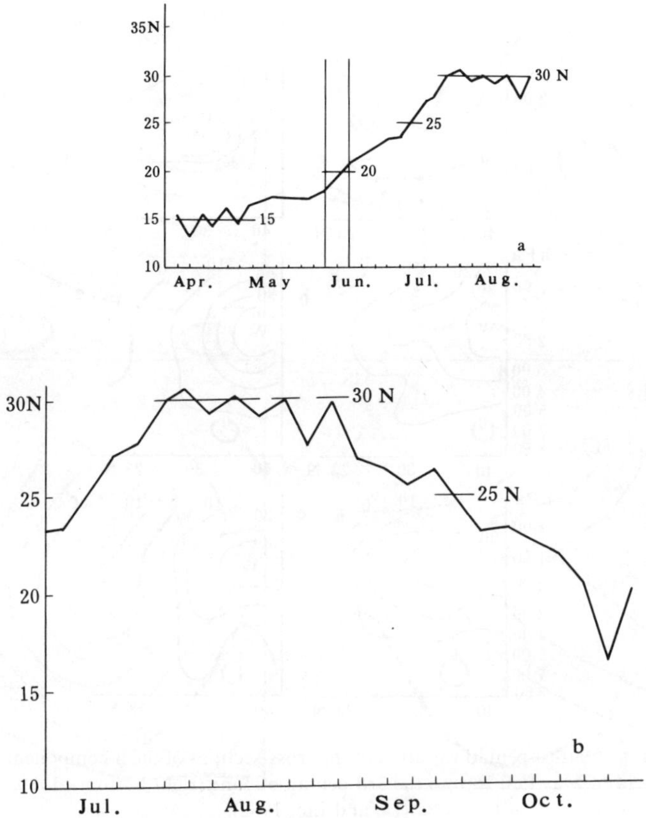

Figure 1.20. Seasonal variations of the average ridge-line position of the subtropical high (125 – 140° E). The position is averaged for 1971–1980.

first southward shift occurs at the end of August or at the beginning of September, indicating the withdrawal of the summer monsoon in northern China. Then in the 4–6th pentads of September the ridge line retreats southward to 25° N, corresponding to the onset of the winter monsoon over the Yangtze River Valley. Finally, during the middle 10 days of October, the ridge line comes down to 15–18° N, with the summer monsoon completely withdrawing from the China Main and the winter monsoon prevailing.

The associated planetary frontal zone and jet stream also show similar changes (Fig. 1.21) (Lu and Lin, 1982). In the 5th and 6th pentads of September, the maximum westerly wind rapidly moves southward by about three latitudes and then, in the 1st pentad of October, again three latitudes, which corresponds to the second and third southward jumps of the ridge line of the subtropical high and the summer monsoon.

Next, we shall address the regional climatology of the summer monsoon in China

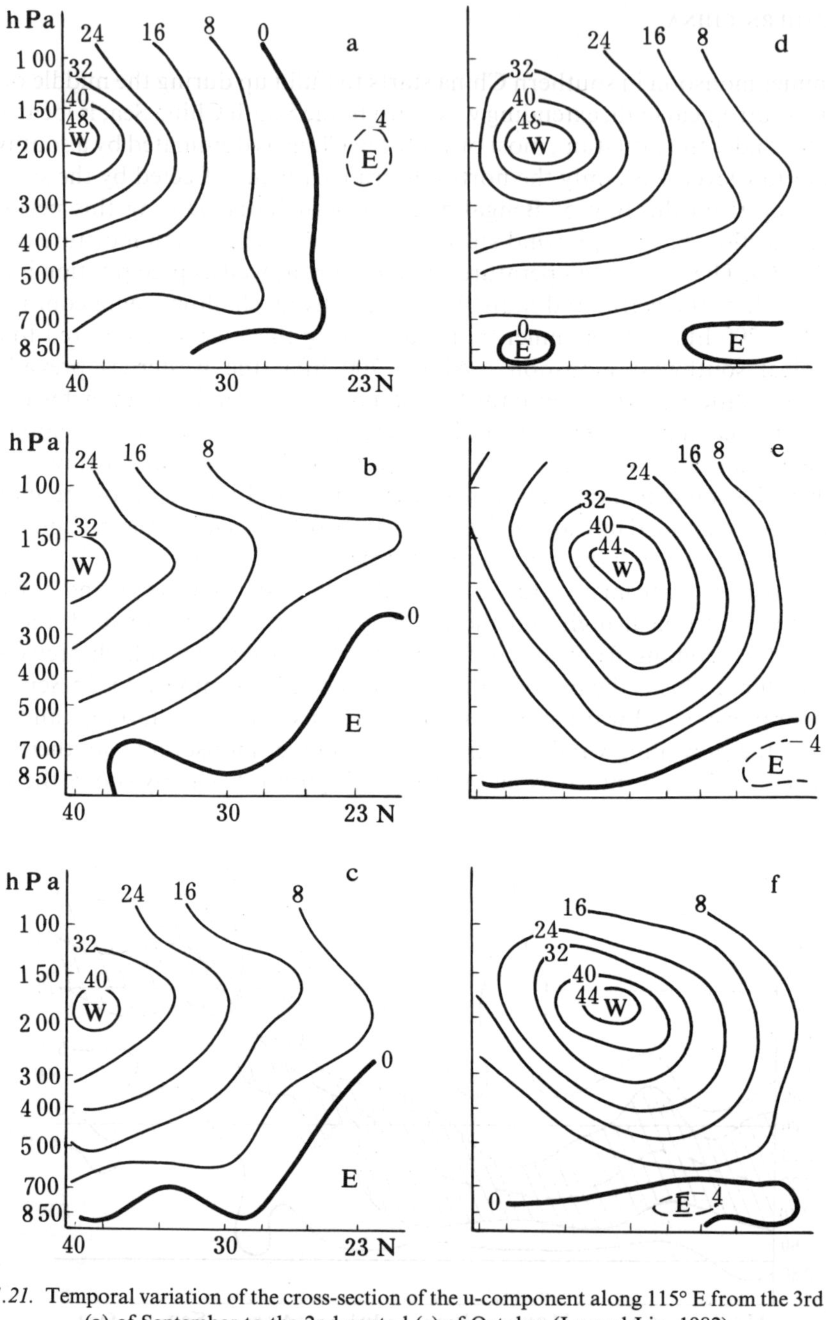

Figure 1.21. Temporal variation of the cross-section of the u-component along 115° E from the 3rd pentad (a) of September to the 2nd pentad (e) of October (Lu and Lin, 1982).

(1) SOUTHERN CHINA

The summer monsoon in southern China starts to build up during the middle of April when the subtropical high, extending westward to the South China Sea, is located near 17–18° N. Under this situation, most of southern China is dominated by a southwestly airflow at low level, with only the northwestern part being affected by the southwest flow coming from the Bay of Bengal and northern India. During this period, the southwest airflow has a high wind steadiness (S > 60%) and high moisture contents ($T-T_d < 4$ °C). The surface temperature and moisture as well as precipitation increase, but the depth of the warm and moist layer is relatively shallow, being confined in a layer below 700 hPa. The summer monsoon during this period is referred to as the "subtropical southwest monsoon". Above 700 hPa, the westerlies prevail. This monsoon condition continues into the first half of June when the first northward jump of the subtropical high, onset of the Indian summer monsoon and the development of cross-equatorial airflow take place. These three branches of airflows rapidly strengthen the summer monsoon in southern China and the depth of the monsoon airflow layer is extended upward to 400 hPa. Meteorologists in southern China call the summer monsoon during this period the "tropical southwest monsoon".

During the period from mid-July to the end of August, after the second northward jump of the subtropical high and the ITCZ greatly moving northward, the southeasterly airflow coming from the southern side of the subtropical high prevails over southern China. This is the period of the tropical southeast monsoon. In September, the cold air moves southward and the winter northeast monsoon develops and sets in. Therefore, the prevailing wind direction of the summer monsoon in southern China experiences an obvious intraseasonal variation. Figure 1.22 clearly shows the shift of

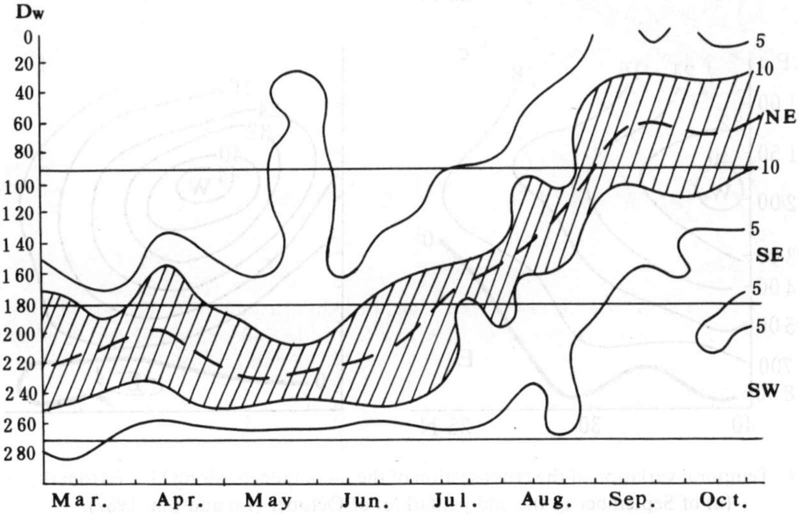

Figure 1.22. Long-term (1976–1980) seasonal variations of average wind direction frequencies at 1500 m at Guangzhou, southern China. Units: degrees. The hatched area denotes high frequency (Chen and Gao, 1980).

wind direction from southwest to southeast, and further northeast as the season progresses (Chen and Gao, 1980). An analysis of the power spectrum of the precipitation data during the period from March through November reveals two types of oscillations: one with a period of about two weeks and other with a period of 6–7 weeks. Oscillations of similar periods for wind data can also be found in Fig. 1.22. Much works on these oscillations has been carried out by meteorologists in India, China, the U.S.A., and Japan. These problems will be discussed in Chapters 3 and 4.

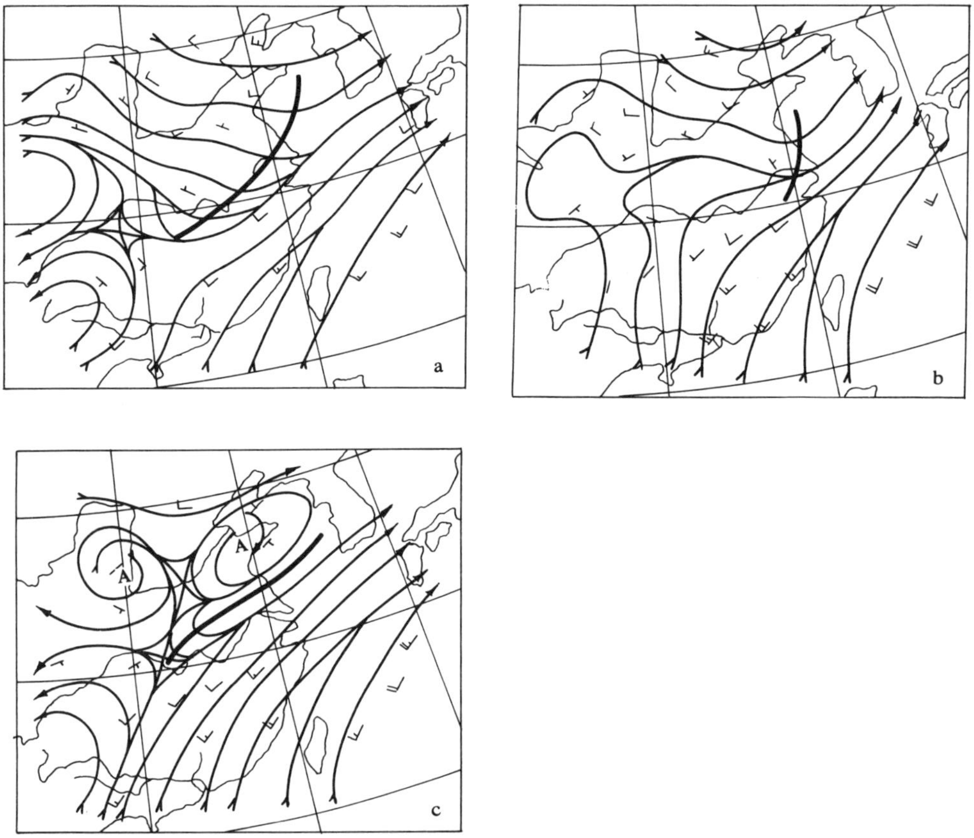

Figure 1.23. Average flow patterns in eastern China at 850 hPa, averaged for 1971–1980, (a) prior to the 1st pentad of June, (b) for the 1st pentad of June, and (c) for the 2nd pentad of June. Wind symbols: full barb: 4 m s^{-1}; half barb: 2 m s^{-1} (Lu and Lin, 1982).

(2) THE YANGTZE RIVER VALLEY (CENTRAL CHINA)

Prior to the 4th pentad of June, the summer monsoon coming from southern China can occasionally appear in this region, but the southwest wind is not very steady (Fig. 1.23a). During this period, the winter monsoon is dominant, although at the beginning of June the winter monsoon (mainly north winds) and the summer

Figure 1.24. Average flow patterns in eastern China at 850 hPa, averaged for 1971–1980: (a) for the 2nd pentad of August, (b) the 5th pentad of August, (c) the 6th pentad of August, (d) the 2nd pentad of September, (e) the 3rd pentad of September, and (f) the 5th pentad of September. Wind symbols are the same as in Fig. 1.23 (Lu and Lin, 1982).

monsoon very often meet in this region (Fig. 1.23b). Until the 4th pentad of June, the steady southwest airflow begins to prevail over the Yangtze River Valley, which indicates the ending of the rainy season in southern China and the beginning of Meiyu in the Yangtze River Valley (Fig. 1.23c). Starting from the 6th pentad of July, after the Meiyu has ended, the southeasterly airflow dominates this region due to the the subtropical high extending westward. Since the 4th pentad of August, the summer monsoon in the Yangtze River Valley weakens with a variable northwestly wind often observed. Until the 1st pentad of October, the winter monsoon eventually builds up in this region. Figure 1.24 depicts the mean flow patterns at 850 hPa during the withdrawal period, which apparently shows the process of rapid retreat of the summer monsoon.

(3) THE HUAIHE AND YELLOW RIVER VALLEYS

The onset of the summer monsoon in this region is associated with the second northward shift of the subtropical high, with its ridge line moving beyond 20° N. This event generally occurs in the first half of July. Figure 1.25 shows the seasonal mean variations of the v-component at 850 hPa and the isoline of θ_{se} = 340 K (Mao and Xu, 1982). During this period, the southerly wind and the air mass of high temperature and moisture dominate this region and the rainy season starts (Fig. 1.26). When the third northward jump of the subtropical high occurs, with its ridge line moving to 30° N or further, the summer monsoon in this region reaches its culmination, with θ_{se} being greater than 347 K, the southerly wind increasing and the precipitation intensity having its maximum. At this time, the summer monsoon may originate in two components: the southwest monsoon coming from the Bay of Bengal and the southeast monsoon from the southern side of the subtropical high, which then turns into the southwest airflow along the western flank of the high. This implies that the Indian southwest monsoon can reach this region. During the first 10 days of August, the summer monsoon in this region begins to weaken, and the amount of precipitation significantly reduces. During the first 10 days of September, the summer monsoon completely retreats from this region.

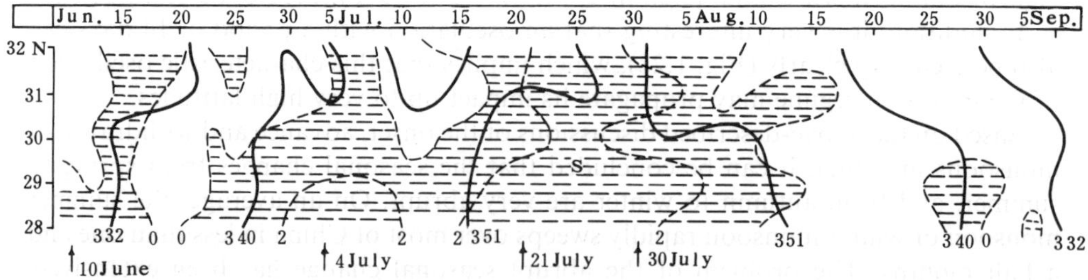

Figure 1.25. Average time-latitude cross-section of the v-component and the selected characteristic θ_{se} isolines at 850 hPa for 1971–1980. The southerly wind area is shaded. Dashed lines denote isotachs; solid lines isopleths of equivalent potential temperature (Mao and Xu, 1982).

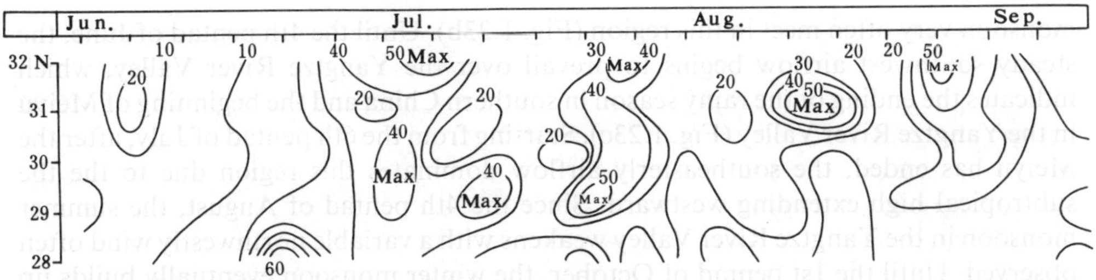

Figure 1.26. Average time-latitude cross-section of pentad rainfall amounts for 1971–1980. Units: mm.

(4) NORTHERN AND NORTHEASTERN CHINA

The summer monsoon in northern China sets in when the subtropical high experiences the third northward jump, which generally occurs in the middle of July. In this situation, the monsoon may go to 45° N, the highest latitude of its northward advance in China. Further north, the synoptic systems in the westerlies predominates. The time span of the summer monsoon prevailing in northern China is much shorter. Starting from 10 August, the summer monsoon begins to weaken and withdraw. Until the end of August, the summer monsoon completely retreats southward to the region of the Huaihe and Yellow Rivers. Thus, the summer monsoon in northern China only lasts for about one month. This time period also corresponds to the main spell of rainfall in this region and in northeastern China. Figure 1.27 clearly indicates the monsoon condition in northern China (Shi and Chao, 1982). The airflows during the period of the summer monsoon in northern China come from the southwest airflow originating in southern China or the Bay of Bengal and the southeast airflow at the southern side of the subtropical high, which then turns into the southwest airflow in the northwestern flank of the high.

The precipitation in northern China is closely related to the activity of the summer monsoon. If the summer monsoon is too weak to reach northern China or has a very short duration there (e.g., in 1972, 1974 and 1979), droughts may often occur due to a deficiency in rainfall. On the contrary, when the summer southwest monsoon is vigorous in northern China, (e.g., in 1973, 1977 and 1978), the floods often occur.

In addition, it is very interesting that an oscillation with a period of 14 days can also be seen very clearly (Fig. 1.28), which implies that this characteristic oscillation of the monsoon regime may propagate its impact up to very high latitudes.

Based on the above-described discussions of the onset, advance and withdrawal of monsoons in China, it can be concluded that the seasonal changes from spring to summer, and from autumn to winter are very abrupt. On an average, the summer monsoon or winter monsoon rapidly sweeps over most of China in less than one and a half months. The problem of the abrupt seasonal change has been extensively studied by Chinese and Western meteorologists over the last 40 years. Their studies have indicated that there are four seasons from the astronomical viewpoint, but for meteorological natural seasons, the spring and autumn are rapid transition seasons, with only the winter and summer circulations being obviously different.

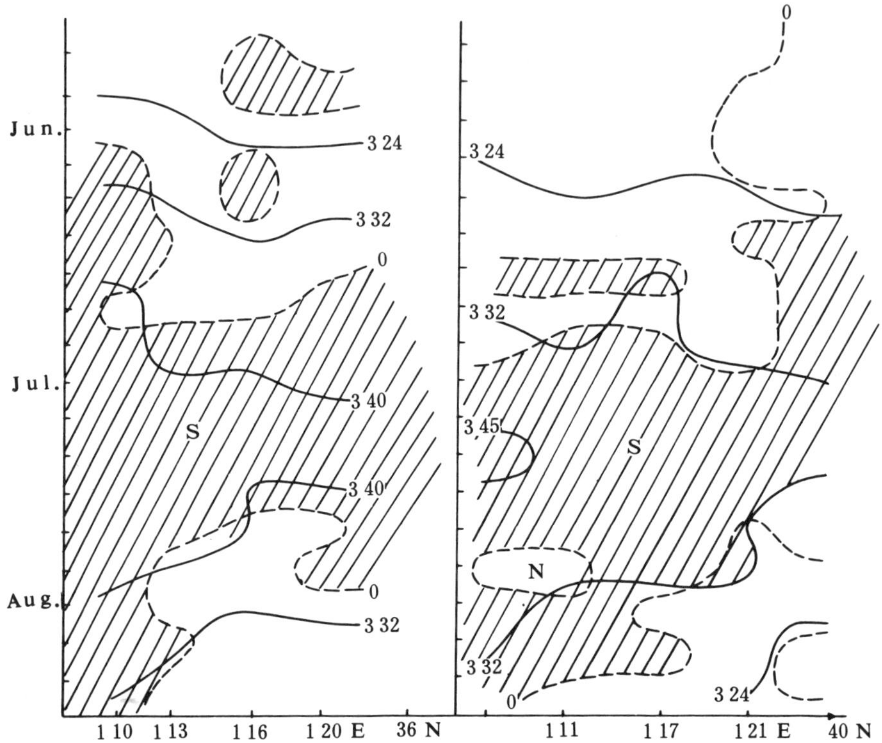

Figure 1.27. Average time-longitude cross-sections of the v-component and θ_{se} at 850 hPa for 1971–1980, (a) along 36° N (left), and (b) along 40° N (right). The area of the southerly wind is shaded. Solid lines denote isopleths of θ_{se}; dashed lines wind isotachs. Units: m s^{-1} for wind; K for θ_{se} (Shi and Chao, 1982).

In 1954, in a discussion on the beginning of the transition season in East Asia, Yeh and Zhu (1954) stressed that the beginning of the East Asian transition season is a process of sudden change. In the same year, Sutcliffe and Bannon (1954) demonstrated the presence of sudden changes in the upper-level wind over the Mediterranean Sea and the central Asia. Two years later, Tao and Chen (1957) made a diagnostic analysis of changes in the atmospheric general circulation in South Asia for the early summer of 1956 and drew the conclusion that the atmospheric general circulation in Asia underwent a jumping change for this time. Inspired by this observational evidence of abrupt changes in the regional circulation, Yeh, Tao and Li (1959) further studied the annual variation of global general circulation and revealed that there are abrupt seasonal changes in June and October every year. The above results clearly indicate the bimodality and seasonal sudden changes of the atmospheric general circulation.

The bimodality of the atmospheric general circulation is on a global scale, but especially shows up in the region of the Asian monsoon (South Asian and East Asian monsoon regions). As indicated here and in Chapter Five, in the middle of June, the subtropical upper-level jet, situated over the southern foothills of the Himalayas in

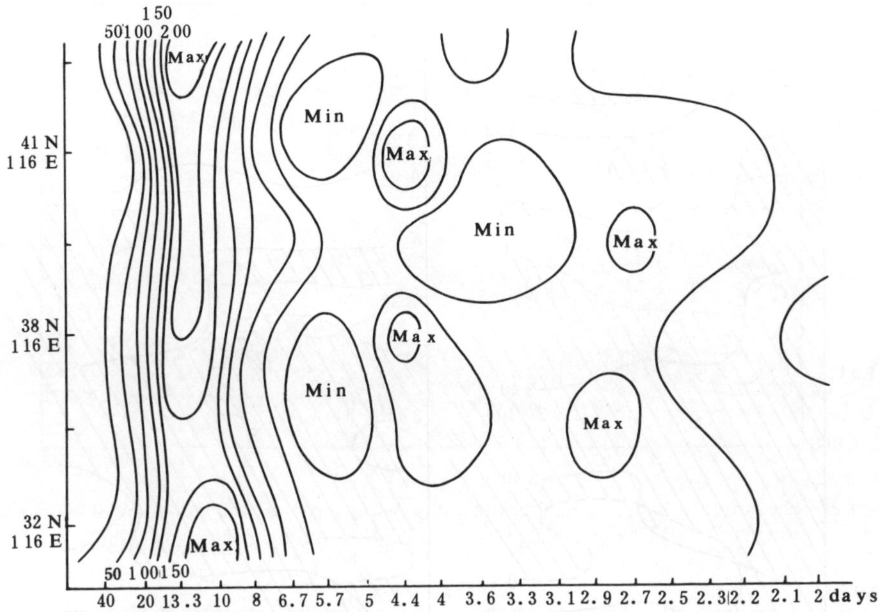

Figure 1.28. Time-longitude cross-section of the power spectrum of θ_{se} at 850 hPa for July and August 1972 (Shi and Chao, 1982).

the winter half-year, rapidly retreats northward and the huge Tibetan high in the upper troposphere and the upper-level easterly jet are established. At the same time, the Indian monsoon sets in and the Meiyu begins in the Yangtze River Valley. In the middle of October, the tropical easterlies suddenly turn into westerlies and the westerly jet in the high troposphere withdraws southward down to 30° N. At the same time, the winter monsoon is established in the Yangtze River Valley and southern China and the Indian southwest monsoon withdraws.

This seasonal sudden change in the general circulation has been well simulated by using the GCM numerical experiment. Utilizing IAP-GCM (Institute of Atmospheric Physics, Academia Sinica), Zhen, Lian and Zhang (1988) simulated the two sudden changes in the zonal circulation in June and October, the seasonal shift of the ITCZ and the meridional movement of the East Asian rain belt (Fig. 1.29). They suggest that the seasonal sudden change of circulation is closely associated with atmospheric heat sources, especially condensational heating, and possibly to a considerable extent, it is also affected by land-sea contrast and topography. Based on further experiments with and without taking the topography into account, it can be found that there is no sudden northward shift of the wind belt in the experiments excluding the topography. When the topography is included in the model, a sudden change may be observed (Hahn and Manabe, 1975).

Although the seasonal abrupt change may be simulated by using GCM, the dynamic mechanism of the abrupt change remains a difficult problem to be solved.

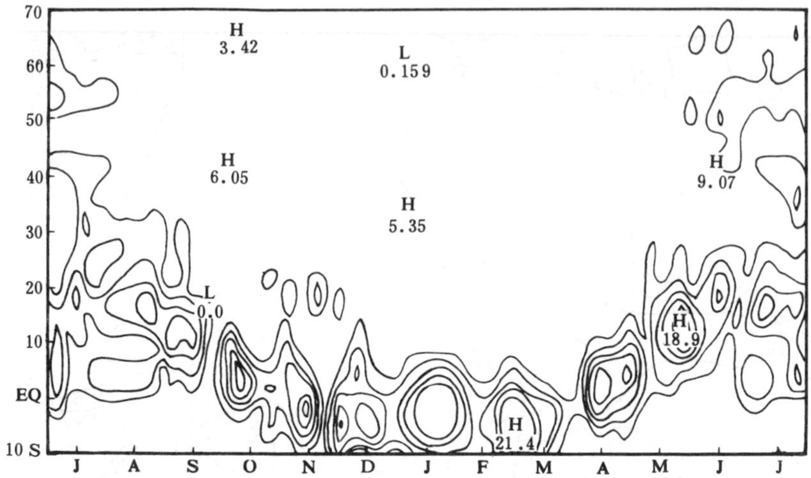

Figure 1.29. The latitudinal and seasonal changes of a 5-day mean rainfall amount at the longitudinal range of 110–125° E, simulated by IAP-GCM. Units: mm (Zhen *et al.* 1988).

Further discussion will be presented in Chapter Five in connection with the thermal and dynamic effects of the Tibetan Plateau.

1.3 The Seasonal Variation of Monsoon Rain Belts in China

Precipitation patterns in China are greatly dependent upon monsoon activity. If one inspects the long-term annual mean precipitation distribution in China (Fig. 1.30), the immediate impression is that the rainfall amount declines from southeast toward northeast. In southern China, the annual mean rainfall amount is greater than 1000 mm. In the southeastern coastal region and the southern part of Yunnan Province, it comes to 1500–2000 mm or even more. In most of northern and northeastern China, the rainfall amounts to 500–800 mm, with some local areas exceeding 1000 mm. Note the 400 mm isohyet running from the northwestern part of northeastern China across Huhohot (Inner Mongolia) and Lanzhou to Lahsa (the Tibetan Plateau), which divides the precipitation pattern of China into two types. To the northwest of this demarcation line, the rainfall amount is lower than 400 mm and even less than 100 mm in the region to the west of Ningxia Province. In extreme conditions, some places only receive an annual rainfall amount of less than 25 mm (e.g. in Turba only 16 mm). To the southeast of this line of demarcation, the annual precipitation amount rapidly increase toward the coastal region. The maximum annual rainfall amount in China is observed in Taiwan, with an annual mean value of 6557.8 mm at the northern tip of the island.

This characteristic pattern of precipitation in China is related to two factors. One of them, as pointed out at the beginning of this section, is due to the summer monsoon. The other one is the effect of large-scale topography which slopes overall

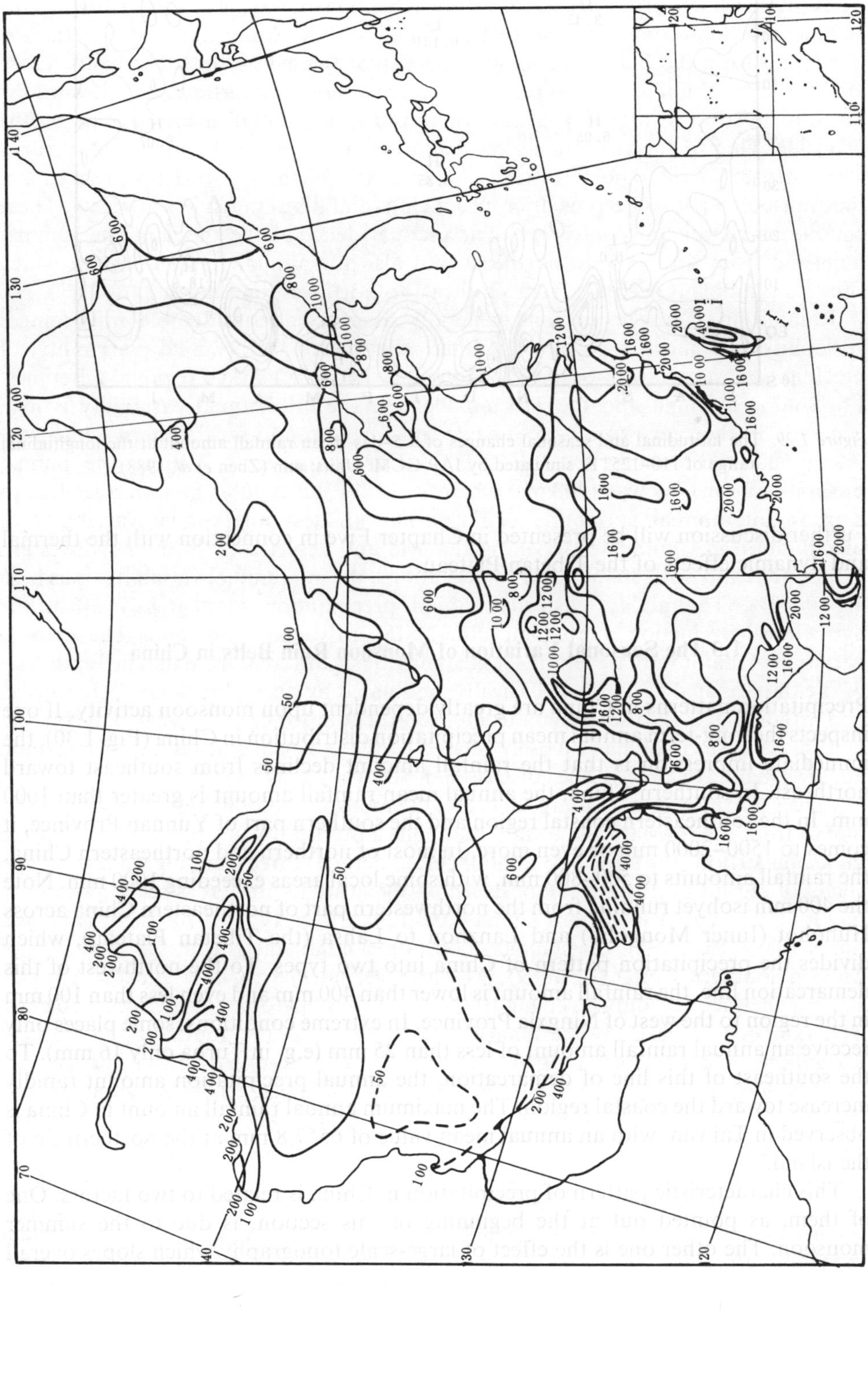
140
130
120
110
100
90
80
70
40
30
20

from southeast to northwest. As discussed in Sections 1.1 and 1.2, the airflows of the summer monsoon come from the western Pacific and the South China Sea, on the one hand, and from the Indian Ocean and the Bay of Bengal on the other hand. These airflows carry abundant moisture. As the summer monsoon airflows flow toward the north or northwest, lifted by large-scale terrains, the rainfall mainly occurs on the windside. Due to the continuous condensation of vapor in the monsoon airflows when they go inland to northwestern China and the northern part of northern China, the monsoon airflows become drier, producing rainfalls of much less intensity. These conditions cannot only interpret their gradual delay of the onset of the summer monsoon from southeast to northeast, but also the annual rainfall pattern which shows a significant decrease from southeast to northwest and north.

Rainfall in monsoon regions generally assumes a considerable inter-annual variability. In most regions of China, the mean inter-annual variability of precipitation ranges from 10 to 30%. However, there is a marked regional difference. For southern China and the Tibetan Plateau, the interannual variability is low, generally in the range of 10–15%, while for northern and northeastern China, the interannual variability rises to 15–30%. For example, in Beijing the annual precipitation of 1959 was 1408 mm whereas in 1891 it was 168.5 mm, thus leading to a mean interannual variability of 31%. The greatest interannual variability of precipitation in China is observed in the arid and desert zone in the northwestern China, generally with 30–50%. At some stations, the interannual variability is as high as 50–60%. The highly unsteady annual precipitation may greatly affect the agricultural production, due to high vulnerability to droughts and floods. The seasonal variation of major rainfall belts in China is obviously different from that in South Asia where the seasonal variation is basically modulated by onset, break and active cycle, and withdrawal of the Indian summer monsoon. The advance and retreat of the rainfall belt in the eastern China is closely related to the advance and retreat of monsoon airflows that have already been discussed in Section 1.2. When the summer monsoon advances northward, its leading zone, i.e., the monsoon rainfall belt correspondingly moves from the low-latitude to mid and high latitudes. In this process, as does the monsoon airflow, the monsoon rainfall belt undergoes three standing stages and two stages of abrupt northward jump. Based on the latitude-time cross-section of the 10 day rainfall amount in eastern China (105–120° E) for 1961–1970 (Lau, Yang and Shen, 1988) (Fig. 1.31), the summer monsoon rainfall starts from the early part of May in southern China (<30° N). At this time, northern China and the Yangtze River Valley are dry, only receiving 10–20 mm of rainfall for a time period of 10 days. The rain belt is basically stationary in the region to the south of the Nanling Mountains, which corresponds to the rainy season of southern China (presummer rainy season). This first standing stage of the major rain belt generally continues until mid-June when it rapidly shifts to the Yangtze River Valley. The second standing stage starts the rainy season there, the so-called "Plum rains" (Meiyu). The time span of the Plum rain season lasts on an average for one month (10th June–10th July). From mid-July, the rain belt rapidly moves over northern China and, in August further over northeastern China, which is the northernmost position of the summer monsoon rainfalls. This third standing stage of the rain belt causes a rainy season in

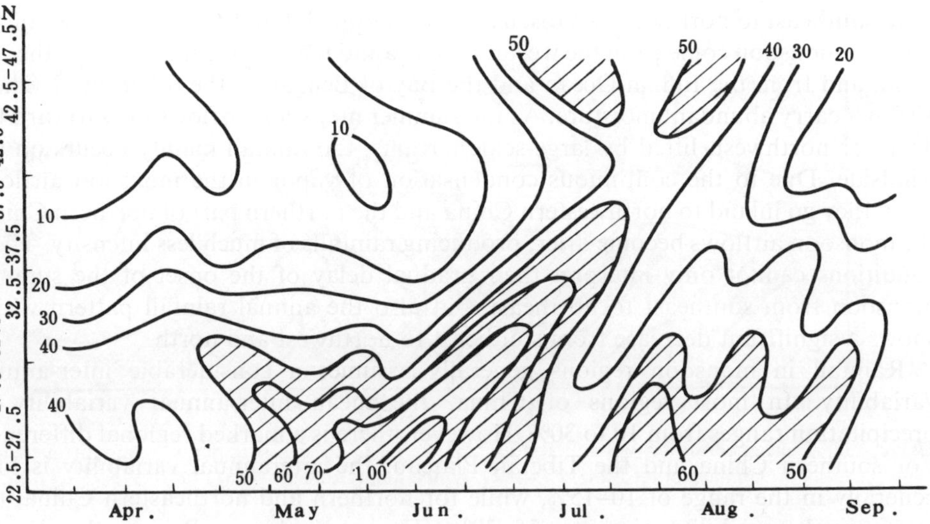

Figure 1.31. The latitude-time cross-section of the 10-day mean rainfall amount in the eastern part of China (105–120° E) for 1961–1970. Units: mm (Lau, Yang and Shen, 1988).

the northern part of China that generally lasts for one month. In the last part of August the rainy season of northern China ends, with the major rain belt retreating southward. At the end of August, it crosses the Yellow River and, early in September, it moves back to southern China. At such times, most of the eastern part of China is dominated by a dry spell. Note that from mid-July the second rainy season predominates over southern China, with a gap of one month between the presummer rainy season and this rainy season which mainly is caused by typhoons, ITCZ and other tropical disturbances. Around the beginning of October, the rainy season over the eastern part of China comes to an end and rapidly makes way for the winter monsoon.

Guo and Wang (1981) analyzed the pattern of rainfalls during the summer monsoon by using a longer record of precipitation in China (1950–1979). Similarly, they pointed out that the summer monsoon rain belt assumed three relatively stable periods, with the first period (the rainy season in southern China) having the longest persistence, the third period (the rainy season in northern China) the second longest persistence, and the second period (the Meiyu season) the shortest. Figure 1.32 is the 30-year mean precipitation pattern, showing that the axis of the maximum rainfalls (shaded area) runs nearly vertically rather than parallel to the abscissa. This implies that the Meiyu over the Yangtze River Valley are not very stable, with the position of rain belt significantly swinging southward and northward, and the duration of the rain period becoming shorter.

Table 1.1 presents the onset and ending dates, and the mean durations of persistence and break for rainy seasons in northern China, the Yangtze River Valley and southern China (Zhang and Lin, 1985). It can be seen that the rainy period of the Yangtze River Valley has only 7 pentads, while the rainy period of southern China

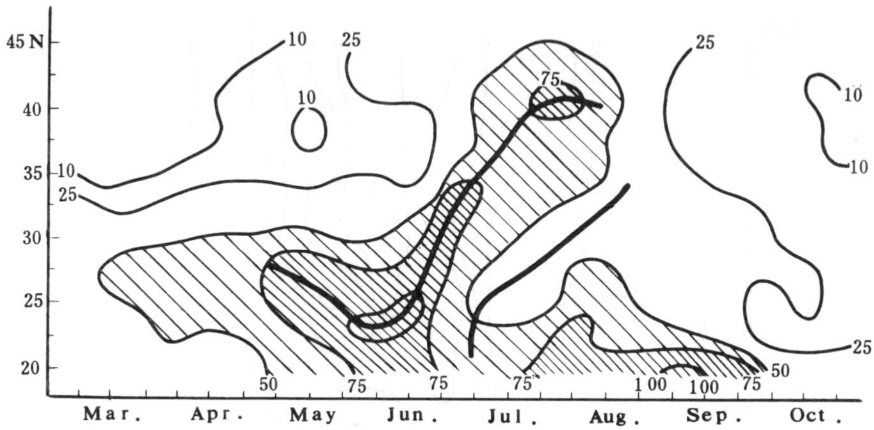

Figure 1.32. The temporal and spatial variations of the mean 10–day precipitation amount in eastern China averaged for 1950–1979. Units: mm (Guo and Wang, 1981).

has 13 pentads. Another interesting point in Table 1.1 is the presence of the active-break cycle that is broadly similar to the situation for the Indian monsoon rainfalls. On an average, the break duration for southern China is the longest, northern China has the second longest, and the Yangtze River Valley has the shortest.

The interannual variability of rainy periods for various regions is also different. The date of the onset of monsoon rains in southern China has maximum variability, with the earliest onset in the 3rd pentad of March and the latest onset in the 1st pentad of June. The RMS (root mean square) error is 4.3 pentads. The time difference between the earliest and latest onset of monsoon rains over the Yangtze River Valley spans 16 pentads. For most years, the rainy season in this region starts during the last part of June, with the RMS errors being 3.6 pentads. The RMS of the onset date is at the minimum, being 3.2 pentads. The time difference between the earliest and latest onset is only 10 pentads, with most years starting the rainy season betweem the last part of June and the early part of July. The interannual variability of the ending date of the monsoon rainy season is smaller than that of the onset date. Southern China has minimum variability, spanning 9 pentads, with the RMS error being 2.3 pentads. For most years, the rainy season ends in the middle or last part of June. The RMS error of the Yangtze River Valley is 3.1 pendads and the time difference between the earliest and latest ending dates spans 10 pentads. For most year, the monsoon rains

Table 1.1. The onset and ending dates, and the duration of persistence and break of rainy seasons for northern China, the Yangtze River Valley and southern China (Zhang and Lin, 1985)

Regions	Onset date	Ending date	Duration of persistence	Duration of break
Northern China	2nd pentad, July	6th pentad, August	10 pentads	1.4 pentads
Yangtze River	3rd pentad, June	4th pentad, July	7 pentads	0.8 pentads
Southern China	5th pentad, April	6th pentad, June	13 pentads	3.4 pendads

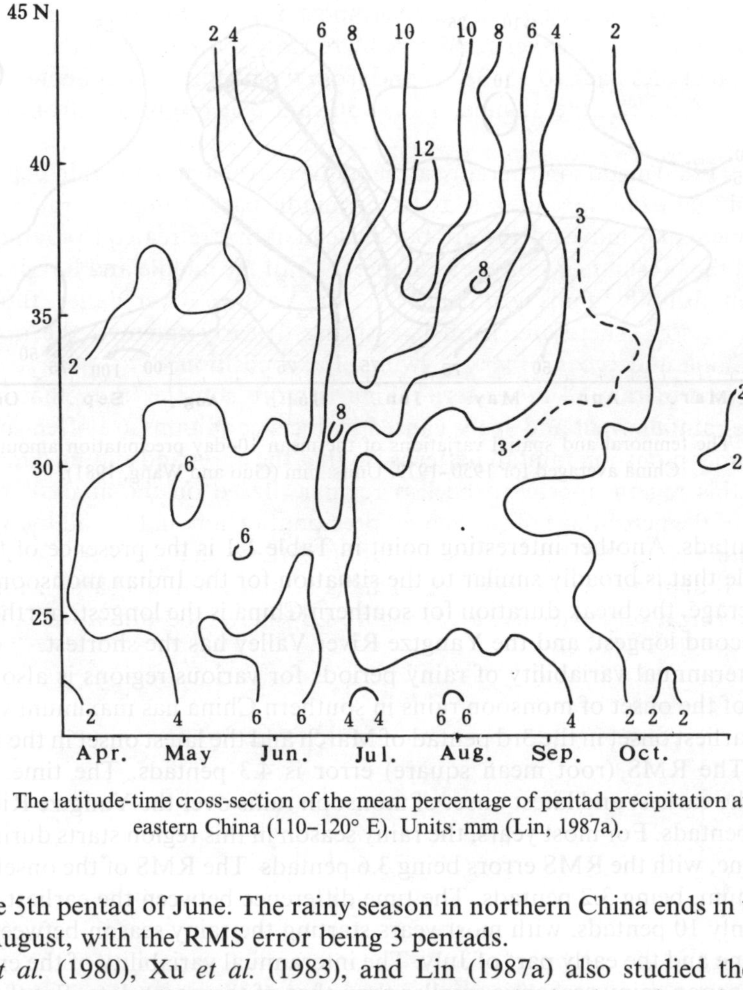

Figure 1.33. The latitude-time cross-section of the mean percentage of pentad precipitation amount (%) in eastern China (110–120° E). Units: mm (Lin, 1987a).

end in the 5th pentad of June. The rainy season in northern China ends in the middle part of August, with the RMS error being 3 pentads.

Tao *et al.* (1980), Xu *et al.* (1983), and Lin (1987a) also studied the seasonal variations of the rainfall belt in eastern China. They obtained results quite similar to those described above. Lin, and Tao and co-workers used the percentages of a particular pentad precipitation amount relative to the annual precipitation or summer precipitation (May-July) to characterize the features of the distribution and variation of the rain belt. This index of monsoon rainfall is relatively objective when one hopes to make a comparison of movement, and onset and ending dates of monsoon rainfalls for various regions, because of their quite significant regional differences. Figure 1.33 is the latitude-time cross-section of the mean percentage of pentad pricipitation amount obtained by Lin (1987a).

It can be seen that prior to the middle part of June the major rain belt is located in southern China (to the south of 25° N), during the last part of June it jumps to the middle and lower the Yangtze River Valleys (29–32° N), in the early and middle part of July, it continues to advance northward, being located in the region of 33–37° N and 36–38° N, respectively, and during the period from the last part of July to the middle part of August, it reaches its northernmost position. Starting from the last part of August, the rain belt weakens and rapidly retreats southward. In the early part of September, it comes back to 35–38° N and from the middle part of September to the early part of October, the summer monsoon rain belt disappears from the China Main.

Xu *et al.* (1983) made some modification to the movement of the rain belt in China. They found that from the end of May to the middle part of June the rain belt in fact did not immediately move northward but, instead, it first retreated to southern China and caused the presummer rainy season there. Until the middle and last part of June the rain belt abruptly jumps northward over the Yangtze River Valley, thus starting the Meiyu season. Therefore, from late spring to early summer, the rain belt in southern China undergoes a marked swing of axis position.

The pattern of rainfall in western China is more complex than that in eastern China. The rainfall event and spells caused by local conditions are often superposed on the summer monsoon rainfall patterns, so that the onset and ending dates of rainy season in this region appear to be less regular. Based on the analysis of annual variations of the percentage of the pentad precipitation amount for all the stations in western China (Lin, 1987b), four precipitation patterns may be identified:

(1) The Indian monsoon rain type. This is a major rainfall pattern for western China, covering a very extensive area. Figure 1.34 shows the annual variation of the

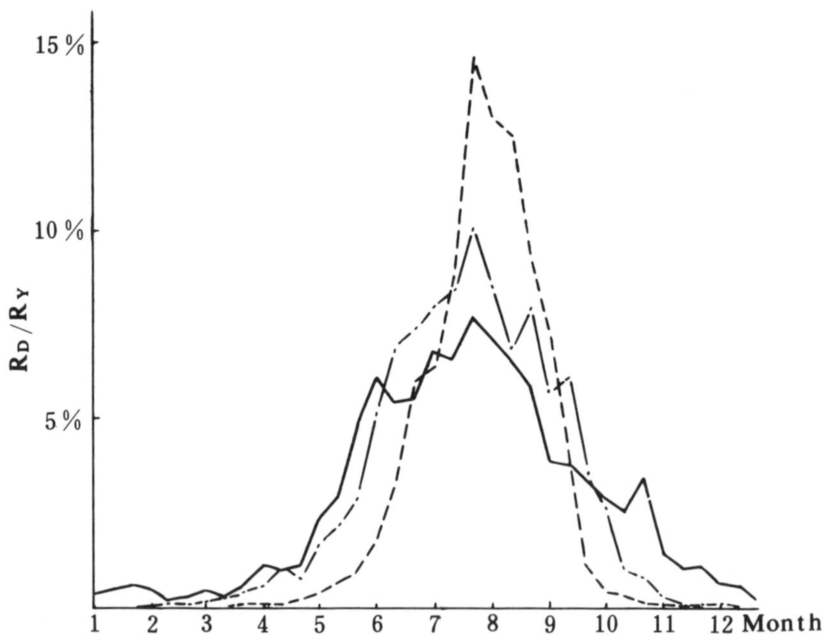

Figure 1.34. The Indian monsoon rain type, depicted by the percentage curve of the ratio of pentad to annual precipitation amount. The solid line represents the annual variation for the subregion of the western part, the dashed line the subregion of the middle part and the dashed-dotted line the southern part (Lin, 1987b).

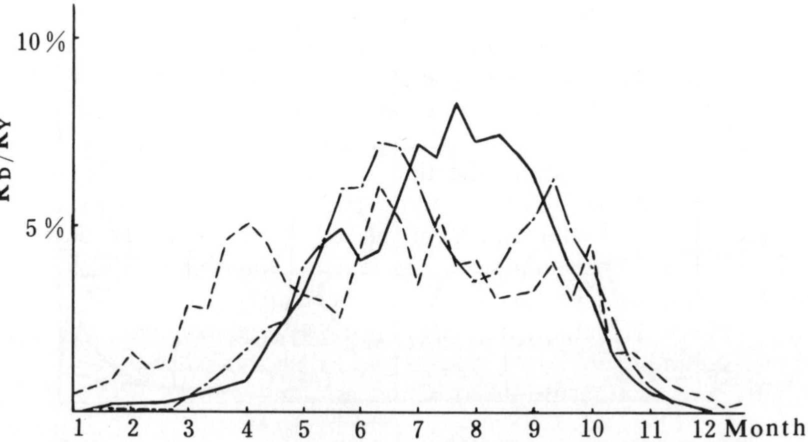

Figure 1.35. The type of spring rain in the northeastern part of the Tibetan Plateau (solid line), the type of mid-summer drought in western Province and the eastern plateau (dashed line), and the type of the dual rainy season in western Yunnan Province and the southern plateau (dashed-dotted line) (Lin, 1987b).

pentad precipitation amount for three groups of representative stations (the western subregion, the central subregion and the southern subregion). There is only one rainy season for the whole year, with the maximum rainfall amount occurring in July. The duration of the rainy season and the focusing degree of rainfall amount are different for various subregions. In the southern part, the rainy season sets in earlier and ends later, the rainy season is long and the focusing degree of rainfall amount is low, whereas in the western part, the case is just the opposite.

(2) The type of spring rain in the northeastern part of the Tibetan Plateau (Fig. 1.35). There are two peaks on the annual curve, i.e., a secondary peak occurs during the middle and last part of May before the primary peak caused by the Indian summer monsoon rain season. The spring rainfall peak may be caused by the weather systems in the westerlies.

(3) The type of mid-summer drought in western Sichuan and the eastern plateau. There are two rainy seasons for the whole year, occurring in June and July, and in September, respectively, and a marked dry season in between. The area affected by this type is relatively limited.

(4) The type of dual rainy season in western Yunnan Province and the southern plateau. This is a very unique rainfall pattern in western China. The first rainy season, mostly local, begins in the early spring (March and April) and is believed to be caused by amplifying wave troughs. The characteristics of the spring rain also greatly depend on the local terrain. The second rainy season starts during the early part of June, at the same time as the extensive Indian summer monsoon rain. Figure 1.36 indicates the dates of the onset of rainy seasons in western China. Prior to the middle part of May, for most of western China there are no significant ranfall areas where the percentage of the mean pentad precipitation amount exceeds 4% and which can be defined as the index of the rainy season. Starting from the middle and last part of May, the rainy

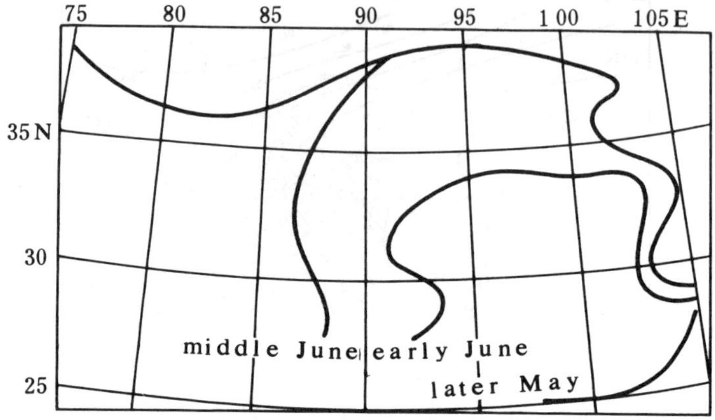

Figure 1.36. The dates of the onset of summer monsoon rainy season (> 4% isoline, see the text) for May and June (Lin, 1987b).

season, which is basically of localized rainfalls, first occurs in the northeastern part of Tsinghan Province and western Sichuan Province. Up to the early part of June, the rainy season advances northwestward, and during the middle of that month, the summer monsoon rainy season dominates the region to the east of 86–87° E. Until the last part of June, the whole Tibetan Plateau and its surrounding areas are subjected to heavy rain. The rainy season in this region may persist until the middle part of September, but with some active-break cycles. The rain area during this period is closely linked to a west-east oriented low-level wind shear-line located along 33° N remaining presence there for the whole summer (Fig. 1.37) (Xu *et al.*, 1983). This condition is in obvious contrast to that in eastern China where the low-level shear-line associated with the major frontal zone migrates seasonally. The retreat and weakening of the summer monsoon season begins from the last part of August in the central Tsinghai Province. Then the rainfall areas with an isoline greater than 4% continuously withdraw southeastward, with the rainy season ending during the last part of September for western China. Therefore, the characteristics of the summer monsoon rainy season in this region are similar to those in India rather than to those in eastern China.

Lau *et al.*(1988), however, have indicated that the monsoon rainfall for western China, including Tibet, is a much simpler picture compared to the east. The rainfall season is quite well defined around 30° N. The heavy monsoon rain comes late in the season around 20–31, July and continues only for about a month without a break. There is a persistent minimum in the rainfall around 40° N. Farther north, rainfall appears to sporadic and ill-organized.

Finally, we shall discuss the relationship between the Indian summer monsoon rainfall and the summer monsoon rainfall in China. In the 30's, Zhu (1934), the founder of modern meteorology in China, in his pioneering work “Monsoon in Southeast Asia and rainfall amount in China” pointed out: “Our country is located in proximity to India. In winter, the weather is controlled by the Siberian high and in

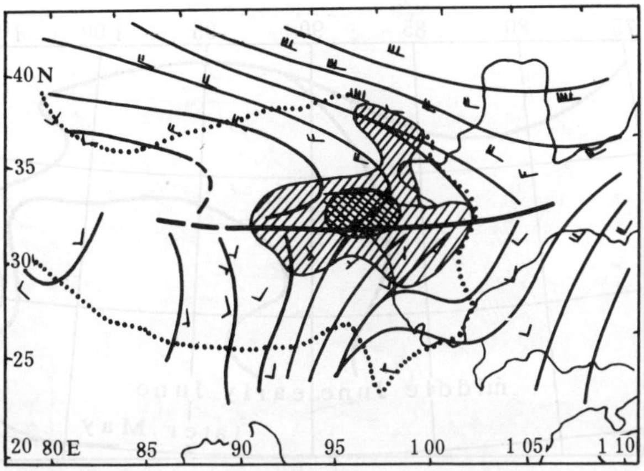

Figure 1.37. The mean streamlines at 500 m for July. The shaded area is the rain area (Xu *et al.*, 1983).

summer by the central Asian low pressure. For both these countries, therefore, the behavior of monsoons is very similar". He further pointed out: "Due to the differences in geographical location and terrain features, the monsoon in our country is obviously different from that in India." These statements emphasize both similarity and difference between the Indian monsoon rainfall and the monsoon rainfall in China. Subsequent studies have broadly followed this line. Shen and Zhu (1980) found that while the Indian southwest monsoon is weak, the rainfall over the Yangtze River Valley is below normal, and vice versa, indicating positive correlation between them. But this relationship is not true for the Tibetan Plateau, which generally shows a plausible negative correlation with the Indian monsoon rainfall, i.e., an active monsoon corresponds to the break monsoon over the Tibetan Plateau (Huang *et al.*, 1980; Chang, 1981). But a recent study of the relationship of summer monsoon rainfall between the central plateau and the central Indian Peninsula for May-August 1979 has cast some doubts on the above negative relationship (The Second Group of QXPMEX, 1988). They have indicated that the negative correlation relationship was only true for a half of the cases for this period, with the other half indicating a positive correlation relationship. From the above discussions, it is difficult to draw some definite conclusions with confidence concerning the relationship of summer monsoon rainfalls between India and China. This is greatly dependent upon the regions and time periods under study. Recently, Guo and Wang (1988) used a longer set of data (1951–1980) for 110 stations in China and 31 subregions in the Indian peninsula to further study this problem, and have, to some extent, clarified the relationship between the Indian summer (June-September) monsoon rainfall and that in China. Figure 1.38 shows the correlative distribution between the total amount of rainfall in the Indian peninsula and the rainfall amount at 110 stations in China for June-September. The most significant correlative region with a significance level of 0.95 is northern China (designated as region a) which demonstrates a positive correlation. The extensive region to the west of 100° E, including the Tibetan Plateau and southern Xinjiang Province, shows a negative correlation (region b), with the maximum

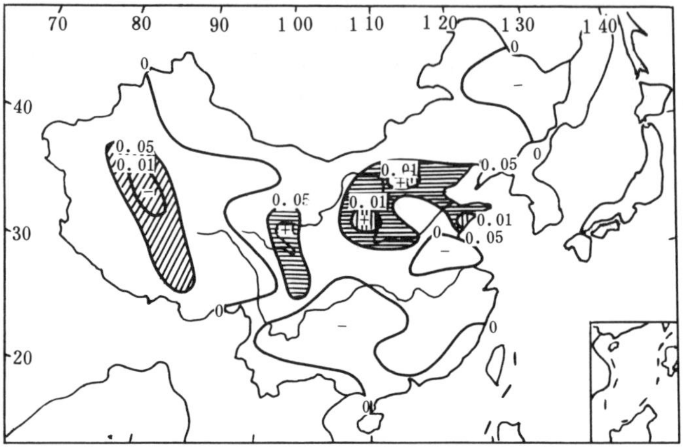

Figure 1.38. The correlation coefficients between summer rainfalls at 110 Chinese stations and rainfall averages for the whole Indian peninsula during June to September 1951–1980. The isopleths of the correlation coefficients are represented by the confidence level, with the regions at 0.05 or 0.01 confidence level shaded (Guo and Wang, 1988).

correlation observed in Lahsa and Qiemo. In contrast to the previous results, over the Yangtze River Valley (region c), the correlation coefficient is very small, with no station attaining the confidence level of 0.90. Figure 1.39 is the multi-year variation of summer monsoon rainfall for the Indian peninsula, and regions a, b and c. It can be seen that the curves for northern China and the Indian peninsula has similar trends of variation, with their correlation coefficient being 0.65 (the significance level exceeding 99.9%). The general trend of variation for the Indian peninsula and region B is the opposite, with their correlation coefficient being −0.50. For the Yangtze River Valley, there is no same or opposite trend of variation between it and the Indian peninsula, with the correlation coefficient being 0.15. This result is believed to present a more confident correlation relationship than that obtained based on shorter-term data sets.

1.4 Subtropical High over the Western Pacific Ocean

In the surtropical region to the south of mid-latitude westerlies, there are extensive belts of high pressure nearly encircling the globe in which are embedded six or seven high cells. The subtropical high over the western Pacific is one of these cells. As depicted in Fig. 1.14, this high is one important component of the East Asian monsoon system. To the western flank of the high, the southeasterly or southwesterly monsoon wind often apparently develops in summer. The subtropical high over the western Pacific assumes a marked seasonal variation in respect to its position. Table 1.2 shows the long-term mean position of the ridge line of the subtropical high over the western Pacific Ocean (The Second Group, 1988). As the summer monsoon develops from spring to summer, the subtropical high continuously moves

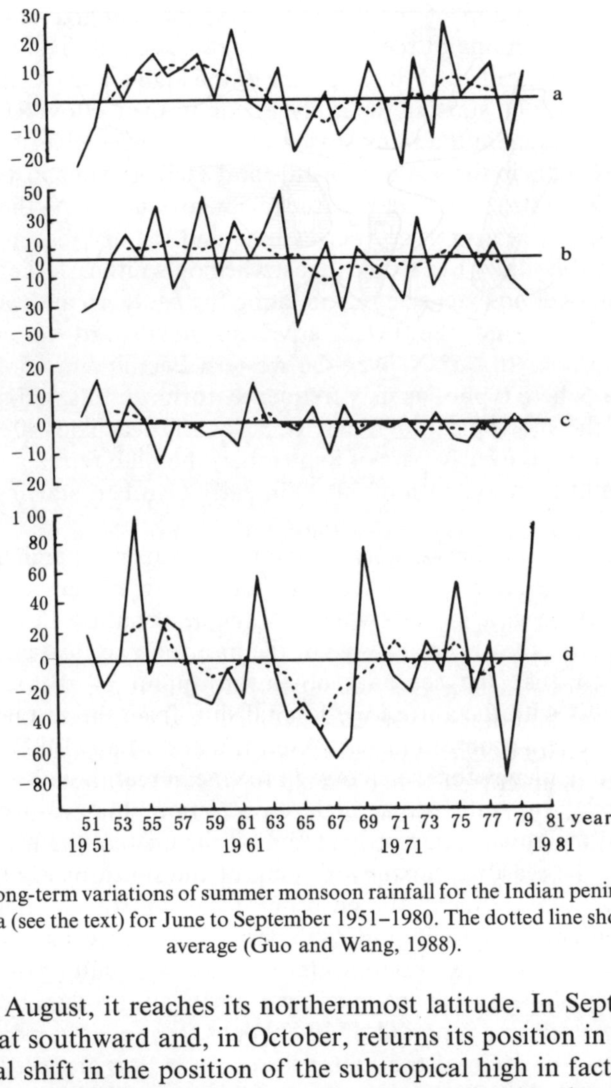

Figure 1.39. The long-term variations of summer monsoon rainfall for the Indian peninsula (d) and regions a, b, and c in China (see the text) for June to September 1951–1980. The dotted line shows a 5-year running average (Guo and Wang, 1988).

northward. In August, it reaches its northernmost latitude. In September, the high begins to retreat southward and, in October, returns its position in June.

The seasonal shift in the position of the subtropical high in fact shows up as an alternation of gradual shift and abrupt jump, which is closely related to the advance and retreat of major rain belts in eastern China. In general, three abrupt northward jumps may be observed. The first northward jump occurs in the middle or last part of June, with the ridge line of the high moving over the region of 20–25° N. This shift

Table 1.2. Long-term mean position of the ridge line of the subtropical high over the western Pacific at the longitudinal range 110–115° E

Month	May	June	July	Aug.	Sept.	Oct.
Latitude	19.3	21.3	24.4	27.8	25.4	21.8

indicates a northward jump of the rain belt over the Yangtze and Huaihe River Valleys, heralding the ending of the pre-summer rainy season in southern China. In the early and middle part of July, the ridge line of the high assumes second northward jump, reaching 25–30° N, starting the rainy season in the Yellow River and Huaihe River Valleys and ending in the rainy season in the Yangtze River Valley. The third northward jump occurs in the last part of July and early in August. This jump brings the ridge line of the subtropical high over the region around 30° N, thus leading to the beginning of the rainy season in northern China and the dry season in the Yangtze River Valley. At this time, the tropical easterlies or southeasterlies (southeasterly monsoon) prevail over an extensive region along the southern and western flanks of the subtropical high, and the ITCZ advances northward up to its seasonal northernmost latitude, 10–15° N over the western Pacific and 15–20° N over the South China Sea, where typhoons may frequently form. In this mid-summer season, the ridge line of the subtropical high may occasionally reach 35–40° N on the daily weather map. During the early part of September, the high retreats to the region to south of 25° N and then to south of 20° N in early October, starting the transition toward winter.

The cause of the seasonal northward jump of the subtropical high has not yet been clarified, although arguments have been put forward to explain this unique event. Some investigators suggest the association of the seasonal northward jump of the subtropical high with the abrupt change of the monsoon vertical circulation cell. It has been revealed that the seasonal northward jump of the subtropical high concurrently occurs with the abrupt meridional shift from the south to north of the monsoon vertical circulation cell in East Asia. Lin and Miao (1981) further pointed out that there is a tendency for the change in the mean meridional cell to lead to the northward jump of the subtropical high. In addition, they also emphasized the important role of the downstream propagation of the unstable growth of long waves in the process of the seasonal northward jump of the subtropical high. Besides the seasonal north-south movement, the position of the subtropical high demonstrates the short and medium-term oscillations with periods of 1–2 weeks (Tang *et al.*, 1982). Figure 1.40 is the curves of the variations in the three-day running mean position for four years, clearly showing the medium-term north-south oscillation of the mean position of the high during a period of 10–20 days. It should be pointed out that a change in the mean position of the subtropical high is first observed in the western part of the high and then in the eastern part. Their time difference or delay spans about 15–20 days. For a particular site, this consecutive change in the mean position may cause a local oscillation of nearly the same period. Figure 1.41 shows some evidence for this change. It can be seen that the change in the mean position of the high propagates westward from the central Pacific.

The east-west movement of the subtropical high also shows the complex seasonal and medium-term oscillations. Figure 1.42 indicates the case for 1979. Before June, the subtropical high was stationary at the longitudinal range 170° E–170° W. From June onward, the high started to move westward and then stalled over the region 140–170° E. In July through September, the high moved further westward to the range of 120–140° E. In the process of westward movement of the subtropical high, one may

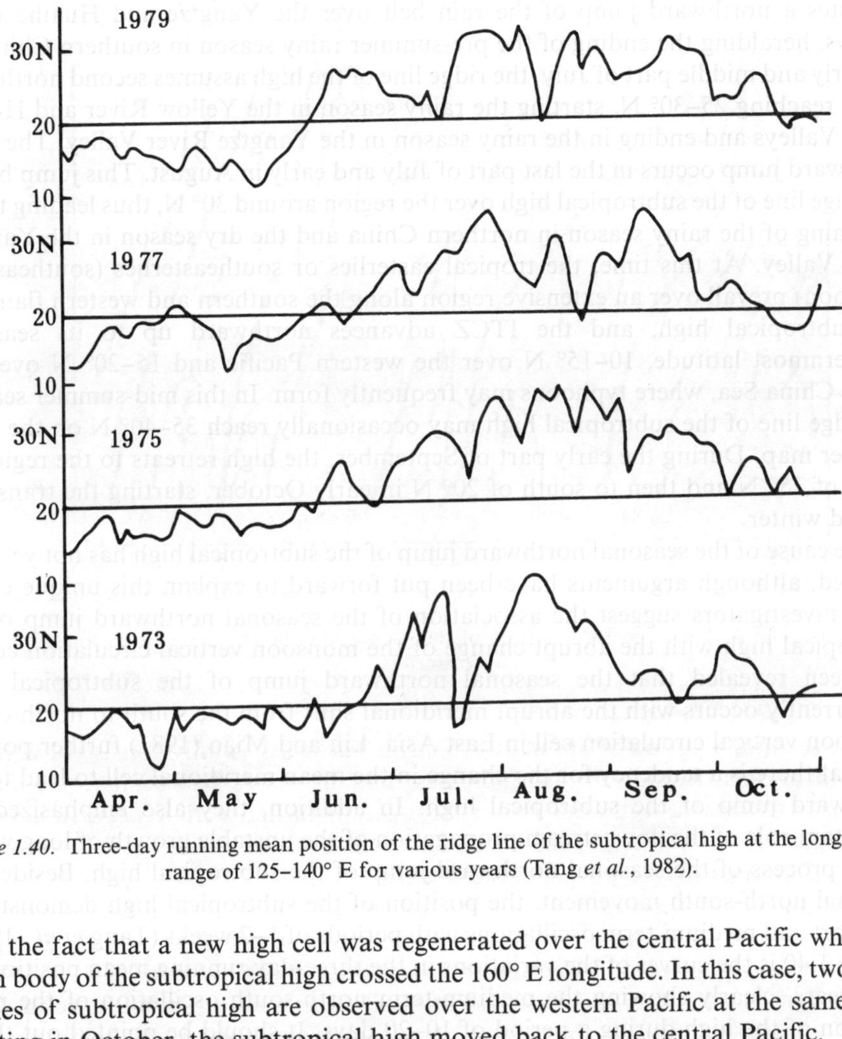

Figure 1.40. Three-day running mean position of the ridge line of the subtropical high at the longitudinal range of 125–140° E for various years (Tang *et al.*, 1982).

note the fact that a new high cell was regenerated over the central Pacific when the main body of the subtropical high crossed the 160° E longitude. In this case, two main bodies of subtropical high are observed over the western Pacific at the same time. Starting in October, the subtropical high moved back to the central Pacific.

It is worthwhile noting that while the cell of the subtropical high was stable at a certain longitudinal range, it demonstrates east-west oscillatory behavior, with a mean period of 7–10 days and a mean amplitude of 10 longitudes or so. Further, when a main cell moves away from the central Pacific and arrives at the longitudinal range 120–130° E, another new cell may form again in the central Pacific. This process of regeneration assumes a quasi-periodic character, with the period of 10–15 days. Utilizing the complex empirical orthogonal function analysis, Bi (1990) has studied the medium-term oscillation and propagation of the subtropical high in summer. She found that the subtropical high has two types of oscillation: 10–15 and 20–30 days (Fig. 1.43). It can be seen that the phase function for both the first and second modes demonstrates a linearly increasing trend with time and the phase shift from 2_π to zero assumes a periodic oscillation, with the first mode having period of about 20–30 days

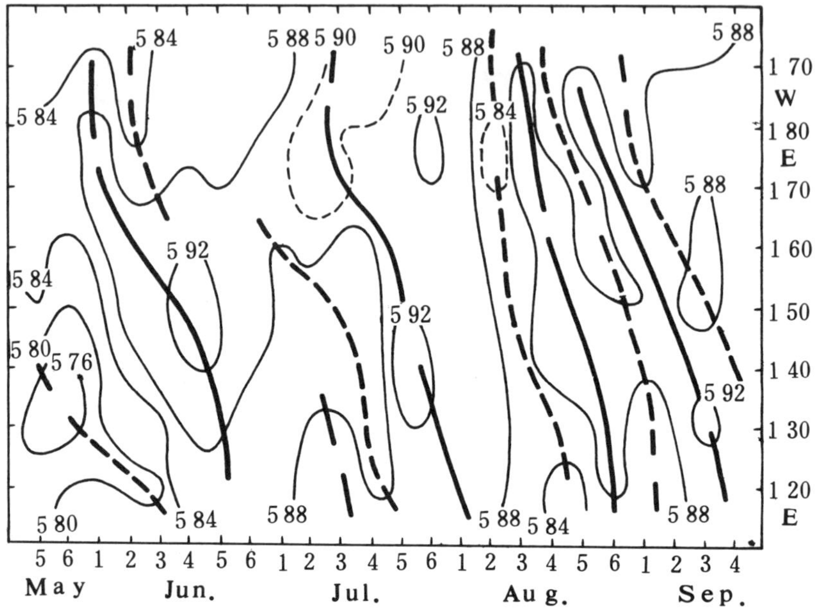

Figure 1.41. The pentad to pentad variation of the geopotential height along 30° N for 1979. Units: decameters.

and the second, a period of about 10–15 days. The latter is fully consistent with the above-mentioned results. Bi also indicates that the above two oscillations mainly propagate westward, with the initial stronger disturbance first occurring in the eastern part of the western Pacific. But, for oscillations of shorter periods (less than 10 days), the disturbance propagates eastward.

The intensity of the subtropical high is generally characterized by the area and intensity indexes. The former is defined as the total number of grid points that have geopotental heights of 588 decameters or greater in the region of 110–180° E, poleward of 10° N. Thus, this index may describe the areal extent of the subtropical high. The latter is defined as the accumulative sum of the geopotential height equal to or greater than 588 decameters in the region of 110–180° E, poleward 10° N, with 588 (decameter) = 1, 589 = 2, 590 = 3, etc. Figure 1.44 is the annual variations of the long-term mean area and intensity indexes of the subtropical high over the western Pacific. It is very evident that these two indices have their peaks during the months of June to September, while they are markedly weak in winter.

Numerous works have indicated that the subtropical high over the western Pacific is a rather stable circulation feature. It may often maintain above or below-normal intensity for several months, or for even a couple of years in individual case. For instance, the intensity of the subtropical high turned into the stronger phase in the second half of 1957, and since then it persisted with a positive anomaly until the end of 1962, lasting for about 5 years. Figure 1.45 shows the persistence of the subtropical high in terms of the correlation coefficient of the area index for consecutive months

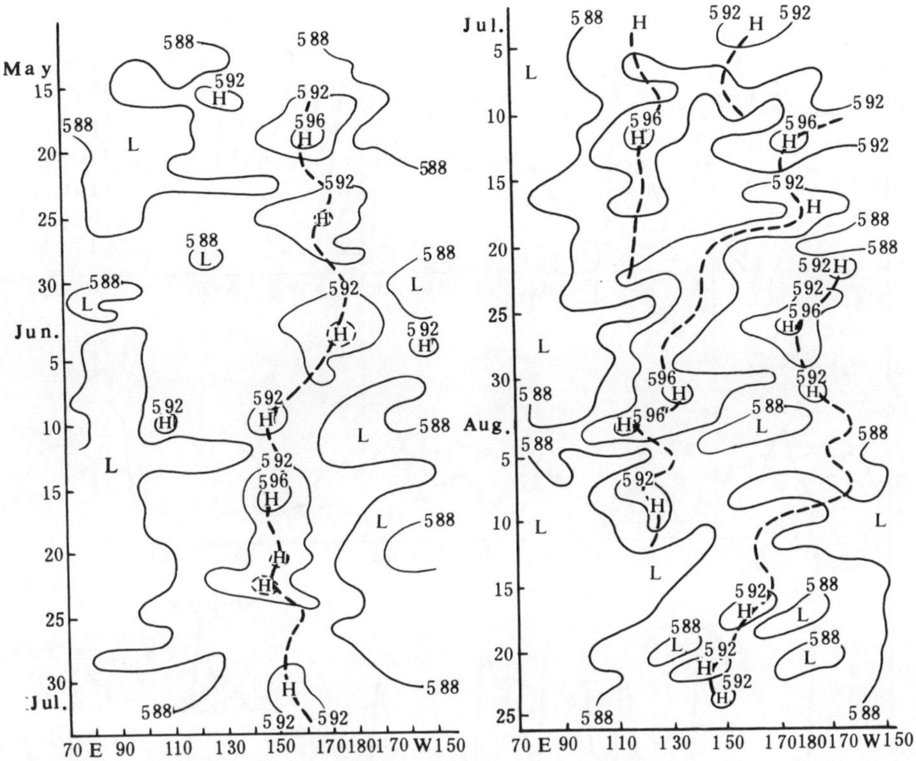

Figure 1.42. The longitude-time cross-section of the geopotential height of ridge line of the subtropical high for May through August 1979. Units: decameters.

(Chen and Yang, 1987). It can be seen that the month-to-month correlation coefficients generally have values greater than +0.60, an indication of marked self-persistence of the subtropical high. In September and October, however, the correlation is very low, implying that the breakdown or destabilization of the subtropical high often occurs during the fall. Recently, Yang and Chen (1990) pointed out that the persistence of the ridge line of the subtropical high is much lower than that of its area index. This persistent character is very useful for long-term forecasting of large-scale circulation patterns.

Another prominent feature of the area index of the subtropical high shows up in its climatic oscillations. The spectral analysis of the time series of the area index has revealed the existence of some long-term oscillations with periods of 2.5 and 3.5 months, and 3.5, 11–12, and 22.5 years.

Among them, the 3.5 year oscillation is most significant (Fig. 1.46). Chen (1982) and Zhang (1988) related this oscillation to the sea surface temperature over the equatorial eastern Pacific, indicating that they have a similar trend of variation.

The variation of the subtropical high is closely related to the activity of the summer monsoon in China. Table 1.3 indicates the correlation coefficients between SMI (summer monsoon intensity index) and the characters of the subtropical high. SMI is

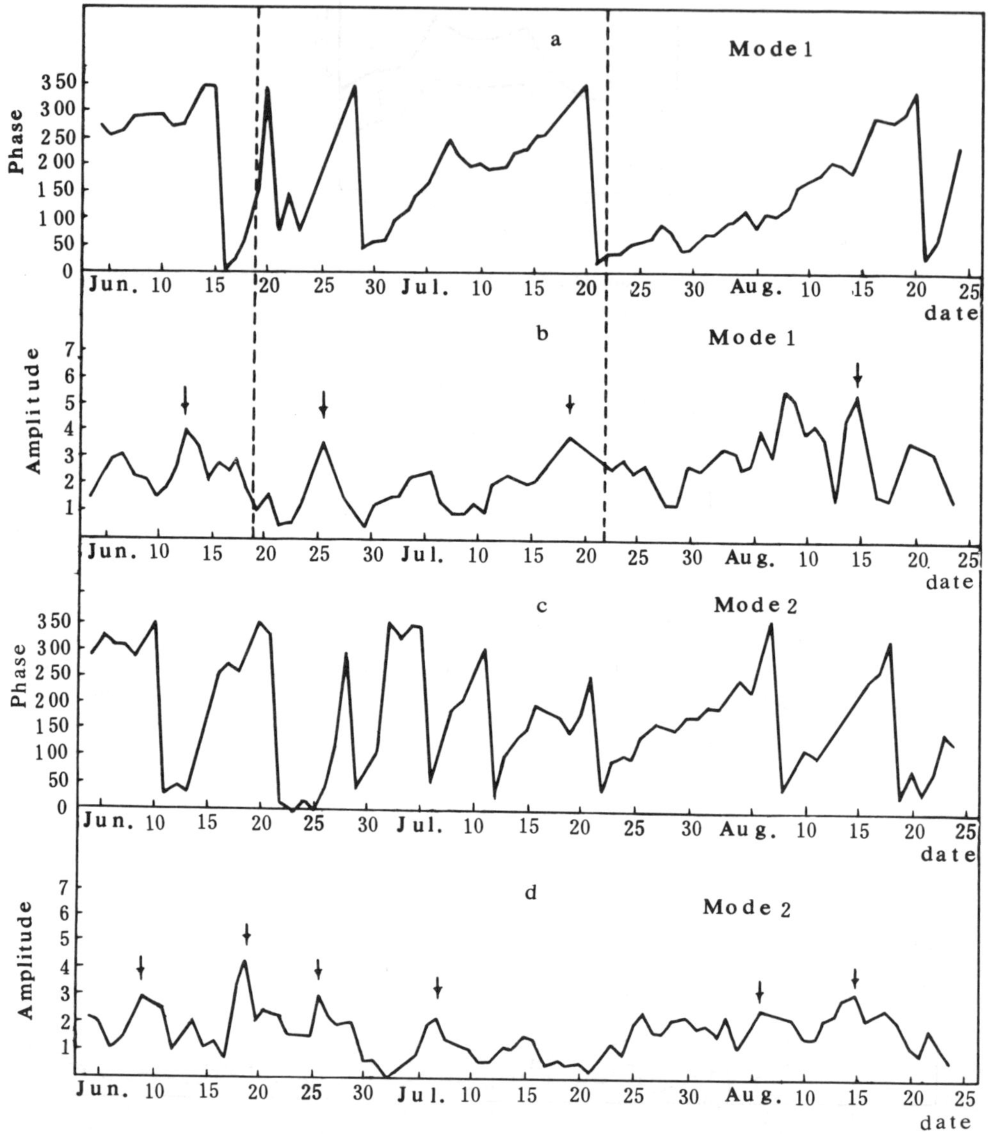

Figure 1.43. The phase and amplitude functions of the subtropical high in the summer 1979: (a) the phase function for the first mode, (b) the amplitude function for the first mode, (c) the phase function for the second mode, and (d) the amplitude function for the second mode (Bi, 1990).

defined as $\Sigma_i \Sigma_j \Delta p$, when $\Delta p \leq -5$ hPa, where $i = 1, 2 \ldots 12$ (months), $j = 10, 20, \ldots$, 50° N, $\Delta p = p_{110^\circ \text{ E}} - p_{160^\circ \text{ E}}$, and p is the sea-level pressure. The characters of the subtropical high mainly include five quantities: the position of the westernmost point of the ridge line, the position of the ridge line, the northern boundary of the subtropical high, and the intensity and area indexes. It can be seen that the intensity

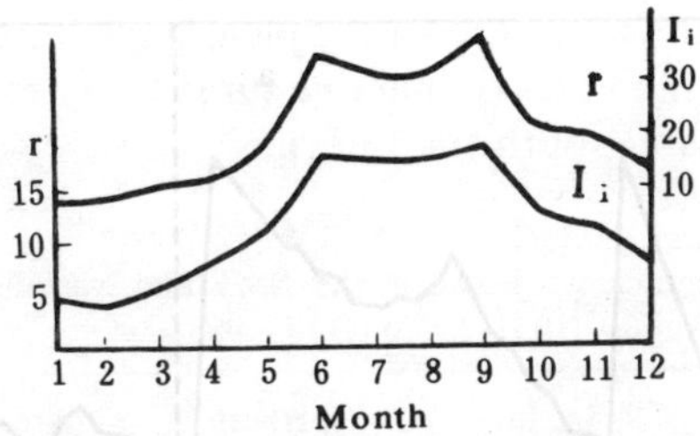

Figure 1.44. Annual variations of the long-term mean area (r) and intensity (I_i) indexes of the subtropical high over the western Pacific.

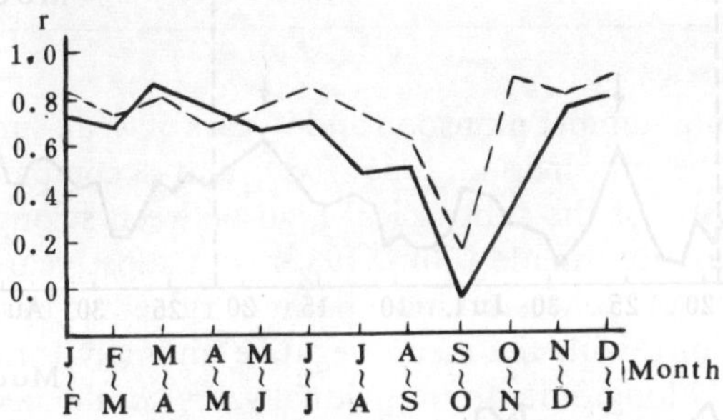

Figure 1.45. The persistent correlation coefficients of the area index of subtropical highs over the western Pacific (solid line) and the Northern Hemisphere (dashed line) for consecutive months (Chen and Yang, 1987).

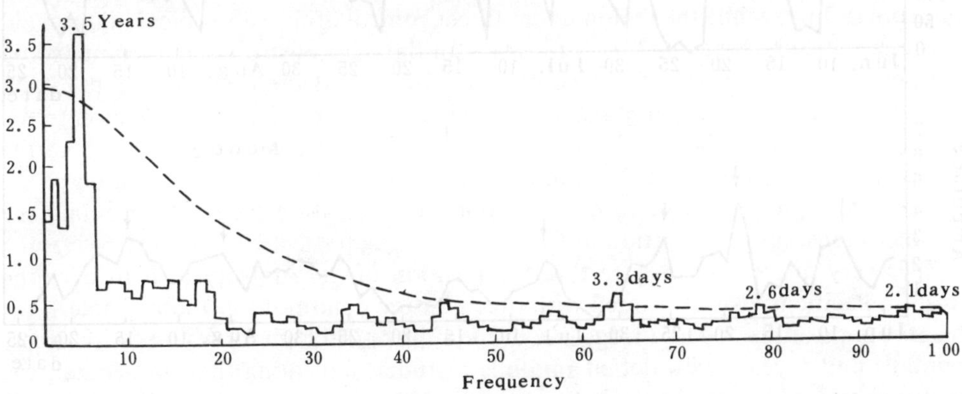

Figure 1.46. The power spectrum of the long-term mean area index of the subtropical high (Long-term Weather Prediction Group, 1976).

variation of the summer monsoon has the best correlation with the position of the westernmost point of the ridge line, and a certain degree of correlation with other characteristic quantities. For example, in July, the SMI has a better correlation with the position of the ridge line and the northern boundary of the ridge line and, in August, a better correlation with the area and intensity indexes. In summary, for the years of active summer monsoon, the subtropical high shifts farther north and west than normal, whereas for the years of inactive summer monsoon the subtropical high moves down farther east and south than normal.

Table 1.3. The correlation coefficients between the SMI and various characteristic quantities of subtropical highs for July and August (1951–1980) (Guo, 1985)

Month	Westernmost point	Ridge line	Northern boundary	Area index	Intensity index
July	−0.50	0.46	0.33	0.07	0.15
August	−0.34	0.11	0.10	0.24	0.28

Eight years of strong summer monsoon and 9 years of weak summer monsoon have been identified for the time period 1951–1980, and a comparison of the various quantities characteristic of the subtropical high between strong and weak summer monsoon regimes has been made (Guo, 1985). For most years of strong summer monsoon, the position of the westernmost point of ridge line has a negative anomaly (more western position), with the mean negative anomaly (more western) being 9 longitudes and even 17 longitudes for individual years. At the same time, the position of the ridge line has a positive anomaly (more northern). In contrast, for the years of weak summer monsoon, the behavior of the subtropical high is nearly opposite to that for the years of strong summer monsoon. It is known that the major rain belt is usually located in the region at about 6 latitudes poleward of the ridge line of the subtropical high, while the region under direct control of the subtropical high is void of rainfall. In July, based on Guo's statistics for the years of strong summer monsoon, the position of the westernmost point is at 110° E and the ridge line is at 27° N. Therefore, above-normal rainfalls occur over the Huaihe Valley or the region to the north of it, whereas the rainfall deficit occurs over the region to the south of the Yangtze River. These results have indicated that the variation in the position of the subtropical high is deemed to be an important circulation factor determining the summer monsoon precipitation in China.

It is very interesting to note that the position of the subtropical high over the western Pacific is also associated with precipitation conditions in India. Table 1.4 is the locations of the 500 hPa ridge line of the subtropical high for 5 years with excess rainfalls and 5 years with deficit rainfalls in India, indicating that, in general, the latitude of the ridge line of the subtropical high for the years with excess rainfalls is higher than that for the years with deficit rainfalls (Guo and Wang, 1988).

The variation of the subtropical high (location and intensity) is related to several atmospheric and oceanic factors, including the tropical and subtropical air flows, the longwaves at mid and high-latitudes, the convective activity in the tropics and the sea-surface temperature. Many Chinese forecasters have indicated that the short-term variation and activity of the subtropical high are frequently affected by disturbances in the westerlies at the middle latitude, in particular the adjustment process of long-waves (Tao *et al.*, 1963). It has been often observed that if a new longwave trough develops at 80° E upstream and at the same time extends southward, a longwave ridge would then generate at 110–120° E downstream, at distance of about half a wavelength. At this time, the subtropical high over the western Pacific would move northward and westward and establish a cell in the eastern part of China. In contrast, if a new long-wave ridge grows at about 80° E, a longwave trough would develop at 110–120° E and force the subtropical high to move southward down to the oceanic area around 140° E. This adjustment process is usually accomplished in 3–5 days, with a considerable shift of position of the induced subtropical high.

Table 1.4. The long-term locations of the ridge line of the subtropical high at 500 hPa for years with excess and deficit rainfalls in India, respectively (Guo and Wang, 1988)

	Years with excess rains (June-September)						Years with deficit rains (June-September)					
Year	1953	1956	1959	1961	1975	Aver.	1951	1965	1972	1974	1979	Aver.
Average latitude of ridge line (° N)	25.3	26.0	23.8	28.0	26.5	25.9	24.5	24.0	23.0	23.3	23.8	23.7

The blocking situation in East Asia may affect the activity of the subtropical high, especially the position of its ridge line (Liao and Zhao, 1990). If a stable and huge high ridge or blocking situation persists over East Siberia in summer, the subtropical high in most cases assumes a more southern position, thus shifting the rain belt further southward than the normal condition. A statistical analysis has indicated that for 60–70% of the years when the position of the subtropical high shifted farther southward than normal, one may observe the persistent existence of blocking situations in East Asia.

A few numerical simulations of the effects of disturbances at the middle latitude on the evolution of the subtropical high, have been undertaken (Department of Meteorology, 1976). They have justified the possibility of the effect coming from the middle latitude, especially for the longwave disturbances.

The variation in intensity of tropical easterlies to the south of the subtropical high may exert an important effect on the movement of the high itself. For example, when the patches of strong easterlies propagate westward from the east, which may be caused by the development of easterly disturbances over the western Pacific in summer, a westward movement of the center of the subtropical high is observed, thus leading to the intensification of the Pacific subtropical high. When a typhoon develops and moves westward to the south of the subtropical high, the high would move northward and westward simultaneously, due to the intensification of easterlies to the south. However, if the typhoon moves northward and is located to the southwest of the soubtropical high, the main body of the high often retreats southward. Sometimes, the near-equatorial anticyclonic system or buffer zone may move northward up to a higher latitude (10–15° N), thus favoring a westward shift and intensification of the subtropical high and, in some cases, it may merge with the subtropical high.

Activity of the warm high (or the continental high) at 500 hPa and the South Asian high at 100 hPa over the Tibetan Plateau, may also affect the behavior of the subtropical high. Statisitics have shown that 23 out of the 30 cases where the subtropical high demonstrated an evident westward shift for summers of 1957–1963, were possibly caused by warm air advection (or splitting warm high cells) originating over the Tibetan Plateau and being transported eastward, while 7 other cases were caused by disturbances in the westerlies.

A number of works have been directed to the relationship of the Pacific subtropical high and the equatorial Pacific sea-surface temperature (Chen, 1982; Zhan, 1988). Some investigators have indicated that in the case of high equatorial sea-surface temperature (SST), the axis line of the subtropical high slightly shifts southward, and assumes a east-west elongated distribution and a stronger intensity. In the case of low equatorial SST, the opposite condition is true. This evidence was argued by other investigators who had found a marked positive correlative relationship, i.e., the axis line of the subtropical high shifts northward rather than southward for the condition of high equatorial SST, and a lagging by 1–2 months response of the subtropical high to the equatorial SST. The above argument demonstrates a complex relationship, possibly depending upon the season, different temporal and spatial scale, and interactive process. Chen (1982) realized that the relationship between the subtropical high and the SST is in fact, an interactive, feedback process. The adjustment time duration of the SST in the equatorial region to the subtropical high on average comes to about 2 months, while the adjustment time of the latter to the former is about 4

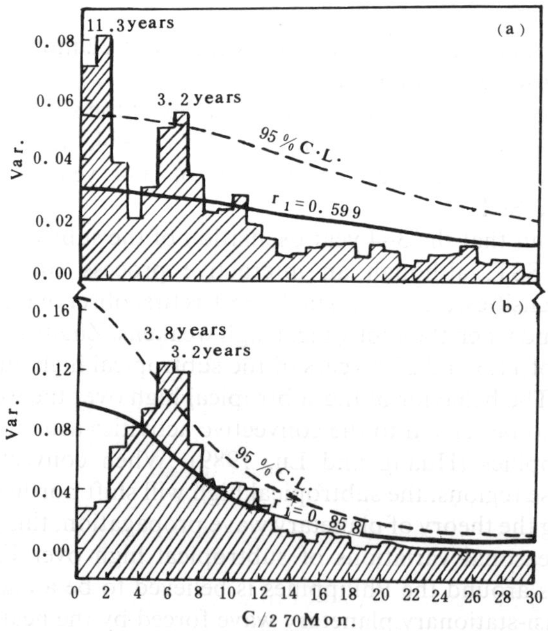

Figure 1.47. The power spectrum of the mean area index of the subtropical high over the western Pacific (a) and the SST (b) over the equatorial eastern Pacific Ocean (Zhang, 1988).

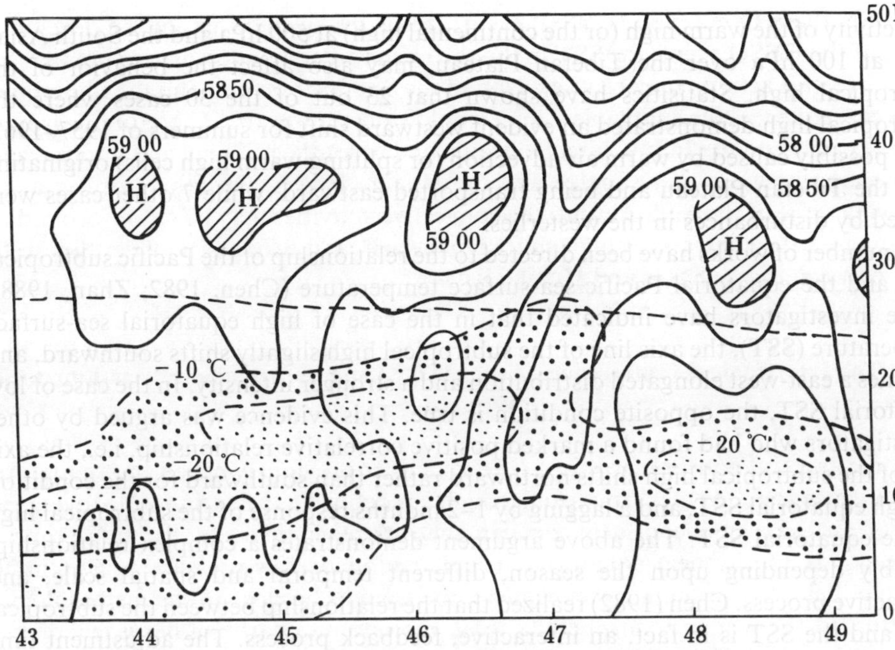

Figure 1.48. Latitude-time cross-section of 500 hPa height (m) averaged for 128–145° E and 5-day mean TBB below−10° C (dotted areas) along 130° E in August, 1984 (Huang and Lu, 1989).

months, with the whole interactive process taking about 22 months, nearly equal to half of their major oscillation of 3.5 years.

Based on data sets of the monthly mean area index of the subtropical high and the SST in the equatorial eastern Pacific for 1951–1984, Zhang (1988) further documented a very low frequency oscillation of 3–4 year (Fig. 1.47) for the subtropical high and SST.

But, he pointed out that the 3–4 year oscillation of the subtropical high is mainly controlled by the SST, although an interactive process still exists. When the subtropical high intensifies, a variation in the SST is first observed, while the variation in the SST lags behind when the subtropical high weakens. Zhang also argued that the longer oscillations of 11.3 and 22.5 years of the subtropical high might be associated with solar activity. The behavior of the subtropical high over the western Pacific and East Asia is found to be related to the convective activities around the South China Sea and the Philippines (Huang and Lu, 1989). When convective activities are intensified over these regions, the subtropical high will shift northward and intensify (Fig. 1.48). By using the theory of planetary wave propagation, this close relationship between the intraseasonal variations of subtropical high over East Asia and the convective activities around the Philippines is believed to be a manifestation of the propagation of quasi-stationary planetary wave forced by the heat source produced by vigorous convective activity around the Philippines.

The convective activity around the Philippines may be related to an El

Ñino/Southern Oscillation (ENSO) event. It was discovered that, at the developing stage of the ENSO event, the SST in the the western Pacific is colder in summer, and the convective activity may be weak around the South China Sea and the Philippines. As a consequence, the subtropical high would shift southward. Therefore, a drought may be caused on the Indo-China Peninsula and in southern China and floods occurs in the area between the Yangtze and the Huaihe Rivers. On the contrary, at the decaying stage of the ENSO event, the convective activities may be strong around the Philippines, the subtropical high shifts northward, and drought may develop in the area between the Yangtze and Huaihe Rivers.

1.5. The Intertropical Convergence Zone (ITCZ) and its Attendant Near-Equatorial Westerlies

The intertropical convergence zone (ITCZ), either previously called the "equatorial convergence zone" or the "equatorial trough", is defined as a continuous convergence line of airflows or a confluent zone in the airstream field. At the same time, it corresponds to the near-equatorial trough region or low-pressure region in the pressure field. In the summer hemisphere, the troughs lie close to the meteorological equator. In the winter hemisphere, a weak discontinuous trough lies a few degrees from the equator. The airflows converging into the equatorial trough mainly come from the easterly trade winds in the Northern and Southern Hemispheres. On cloud images taken by weather satellite, the ITCZ is apparently identified as a single or several nearly continuous and elongated convective cloud bands. Other satellite data, including the outgoing longwave radiation (OLR) data, may also be used to define the intertropical convergence zone in terms of the latitude of the minimum of the OLR in the zonal belt between 30° N and 30° S (Jiang, 1988). Figure 1.49 and Table 1.5 show the long-term mean position and intensity of the ITCZ. Over the oceanic basins, the ITCZ runs almost parallel to the latitude circle, with a mean position of 5–6° N over the Atlantic, 6° N over the Pacific, and 5° S over the Indian Ocean. Note another ITCZ over the southeastern Pacific which is located at about 5° S. Therefore, over the western Pacific, one may observe a double ITCZ (Hebert, 1969). From the seasonal distributions of OLR (Fig. 1.50a-b), the double ITCZ is seen to show obvious seasonal oscillations, i.e., an east-west oscillation for its position and a north-south oscillation for its intensity. As a matter of fact, the ITCZ over the southwestern

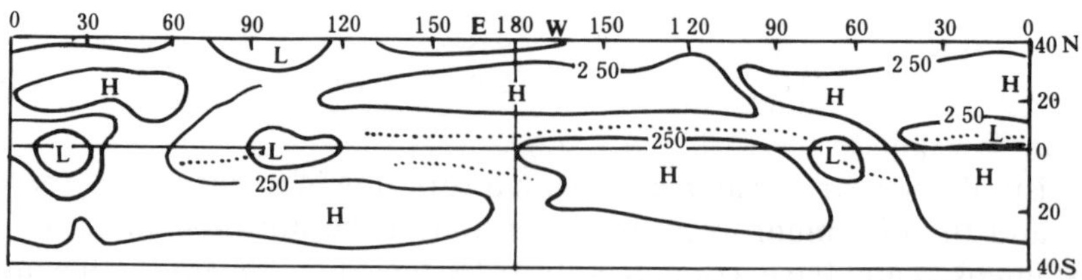

Figure 1.49. The long-term mean distribution of the OLR and ITCZ (1974–1984). Solid lines denote the isopleths of OLR, with an interval of 25 W m^{-2}; the dashed lines denote the ITCZ axis lines (Jiang, 1988).

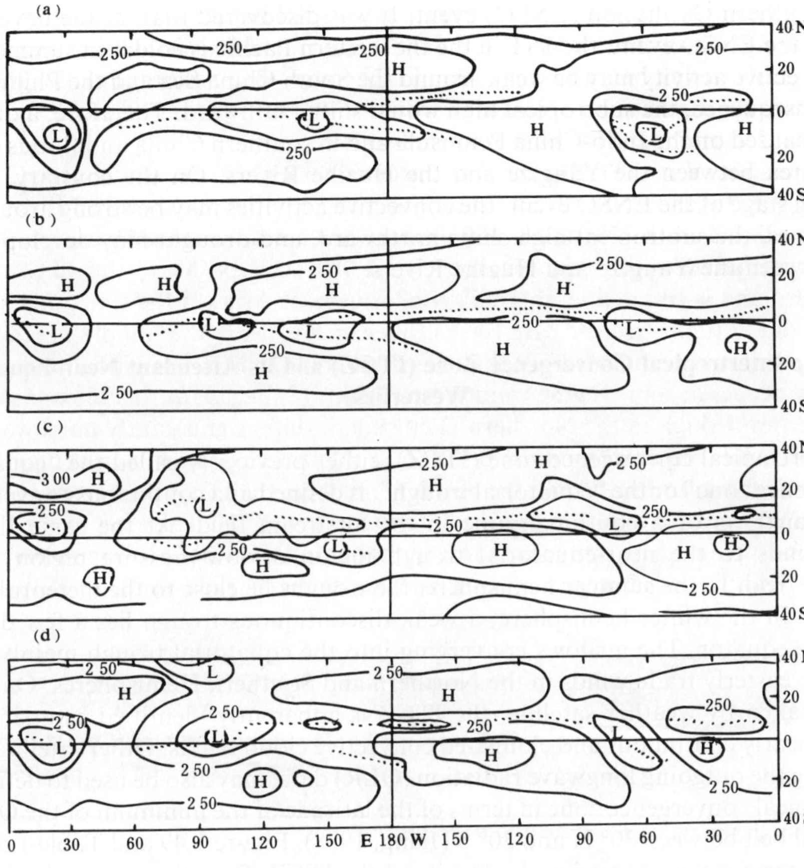

Figure 1.50. The same as Fig. 1.49, but for the seasonal mean distributions of the OLR and ITCZ, (a): winter (December-February), (b): spring (March-May); (c) summer (June-August), and (d) autumn (September-November) (Jiang, 1988).

Table 1.5. The long-term mean position and the OLR intensity (W m^{-2}) of the ITCZ in various tropical regions (Jiang, 1988)

Regions	Atlantic Ocean 10–40° W	East-Pacific Ocean 172.5–890° W	West-Pacific Ocean 120° E–170° W		Indian Ocean 65–90° E	Africa 5–35° E	South America 52.5–80° W
Position	5.6° N	6° N	6° N	5° S	5° S	2° N	2° S
Intensity	237	234	226	220	230	221	221

Pacific exists all the year round, mainly with a variation in intensity, i.e., the weakest intensity and the maximum westward extension in the northern summer, and the strongest intensity and the maximum eastward extension in the northern winter. The ITCZ in the Northern Hemisphere always persists over the tropical central and eastern Pacific. From winter to summer, its intensity continuously enhances and extends westward. In June, the ITCZ establishes itself in the tropical western Pacific, thus creating a long continuous ITCZ over the northern Pacific. This condition may persist until November and then the ITCZ over the western Pacific gradually disappears. As a result, an east-west oscillation of the position of the ITCZ can be observed, due mainly to its seasonal zonal shift.

The position of the ITCZ in the flow patterns has a greater seasonal variation than that in OLR fields, especially in South Asia and Southeast Asia. In January (Fig. 1.51), the ITCZ in the Northern Hemisphere has a southernmost position, located somewhere between the equator and 5° N. In April or May, the ITCZ moves northward and is situated at about 5° N in Southeast Asia. In the transition seasons (spring and autumn, see the case for October in Fig. 1.51) the double ITCZ is very evident. It should be pointed out that this double ITCZ is not of global phenomena, and only occurs in some regions and some seasons (Saha, 1970). In July and August, the monsoon trough in the Northern Hemisphere shifts significantly northward and develops. It has the northernmost position in the Indian peninsula, reaching 20–25° N. Over the western Pacific and the South China Sea, the ITCZ runs from northwest to southeast, with an averaging position found at 10–15° N. This part of the ITCZ is very active, being the birthplace of most typhoons in this region. It is the convergence zone between the southwest monsoon and the easterly trade wind. In autumn, the

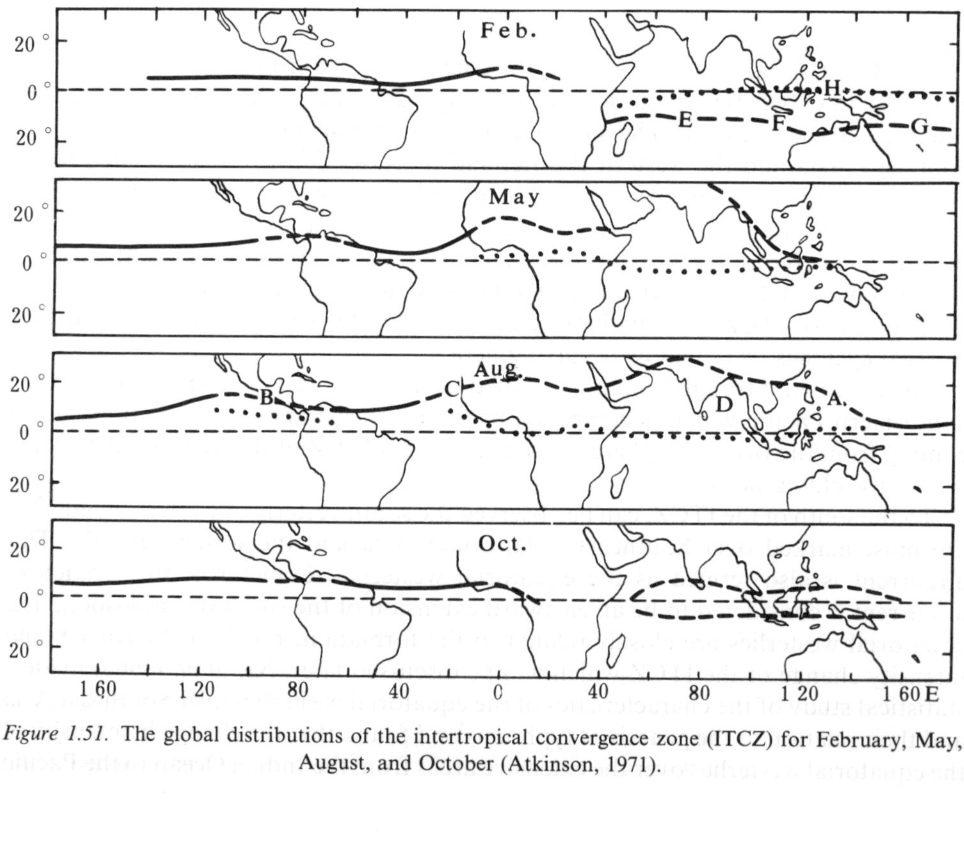

Figure 1.51. The global distributions of the intertropical convergence zone (ITCZ) for February, May, August, and October (Atkinson, 1971).

Asian monsoon trough retreats southward more rapidly than over the western Pacific.

The discrepancy of the positions of the ITCZ between the OLR field and the flow field is readily interpreted in terms of the mutual relationship between the convergence zone of airflows and cloud bands associated with the ITCZ. In general, the latter is located to the south of the former, in the southwest monsoon flow. Figure 1.52 schematically illustrates this situation (Gray, 1979). Therefore, the latitude of the ITCZ in the OLR field that reflects the distribution of cloud system, is lower than that in the flow field.

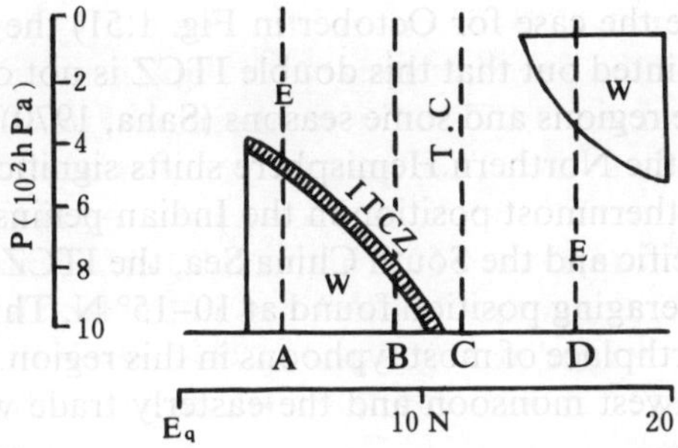

Figure 1.52. North-south cross-section of wind across the monsooon trough (Gray, 1979).

Based on the different character of the convergence of airflows, the ITCZ may be classified into two types: the monsoon trough and the trade wind trough or the equatorial trade wind belt. The monsoon trough is usually observed over South Asia, Southeast Asia and the western Pacific, and its formation is closely related to the monsoon. The monsoon trough is characterized by the great horizontal wind shear, with the westerlies or southwesterly winds to the south and the easterly trade wind to the north. Therefore, in the transition zone from westerlies to easterlies, there exists a relatively calm doldrum where the wind is weak and highly variable. This doldrum is of the active ITCZ type, with large areas of cloud systems or clusters, and even tropical cyclones or typhoons occurring here.

The trade wind trough is mainly situated over the North Atlantic, and the central and eastern Pacific which is composed of the northeast and southeast trade winds coming from the two hemispheres, respectively. An ITCZ of this type is inactive and typhoons seldom occur.

To the south of the ITCZ, can be observed the southwesterly or westerly winds that are most marked over Southeast Asia, South Asia and the western Pacific. This aircurrent is also termed as the equatorial westerlies. Sometimes, this branch of aircurrent is considered to be an eastward extension of the southwest monsoon. The equatorial westerlies are closely related to the formation, northward advance, and intensity change of the ITCZ which will be discussed later. Xie *et al.* (1963) made a statistical study of the characteristics of the equatorial westerlies over Southeast Asia and their relation to the genesis of typhoons and found that the change in intensity of the equatorial westerlies over the extensive areas from the Indian Ocean to the Pacific shows a considerable degree of consistency, i.e., when the westerlies over the southern part of the Indian peninsula intensify, the westerlies over Southeast Asia also intensify, but with a time lagging from the west to the east. Most typhoons in summer occur during the time period of the eastward intensification of the equatorial westerlies, especially after they have passed over the southern part of the Philippines.

The ITCZ over the South China Sea is rather unique for its long persistence and significant seasonal variation, as revealed by numerous recent studies (Jiang and Yu, 1985; Institute of Tropical and Marine Meteorology, 1976). The activity of the ITCZ in this region is mainly observed in July-November, with a monthly mean duration of persistence of 14–20 days. The ITCZ very seldom occurs in the cold season. Figure 1.53 indicates the seasonal variation in the monthly mean position of the ITCZ over the South China Sea. In January, the northeasterly wind or cold surge prevails in this region and the ITCZ has its southernmost position, located at 15–17° S. As the season progresses, the ITCZ gradually moves northward at a rate of 1–2 latitude degrees per month. In March, in a few cases, the ITCZ may pulse into the Northern Hemisphere. In April, the ITCZ in the Southern Hemisphere rapidly moves northward and, meanwhile, another ITCZ forms in the Northern Hemisphere, thus creating a double ITCZ, located at 6–9° S and 5–7° N, respectively. Starting from the middle part of June, the ITCZ in the Southern Hemisphere disappears and only that in the Northern Hemisphere persists. In June, the western part of the ITCZ

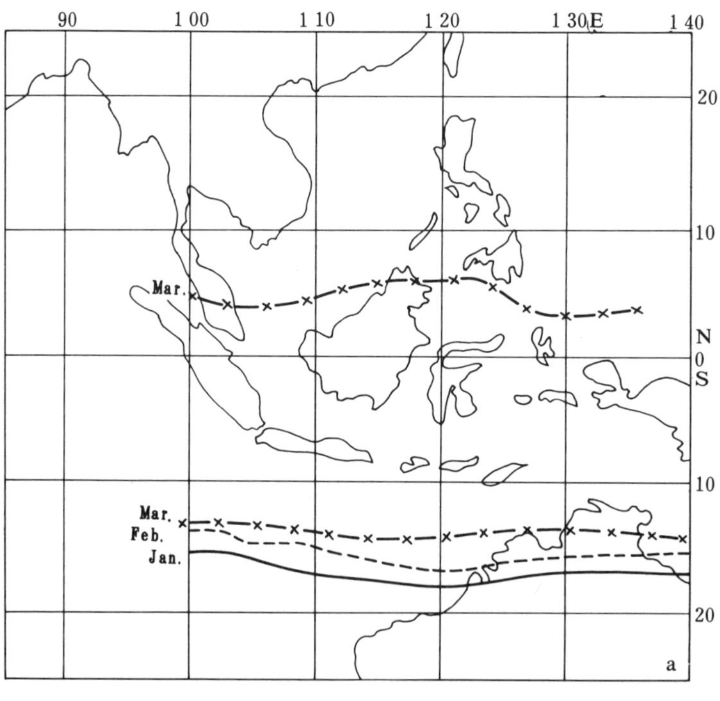

Figure 1.53a. Monthly mean positions of the ITCZ over the South China Sea (1973–1979), (a) for January-March.

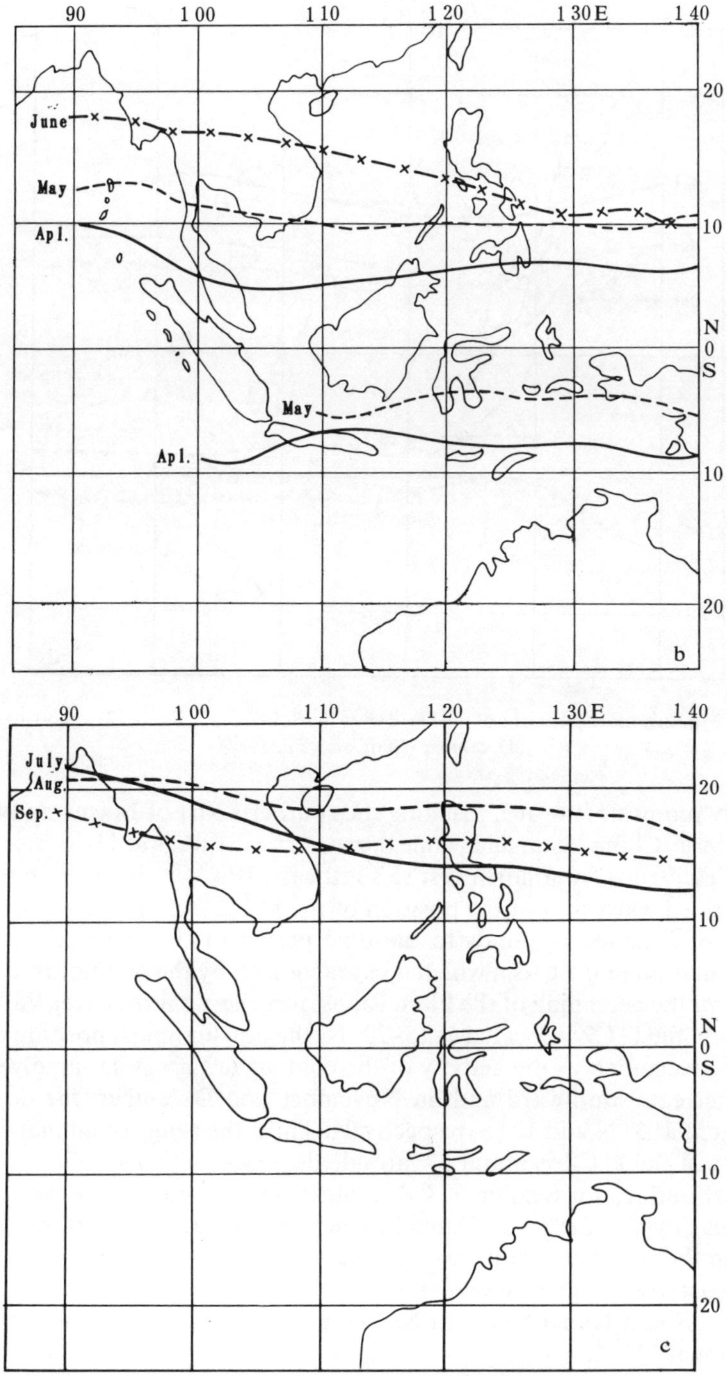

Figure 1.53b-c. Monthly mean positions of the ITCZ over the South China Sea (1973–1979), (b) for April-June, (c) for July-September.

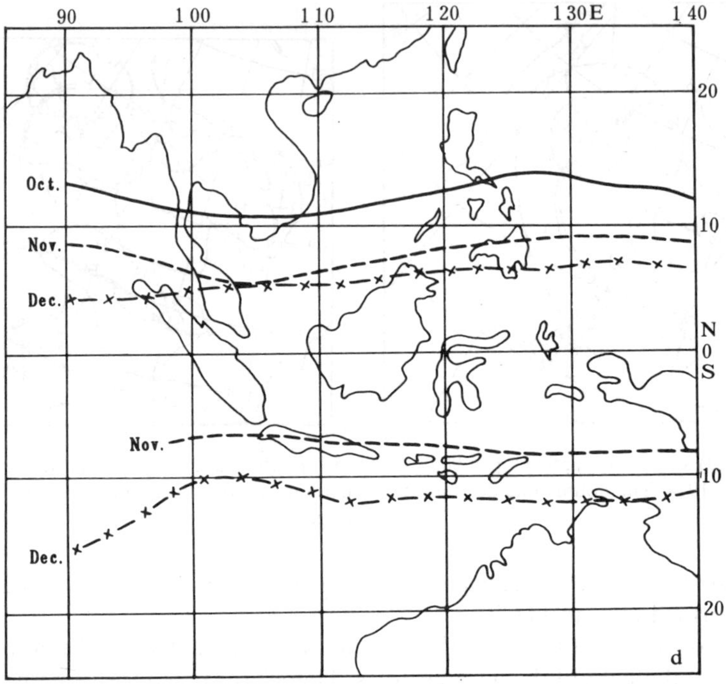

Figure 1.53d. Monthly mean positions of the ITCZ over the South China Sea (1973–1979), (d) for October-December (Jiang and Yu, 1984).

significantly jumps northward, reaching the southern part of Burma and the central part of the Indo-China Peninsula, while the eastern part stalls at 11–12° N. Therefore, the entire ITCZ runs from northwest to southeast. It is very interesting to note that the abrupt northward shift of the position of the ITCZ from the first part of May to the middle of June, corresponds to the time period of the seasonal change in the general circulation in East Asia which is characterized by the onset of Indian summer monsoon and the beginning of the Plum Rains over the Yangtze River Valley. In July to September, the ITCZ is located at 15–20° N, the northernmost position of the year. In October-December, as the activity of the cold air in East Asia rapidly intensifies, the ITCZ retreats southward and, in November and December, the double ITCZ forms, located at 5° N and 12° S, respectively. Thus, the range of annual variation in the position of the ITCZ is about 37 latitude degrees.

Over the South China Sea, the ITCZ results from the cold front or wind shear line which comes down to the South China Sea and assumes a significant transformation, especially in the relatively cold season or the transition season. In this case, the low-level northeasterly wind accompanying cold surges may penetrate into the central or the northern South China Sea, and converge with the equatorial westerly wind to form a regional ITCZ.

Figure 1.54 is an example. On 5 June, a wind shear line associated with a cold front moved down to 20° N. The next day, the low over Japan moved northeastward and

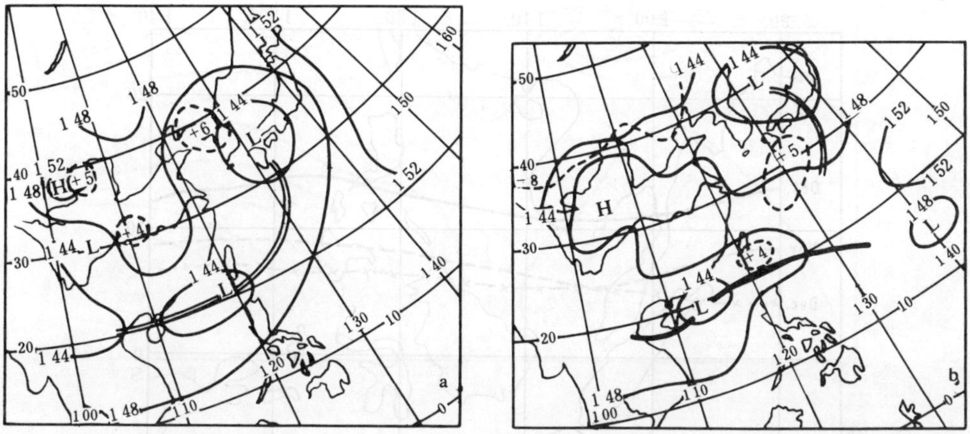

Figure 1.54. (a) 850 hPa synoptic map at 00:00 GMT, 5 June 1974, and (b) 850 hPa synoptic map at 00:00 GMT, 6 June 1974. The double line denotes the trough line or shear line, and bold solid line, the ITCZ (Chen and Ding, 1979).

the associated wind shear line was left over the northern part of the South China Sea and transformed into an ITCZ.

Analysis of transient features of the structure of the ITCZ has been hindered due to insufficient data about the oceanic areas. Therefore, a composite method is used to obtain an average picture of the structure of the ITCZ, as was done previously by Gray (1967). Figure 1.55a-c show some aspects of its structure (Group of Tropical Meteorology, 1976).

To the south of the ITCZ, there is the westerly wind below 500 hPa and the easterly wind above 500 hPa, whereas to the north of the ITCZ, there is a consistent easterly wind. Thus, the convergence of the two kinds of aircurrents is only a phenomenon in the lower troposphere. The maximum (10 and 6 m s^{-1}) of the westerlies and easterlies are both observed at about 850 hPa, 10 and 6 latitude degrees away from the ITCZ, respectively.

The axis line is nearly vertical from the surface to 700 hPa and then shows a marked tilting southward from 700 to 500 hPa.

Figure 1.56 is a streamline analyses at 850 and 200 hPa, showing a disturbance with the wavelength of 1800 km embedded in the ITCZ at low-level and an upper level anticyclonic circulation located to the north of the disturbance at 200 hPa. The divergence field in Fig. 1.56 is consistent with the flow fields, with convergence found along the ITCZ and to its south, and divergence to the north of the ITCZ at low-level. It is opposite at high-level, with the divergence center corresponding to the anticyclonic center at 200 hPa.

From Figure 1.55 (b), it can be seen that the relatively moist air is distributed within the ITCZ and to its south. Moist air is located at a lower level to the south of the ITCZ, which may be a reason for the occurrence of the major cloud system in this region. The moisture convergence is concentrated in the southern and northeastern sectors of the ITCZ, which is consistent with the distribution of low clouds.

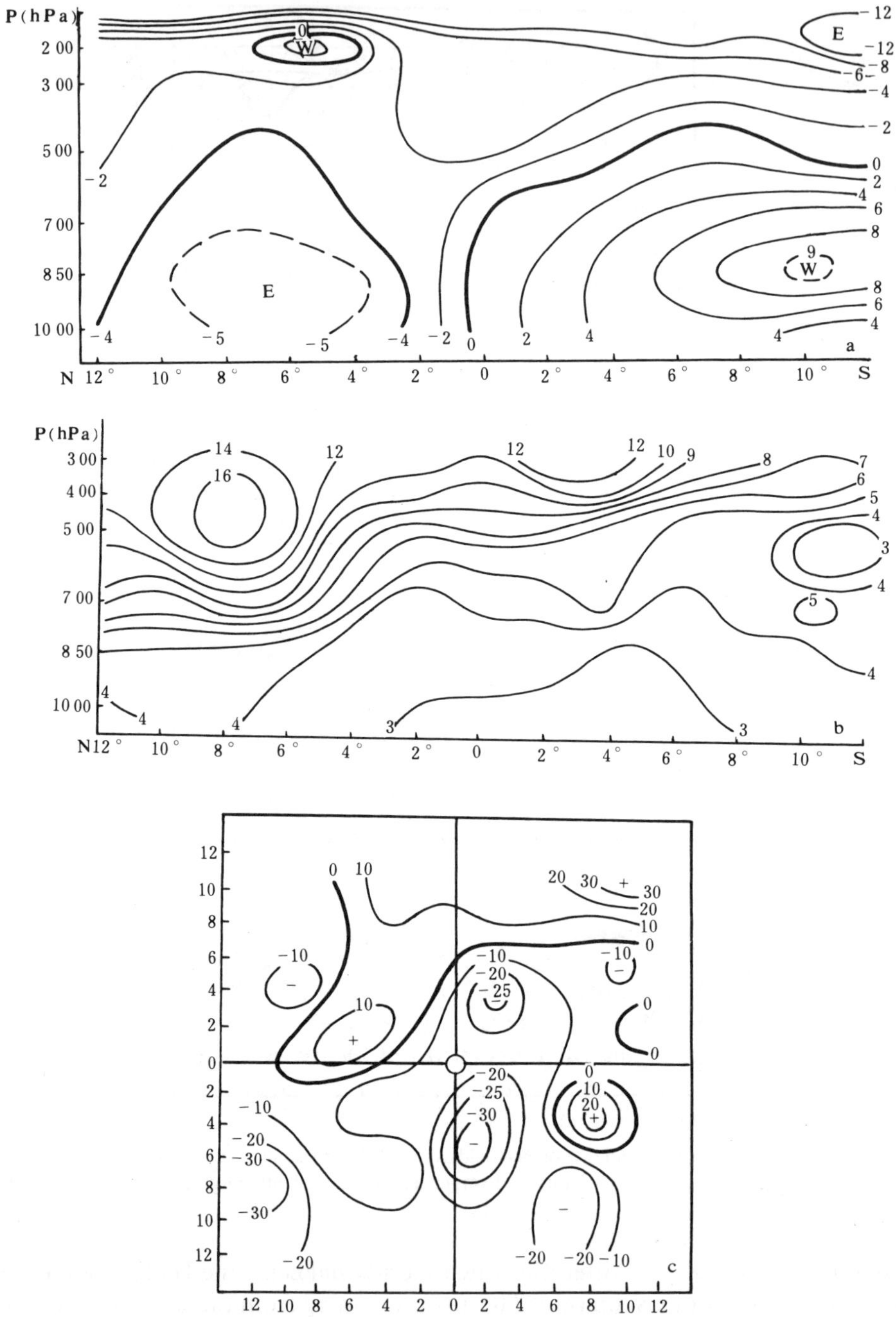

Figure 1.55. (a) The meridional cross-section of the zonal wind component across the the ITCZ. Units: m s^{-1}; (b) same as (a), but for dewpoint depression, Units: °C ; (c) the integrated convergence of moisture from the surface to 300 hPa. Units: mm day^{-1} (Group of Tropical Meteorology, 1986).

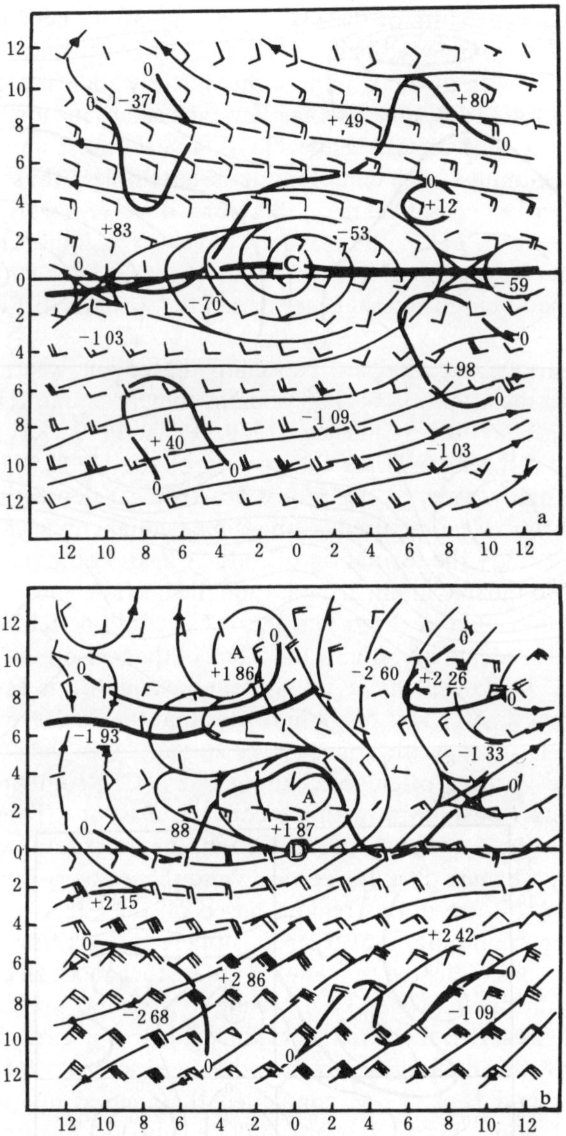

Figure 1.56. (a) The mean flow and divergence fields at 850 hPa; (b) same as (a), but for 200 hPa. The long barb of the wind symbol is 4 m s^{-1}. Units of divergence: 10^{-7} s^{-1}. Units of grid length: latitude degree.

Around the ITCZ, one often finds two meridional cells: the Hadley cell and the anti-Hadley cell (or monsoon cell). The airflow coming from the south ascends while it approaches the ITCZ and at the upper level splits into two branches: one branch of airflow returns back to the south, thus constituting an anti-Hadley cell; the other flows further northward and descends at the middle latitude, thus forming a Hadley-type cell. Finally, the structure of the ITCZ over the South China Sea and western Pacific Ocean will be summarized below. The ITCZ is really a wind shear line in the horizontal wind field, with the maximum of the westerly and easterly wind located at either side. Around the ITCZ, there is a Hadley-type cell to its north and a monsoon-type cell to its south. The air along the ITCZ is very moist and a strong vertical velocity produces cumulus and cumulonimbus ensembles, thus releasing a large amount of condensation heat. One may, therefore, observe a warm core at the mid- and upper levels. Near the ITCZ, the vertical wind shear is slight. The structure of the ITCZ shows a very asymmetric feature about the axis line, with the axis tilting southward and a major cloud system generated in the westerly airflow to the south of the ITCZ.

Careful observation and analysis, especially based on weather-satellite cloud images, have drawn the attention of theoreticians to near-equatorial troughs and the associated convergence lines (Charney, 1969; Pike, 1970; Krishnamurti, 1969). Charney suggested a mechanism for the growth of the ITCZ, based on the CISK theory, similar in many ways to the process of tropical depression throuth CISK instability mechanism. He first used a one-dimensional model and then a global nonlinear model to study the formation of the ITCZ and obtained a rather realistic result with respect to the maximum growth rate, the latitude and width of occurrence and cloud distribution. Pike realized that the ITCZ could be related to the SST. He used the multi-level model to investigate the effect of the SST on the convergence zone and found that the ITCZ often moves away from the equator, in association with the latitude of the warmest SST. This result indicates that the SST is an important factor in determining the latitude of the convergence zone.

Several cyclonic vortices often form along the ITCZ; sometimes three or four vortices may simultaneously form within a convergence zone (Ding *et al.*, 1975). At 2–4° latitude to the north of the monsoon trough, the horizontal wind shear reaches a maximum, and at the same time, the vertical wind shear is very weak, so the cyclonic vortex is quite easy to form in this region, based on the CISK mechanism (friction turning, convection, pumping convergence, upper-level outflow, and so on). An observational study has indicated that 80–85% of disturbances all occur in the region at 2–4° latitude to the north of the ITCZ. Owing to the important effect of the strong horizontal wind shear across the ITCZ, some investigators argued that the barotropic instability may provide an energy source for the development of disturbances. Nitta and Yanai (1969), and Lips (1970) concluded that, based on their observational analysis and calculations, this mechanism could be at work. The unstable wavelength is about 2000 km which is in basic agreement with the realistic condition. Later, Bates (1970) studied the growth of disturbances on the ITCZ under the combined effect of barotropic instability and the CISK mechanism. He also obtained a most unstable wavelength of 2000 km. At the initial stage, most of vortices acquire their kinetic energy from the mean flow through the Reynolds stress, then grow under the combined effects of barotropical instability and the CISK mechanism and finally at the mature stage, the CISK mechanism provides the primary energy source through the direct conversion of the available potential energy generated by the condensation latent heat. At this time, the thus provided energy supply is one order of magnitude greater than the barotropic conversion and two orders of magnitude greater than the baroclinic conversion at the expense of the mean temperature gradient.

The activity of the ITCZ over the tropical western Pacific shows a marked intraseasonal variation which is mainly characterized by the degree of eastward extension. It is often observed that in some spells of summer, the ITCZ is well developed, extending itself eastward down to 150° E or even further, while in other spells, it is confined to the region to the west of 140° E and consists of easterly trades. This variation in the activity of the ITCZ depends greatly upon large-scale circulation conditions. For the active spell of the ITCZ, the position of the ridge line of the subtropical high is at 30–35° N, further north than normal; the Australian region is dominated by a stronger cold high and by the southeasterly trade and cross-equatorial aircurrent. At 200 hPa, the TUTT (Tropical Upper Tropospheric Trough) is shifted farther northward and eastward (Ding and Reiter, 1983; Group of Tropical Meteorology, 1976). For the inactive ITCZ spell, the large-scale circulation conditions are generally opposite to the active ITCZ spell (Zhou and Chen, 1984). The above intraseasonal variation in the activity of the ITCZ may exert a very significant effect on the weather events over the South China Sea and the western Pacific, especially rainfall events and typhoon activities. A brief discussion of this matter will be presented below. Based on 30 years of typhoon data over the western Pacific Ocean and on monthly upper air wind data (1949–1978), Ding and Reiter (1983) studied some climatological aspects of the variability of typhoon formation frequency and its relationship to the tropical general circulation in the Asian monsoon region and the central and eastern Pacific. Some emphasis is placed on a contrasting study of the circulation features responsible for months with many typhoons and months with few typhoons. The difference in typhoon numbers between months with many typhoons and few typhoons is quite significant, the former being 2–3 times larger than the latter. One of the major differences in the circulation features between the above two cases shows up in the condition of the ITCZ.

Figure 1.57 is the flow patterns at 850 hPa for months with many and few typhoons. The significant differences between these patterns are as follows:

(1) For months with many typhoons the ITCZ is located near 20° N, whereas for months with few typhoons the ITCZ is found near 10° N. For months with many typhoons, the ITCZ extends eastward to 160° E, whereas for months with few typhoons, the ITCZ ends at about 130° E.

(2) For months with many typhoons, conditions characteristic of a break in the monsoon prevail over the Indian subcontinent, with the monsoon trough there advancing to the foothills of the Himalayas. In contrast, during months with few typhoons, the monsoon is very active over India, with the monsoon trough lying over the northern part of the Indian Peninsula.

(3) During months with many typhoons, the ridge line of the subtropical high over the western and central Pacific Ocean assumes a position further to the north than during months with few typhoons, near 30° N in the former case and near 25° N in the latter situation.

(4) The pattern of the locations of typhoons formation also shows a characteristic difference. During months with many typhoons, most tropical storms occur close to

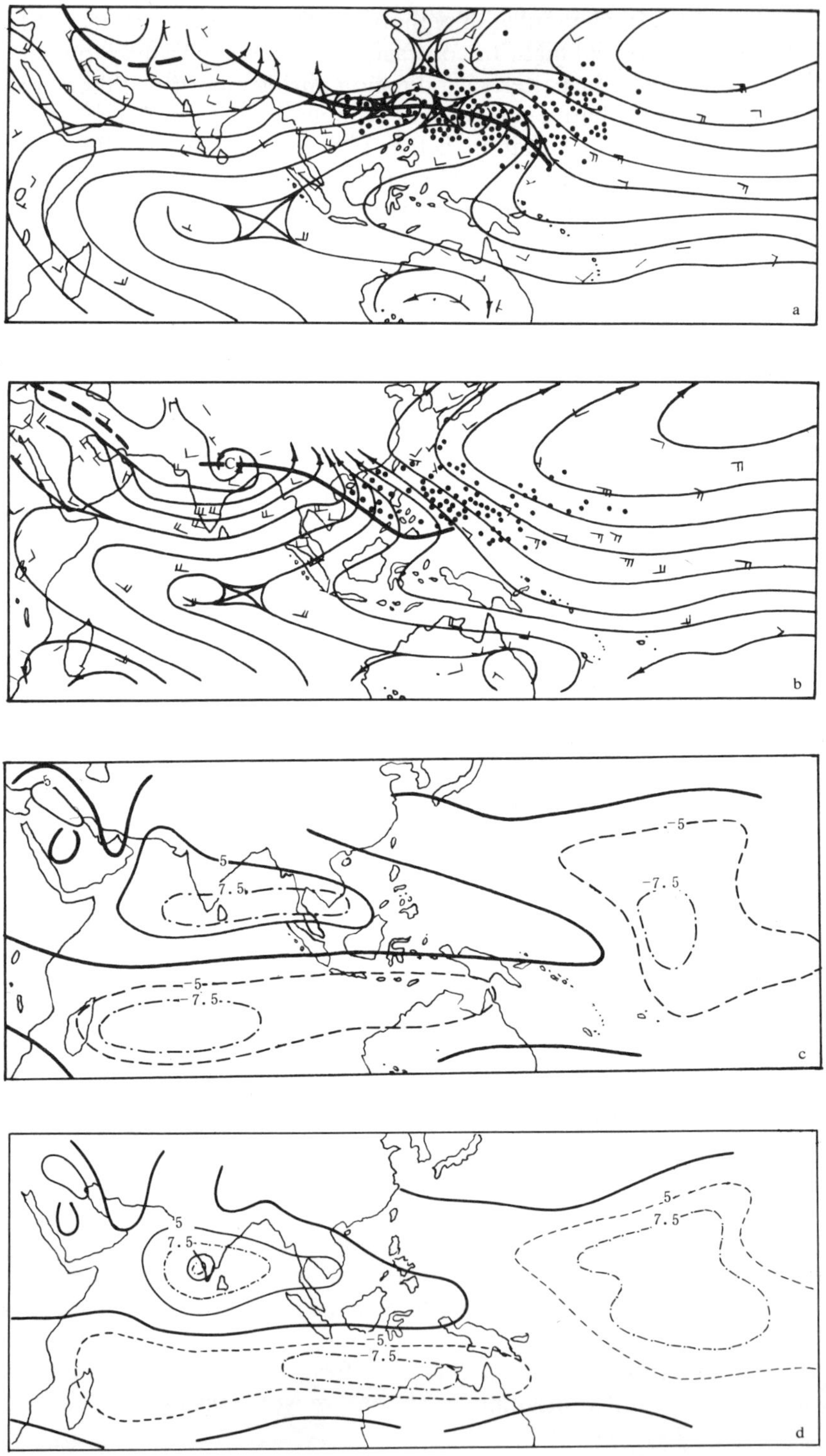
a
C
b
5
7.5
−5
−7.5
c
d

the ITCZ, except for a small number of typhoons in the region of easterly winds to the southwest of the subtropical high. In contrast, only a few typhoons form within the ITCZ and many occur in the easterly wind region during months with few typhoons. This suggests that the formative processes of typhoons may be different in each case. Ding *et al.* (1975) indicated that in active typhoon seasons, the ITCZ is well-established and well-extended eastward and typhoon activity tends to cluster in time and space (Gray, 1979). For seasons or years with many typhoons during periods of clustering, on an average, 26 typhoons are generated, which accounts for 69% of the typhoon total for one year, whereas 10 typhoons, on an average, are contained in clusters during years with few typhoons, which accounts for 44% of the total typhoon number for these years.

(5) There is some evidence that the lower tropospheric high pressure cell in the Australian region assumes a position further the west during months with many typhoons than during those months with few typhoons. As previously indicated, the position of the ITCZ is closely related to the activity of the monsoon aircurrent to its south. Figure 1.57c and d clearly demonstrate this point. For months with many typhoons, the westerly winds at 850 hPa extend to 170° E, whereas during months with few typhoons they terminate at 135° E. Months with many typhoons have a weaker easterly wind regime over the central and eastern Pacific and a weaker southwest monsoon over the Indian Peninsula. The easterly winds over northern Australia and its neighborhood are stronger during months with few typhoons.

The effect of the ITCZ on the rainfalls in China has been studied by numerous investigators. They have found that for many cases of heavy rainfalls in central and northern China, one may often find northward advances of the ITCZ, thus enhancing the transport of moisture into the regions of heavy rainfall. A more detailed discussion will be given in Chapter 4. Furthermore, Zhu *et al.* (1987) have pointed out another situation that when the continuous ITCZ breaks in the South China Sea and the Indo-China Peninsula the low-level southwest monsoon may advance northward and merge with the westerlies at the middle latitude, thus causing active cloud surge and its merging with the cloud systems of the Meiyu. In this case, the rainfall amount of the Meiyu would be enhanced and the rain belt would be shifted northward.

Recently, considerable attention has been paid to the diagnostic study of the ITCZ, including the estimate of the budgets of kinetic energy and vorticity, to better understand the mechanism of the maintenance of the ITCZ. Ding and Reiter (1983) calculated the kinetic energy budget over the western Pacific during dominant trade wind and active monsoon regimes. For the latter case (July 1972), a marked ITCZ dominated much of the tropical western Pacific and five typhoons and a depression formed in the area under study. For the former case (July 1975), strong southeast

---

←

Figure 1.57. Mean streamline and wind isotach analysis at 850 hPa for summer (July-Sept.). (a) Mean streamline analysis for months with many typhoons, (b) same as (a), but for months with few typhoons, (c) 850 hPa u-components of wind for months with many typhoons (solid line: west wind; dashed-dotted line: east wind), and (d) the same as (c), but for months with few typhoons. Wind symbols: triangle denotes 20 m s^{-1} speed, full bar 4 m s^{-1} and half barb 2 m s^{-1}. Black circle denote the location of first detection of typhoons (Ding and Reiter, 1983).

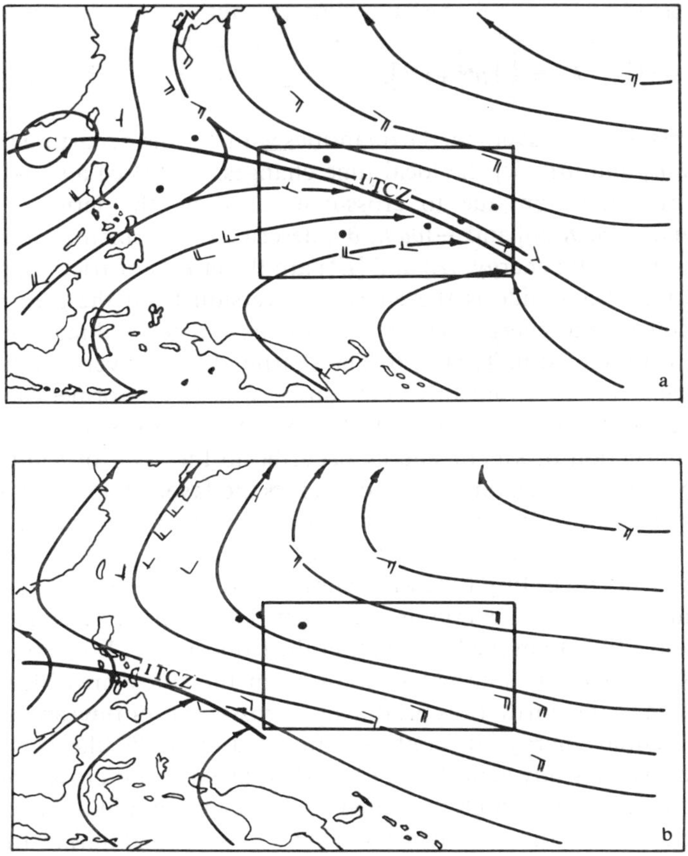

Figure 1.58. Monthly mean streamlines at 850 hPa, (a) for July 1972 and (b) for July 1975. Black circles indicate the positions of the first reported detection points of the typhoons. Open circles indicate the upper air stations used in this analysis. The rectangle represents the studied area. Wind symbols: half bar = 2 m s^{-1}, full bar = 4 m s^{-1}, and triangle = 20 m s^{-1} (Ding and Reiter, 1983).

trade winds prevailed over most of the western Pacific and no typhoons formed in that area (Fig. 1.58)

The time-averaged budgets of kinetic energy in the general flow and large-scale disturbance, according to Kung and Merritt (1974), were estimated by use of the following equations, respectively:

$$\left[\frac{\partial K}{\partial t}\right] = -[\nabla \cdot VK] - \left[\frac{\partial \omega K}{\partial p}\right] - [V \cdot \nabla \phi] - [E] \quad (1.1)$$

$$\left[\frac{\partial K_e}{\partial t}\right] = -[\nabla \cdot VK_e] - \left[\frac{\partial \omega K_e}{\partial p}\right] - \{[v^* \omega^*] \frac{\partial [v]}{\partial p} \quad (1.2)$$

$$+ [v^* \omega^*] \frac{\partial [v]}{\partial p}\} - [V^* \cdot \nabla \phi^*] - [E^*]$$

where

$$K = \frac{1}{2}[u^2 + v^2], \quad K_e = \frac{1}{2}[u^{*2} + v^{*2}], \tag{1.3}$$

[ ] represents the areal average, the quantities with "*" the departure from the areal average, $[\partial K/\partial t]$ and $[\partial K_e/\partial t]$ the local time changes, $-[V \cdot \nabla\phi]$ and $-[V^* \cdot \nabla\phi^*]$ the kinetic energy generation due to cross-contour flow, the terms $-[\nabla \cdot VK]$ and $-[\nabla \cdot VK_e]$, and $-[\partial \omega K/\partial p]$ and $-[\partial \omega K_e/\partial p]$ describe the horizontal and vertical flux convergences of kinetic energy, $[E]$ and $[E^*]$ are the effects of friction, and the third term on the right-hand side is the energy conversion from the area-mean kinetic energy to eddy kinetic energy. The results of the budget of kinetic energy in the general flow is presented in Table 1.6. The sharply different vertical distribution of $-[V \cdot \nabla\phi]$ during July 1972 and July 1975 can be seen. In the former case, the kinetic energy was destroyed in the whole tropical troposphere of the studied area. The most significant destruction of kinetic energy occurred in the layer of 850 to 500 hPa in which the southwest monsoon prevailed, and in the layer of 150 to 100 hPa where strong upper westerlies dominated. The generation of kinetic energy in July 1975 showed its maximum in the upper troposphere where a deep layer of easterlies was found. The destruction in July 1972 was mainly determined by the large divergence of horizontal flux of potential energy in the upper half of the troposphere and by the convergence of $-[\phi\nabla \cdot V]$ in the lower half of the troposphere. Inspecting Figure 1.57 (a), one may see that the southwest monsoon in the study area blew toward the subtropical high to the north, thus producing a cross-contour flow and conversion of kinetic to potential energy. In contrast, during July 1975, the subtropical high dominated the whole study area (Figure 1.57 (b)). The obvious cross-isobaric flow to the south of the subtropical high generated a significant amount of kinetic energy. In the upper troposphere, the large negative values of $-[\nabla \cdot V\phi]$ during July 1972 were mainly caused by a strong mass divergence associated with the subequatorial ridge. The magnitudes of the transport terms $-[\nabla \cdot VK]$ and $-[\partial \omega K/\partial p]$ in Table 1.6 were rather small in comparison with the generation term $-[V \cdot \nabla\phi]$ and the dissipative sink

Table 1.6. Kinetic energy balance of the general flow over the western Pacific area within specified layers, for 00:00 GMT, July 1972 and July 1975. Units: W m^{-2} (Ding and Reiter, 1983)

Pressure layer	$-[V \cdot \nabla\phi]$		$-[\nabla \cdot VK]$		$-[\partial \omega K/\partial p]$		$[E]$	
(hPa)	1972	1975	1972	1975	1972	1975	1972	1975
100–150	− 31.50	82.92	−0.63	−0.04	0.17	0.07	−31.96	82.95
150–200	− 19.20	38.62	−0.50	0.00	0.05	0.04	−19.65	38.96
200–300	− 18.00	46.59	−0.29	0.04	−0.03	−0.02	−18.32	46.61
300–500	− 14.03	35.91	−0.13	0.06	−0.09	−0.05	−14.25	35.92
500–700	− 37.52	13.79	−0.04	0.04	0.00	−0.01	−37.56	13.82
700–850	− 21.49	4.44	−0.01	0.02	−0.02	−0.02	−21.52	4.44
850–1000	− 2.49	0.69	0.08	0.03	−0.02	−0.02	− 2.50	0.70
Total 100–1000	−144.2	222.98	−1.52	0.15	−0.01	−0.01	−145.8	223.1

[*E*]. This result suggests that the observed mean flow kinetic energy sources and sinks on the grid-scale were nearly balanced in the study area. During July 1972, the large negative values [*E*] in all layers seem to indicate that the conversion of subgrid-scale energy to that of the grid-scale was a very important process for compensating for the destruction of kinetic energy produced by cross-contour flow. This result demonstrates the vital importance of convective activity in maintaining the budget of kinetic energy in the general flow. In contrast, in the trade wind case, almost all general flow kinetic energy generated through the cross-isobaric flow was dissipated by a frictional effect.

The time averaged eddy kinetic energy budgets in the ITCZ was computed, and is presented in Table 1.7. More eddy kinetic energy was generated in the upper troposphere between 300 and 100 hPa in July 1972 than in July 1975. This comparison suggests that there is a stronger ageostrophic acceleration of the flow at upper levels, possibly caused by the heating aloft in the storm areas through convective activity rather than during quiscent periods. A weak adiabatic destruction at middle levels occurred between 700 and 300 hPa. A secondary maximum of the eddy kinetic energy generation was found in the lower troposphere below 700 hPa. In July 1975, there

Table 1.7. Kinetic energy balance of large-scale disturbances over the western Pacific area within specified layers of 00:00 GMT, July 1972 and July 1975. Units: W m^{-2} (Ding and Reiter, 1983)

Pressure layer (hPa)	$-[V^{*}\cdot\nabla\phi^{*}]$		$-[\nabla\cdot VK_e]$		$-[\partial\omega K_e/\partial p]$	
	1972	1975	1972	1975	1972	1975
100–150	0.525	0.223	−0.311	−0.003	0.028	0.004
150–200	0.346	0.121	−0.264	0.008	0.141	0.034
200–300	0.327	0.181	−0.198	0.035	−0.041	−0.034
300–500	−0.017	0.062	−0.088	0.044	−0.098	−0.035
500–700	−0.003	−0.127	−0.017	0.017	0.012	−0.006
700–850	0.244	−0.128	0.022	0.003	−0.012	−0.007
850–1000	0.370	0.037	0.088	0.002	−0.082	−0.009
Total 100–1000	1.792	0.369	−0.769	0.106	−0.052	0.013

Pressure layer (hPa)	$-[u^{*}\omega^{*}]\,\dfrac{\partial[u]}{\partial p}$		$-[v^{*}\omega^{*}]\,\dfrac{\partial[v]}{\partial p}$		$[E^{*}]$	
	1972	1975	1972	1975	1972	1975
100–150	0.008	0.003	−0.002	−0.001	0.252	0.268
150–200	−0.006	−0.002	0.007	0.002	0.224	0.163
200–300	−0.025	−0.009	0.014	0.005	0.077	0.178
300–500	0.015	−0.003	−0.008	0.002	−0.196	0.076
500–700	−0.005	0.000	−0.006	0.000	−0.019	−0.116
700–850	0.000	0.000	0.002	0.000	0.256	−0.132
850–1000	−0.001	0.000	0.002	0.000	0.379	0.030
Total 100–1000	−0.012	−0.004	0.013	0.106	0.010	0.468

were two similar maxima in the upper and lower troposphere, but their magnitudes were much smaller than those observed during July 1972.

These computations confirm the fact that the upper and lower troposphere in the tropics contain active layers for the potential growth of disturbances, whereas the middle tropical troposphere is relatively inactive. The horizontal transport term, $-[\nabla \cdot VK_e]$, also revealed significant differences. In July 1972, a large part of the generated eddy kinetic energy (43%) was exported from the area in the middle and upper troposphere between 700 and 100 hPa. Therefore, the tropical atmosphere with strong convective activity supplies a considerable amount of kinetic energy to the surrounding atmosphere, mainly at upper levels. This term showed a significant flux convergence of eddy kinetic energy in the lower troposphere in July 1975. Almost all vertical layers gained eddy kinetic energy by import from outside the area during that month, the exception being the top layer.

Generally, the values of vertical transport in the total column were small as compared to other terms of the energy budget. Also, the conversion between the area-averaged kinetic energy and the eddy kinetic energy contributed only insignificantly to the budget. The term $[E^*]$, the friction dissipation of eddy kinetic energy, indicates that in both cases, a significant part of the generated and/or imported eddy kinetic energy of the synoptic disturbances was cascaded into a smaller scale turbulence at jet-stream level. The negative values of $[E^*]$ observed in the middle troposphere should be interpreted as the feeding of eddy kinetic energy from small to large-scale disturbances.

The generation terms can be written as

$$-[V^* \cdot \nabla\phi^*] = -[\nabla \cdot V^*\phi^*] - [\frac{\partial \omega^*\phi^*}{\partial p}] - [\omega^*\alpha^*] \qquad (1.4)$$

It is interesting to note that the largest contribution to the budget of eddy kinetic energy did not come from $-[\omega^*\alpha^*]$, the conversion between available potential energy and kinetic energy, but from the horizontal eddy flux of potential energy. Therefore, $-[\nabla \cdot V^*\phi^*]$, rather than $-[\omega^*\alpha^*]$, made the largest contribution to the eddy kinetic energy generation, especially in 1972. Kung and Merritt (1974) obtained a similar result, finding that the general distribution of $-[V^* \cdot \nabla\phi^*]$ followed more closely that of $-[\nabla \cdot V^*\phi^*]$ rather than that of $-[\omega^*\alpha^*]$ and had an apparent relationship to the shift of the position of the ITCZ. When the ITCZ was south of, or near, the equator, which was the southern boundary of their analysis area, a large export of the eddy potential energy from the area was observed in terms of large negative values of $-[\nabla \cdot V^*\phi^*]$. On the other hand, when the ITCZ migrated northward into the analysis area, $-[\nabla \cdot V^*\phi^*]$ showed an oscillation between positive and negative values as the ITCZ migrated within the area, but on an average, yielded positive values, indicating an import of eddy potential energy into the area. This finding can be interpreted as a positive generation by the virtue of $-[V^* \cdot \nabla\phi^*]$. In short, the area close to the ITCZ received eddy potential energy through the horizontal flux $-[\nabla \cdot V^*\phi^*]$ while other areas served as suppliers.

A case study of the budget of the ITCZ's kinetic energy was made by Jian *et al.* (1987). They studied the budget of eddy kinetic energy of the ITCZ or the monsoon trough from 27 July–14 August, 1979. This monsoon trough extended from the Arabian Sea eastward down to 140 E, lasting for 20 days, during which time period three tropical depressions and three typhoons successively formed. The major results are basically the same as those obtained by Ding and Reiter (1983a) for the monthly mean study. A further analysis of the production term of kinetic energy has shown that its major part comes from the contribution of the production of kinetic energy by the divergent wind that was mainly caused by the condensational heating effect.

The estimate of the budget of vorticity is another important aspect of the diagnostic study of the ITCZ. Xue (1989) estimated the budget of vorticity for a strong ITCZ that occurred in August 1979 along 20° N, extending from the Indian subcontinent to the western Pacific Ocean, with a special emphasis on the cumulus friction effect. He obtained that the cumulus activity within the ITCZ creates the apparent vorticity sink at the low level and the apparent vorticity source at the upper level. This indicates that residual vorticity change would decrease the vorticity produced by large scale motion at both high and low levels. Thus, the vertical transport of the vorticity by cumulus convection would weaken the positive vorticity at the low level and enhance the positive vorticity at the upper level, thus maintaining the ITCZ development.

1.6. The Cross-Equatorial Flow in the East Asian Monsoon Region

Late in the sixties, Findlater (1969) found the existence of the Somali low-level jet anchoring off eastern Africa. This jet in fact originates in the South Indian Ocean (around the Mascrene high), then flows northwestward with the southeast trade wind and turns into the southwest monsoonal flow in the Northern Hemisphere while crossing the equator. Therefore, the summer monsoon in South Asia is believed to be closely related to the cross-equatorial airflows. Chinese meteorologists have, for a long time, realized the importance of the cross-equatorial airflows coming from Australia in the development of synoptic systems and events in the tropical western Pacific and East Asia. Previously, Li (1956) revealed the relationship between the typhoon genesis over the tropical western Pacific and the activity of cold airs over Australia. Early in the sixties, Tao *et al.* (1962) pointed out that when meridional type of circulation prevailed over East Asia, the probability of a northward transport of the air of the Southern Hemisphere near the equator is very high. This strong southerly wind component is related to the activity of cold airs in Australia. Later, Ramage (1972) also pointed out that if there is the transequatorial flow coming from the Southern Hemisphere, it would cause an active spell of typhoons off the Seaboard of China. As weather-satellite data and data from other sources become available from the tropics, especially over the tropical oceans, it is feasible that the behavior of the cross-equatorial airflows may be further studied. Some studies have revealed that, from the East African coast to the central Pacific, there are five cross-equatorial airflows. Of them, the most obvious is the Somali jet. Further east, in the eastern Indian Ocean (80–85° E), there is likely to be another cross-equatorial airflow, although its steadiness is relatively low. Over the South China Sea and the western

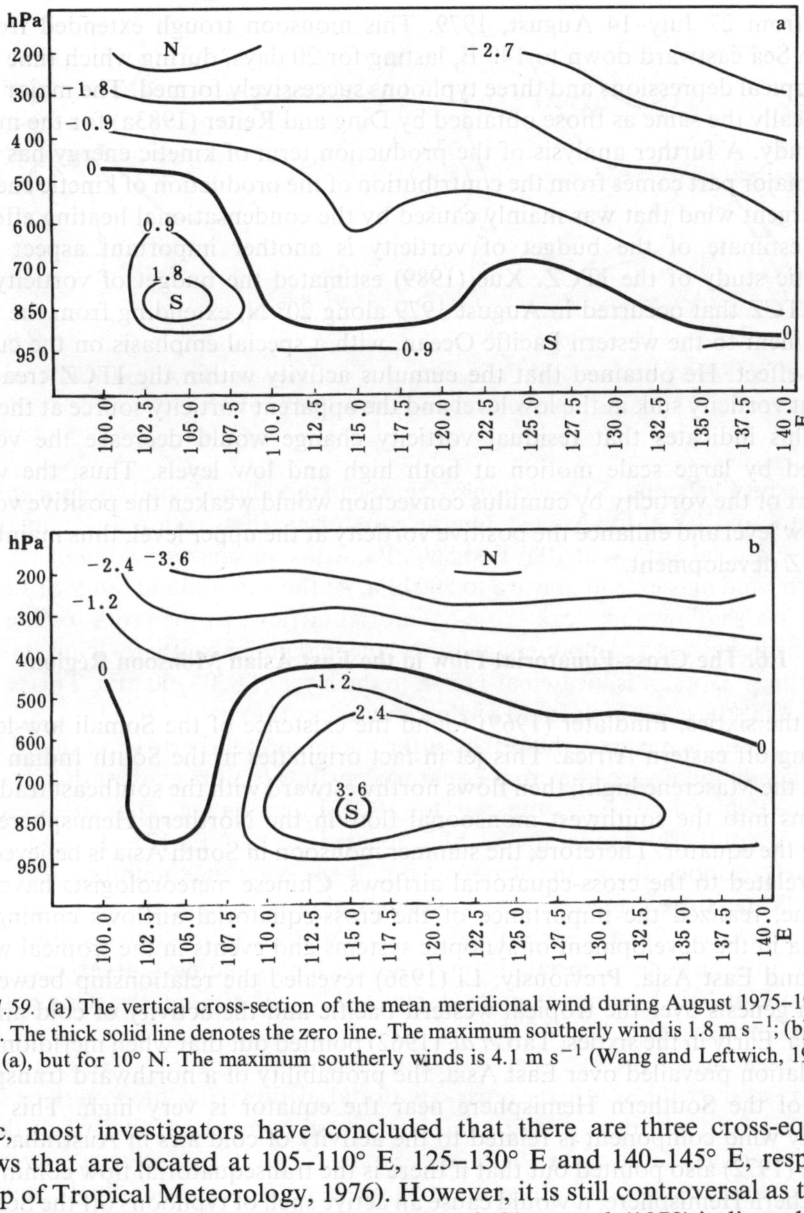

Figure 1.59. (a) The vertical cross-section of the mean meridional wind during August 1975–1981 at the equator. The thick solid line denotes the zero line. The maximum southerly wind is 1.8 m s^{-1}; (b) the same as (a), but for 10° N. The maximum southerly winds is 4.1 m s^{-1} (Wang and Leftwich, 1984).

Pacific, most investigators have concluded that there are three cross-equatorial airflows that are located at 105–110° E, 125–130° E and 140–145° E, respectively (Group of Tropical Meteorology, 1976). However, it is still controversal as to which one is the major cross-equatorial airflow channel. Chen *et al.* (1979) indicated that the airflow coming from the northeastern part of Australia crosses the equator mainly around 120° E. Liang *et al.* (1981) obtained that there is a marked cross-equatorial airflow at about 105° E. But, Wang and He (1979) argued that no significant cross-equatorial airflow was observed in the southern part of the South China Sea. Later, Wang and Leftwich (1984) documented the existence of the cross-equatorial airflow in this region, based on the wind data for June-October of 1975–1981. Figure 1.59 (a),

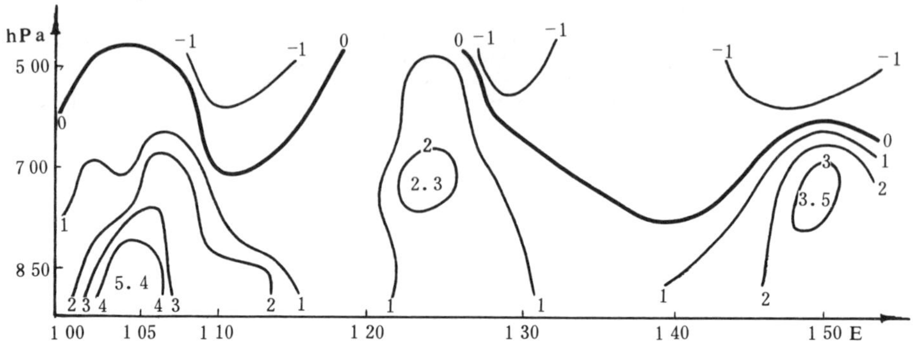

Figure 1.60. The zonal cross-section of the meridional wind component for August 1979. Units: m s^{-1} (Wang and Leftwich, 1984).

(b) are, respectively, the cross-sections of the mean meridional wind component along the equator and 10° N for August. On the equatorial cross-section, one may observe a cross-equatorial airflow at 105° E at 850 hPa, with a mean south wind of 1.8 m s^{-1}. The south wind may extend upward to 500 hPa. At the longitudinal range of 120–130° E, there is a shallow and weak cross-equatorial airflow. In the cross-section of the mean meridional wind component at 10° N, the maximum southerly wind intensifies up to 4.1 m s^{-1}, but, it is located at 115° E in the layer of 850–700 hPa. This evidence indicates that the cross-equatorial airflow moves northeastward from the equator to 10° N and, at the same time, the height of the maximum southerly wind slightly rises.

For an individual year, the cross-equatorial airflow is more evident than the mean condition. Figure 1.60 shows the case for 1979. It can be seen that there are three maxima of southerly winds at 850 hPa, located at 105° E, 125° E, and 150° E, respectively. Among them, the strongest one is the wind maximum at 105° E. This result indicates the existence of three channels of cross-equatorial airflows over the South China Sea and the western Pacific. The statistics of the wind steadiness shows that these three cross-equatorial airflows all have high steadiness, all exceeding 40%, with the one at 105° E being the most stable (> 70%). Therefore, the strongest cross-equatorial airflow exists near 105° E.

It has been observed that an easterly return flow at the upper level (for example, at 200 hPa) lies over the southerly cross-equatorial airflow, and there may be a close relationship between them. Liang *et al.* (1981) also pointed out that when the low-level cross-equatorial airflow from the Southern to the Northern Hemisphere enhances, the upper-level return flow lying over it correspondingly enhances, especially during the active monsoon spell.

When the cross-equatorial airflow arrives at the northern part of the South China Sea and at the coastal area of the Guangdong Province of China, the southerly wind component may often accelerate to reach the intensity of a low-level jet. Figure 1.61 is the cross-section of the long-term mean meridional wind component with the wind direction between 170 and 189° for July along 20° N (Wang and Li, 1982). It can be seen that there is a 8 m s^{-1} maximum of low-level southerly wind in the layer

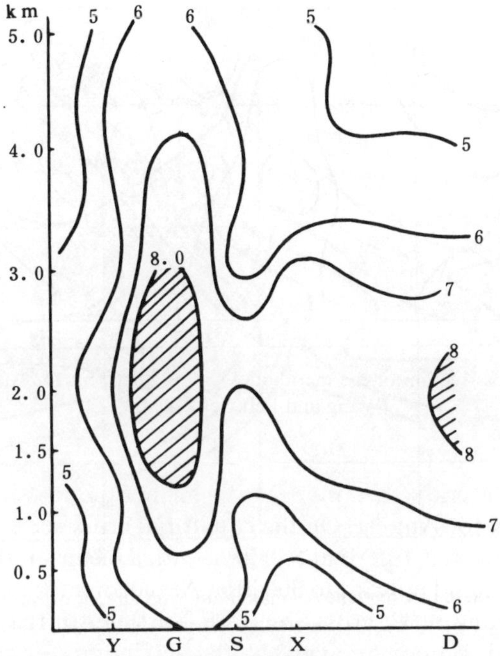

Figure 1.61. The cross-section of the long-term mean wind speed with the wind direction between 170 and 189° for July at 20° N. Y, G, S, X and D represent locations of upper air stations, arranged from west to east. Units: m s^{-1} (Wang and Li, 1982).

between 1000 and 3000 m. Figure 1.62 clearly shows that a cross-equatorial airflow, having its origin in Australia, merges with the southwest monsoonal flow to form a marked low-level jet with a maximum wind speed of 30 m s^{-1} at 850 hPa in the northern part of the South China Sea. This result is in good agreement with the position indicated in Figure 1.61. From Figure 1.62 one can also see the northeastward flow of the cross-equatorial airflow from the equator.

The effect of the cross-equatorial airflow on weather events in the Northern Hemisphere has long been realized by Chinese meteorologists. This effect generally includes three aspect: the genesis and development of typhoons, monsoon activities, and the precipitation. Below, a brief discussion will be devoted to them. In the fifties, Li (1956) proposed a triggering effect of the cold air coming from the Southern Hemisphere on the genesis of typhoons over the western Pacific. He pointed out that the strong enough invasion of cold air originating in the other hemisphere may exert a forced lift to the moist unstable air, so that a large amount of latent heat could be released, thus leading to the genesis and development of typhoons. Li believed that this invasion of cold air is a nesessary condition for the genesis of typhoon. Later, Shanghai Observatory verified Li's proposition based on tropical analyses for July and August 1958 and basically confirmed his idea (Chen and Ding, 1979). A later study further documented this association between the genesis of typhoons over the western Pacific and the cross-equatorial flow (Wang and Leftwich, 1984). For the

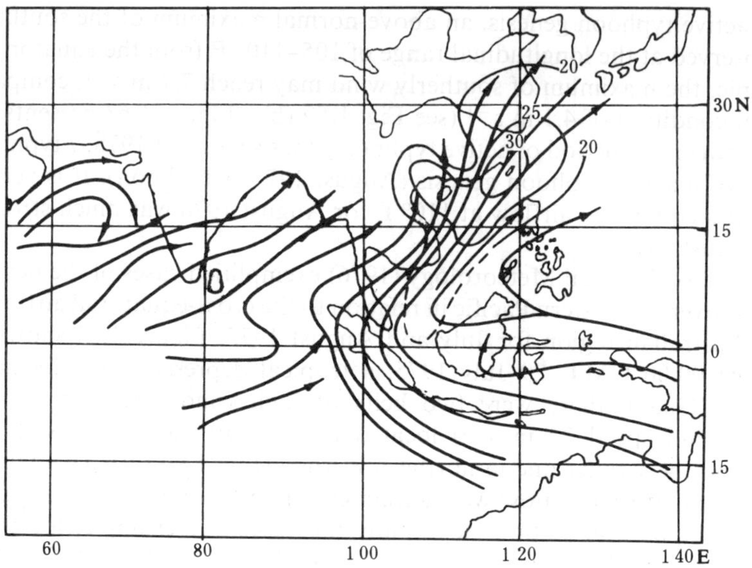

Figure 1.62. The distribution of the frequency of maximum wind direction and the maximum of wind speed for July at 850 hPa. The full line denotes the isotach. Units: m s^{-1} (Wang and Li, 1982).

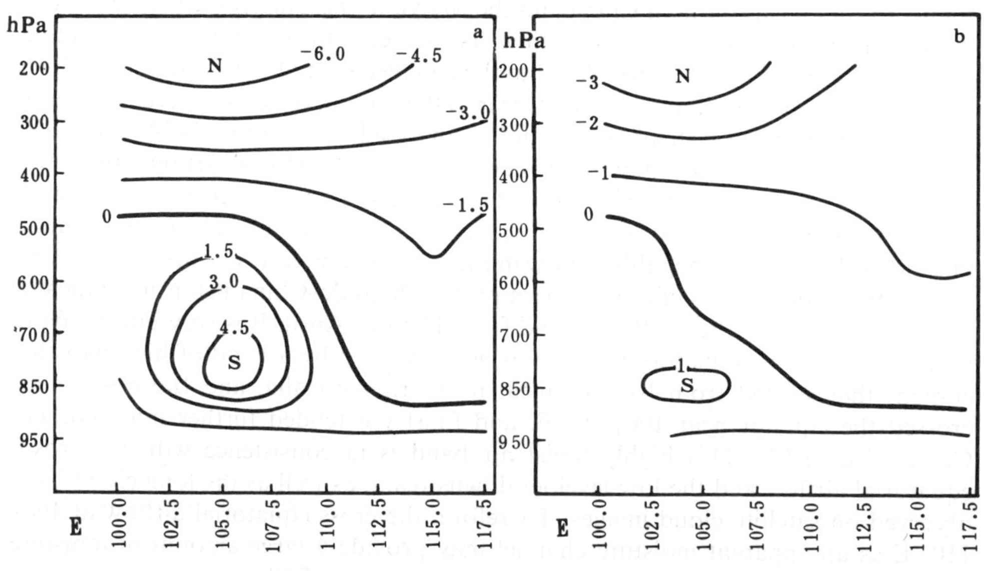

Figure 1.63. A comparison of the monthly mean meridional wind between August 1978 (the month of active typhoon genesis) (a) and August 1977 (the month of inactive typhoon genesis) (b). Units: m s^{-1} (Wang and Leftwich, 1984).

period of active typhoon genesis, an above-normal maximum of the southerly wind may be observed at the longitudinal range of 105–110° E from the equator to 30° N. For example, the maximum of southerly wind may reach 7.7 m s^{-1}, compared with the normal condition of 4.1 m s^{-1} (see Fig. 1.59 (b)). Figure 1.63 presents a clearer contrast between the period of active typhoon genesis (August 1978, 7 typhoons) and the period of inactive typhoon genesis (August 1977, 1 typhoon). It is very evident that the cross-equatorial airflow at 110° E for August 1978 was much stronger than that for August 1977.

The Group of Tropical Meteorology (1976) exemplified cases of the development of typhoons over the western Pacific in relation to the cross-equatorial airflow coming from the Australian region for July and August 1975. Figure 1.64 shows such an example. At 00:00 GMT, 29 July 1975, a tropical depression was located to the southeast of Guam; there were two belts of high wind, with the stronger one (maximum wind speed 22 m s^{-1}) located in the northeastern coastal region of Australia and the weaker one in northern Borneo (Fig. 1.64 (a)). The next day, the strong cross-equatorial airflow was getting closer to the equator, causing the wind speed at the equatorial stations to rapidly increase. For example, for the station 97014, the wind speed increased from 4 m s^{-1} for 29 July to 10 m s^{-1} for 31 July. At the same time, a long cloud band following the cross-equatorial airflow extended into the formative typhoon, possibly causing the rapid development of this typhoon.

Wang and Li (1982) and Tao *et al.* (1983) have studied the association of the pre-summer rainfalls in southern China and the Meiyu in the Yangtze River Valley with the activity of the cross-equatorial airflow. The effect of the latter on the pre-summer rainfalls is often observed to be related to the enhancement of the low-level jet which, in turn, may transport abundant moisture into the rainfall region or exert a triggering effect on rain-bearing systems. Wang and Li analysed the case for 12–14 April 1980, indicating that at the early stage of this event, a cross-equatorial airflow originating in Australia came to southern China and caused the development of a low-level jet with a maximum wind speed of 18 m s^{-1} at 850 hPa. Corrspondingly, a long cumulus line or band extended along this airflow in the South China Sea. The upper-level wind observations for stations of the southern part of the Indo-China Peninsula indicated an intensification of the southerly wind from 11 to 13 April. It is very interesting to note that this cross-equatorial airflow is accompanied by a band of high moisture content that started from New Guinea, northern Australia, and Indonesia, then crossed the equator near 100-110° E, and finally extended further into southern China, (Fig. 1.65). This highly moist air band is in consistence with the cross-equatorial airflow and the low-level jet downstream, as well as the long cloud band observed on satellite cloud images. Therefore, the cross-equatorial airflow at 100–110° E as an apparent moisture channel may provide a large amount of moisture supply for the summer precipitation in the eastern part of China.

But, Tao *et al.* (1983) argued that the origin of the cross-equatorial airflow that may have influenced pre-summer rainfalls in southern China in 1979 is likely to have come from the Indian Ocean, then crossed the equator at 80–90° E and merged with the westerly monsoon airflow to form a strong southwest airflow, and finally flowed into southern China passing over the Indo-China Peninsula. This difference may

Figure 1.64. (a) Streamlines (solid lines) at 850 hPa for 00:00 GMT, 29 July 1975. The heavy full line denotes isotachs; (b) the same as (a) but for 00:00 GMT, 30 July 1975 (Group of Tropical Meteorology, 1976).

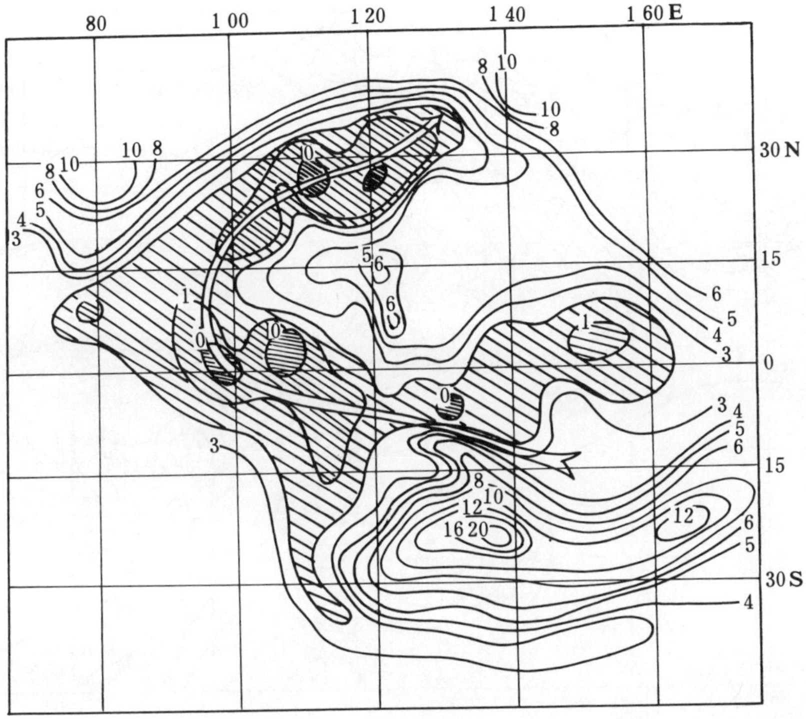

Figure 1.65. The field of dew-point depression at 00:00 GMT 12 April 1980. Units: °C (Wang and Li, 1982).

result from the different cases and interannual variability. A definite picture cannot be yet derived.

Owing to the fact that the Meiyu over the Yangtze River Valley is a result of the northward advance of the rain belt in southern China, it is reasonable to think that the Meiyu may also be affected by the cross-equatorial airflow just as the pre-summer rainfalls in southern China may be. The case study for the 1979 Meiyu season illustrates this point. From Figure 1.66, it can be seen that the southeasterly wind flowing out of the Australian cold high reached speed of 12–16 m s^{-1}, with width of about 1000 km, then turned into the southwest wind after crossing the equator, with the wind speed increasing up to 18 m s^{-1} and finally flowed into the southeastern part of China along the western flank of the western Pacific subtropical high, as a low-level jet. Around 28 June, one may observe the enhancement of the heavy rainfall in the middle Yangtze River Valley, with a daily rainfall amount of 80–100 mm. As previously pointed out by Qiu and Ding (1979), this low-level jet constitutes the low-level branch of the monsoon meridional cell.

Based on the analysis of the development of the southwest monsoon over the western Pacific in the summer of 1979 (Tao *et al.*, 1984), it was found that prior to the onset of summer monsoon in this region (23 June 1979), an intensification of the upper westerly jet off the eastern coast of Australia could be observed. Its maximum wind speed increased from 41 m s^{-1} in the first pentad of June to 66 m s^{-1} in the

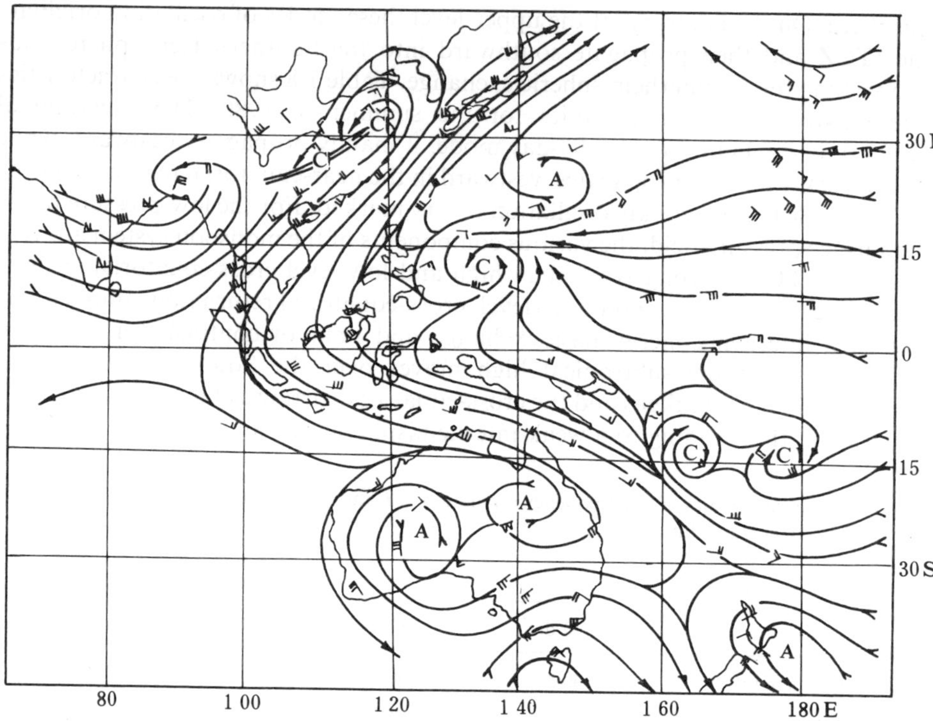

Figure 1.66. The streamline analysis at 850 hPa at 00:00 GMT, 28 June 1979. A: anticyclonic circulation; C: cyclonic circulation (Wang and Li, 1982).

second pentad of that month. At nearly the same time, frontal activity occurred in the region along 160° E. The cloud band associated with the front extended down to the low-latitude area and crossed the equator. When it connected with the ITCZ cloud band in the Northern Hemisphere, the latter rapidly moved to a region to the north of 10° N and, consequently, the southwest monsoonal flow to the south of the ITCZ also moved northward, thus leading to the establishment of the summer monsoon over the western Pacific. This case study provides some evidence of the possible association between the onset of the summer monsoon over the western Pacific and the cross-equatorial airflow and, further, the mid-latitude cold air activity.

Krishnamurti and Bhalme (1976) pointed out the inherent oscillation characteristics of the South Asian monsoon system, with a biweekly period of oscillation as its major mode. Numerous studies of the East Asian monsoon system have also revealed similar modes of oscillation in the Northern Hemisphere (5-day, biweekly, and 40–60 day oscillations). Likewise, these oscillations show up in the activity of the cross-equatorial airflow. Chen and Jin (1983), by use of the station data for May-August 1980 over the South China Sea and the western Pacific, analysed the oscillatory features of the cross-equatorial airflow in this region. They have found that there are 13.4-day and 40-day low-frequency oscillations that both originate in Australian region at low-level and propagate northward into the Northern

Hemisphere. On the contrary, at the upper-level these modes of oscillation originate in the ITCZ and then propagate southward into the Southern Hemisphere, thus accomplishing the air interhemispheric exchange. Table 1.8 shows the characteristics of these two low-frequency modes for the summer of 1980. Their meridional wavelength is 4000 km. For the 40–day mode, it mainly propagates eastward, while for the biweekly mode it propagates westward in the near-equatorial region. Over the South China Sea, the biweekly oscillation is very marked. For the low-pressure phase of the surface pressure field, the various components of the monsoon system are very active: active ITCZ, strong Australian high and low-level cross-equatorial airflow, strong tropical easterly jet and upper-level, cross-equatorial return flow, thus leading to vigorous interhemispheric mass exchange and link. In contrast, for the high-pressure phase, as the subtropical high moves over the South China Sea, the Australian cold high weakens and the cross-equatorial airflows at low and high levels become weak, thus leading to a weak interhemispheric exchange.

Table 1.8. The characteristics of low-frequency oscillation of *u* and *v* wind components at 850 hPa for the summer of 1980 (Chen and Jin, 1983)

Characters	*u*		*v*	
	40 day	13.4 day	40 day	13.4 day
Meridional wavelength (km)	4000	4000		3900
Direction of meridional propagation	northward	northward		northward
Meridional propagation speed (Lat./day)	0.80–0.95	2.3		2.6
Zonal wavelength over the South China Sea (km)		9700		3900
Zonal direction of propagation over the South China Sea		westward		westward
Zonal speed of propagation over the South China Sea (Long./day)		3.7		2.6
Zonal wavelength at the equator (km)	1300			
Zonal direction of propagation at the equator	eastward	westward	slightly eastward	westward
Zonal speed of propagation at the equator (Long./day)	3.2			
Region of origin	northern costal area of Australia	northern costal area of Australia		northern costal area of Australia

Besides the low frequency modes of 13.4-day and 40-day oscillations, there is another oscillation of about a 5-day period (Zhan and Cheng, 1988) for both the summer precipitation in China and the outbreaks of cold air in Australia, but the phase relationship between them is opposite. This correlative relationship is accomplished through the teleconnection between the outbreaks of cold air in Australia and the planetary-scale system at low and middle latitudes in the Northern Hemisphere, and especially a rapid northward propagation of wind pulsation originating in the Australian region. Therefore, the cross-equatorial airflow may exert a significant effect on synoptic processes in the Northern Hemisphere through three modes of oscillation that are probably excited by outbreaks of cold air in the northern part of Australia. This evidence is also documented in studies using FGGE data (Tao *et al.*, 1983).

1.7. The Tropical Easterly Jet Stream

The tropical easterly jet stream (TEJ) in the upper troposphere is one of the most important components of summer monsoon systems in the Indian and East Asian monsoon regions, and also one of the major circulation features in the tropical upper troposphere in the northern summer. Its establishment and activity are closely associated with the seasonal change in the upper troposphere in the Northern Hemisphere, and summer monsoon and precipitation events in South Asia and Southeast Asia. Numerous investigators have studied the characteristics of activity and the cause of the formation of the TEJ as well as its close relationship to the precipitation in monsoon regions (Koteswaram, 1958; Flohn, 1964; Krishnamurti, 1978). Tanaka (1982) pointed out that the interannual fluctuation of the TEJ at 100 hPa is caused by the severe influence of the middle latitude atmospheric circulation in addition to following the seasonal evolution at low latitudes. Yang (1982) has studied the splitting phenomenon of the TEJ at 100 hPa and indicated that it has two branches over the region from South Asia and Southeast Asia to the tropical western Pacific with the northern branch located over the latitude of 10–20° N at 100 hPa, and the southern branch found near 8° N at 150 hPa, slightly lower than the northern one. The above results have suggested that the intensity and position of the TEJ show apparent intraseasonal and interannual fluctuations. The TEJ has its maximum intensity in the layer of 150–100 hPa. Therefore, it is preferable to use the synoptic maps at 100 or 150 hPa to study the behavior of the TEJ, but the 200 hPa synoptic map has been sometimes utilized due to the sparse data sources at higher levels. From Figure 1.10, one may see the horizontal and vertical extents as well as the intensity of the TEJ. It appears to be a circulation feature from 30° N to 10° S above 500 hPa, with a maximum wind speed of 20 m s^{-1} or greater at 10–15° N at about 150 hPa. Below the TEJ, there is a layer of weaker westerly wind which represents the summer monsoonal aircurrent. This picture is to some extent different from the previous result shown in Figure 1.67 (Koteswaran, 1958). One such difference is the much greater wind maximum (about 40 m s^{-1}) for Koteswaram's cross-section; another difference is the presence of two wind maxima for Figure1.67. The northern branch of the TEJ is stronger, with its wind core at 15° N at about 100 hPa; the southern one is weaker with the wind core near the equator in the layer of 150–200 hPa. These differences may result from the different data sources.

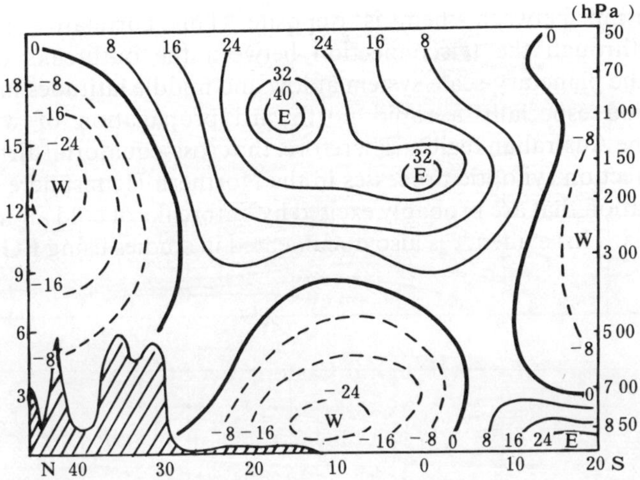

Figure 1.67. A cross-section of the mean zonal wind component along 80° E for July. Solid (dashed) lines denotes isotach of easterly wind (westerly wind). Units: m s^{-1} (Koteswaram, 1958).

The TEJ has a very high steadiness, usually attaining 96–100%. The interannual variability of the position of the jet core is rather small. Therefore, the TEJ is an almost steady and persistent circulation feature during the summer.

Based on the upper-level data for 1966–1975, Zheng and Guo (1982) plotted the long-term mean maps of the TEJ. In June (not shown), the eastern part of the TEJ axis is located at about 15° N and the western part at about 10° N, thus causing the whole axis of the TEJ to be oriented from east-northeast to west-southwest. In July (Fig. 1.68), the axis of the TEJ moves northward by about 5 latitudes, with the eastern part situated at 18–20° N and the western part at 15° N. In August, the TEJ returns back to its June pasition. The greatest intensity of the TEJ is observed in July, with a jet core of 35 m s^{-1} at 15° N, 80° E. The area with strong winds greater than 20 m s^{-1} zonally extends by 110 longitudes (15–125° E).

Based on the wind patterns at 200 hPa, one may also easily identify the features and evolution of the TEJ (Ding, Zhao and Fu, 1988; Ding *et al.*, 1989). From Figure 1.69,

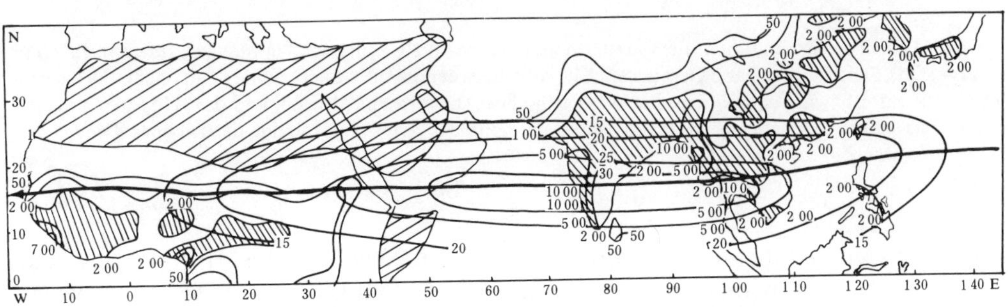

Figure 1.68. The long-term mean TEJ for July 1966–1975. The bold black arrow represents the axis of the TEJ; the heavy full lines denote isotachs (Units: m s^{-1}); the full lines isohyets (Units: mm). Shaded area denotes the regions with rainfall amount greater than 200 mm (Zheng and Guo, 1982).

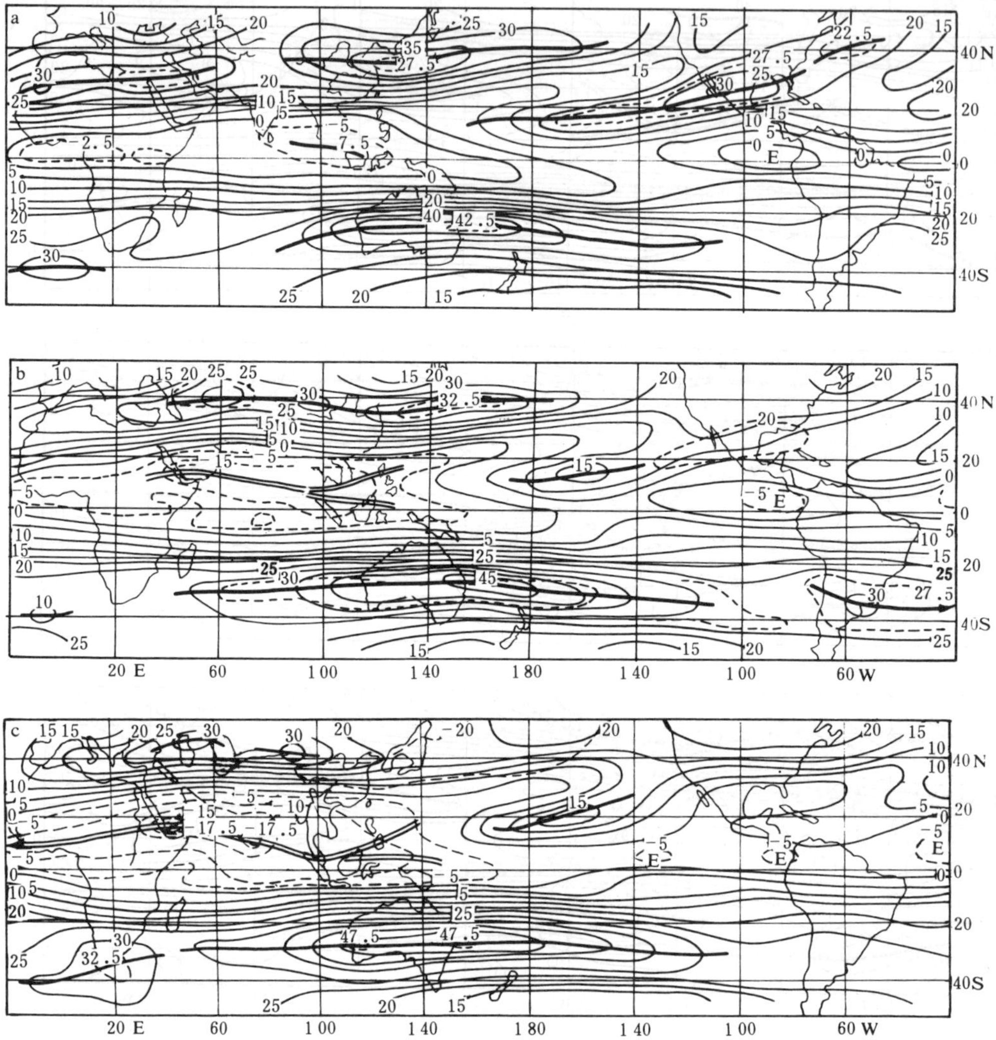

Figure 1.69. The distribution of the long-term mean (1968–1980) zonal wind in May (a), June (b), July (c), August (d), September (e), and October (f). The solid lines denote the westerly wind component, the dashed lines the easterly wind component, the heavy solid lines the zero line of wind, and the bold arrows the jet axis. The change in kinetic energy for two adjacent months is given, with the thin dashed lines (Ding *et al.*, 1988).

it can be seen that the summer tropical easterlies at 200 hPa almost encircle the global tropical zone, but with an asymmetric distribution that is characterized by the wider zone in the Eastern Hemisphere than in the Western Hemisphere. The TEJ forms a narrow latitudinal belt of 5–10° N from 40° W to 160° E. As seen at 100 hPa, the seasonal change in the TEJ is very marked. In May (Fig. 1.69a), the strongest

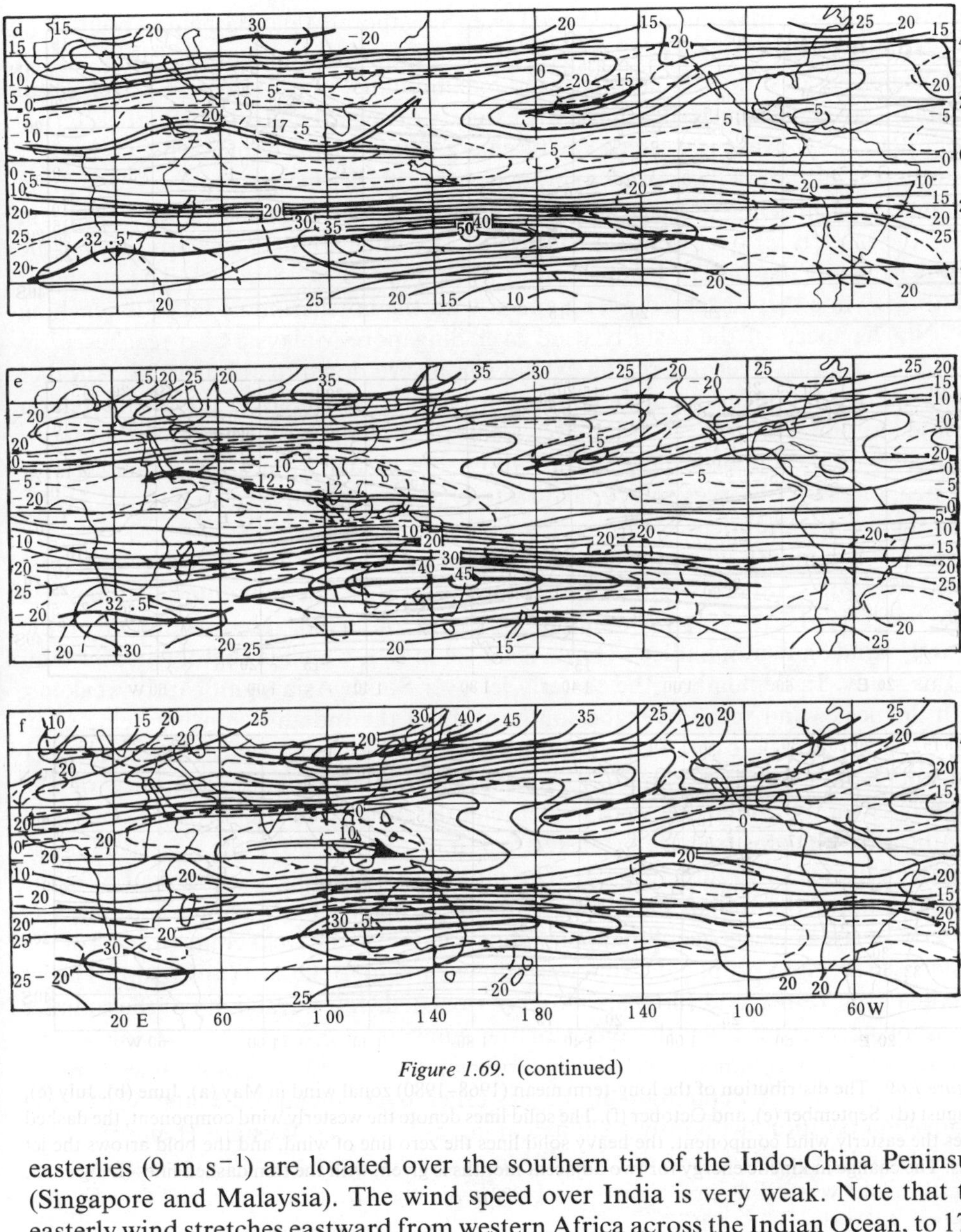

Figure 1.69. (continued)

easterlies (9 m s^{-1}) are located over the southern tip of the Indo-China Peninsula (Singapore and Malaysia). The wind speed over India is very weak. Note that the easterly wind stretches eastward from western Africa across the Indian Ocean, to 175° E along the equator. In addition, over the equatorial region of the eastern Pacific, there is a small region of weak easterlies. Therefore, from the viewpoint of the global tropics, the easterlies in May are interrupted over central and eastern Pacific as well as over the northern South America. In June, as the general circulation has a transition toward summer, and the onset and the northward advance of the South Asian summer monsoon begins, the upper-level easterlies of this region rapidly intensify with the maximum wind speed over Singapore-Malaysia being about 15 m s^{-1}. The strongest easterly wind may be observed over the southern tip of the Indian Peninsula. Besides, there is another wind maximum (14 m s^{-1}) observed over Saudi Arabia-Yemen. The easterlies markedly extend northward and a branch of easterlies from 20 to 25° N over the western Pacific emerges besides one branch which is located at the lower latitude. The above two branches of easterlies merge over the Indo-China Peninsula. The branch of easterlies at the higher latitude is associated with TUTT in July; the easterly jet over South Asia significantly intensifies. The jet streak over the southern tip of the Indian Peninsula has a wind speed of 18 m s^{-1}. The strongest wind of 19 m s^{-1} is observed over Yemen, with the northernmost latitude reaching 15° N. The speed of the easterly wind over Singapore-Malaysia also reaches 17 m s^{-1}. The easterlies over this region extend southward down to about 10° S. Another significant variation shows up in connection of the easterlies over the Pacific region at low latitude with the easterlies over Central America, resulting in a continuous encircling of the upper-level easterlies about the global tropics. Over the Gulf of Mexico, one also observes subtropical easterly airflow which is related to the TUTT over the North Atlantic.

August is a month when the easterly jet reaches its maximum intensity. The wind speed over the southern tip of India reaches 19 m s^{-1}. The remaining two maxima also reach 18–19 m s^{-1}, with similar positions observed. The maximum of the easterly wind shown in Sadler's charts is 20–24 m s^{-1}, which is slightly greater than in this study. In September, the easterly jet over South Asia significantly weakens, with the peak wind speed over the southern tip of the Indian Peninsula being 15 m s^{-1}. The jet streak over Yemen disappears. The easterly wind maximum over Singapore-Malaysia is 15 m s^{-1}. Another change is the disappearance of the easterlies over the Gulf of Mexico. In October, the easterly jet stream over South Asia significantly weakens (see the extensive region of the change in the negative kinetic energy indicated in Figure 1.69 (f)). The most marked change is observed over the southern tip of the Indian Peninsula, with the easterly jet weakening and only the easterly jet streak exists over Singapore-Malaysia. Thus, the weakening process of the easterly jet seems to propagate from west (downstream) to the east (upstream). The jet maxima over Yemen and India are closely associated with the activity of the summer monsoon, occurring after the onset of the summer monsoon, intensifying as the monsoon activity enhances, and disappearing while the summer monsoon withdraws. In contrast, the jet maximum of the easterlies over Singapore-Malaysia is little changed in position, with only a change in intensity. In the Western Hemisphere, in October only a limited area of easterlies may be detected over the tropical eastern Pacific. The westerlies are again observed over the equatorial central Pacific and central Atlantic.

The major intraseasonal variability may be observed in the zone of the TEJ. By the use of space-time spectral analysis and a band-pass filter, Lu and Ding (1989) revealed that there are three modes of low-frequency oscillation: 50-day, 25-day, and quasi-biweekly oscillations for the summers of 1980 and 1982 (Fig. 1.70). However, the interannual variability of and the latitudinal differences among the three oscillations are apparent. The 50 and 25-day period oscillations are mainly caused by the

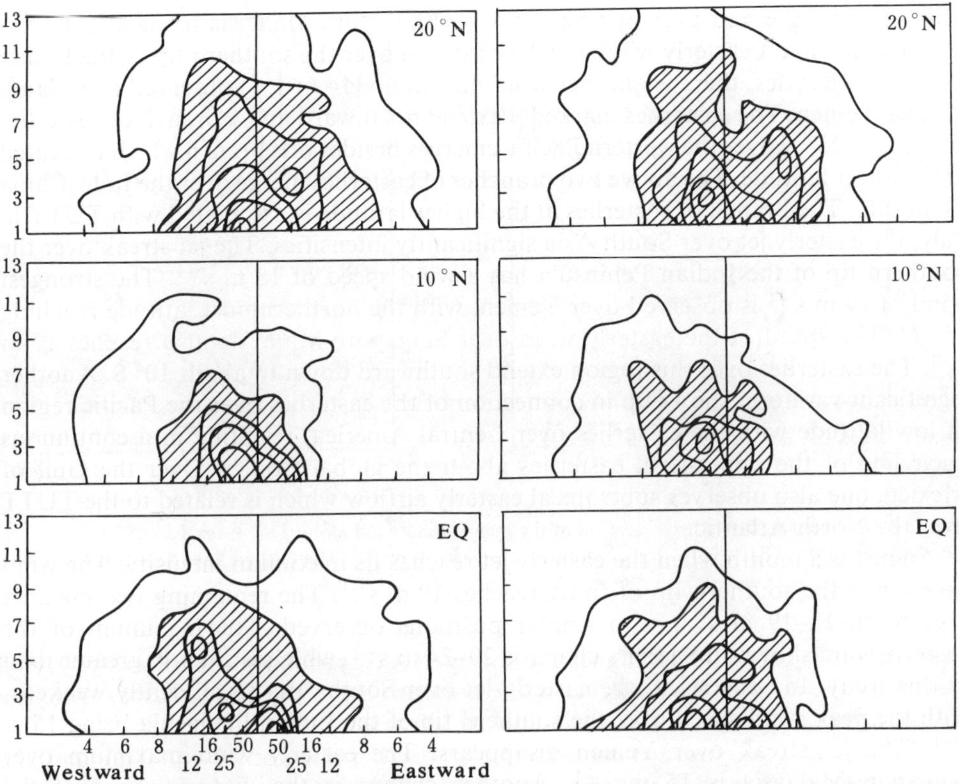

Figure 1.70. Power spectra for the zonal wind in the wavenumber-frequency domain for eastward and westward moving waves with wavenumbers 1 to 13 (ordinate) and periods longer than 4 days. The left panel is for 1982 and right panel is for 1980 with intervals 0.05, 0.1, 0.3, 0.5, 1.0, 2.0 $m^2 s^{-2}$ day; spectrum power in excess of 0.1 units is shaded (Lu and Ding, 1989).

planetary waves of zonal wavenumbers 1–3. At the equator, the 50-day spectrum peak was found in 1980 as an eastward moving wave with wavenumber 1. However, a pronounced peak was found in 1982 as a westward moving wave with wavenumber 1. The obvious westward propagating 50-day oscillation was found at 10° N and 20° N in both those years. The 50 and 25-day oscillations of zonal wind both demonstrate southward phase propagation over the region of the South Asian region. The quasi-biweekly oscillation is mainly related to synoptic waves 4–6.

As previously pointed out by some researchers, the TEJ contains two branches of the jet, especially in the eastern part of the TEJ (75–150° E). Figure 1.69 clearly indicates this point. Xiao and Xie (1983) analysed the structure of the TEJ particularly in connection with the condition of the two branches of the jet in Southeast Asia, which is an important characteristic feature for the East Asian summer monsoon system. Figure 1.71 is the meridional cross-section of mean zonal wind along 115° E for July 1979, indicating the northern branch of the jet located at about 17° N at 100 hPa, with maximum wind speed greater than 20 m s^{-1}. Therefore,

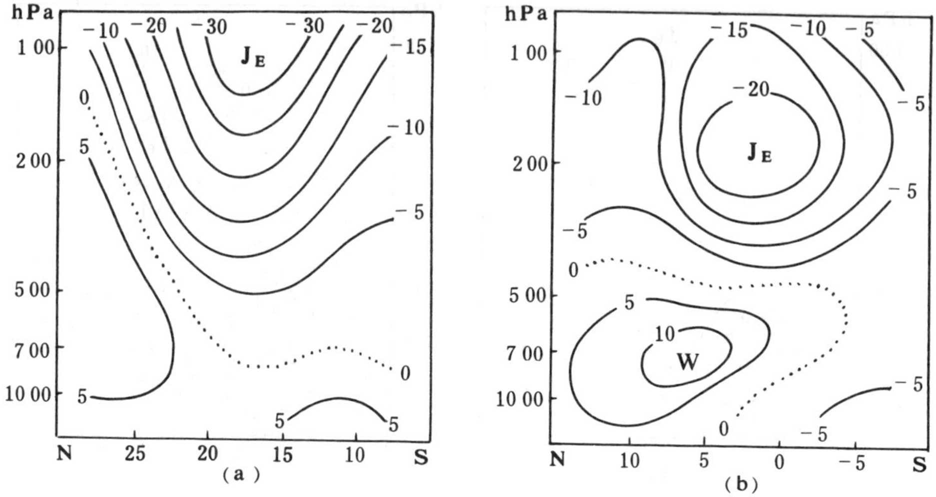

Figure 1.71. The meridional cross-section of the mean zonal wind along 115° E for July 1979: (a) for the northern leg of the TEJ, and (b) for the southern leg. Units: m s^{-1}. The dashed lines denote the zero-line of isotach (Xiao and Xie, 1983).

the northern leg of the TEJ is higher and stronger. A very shallow layer of westerly wind lies below the layer of the easterlies in the northern leg of the TEJ, whereas there is a deeper layer of stronger westerly wind below 500 hPa for the southern leg of the TEJ.

There is a weak upward motion in the southern sector of the TEJ and the downward motion in the northern sector (Fig. 1.72). In contrast, for the southern leg of the TEJ, there is a deep layer of a strong upward motion in the southern sector and a downward motion in the southern sector. The magnitude of vertical velocity in this leg is nearly one order of magnitude greater than that for the northern leg.

It is worthwhile showing the mean meridional cell nearly normal to the axis of the TEJ (Fig. 1.73). In the northern part of this cell there is the very strong upward motion, while in the southern part, the air descends, thus producing a marked negative meridional cell. The sense of this kind of the meridional cell is in good agreement with the result obtained by Flohn (1964) for the condition of the entrance region of the TEJ (90–130° E). He indicates that the meridional cell causes an extensive ascending motion and convective activity over Southeast Asia and the South China Sea (see the bottom panel of Fig. 1.73). The tropical disturbances at low-level develop while they are entering the northern sector of the entrance region of the southern branch of the TEJ. This situation is very similar to that for the westerly jet-stream at the middle latitude.

The TEJ is one of the most important circulation features that affect the monsoon activity and precipitation in the tropical African and Asian regions. Over the Indian monsoon region, on the one hand, the active and break cycles of the summer monsoon, the shift of the monsoon rain belt, and the periodic fluctuation of the intensity of monsoonal airflows are all related to the position and change in the intensity of the TEJ (Koteswaram, 1958). On the other hand, the activity of the TEJ

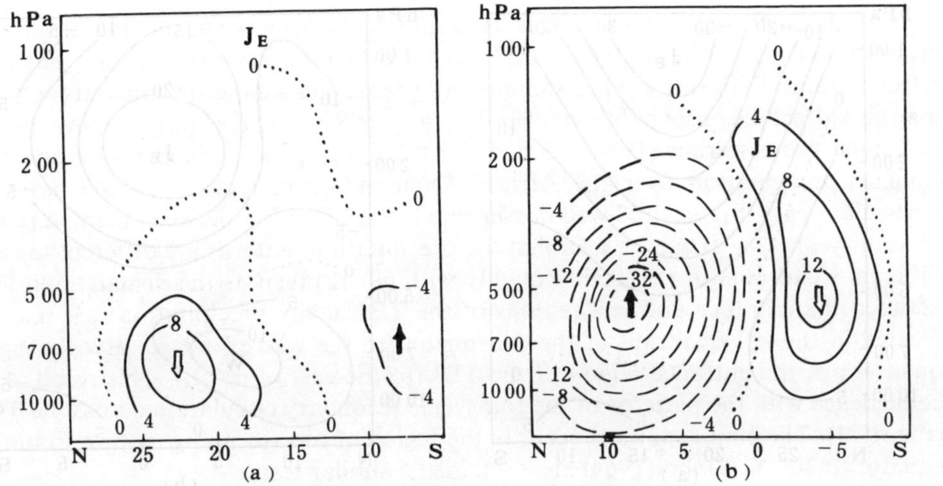

Figure 1.72. The same as Fig. 1.71, but for the vertical velocity ($\omega = dp/dt$). Units: 10^{-3} hPa s^{-1} (Xiao and Xie, 1983).

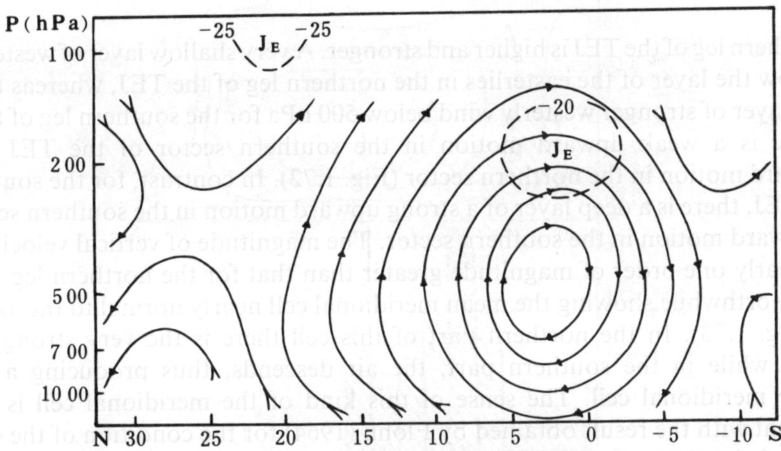

Figure 1.73. The mean meridional circulation across the TEJ (top panel) and the distribution of the total cloudiness (bottom panel).

is also related to the shift of rain belts in China. In May when the TEJ begins to establish itself over the southern part of the Indo-China Peninsula, it may have an effect on rainfalls during the pre-summer rainy season in southern China . Its activity may also affect the onset of Meiyu over the Yangtze River Valley and the rainy season in China. Koteswaram (1958) studied the relationship between the TEJ and the summer precipitation over the African-Asian monsoon regions on the basis of particular years. Figure 1.68 well documents the relationship between them in terms of multi-year data. It can be seen that the precipitation patterns are different for the different quadrants of the TEJ. Roughly with 60° E taken as the demarcation line between the entrance and exit regions of the TEJ, many precipitations are mainly located in the left quadrant of the exit region (to the west of 60° E) and the right quadrant of the entrance region of the TEJ (to the east of 60° E). This result is in coincidence with the patterns of the transverse secondary circulation across the TEJ (Fig. 1.74). The major rainfalls occur in the region of the ascending motion from the equator to 30° N. Flohn (1964) also obtained a similar result.

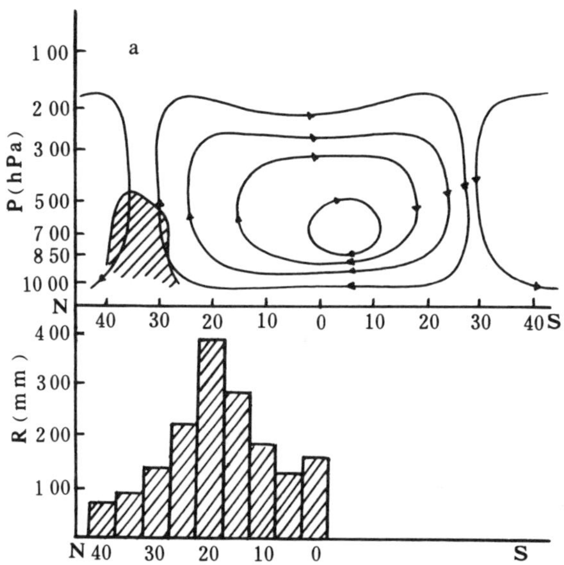

Figure 1.74. The meridional circulation (top panel) and the distribution of precipitation (bottom panel) in the entrance region of the TEJ on the long-term mean basis (Atkinson, 1971).

The interannual variability of monsoon precipitation is associated with the variation in strength of the TEJ. Zeng and Guo, by using data sets for 1966–1975, showed that the years with many monsoon rainfalls, in most cases correspond to a TEJ of above-normal intensity (the more extensive region of easterly wind speed greater than 20 m s^{-1}, more northern position of the axis of the TEJ and the stronger core of jet), whereas the years with deficit monsoon rainfalls correspond to a TEJ of below-normal intensity. In addition, the intensity of the TEJ assumes interannual

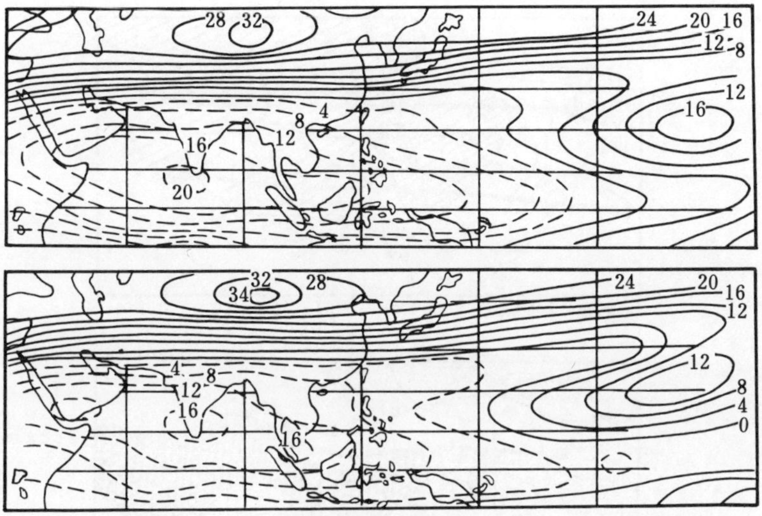

Figure 1.75. The distributions of zonal wind for months with many typhoons (top panel) and few typhoons (bottom panel). The full lines represent the westerly wind and the dashed lines the easterly wind. Units: m s^{-1} (Ding and Reiter, 1983).

variability during a period of 2–3 years. Correspondingly, the precipitation pattern also demonstrates this general tendency.

The behavior of the TEJ is related to the activity of typhoons over the western Pacific. Ding and Reiter (1983b) have studied the large-scale environmental conditions affecting the formation of typhoons there, indicating a significant difference in the TEJ between months with many typhoons and months with few typhoons (Fig. 1.75). For the former case, the strength of the easterly winds at 200 hPa over the tropical western and central Pacific during months with many typhoons is twice that of months with few typhoons.

At about 0–10° N, one may observe a branch of the easterly jet, i.e., the southern branch of the TEJ. During months with few typhoon, this branch is nonexistent. Thus, this may be indicative of the fact that the strong upper-level outflow equatorward of the region of typhoons, caused by the frequent activity of typhoons over the western Pacific, and the subsequent westward acceleration, contributes much to the formation or enhancement of the southern branch of the TEJ.

Xiao and Xie (1983) further identified the difference in the patterns of the TEJ for periods of active and inactive typhoon occurrence in July 1979 (Fig. 1.76). For the latter there were two branches of the TEJ, with the northern branch at about 15° N and the southern branch at about 5° N. The northern branch of the jet was stronger than the southern branch. For the former, however, the opposite case is almost true, i.e., the southern branch was stronger than the northern branch. If the jet is defined here as a maximum wind speed greater than 20 m s^{-1}, one may conclude that there is only the southern branch of TEJ in the region from Southeast Asia to the western Pacific for period with many typhoons, while there is only the northern branch for the

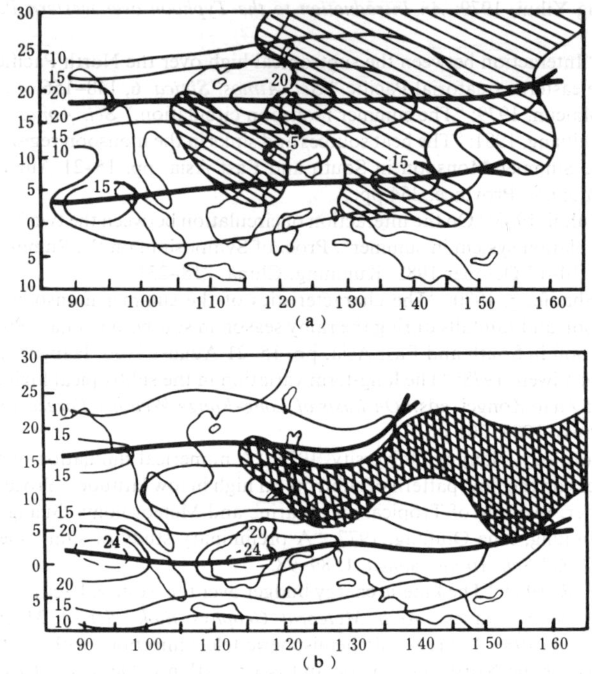

Figure 1.76. The patterns of isotachs at 200 hPa for the periods of no typhoons (shaded area : advection of negative vorticity) (a) and many typhoons(shaded area : advection of positive vorticity) (b) in July 1979. Units: m s^{-1} (Xiao and Xie, 1983).

period with few or no typhoons. Therefore, the above result justifieds the close relationship between the development of typhoons and the existence of the southern branch of the TEJ.

References

Atkinson, G. D., 1971: *Forecaster's Guide to Tropical Meteorology*, AWS, U.S. Air Force.

Bates, J. R., 1970: "Dynamics of disturbances on the intertropical convergence zone". *Quart. J. Roy. Meteorol. Soc.*, **96**, 677–701.

Bi Muying, 1990: "The oscillation and propagation of the subtropical high over the NW Pacific in the summer". Proceeding of Long-Range Weather Forecasting. China Meteorol. Press, pp. 241–247.

Chang, C. C., 1981: "A contrasting study of the rainfall anomalies between central Tibet and central India during the summer monsoon season of 1979". *Bull Amer. Meteorol. Soc.*, **62**, 20–22.

Charney, J. G., 1969: "The intertropical convergence zone and the Hadley circulation of the atmosphere". Proceedings of WMO/IUGG Symposium on Numerical Weather Prediction in Tokyo, November 26, 1968.

Chen Qiushi, Li Weiliang, Miao Jinhai, 1964: "The mean flow fields and meridional cells in the regions of southwesterly monsoon over Southeast Asia nd trade wind over the Pacific." *Acta Meteorol. Sinica*, **34**, 51–61.

Chen Lianshou, Ding Yihui, 1979: *An Introduction to the Typhoon over western Pacific*. Science Press, Beijing. 491 pp.

Chen Lieting, 1982: "Interaction between the subtropical high over the North Pacific and the sea surface temperature of the eastern equatorial Pacific." *Sci. Atmos. Sinica.* **6**, 148–156.

Chen Longxun, Li Maicun, 1979: "The summer monsoon circulation." *Sci. Atmos. Sinica*, **3**, 86–98.

Chen Longxun, Li Weiliang, 1981: "The heat sources and sinks in the monsoon region of Asia". Pro. of the Symposium on the Summer Monsoon in South and East Asia, pp. 15–21 August, 1981, Hongzhou, People's Press of Yunnan Province, 254 pp.

Chen Longxun, Jin Zuhui, 1983: "On the interaction of circulation between the two hemispheres in the East Asia monsoon circulation system in summer". Proc. of Symposium on the Summer Monsoon in South and East Asia, pp. 10–15 October 1982. Kunming, China. 218–231.

Chen Shixun, Gao Shaofeng, 1980: "The characteristics of the summer monsoon and the relationship between the monsoon and rainfalls during the rainy season in southern China". Proc. of Symposium on the Summer Monsoon in South and East Asia, pp. 15–21 August 1980, Hangzhou, China, 129–141.

Chen Xingfang, Yang Yiwen, 1978: "The long-term variation in the subtropical high over NW Pacific", in Wang Shaowu and Zhao Zongci, eds., *On Basis of Long-Range Weather Forecasting*, Shanghai Science and Technology Press, 1987, 201 pp.

Department of Meteorology, Nanjing University, 1976: "A numerical simulation of the effect of the mid-latitude disturbances on the flow pattern of subtropical high in low latitude". Proceedings of Conference on Tropical Weather. Institute of Tropical and Marine and Meteorology, Guangdong, 203 pp.

Ding Yihui, Feng Huijung, Xue Qiufang, 1977: "A preliminary study of simultaneous development of typhoon on the ITCZ." *Sci. Atmos. Sinica*, **1**, 89–98.

Ding Yihui, Reiter, E. R. 1983: "The kinetic energy budget over the western Pacific during dominant trade wind and active monsoon regimes." *Arch. Meteorol. Geophys. Biokl. Ser. A*, **32**, 201–229.

Ding Yihui, Reiter, E. R. 1983: "Large-scale hemispheric teleconnections with the frequency of tropical cyclone formation over the Northwest Pacific and North Atlantic Oceans." *Arch. Meteorol. Geophys. Biokl.*, Ser. A, **32**, 311–337.

Ding Yihui, Zhao Shenming, He Shixiu, 1984a: "Some new aspects in the study on global mean circulation at 200 hPa in summer." *Kexue Tongbao*, **29**, 1506–1509.

Ding Yihui, Zhao Shenming, He Shixiu, 1984b: "A study of the long-termmean circulation at 200 hPa over the global tropics and subtropics during the Northern summer. Part I: Planetary-scale circulation features". *Sci. Atmos. Sinica*, **12**, 221–231 (English edition).

Ding Yihui, Zhao Shenming, He Shixiu, 1984: "A study of the long-term mean circulation at 200 hPa over the global tropics and subtropics during the Northern summer. Part II: Planetary-scale wind systems". *Sci. Atmos. Sinica*, **12**, 242–249.

Ding Yihui, Zhao Shenming, Zhang Jian, 1987: "A study of the global 200 hPa divergent ciculations in the tropics and subtropics in summer". *Acta Meteorol. Sinica*, **45**, 120–127.

Findlater, J., 1969: "A major low level aircurrent over the Indian Ocean during the northern summer". *Quart J. Roy. Meteorol. Soc.*, **95**, 362–380.

Flohn, H., 1964: *Investigations on the tropical easterly jet*. Meteorol. Inst., Univ. of Bonn, Final Report No. 1, contract No. DA-91-591-EVC-2784.

Gray, W. M., 1967: "Global view of the origin of tropical disturbances and storms". *Monthly Wea. Rev.*, **96**, 669–700.

Gray, W. M., 1979: "Hurricanes: their formation, structure and likely role in the tropical circulation", in D. B. Shaw, ed., *Meteorology Over the Tropical Oceans*. Roy. Meteorol. Soc., 155–218.

Group of Tropical Meteorology, Institute of Atmospheric Physics, 1976: "A contrasting study of low-latitude general circulation during active and inactive period of the ITCZ over West-Pacific". Proceedings of Workshop on Tropical Weather, Guangdong, pp. 111–122. Institute of Tropical and Marine and Meteorology.

Guo Qiyun, Wang Jiqin, 1981: "The distribution of precipitation in China during the summer monsoon period for recent 30 years". *Acta Geographica Sinica*, **36**, 187–195.

Guo Qiyun, 1985: "The variations of summer monsoon in East Asia and the rainfall over China". *J. Tropical Meteorol.*, **1**, 44–52.

Guo Qiyun, Wang Jiqin, 1988: "A comparative study on summer monsoon in China and India". *J. Tropical Meteorol.*, **4**, 53–60.

Hahn, D. G., and S. Manabe, 1975: "The role of mountains in the South Asia monsoon circulation". *J. Atmos. Sci.*, **32**, 1515–1541.

Herbert, L. F., 1969: "The double intertropical convergence zone-Fact or fiction". *J. Atmos. Sci.*, **26**, 771–773.

Huang Fujun, Cui Xiumin, Shan Fumin, 1980: *The break and active monsoon rainy spells over the Tibetan Plateau.* Meteorol. Monthly. No. 10, 1–4.

Huang Ronghui, Lu Li, 1989: "Numerical simulation of the relationship between the anomaly of subtropical high over East Asia and the convective activities in the western tropical Pacific". *Adv. Atmos. Sci.*, **6**, 202–214.

Institute of Tropical and Marine Meteorology, Guangdong Province, 1976: "A preliminary study of the intertropical convergence zone over the South China Sea". Proceedings of Workshop on Tropical Weather and System, Ocean Press, Beijing. pp. 41–50.

Jiang Dunchun, Bao Chenglan, Lu Sene, 1987: "An energy analysis on a strong monsoon trough in ITCZ". Collected Papers of Meteorological Science and Technology. Vol. 10, China Meteorol. Press, Beijing, pp. 209–219.

Jiang Quanrong, Yu Zhihao, 1984: "Some aspects of the ITCZ around the South China Sea and a case analysis on its structure". Proceedings of Symposium on Tropical Circulation and Systems. Ocean Press, Beijing, pp. 119–127.

Jiang Shangcheng, 1988: "The climatic characteristics of ITCZ all over the globe". *Acta Meteorol. Sinica.*, **46**, 241–245.

Johnson, D. R., 1985: "On the global distribution of heat sources and sinks and their relation to mass and energy transport". GARP Special Report. No. 44. III, pp. 1–III 13.

Koteswaram, R., 1958: "The easterly jet stream in the tropics". *Tellus*, **10**, 43–57.

Krishnamurti, T. N., 1969: "An experiment in numerical prediction in equatorial latitudes". *Quart J. Roy. Meteorol. Soc.*, **95**, 594–620.

Krishnamurti, T. N., 1971: "Tropical east-west circulations during northern summer". *J. Atmos. Sci.*, **28**, 1342–1347.

Krishnamurti, T. N., Bhalme, H. N. 1976: "Oscillation of a monsoon system. Part I: Observational aspects". *J. Atmos. Sci.*, **33**, 1937– 1954.

Krishnamurti, T. N., 1978: "Northern summer tropical circulations during drought and normal rainfall months". *Mon. Wea. Rev.*, **106**, 331–347.

Krishnamurti, T. N., 1985: "Summer monsoon experiment – A review". *Mon. Wea. Rev.*, **113**, 1590–1626.

Kung, E. C., Merritt, L. P. 1974: "Kinetic energy sources in large-scale tropical disturbance over the Marshall Islands area". *Mon. Wea. Rev.*, **102**, 489–502.

Lau, K-M., Yang, G. J., Shen, S. H. 1988: "Seasonal and intraseasonal climatology of summer monsoon rainfall over East Asia". *Mon. Wea. Rev.*, **116**, 18–37.

Li Xianzhi, 1956: "A synthetic theory of formation of typhoons". *Acta Meteorol. Sinica.*, **27**, 87–100.

Liao Quasun, Zhao Zhenguo 1990: "Relationship between the blocking pattern in East Asia and the West Pacific subtropical high and its influence on the precipitation in China". Proceedings of Long-Range Weather Forecasting. China Meteorol. Press, Beijing, pp. 125–135.

Liang Biqi, Liang Mengxuan, Xu Xiaoying, 1981: "The cross-equatorial current in the lower troposphere and the summer monsoon in the South China Sea". Proc. of the Symposium on the Summer Monsoon in South and East Asia, 15–21 August 1980, Hangzhou, China, pp. 39–48.

Lin Benda, Miao Jinhai, 1981: "The dynamic analysis of the sudden northward shift of subtropical high at the end of Mei-Yu in 1977". *Sci. Atmos. Sinica*, **5**, 255–266.

Lin Chunyu, 1982: "The stability of summer monsoon over the middle and lower reaches of the Yangtze River and the seasonal transition of circulation over East Asia in early summer". Proc. of the Symposium on the Summer Monsoon in South and East Asia, 10–15 October, 1982, Kunming, China, pp. 204–217.

Lin Zhiguang, 1987a: "The continual research of the movement of summer monsoon rain belt in the east part of China". In *Collected Papers of Meteorological Science and Technology*, Vol. 10, Monsoon Meteorology, China Meteorol. Press, Beijing, 235 pp.

Lin Zhiguang, 1987b: "The advance of the monsoon rainfall area in the western part of China". In *Collected Papers of Meteorological Science and Technology*, Vol. 11, Summer Monsoon and Droughts/Floods, China Meteorol. Press, Beijing 142 pp.

Lips, F. B., 1970: "Barotropic stability and tropical disturbances". *Mon. Wea. Rev.*, **98**, 122–131.

Long-term Weather Forecasting Group, Institute of Geography, Academia Sinica, 1976: The effect of the tropical Ocean on the subtropical high and its prediction experiment. In: Wang Shaowu, Zhao Zongci, eds., *On Basis of Long-Range Weather Forecasting*, Shanghai Science and Technology Press, 1987, 201 pp.

Lu Jingxi, Ding Yihui, 1989: "Climatic study on the summer tropical easterly jet at 200 hPa". *Adv. Atmos. Sci.*, **6**, 215–226.

Lu Jingxi, Ding Yihui, 1989: "Medium-range oscillations in summer tropical easterlies at 200 hPa". *Adv. Atmos. Sci.*, **6**, 301–312.

Lu Juzhong, Lin Chunyu, 1982: "The relationship between the retreat of summer monsoon in East Asia and seasonal variation of circulation over East Asia". Proc. of the Symposium on the Summer Monsoon in South and East Asia, 10–15 October 1982, Kunming, China, pp. 45–55.

Mao Tianson, Xu Naiyou, 1982: "A preliminary study on the characteristics of summer monsoon activity over Huanghuai area". Proc. of the Symposium on the Summer Monsoon in South and East Asia. 10–15 October 1982, Kunming, China, pp. 80–96.

Nitta, T., Yanai, M. 1969: "A note on barotropic instability of the tropical easterly current". *J. Meteorol. Soc. Japan*, **47**, 127–130.

Oort, A. H., Rasmussen, E. M. 1971: "Atmospheric Circulation Statistics". NOAA Professional Paper 5, 323 pp.

Pike, A. C., 1970: "Seasonal variation of the ITCZ studied with an interacting atmosphere and ocean model". Proc. of the Symposium on Tropical Meteorology, 2–11 June 1970, Honolulu, 1970.

Qiu Binhuan, Ding Yihui, 1979: "The structure of circulation of the Meiyu for 1973". In: *Formation and Forecasting of the Meiyu Rainfall in Summer*. Science Press, Beijing, pp. 23–32.

Ramage, C. S., 1972: *Monsoon Meteorology*. Academic Press, 296 pp.

Saha, K. R., 1973: "Globle distribution of double cloud bands over tropical oceans". *Quart. J. Roy. Meteorol. Soc.*, **99**, 551–555.

Shen Jianzhu, Zhu Zhihui, 1982: "The intensity of the southwest monsoon and its relationship to the precipitation over the Yangtze River Valley". Proc. of Symposium on Tropical Weather. Science Press, Beijing, pp. 120–126.

Shi Shangwen, Chao Junmin, 1982: "The summer monsoon in North China". Proc. of the Symposium on the Summer Monsoon in South and East Asia. 10–15 October 1982, Kunming, China, pp. 72–85.

Sutcliffe, R. C., Bannon, K. 1954: "Seasonal changes in upper air conditions in the Mediterranean-Middle East Asia". Proceeding of UGGI, Rome, pp. 322–334.

Tanaka, M., 1982: "Interannual fluctuations of the tropical easterly jet and the summer monsoon in the Asia region". *J. Meteorol. Soc. Japan*, **60**, 865–874.

Tang Mingmin, Lu Sene, Huang Shihong, 1984: "The characteristics of the position and intensity of the western Pacific subtropical high". Proc. of Conference on Tropical Circulation and Systems. Ocean Press, Beijing, pp. 41–53.

Tang Mingmin, Huang Shihuong, 1980: "On the advance and retreat of the summer monsoon of eastern China in 1979". Proc. of Symposium of the Summer Monsoon in South and East Asia, 15–21 August 1980, Hangzhou, China, pp. 15–30.

Tao Shiyan, Chen Longxun, 1957: "The structure of the atmospheric general circulation over the Asian continent in summer". *Acta Meteorol. Sinica*, **28**, 234–247.

Tao Shiyan, Xu Suyin, Guo Qiyun, 1962: "The characteristics of meridional and zonal types of circulation in East Asian tropics and subtropics in summer". *Acta Meteorol. Sinica.*, **32**, 91–103.

Tao Shiyan *et al.*, 1963: *A study of some aspects of weather systems in the subtropics in summer in China.* Science Press, Beijing.

Tao Shiyan, He Shixiu, Yang Zhufang, 1983: "An observational study on the onset of the summer monsoon over eastern Asia in 1979". *Sci. Atmos. Sinica.*, **7**, 345–355.

Tao Shiyan, He Shixiu, Yang Zufang, 1984: "A preliminary analysis on the development of southwest monsoon over Western North Pacific during 1979 summer". Proc. of the Symposium on Tropical Circulation and Systems, Ocean Press, Beijing, pp. 1–10.

Tao Shiyan, Chen Longxun, 1987: A review of recent research on the East Asian summer monsoon in China. In: C. P. Chang and T. N. Krishnamurti, eds., *Review of Monsoon Meteorology*, Oxford University Press, 353 pp.

Tao Shiyan, *et al.*, 1980: *The Heavy Rainfalls in China*, Science Press, Beijing. 225 pp.

The Second Group of QXPMEX, 1988: *The Circulation and Monsoon*. Science Press, Beijing, 146 pp.

Wang Zuoshu, He Shixiu, 1979: "A preliminary study of the low-level cross-equatorial airflow and monsoon in summer over the South China Sea and West Pacific". *Acta Meteorol. Sinica*, **37**, 67–78.

Wang Jizhi, Li Maicun, 1982: "Cross-equatorial flow from Australia and monsoon over China". *Sci. Atmos. Sinica.*, **6**, 1–10.

Wang Jizhi, Leftwich, P. W., Jr., 1984: "A major low-level cross-equatorial current at 110° E during the northern summer and its relation to typhoon activities". *Sci. Atmos. Sinica.*, **8**, 443–449.

Wu Yuankang, 1987: *Atlas of Tropical Monsoon*. China Meteorol. Press, Beijing, 740 pp.

Xiao Wenjun, Xie An, 1983: "Two easterly jets in summer monsoon and development of typhoon". Proc. of the Symposium on the Summer Monsoon in South and East Asia, 10–15 October 1982, Kunmimg, China, pp. 285–296.

Xue Feng, 1989: "Cumulus friction effects and diagnostic analysis of ITCZ-tropical low". *J. Tropical Meteorol.*, **5**, 320–329.

Xu Guochang, Li Meifang, Zhang Zhiyin, 1983: "Seasonal variation of the rain belts in China". *Sci. Atmos. Sinica*, **7**, 312–318.

Yang Yiwen, Chen Xingfang, 1980: "Some laws of changes in subtropical high ridge in Northern Hemisphere". Proc. of Long-Range Weather Forecasting. China Meteorol. Press, Beijing, pp. 189–205.

Yang Yazheng, 1982: "A preliminary analysis of the genesis and development of tropical easterlies in the upper troposphere". Proceedings of the Symposium on Tropical Weather for 1980. Science Press, Beijing, pp. 70–81.

Yeh Tucheng, Zhu Baozhen, 1954: "The beginning of the transitional season in East Asia in relation to the change in atmospheric general circulation". *Acta Meteorol. Sinica*, **26**, 71–87.

Yeh Tucheng, Tao Shiyan, Li Maichun 1959: "The abrupt change of circulation over the Northern Hemisphere during June and October". In: B. Bolin, ed., *The Atmosphere and the Sea in Motion*. Rockefeller Inst. Press, New York, pp. 249–267.

Yeh Tucheng, Yang Guangji, 1979: "Mean meridional circulations over East Asia and the Pacific Ocean. I: summer; II: winter". *Sci. Atmos. Sinica.*, **3**, 299–305.

Zeng Zaomei, Guo Qiyun, 1982: "The relationship between the summer precipitation over Asia-Africa monsoon regions and the tropical easterly jet stream (TEJ)". *Sci. Atmos. Sinica*, **6**, 281–192.

Zhang Yuanzhen, Cheng Qiuying, 1988: "The relation of oscillation between the high Pressure of Australian and the summer precipitation of China in 1985". *J. Tropical Meteorol.*, **4**, 146–154.

Zhang Jiacheng, Lin Zhiguang, 1985: *The Climate in China*. Shanghai Science and Technology Press, 603 pp.

Zhang Xiangong, 1988: "The climatic oscillations of the subtropical high over the West-Pacific". *J. Acad. Meteorol. Sci.*, **3**, 1–8.

Zheng Qingcun, Liang Xinzong, Zhang Minhua, 1988: "Numerical simulation of monsoon and the abrupt changes in the atmospheric circulation". *Sci. Atmos. Sinica*, Special Issue, 22–42.

Zhou Qinhua, Chen Longxun, 1985: "An analysis on the circulation structure of strong or weak ITCZ periods in 1979 summer". Proceedings of Symposium on Tropical Circulation and Systems. Ocean Press, Beijing. 54–65.

Zhu Kezhen, 1934: "Monsoon in Southeast Asia and rainfall amount in China". *Acta Geographica Sinica*, **1**, 1–27.

Zhu Qiangen, Wu Hong, Xie Lian, 1987: "The breakdown process and structure characteristics of monsoon trough over Asia in summer". *J. Tropical Meteorol.*, **3**, 1–8.

CHAPTER 2

THE WINTER MONSOON IN EAST ASIA

2.1. Climatology of the Winter Monsoon

The prevailing weather processes in winter over East Asia are outbreaks of cold air which, among the most significant weather events of winter over the Northern Hemisphere, are marked by the invasion of extremely dry and cold air over Mongolia, northern China, and Korea. The associated weather includes high winds, an abrupt temperature drop, severe frost, freezing rain, heavy snowfalls and even sandstorms. The severity of these weathers greatly depend upon the coldness of the air masses, patterns of synoptic systems and circulation and the regional or local terrain features.

In China, the severity of the cold air is generally classified into four grades: (1) nationwide outbreak of the cold air, (2) regional outbreak of cold air, (3) severely cold air, and (4) cold air of moderate intensity. Based on the statistics of the cold air for 1951–1975 (Pan *et al.*, 1985), each winter about 20 cold air outbreaks of different intensities affect different regions of China. On an average, there are 6.3 nationwide and regional outbreaks of cold air, 4.2 severely cold air, and 8.8 cold air of moderate intensity. Of these, about three to five sweep across most of China, and bring temperature drops of the order of 10 to 20 °C within a 24 h period and are associated with strong surface winds of 15–20 m s^{-1}. Frequently, such outbreaks result in serious damage to agriculture, transportation, and other economic activities across China. Figure 2.1 and Figure 2.2 show the mean and maximum temperature fall during the whole period of outbreaks of the cold air, respectively. The mean temperature fall is 10–12 °C in northwestern China and eastern parts of China, and the maximum temperature fall of about 28 °C mainly occurred in southern China. Figure 2.3 is the geographical distribution of mean maximum wind vectors for 123 outbreaks of cold air, indicating that the regions of northwestern China, Inner Mongolia, and the Shangdong, Henan, Hunan and Jinagxi Provinces, vulnerable to the impact of outbreaks of cold air, may have maximum wind speeds of 20–28 m s^{-1} and other regions generally have mean maximum winds of 12–18 m s^{-1}. it should be pointed out that the wind force at any station or site not only depends upon the intensity and tracks of cold air, but also greatly depends on the terrain features. Therefore, the above-described distribution of the maximum wind speed basically reflects the feature of the geographical terrain of China.

The characteristics of cold air outbreaks and the winter monsoon are closely associated with the formation, intensification, southward excursion, transformation and dissipation of the Siberian high. Therefore, it is necessary to examine in more

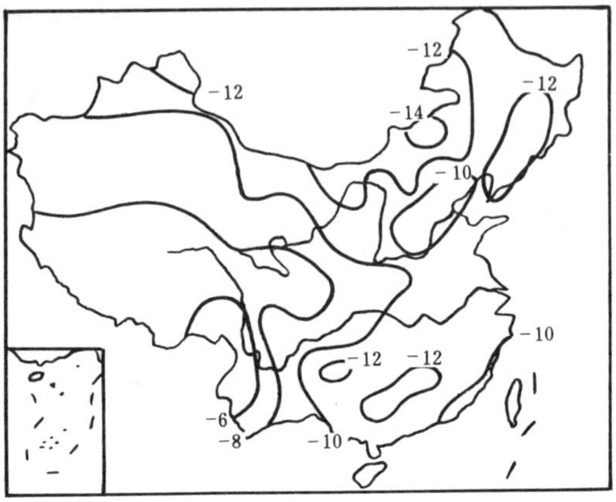

Figure 2.1. The distribution of mean temperature fall for the whole process of cold air outbreaks averaged for 123 cases between 1955 and 1957. Units: °C (Pan *et al.*, 1985).

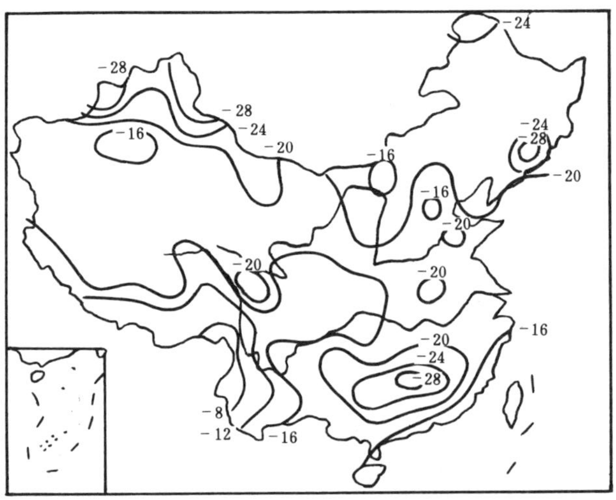

Figure 2.2. The same as Fig. 2.1, but for the maximum temperature fall. Units: °C (Pan *et al.*, 1985).

detail the climatology of that high which acquires its maximum intensity in the winter season and frequently affects China and other regions in East Asia. On an average, a cold high passage occurs over China approximately once every 4 days. The intensity of the Siberian high may reflect to a great extent, the severity of the cold air. Table 2.1 gives the statistics of the central sea level pressure of the Siberian high based on the data of 1958–1979. It can be seen (1) that the high has its maximum intensity in January with a mean sea level pressure (SLP) of 1052 hPa, followed by December and February which have mean maximum pressures of 1051 and 1049 hPa, respectively; (2) highs with maximum sea level pressures exceeding 1080 hPa alone occur in December through February with an extreme maximum sea level pressure of 1082 hPa

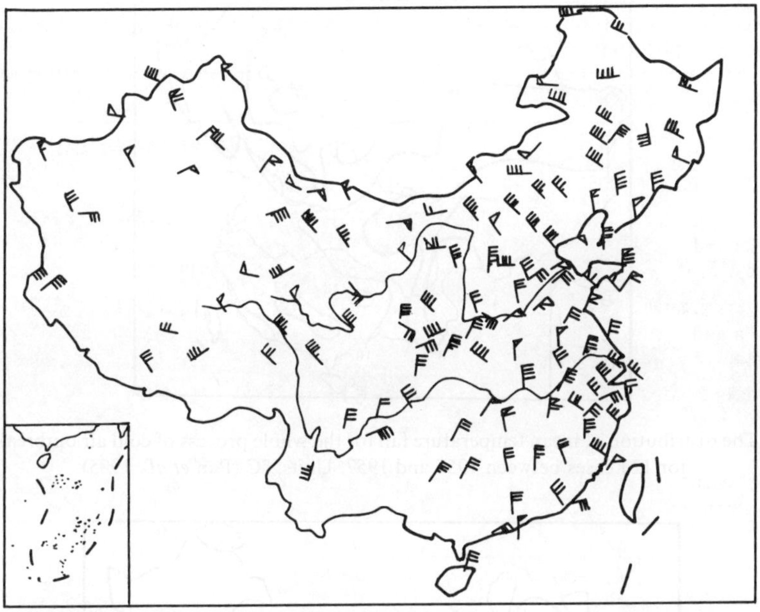

Figure 2.3. The same as Fig. 2.1, but for the maximum wind vectors. Wind symbols: triangle representing 20 m s^{-1}; full barb, 4 m s^{-1} and half barb, 2 m s^{-1} (Pan *et al.*, 1985).

observed in December; (3) the high gradually intensifies from October to January, with the maximum intensity reached in January, and then gradually weakening; (4) the number of days with the maximum sea-level pressure greater than 1050 hPa in January accounts for 51%, in December for 45%, and in October and April for about 3–4%; and (5) in seven months from October to April, the number of days when high centers were observed in the region of 45–55° N, 90–105° E, accounts for over 70%, with January having a maximum of 83%.

There is a close relationship between the intensity of the Siberian high and the severity of the cold airs. As a rule, when the Siberian high builds up to a considerable intensity, one may anticipate an outbreak of cold air. Based on the statistics of 16 cases of cold air outbreaks and strong cold air intrusions, it has been found that the nationwide and regional outbreaks of cold air often occur when the positive departure of the maximum surface pressure is equal to or greater than 18 hPa and 11 hPa, respectively. Further, if the departure of the maximum surface pressure is greater than 10 hPa, the probability of the occurrence of nationwide or regional outbreaks of cold air is very high. The above statistics indicate that the higher the maximum surface pressure of the Siberian high, the greater the possibility of the occurrence of outbreaks of cold air. According to the statistics of Bao (1982), during 1954 through 1979, for nine below-normal (cold) winter months, the central sea-level pressure of the Siberian high was greater than 1050 hPa for 20 days in each month. In nine above-normal (warm) winter months the pressure of the Siberian high exceeded 1050 hPa only for about 5 days in each month. Therefore, the above statistics suggest that the cold-wave

Table 2.1. The statistics of central sea level pressure of the Siberian high in the region of 45–55° N, 90–105° E for the period 1958–1979 (Pan *et al.*, 1985)

Months	Oct.	Nov.	Dec.	Jan.	Feb.	Mar.	Apr.
Mean pressure (hPa)	1036	1045	1051	1052	1049	1041	1032
Maximum SLP (hPa)	1057	1075	1087	1085	1081	1074	1076
Number of days greater than 1050 hPa and percentage (%)	26 (4)	147 (25)	280 (45)	330 (51)	214 (38)	82 (13)	16 (3)
Number of days greater than 1060 hPa and percentage (%)		42 (7)	134 (22)	136 (21)	75 (13)	15 (2)	2 (0.3)
Number of days greater than 1070 hPa and percentage (%)		4 (1)	34 (5)	33 (5)	16 (3)	1 (0.2)	1 (0.2)
Percentage of occurence of centers of highs in this region	71	77	73	83	75	71	59

activity is a basic characteristic associated with the passage of intense Siberian highs over this region.

The tracks of the Siberian high have been studied by numerous investigators (Tao, 1955; Peterssen, 1956; Boyle and Chen, 1987; Pan *et al.*, 1985; Ding and Krishnamurti, 1987). In general, the cold highs associated with outbreaks of cold air come from the region to the west of 110° E, then traverse western Mongolia and arrive in the region to the west of the Lake Baikal (50–55° N, 90–100° E), and eventually more rapidly traverse southeastward across Mainland China once the cold air outbreaks. Figure 2.4 shows the distribution of the frequency of the centers of cold highs for a 5° × 5° longitude/latitude area. It can be seen that the early tracks of cold highs may fall into five types: (1) the northeast track. The highs move from the eastern Siberia southwestward to the region to the east of Lake Baikal and Mongolia (the high of this track seldom occur); (2) the north track. The highs originate in the Tymer Peninsula and move southeastward down to the weatern part of Mongolia; (3) the northwest track (I). The highs come from the Kera Sea and move southeastwards down to the western part of Mongolia; (4) the northwest track (II). The highs have their origin in northern Europe, then traverse the northern Ural region, arrive at the Siberian region and finally merge with the track (3); (5) the west track. The highs come from southern Europe, then move eastward and traverse the Xinjiang Province of China or the western part of Mongolia. Out of the above five tracks, the latter three tracks are quite often observed.

The Siberian high which intrudes into China has two tracks: (1) the major one extends southeastward from the source region to the west of Lake Baikal, passing the middle Yellow River Valley, northern China, and the lower Yangtze River Valley, and eventually extends toward the sea region to the south of Japan or down to southern China; (2) the second track extends eastwards from Mongolia, crossing

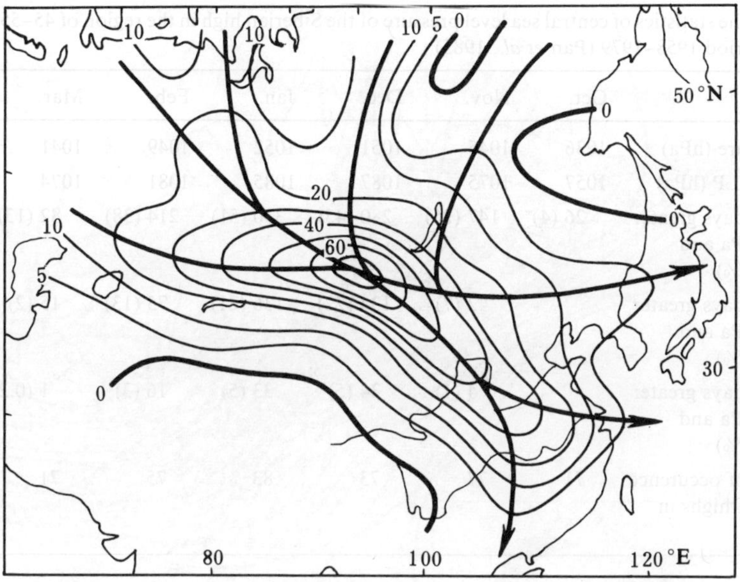

Figure 2.4. The distribution of the frequency of the centers of the Siberian highs in an area of 5° × 5° longitude/latitude for the winter half years of 1955–1979 (Pan *et al.*, 1985).

northeastern China and moving over the Sea of Japan (see Fig. 2.4). Based on an examination of monthly distributions of anticyclone frequencies over eastern Asia and the western Pacific for the period 1957 to 1961, and using a 5° × 5° latitude-longitude grid, Boyle and Chen (1987) obtained a similar result. The major track of anticyclones in November extends from Mongolia to the southwest of the Lake Baikal, southeastward passing the lower Yellow River Valley, and then eastward through the Yellow Sea, southern Korea, and northern Honshu. The major track shifts southward as the winter progresses until it reaches the southernmost latitudes in February, when the anticyclones mainly pass through the lower Yangtze River Valley and then towards the south of Japan.

Ding and Krishnamurti (1987) plotted the tracks of all the Siberian highs that traversed China during the winters of 1980–1984 (Fig. 2.5) based on the position of the maximum sea level pressure. It appear as though most of these tracks cluster along three major paths (see the inserted schematic diagram in the bottom left corner). The first track (identified as the northwestern track) was the most frequently observed. This track accounted for nearly 64% of all the paths of Siberian highs inverstigated by them. A second track is characterized by paths of cold highs that move from west to east south of 50° N. These paths enter the Xinjiang Province of China, moving eastward to the western part of Mongolia and finally arriving over eastern China. A few of these anticyclones move towards the southwestern region of China along the eastern periphery of the Tibetan Plateau. About 27% of the tracks are of this type. A small number of Siberian highs (roughly 10%) moved along a third track shown in Figure 2.5. These mainly affected northeastern China, Korea, and the Sea of Japan.

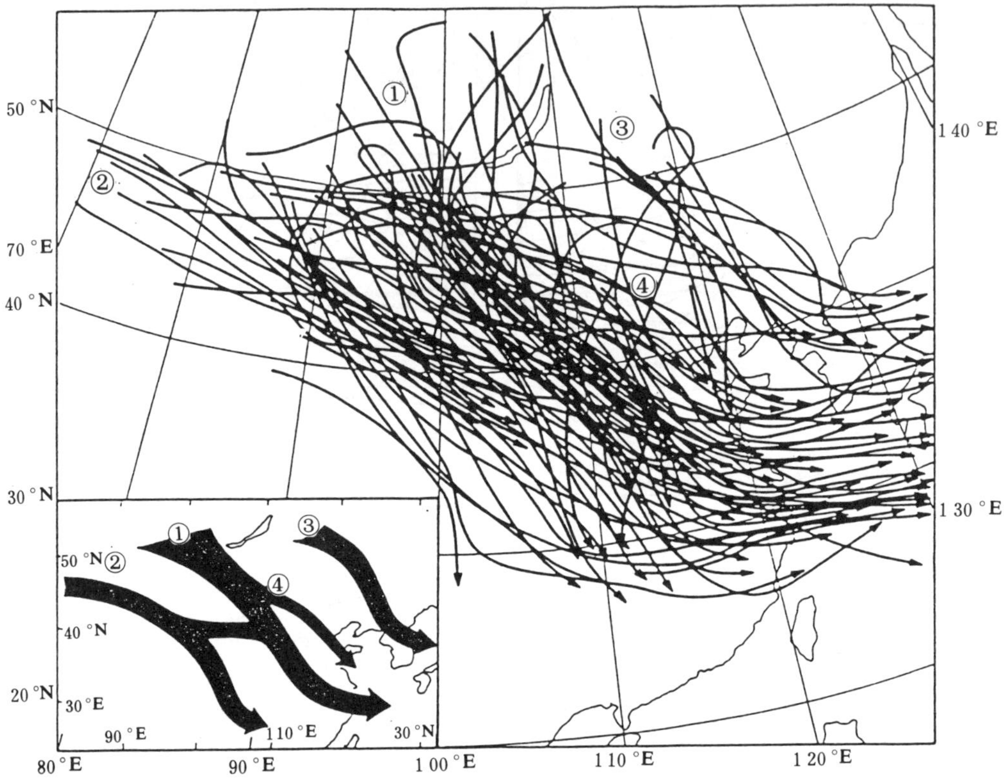

Figure 2.5. Tracks of Siberian highs which crossed into China for the five winters (December-February) of 1980–1984. A schematic map of the tracks of the Siberian high appears at the bottom left (Ding and Krishnamurti, 1987).

Of the Siberian highs moving along the first family of tracks, a subset propagates farther east towards the Beijing area and ultimately passes over the Yellow Sea. These constitute a fourth family of tracks. The above-described first and third paths are similar to those paths indicated by Pan *et al.*

The build-up and development of the Siberian high is closely associated with large-scale circulation and synoptic conditions. Figure 2.6 is representative of the synoptic conditions prior to and after a cold air outbreak. These conditions were obtained from a composition analysis of 19 cases of strong Siberian highs that moved over China from the northwest for five winters (December-February of 1980–1984). Prior to the outbreak of cold air (Fig. 2.6a), an intense composite high was exhibited over western and central Siberia and the northern part of Mongolia. It also appeared to the west of a major upper trough off the eastern coast of East Asia, with a persistent northwesterly flow aloft (Fig. 2.6b). Note that the strongest cloud air advection is found over the southeastern sector of the Siberian high where north-northwesterly winds prevail. Over the western part of the Siberian high, slight warm air advection is descernible. Figure 2.6b represents the composite synoptic situation after cold air

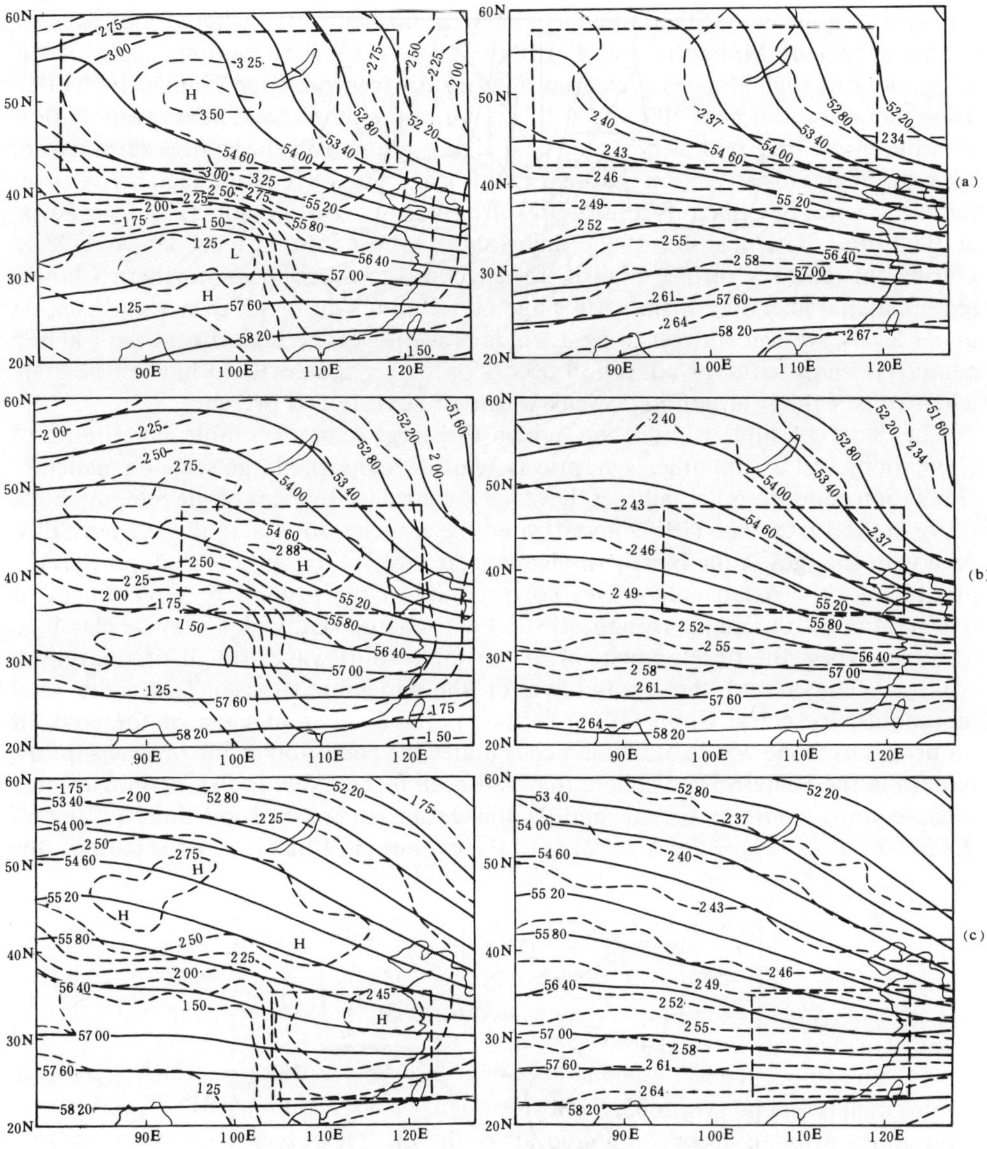

Figure 2.6. Left panel: composite geopotential height fields (in meters) at 1000 hPa (dashed lines in the left panel) and 500 hPa (solid lines in the left panel), and right panel: composite geopotential height (solid line in the right panel) and temperature analysis (dashed lines in the right panel) at 500 hPa. (a) For the stage prior to cold air outbreaks, (b) for the stage post to cold air outbreaks and (c) for the later stage of cold air outbreaks. The above charts were obtained by combining of 19 case of strong outbreaks of cold air for five winters of 1980–1984. Units are meters for height and Kelvins for temperature (Ding and Krishnamurti, 1987).

outbreaks, showing the Siberian highs somewhat farther south, being located over the southern part of Mongolia and northern China. The geopotential height of the anticyclone at 1000 hPa has decreased significantly (central value 294m at 1000 hPa). The synoptic situation at 500 hPa at this point of the time shows northwesterly flow over the entire area, with very strong cold air advection over the southeastern part of the high. Warm air advection is absent at this stage. As the Siberian high arrives over the eastern part of China, its central pressure decreases considerably (Fig. 2.6c; 269 m at 1000 hPa). The position of the high center is now located near 33° N, 117° E. During this stage a surface high is absent over the Siberian or northern Chinese region, a characteristic of the latte stage of a cold air outbreak over East Asia. At upper levels, west-northwest to west winds prevail over the high. From the 500 hPa composite chart, cold air advection occurs only over the northern half of the high, whereas over the southern half, weak warm air advection is present.

The Siberian high is built up under the large-scale favorable conditions of circulation. Just as the other synoptic systems develop, the large-scale or planetary circulation conditions actually set the stage for the development of the Siberian high. Early in the 1950s, Tao (1957) investigated the association of a cold wave over East Asia with changes in the Northern Hemisphere circulation and pointed out that the process of East Asian cold waves corresponds to the process of replacement or regeneration of the major trough off the East Asian coast, which may be classified into two types: the development of an unstable short-wave trough originating in Novaya Zemlya and the breakdown of the blocking situation over the Ural mountains. Recently, Chou (1985) further pointed out that most of the cold air outbreaks over the whole of China occur under the condition of the inverted-Ω flow pattern in the Eastern Hemisphere that had been formed due to the merging of high ridges coming from the Atlantic and Pacific Oceans and at the same time, cutting off the polar vortex (Fig. 2.7). According to the evolution of the inverted-Ω pattern, the

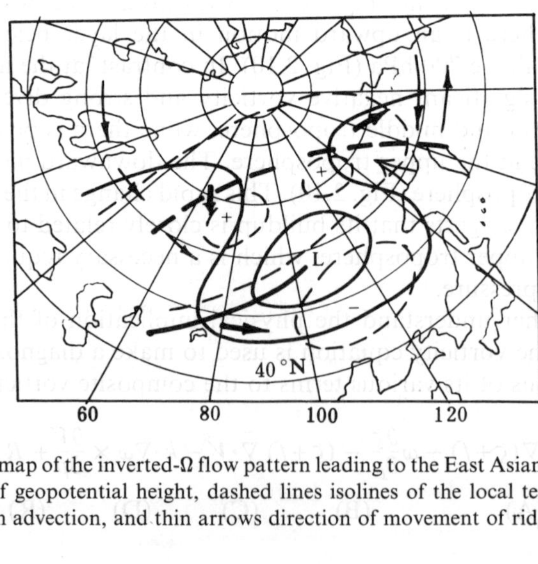

Figure 2.7. Schematic map of the inverted-Ω flow pattern leading to the East Asian cold air outbreak. Thin lines denote isolines of geopotential height, dashed lines isolines of the local tendency of heights, bold arrows cold or warm advection, and thin arrows direction of movement of ridge lines (Chou, 1985).

large-scale process of cold air outbreaks in East Asia may be divided into four stages: initial, building-up, outbreak, and decaying. The initial stage of cold air outbreaks is associated with the blocking situations over Europe and the Ural region and the decaying stage is characterized by the re-establishment of the East Asian major trough. The intense and extensive cold air outbreaks are mostly evolved from the above medium-range weather process. Recent statistics of cold air outbreaks for 10 years have indicated that 85% of them may fall into this process in winter and 70% in spring.

Also, there is another large-scale weather process that may lead to East Asian cold waves. This process is closely related to an eastward system of trough and ridge (Joung and Hitchman, 1982). Before a polar air outbreak, a clear sequence of successive downstream developments is noted. Over the western Atlantic, 6 or 7 days in advance of a cold air outbreak over East Asia, troughs and ridges are seen to form which develop and decay successively downstream of each other across the Eurasian continent until a polar outbreak occurs.

2.2. Build-up and the Dynamic Structure of the Siberian High

The Siberian high usually has very high surface pressure values, generally in the range of 1050–1070 hPa, and even in excess of 1080 hPa in extreme cases. This increase of surface pressure by 50–70 hPa within a few days indicates the presence of a strong net mass convergence in the region of the high, which may be caused by dynamic and/or thermal factors. For this purpose, a study of the dynamic structure and budgets of vorticity and heat of the Siberian high has been made to explore dynamic and thermal factors critical for a local buildup of the high (Ding *et al.*, 1991; Ding and Krishnamurti, 1987). At the initial stage of the formation of the Siberian high, positive vorticity dominates the region of the high in the troposphere (Fig. 2.8). There is convergence at both low and high levels, with divergence in between (Fig. 2.9). Correspondingly, there is an upward motion in the layer below 700 hPa and a downward motion above 700 hPa (Fig. 2.10). In contrast, at the mature stage of the Siberian high, the significant negative vorticity and strong divergent airflows are observed in the lower and middle troposphere, while there is positive vorticity and convergent airflows in the upper troposphere. The downward motion occurs in the whole layer of the troposphere (Fig. 2.10). This rapid change in the dynamic structure of the Siberian high suggests that its buildup is closely related to a significant mass convergence in the upper troposphere, which is a necessary requirement for a rapid increase in surface pressure.

In order to further understand the physical implication of the above-described vorticity profiles, the vorticity equation is used to make a diagnostic analysis of the relative contributions of its various terms to the composite vorticity field,

$$\underset{(G)}{\frac{\partial \zeta}{\partial t}} = \underset{(A)}{- V\cdot\nabla(\zeta+f)} \underset{(B)}{- \omega\frac{\partial \zeta}{\partial p}} \underset{(C)}{- (\zeta+f)\,\nabla\cdot V} \underset{(D)}{- k\cdot\nabla\omega\times\frac{\partial V}{\partial p}} + \underset{(R)}{R} \qquad (2.1)$$

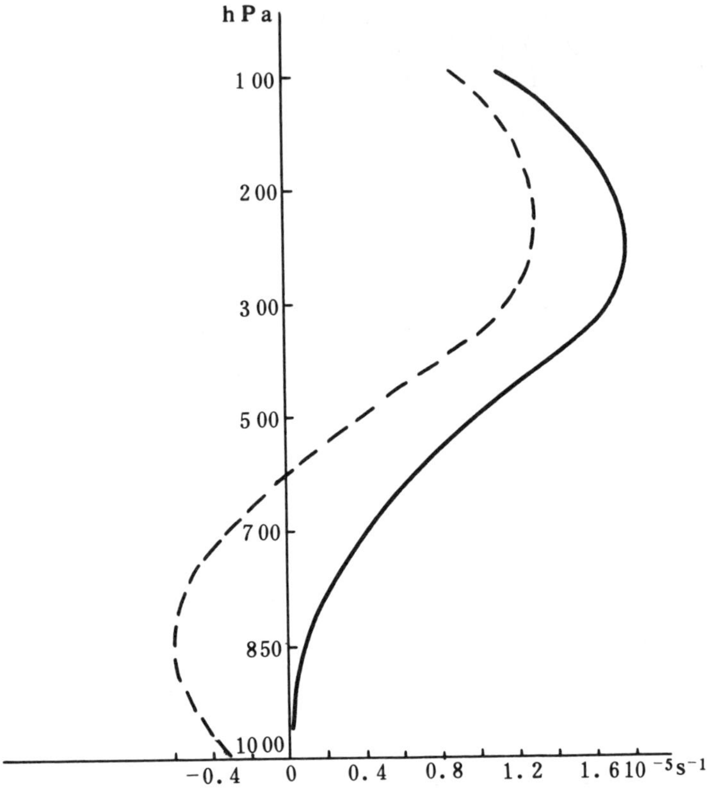

Figure 2.8. The profiles of mean vorticity for the Siberian high, averaged for five cases in the domain of 40–60° N, 77.5–122.5° E. The solid line denotes the curve for the initial stage and the dashed line the curve for the mature stage. Units: $10^{-5}s^{-1}$ (Ding *et al.*, 1991).

Table 2.2. The estimate of various term of the vorticity equation (2.1). Units: $10^{-10}\,s^{-1}$ (Ding *et al.*, 1991)

	Inital stage						Mature stage					
Level (hPa)	G	A	B	C	D	R	G	A	B	C	D	R
100	0.1	0.3	−0.1	0.2	−0.0	−0.3	−0.2	−0.7	−0.0	0.8	−0.0	−0.3
200	0.6	0.8	−0.0	0.5	−0.0	−0.6	−0.3	−1.8	0.0	1.4	0.0	0.1
300	0.8	1.3	0.1	−0.0	−0.0	−0.5	−0.4	−2.4	0.1	1.5	−0.1	0.5
500	0.5	0.9	0.2	−0.3	−0.1	−0.3	−0.4	−0.8	0.3	0.2	−0.2	0.1
700	0.2	0.7	0.1	−0.3	−0.1	−0.3	−0.4	0.2	0.2	−0.4	−0.1	−0.2
850	0.1	−0.1	−0.0	1.1	0.1	−0.9	−0.3	−0.1	−0.0	−1.0	0.1	0.7
1000	−0.1	−0.4	0.0	0.1	−0.0	0.2	−0.1	−0.2	−0.0	−2.8	0.0	2.9

Table 2.2. is the computed results averaged for five cases of the Siberian high. It can be seen that at the initial stage, the local change of the relative vorticity (G) is positive over almost the whole troposphere, except for 1000 hPa. This increase in vorticity is

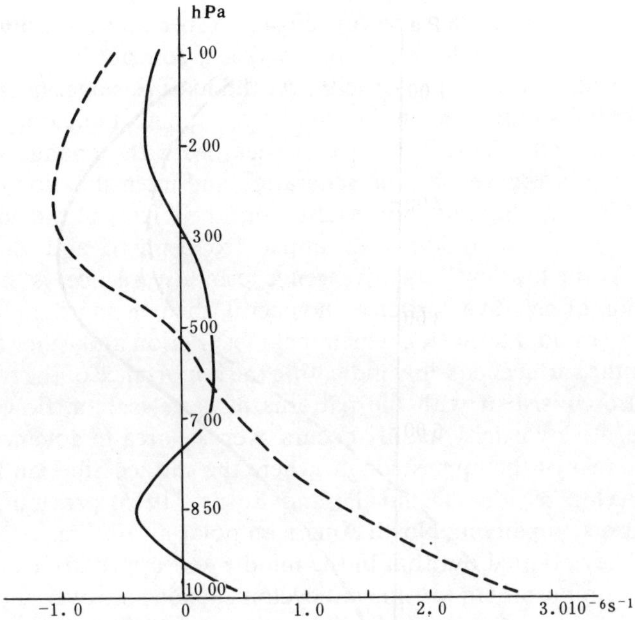

Figure 2.9. Same as Fig. 2.8, but for the divergence (Ding *et al.*,1991).

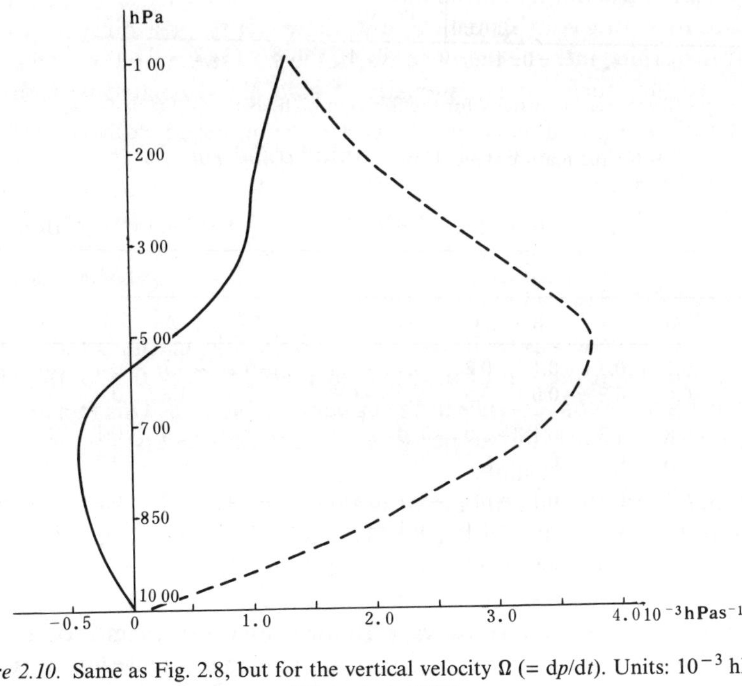

Figure 2.10. Same as Fig. 2.8, but for the vertical velocity Ω (= d*p*/d*t*). Units: 10^{-3} hPa s^{-1} (Ding *et al.*, 1991).

mainly caused by the advection of the vorticity (A), especially in the middle and upper troposphere. The divergence term (C) also makes a contribution, especially at low levels where it produces positive vorticity. At the mature stage, however, the local change of the vorticity is negative in the whole troposphere, indicating the generation of anticyclonic vorticity. This is fully consistent with the change of the relative vorticity, as shown in Figure 2.8. The generation and intensification of the negative vorticity in the Siberian high mainly results from the effects of the advection of the negative vorticity in the middle and upper troposphere and divergence term (convergence). At the low level, the divergence term (divergence) is the major factor for the generation of negative vorticity. In general, the twisting term (D) is of minor importance. The residual term (R), which includes friction and other effects, is most significant near the surface, possibly indicating the importance of the frictional effect. These results are consistent with the patterns of large-scale airflows, because the advection of negative vorticity mainly occurs over the area in advance of the upper ridge and at the rear of the upper trough, where the surface Siberian high is usually observed to develop rapidly. Dallavalle and Bosart (1975) previously studied the anticyclogenesis accompanying North American polar air outbreaks and obtained a similar result. They pointed out that in the middle and upper troposphere, vorticity advection is most important in causing anticyclonic tendencies over and downwind of the surface anticyclone. In contrast, divergence usually dominates in producing similar tendencies in the lower troposphere over the surface high. Subsidence in the anticyclone over North America helps alter the thermal structure to that of classical warm anticyclone. Maximum subsidence, located to the east of the anticyclone, is mainly caused by strong cold advection. But, there is little possibility for the Siberian high to be transformed into the high of a warm thermal structure through subsidence, due possibly to the much greater intensity of cold air associated with the Siberian high.

The advection of negative vorticity in the middle and upper troposphere may cause the upper-level convergence. As an approximation, the equation of vorticity may be simplified into the following form by neglecting the tendency, twisting, and friction terms:

$$V\cdot\nabla(\zeta+f) = -(\zeta+f)\nabla\cdot V \qquad (2.2)$$

This equation is generally believed to be applicable in the middle and upper troposphere. For the advection of negative vorticity, $V\cdot\nabla(\zeta+f) > 0$, therefore, $\nabla\cdot V < 0$, i.e., there is a mass convergence in the upper troposphere. This fact also indicates the major importance of the upper-level advective process for triggering and development of the Siberian high.

On the other hand, the thermal factor is also very important for the formation of the Siberian high. Previously, Wexler (1951) emphasized that an anticyclone can be created through mass convergence, through local radiational cooling, or through the advection of colder, denser air. This is based on the reasoning he suggested in 1937 that if the radiational cooling associated with the formation of continental polar (C_p) air is confined to a restricted area, it will be accompanied by a lowering of the isobaric surfaces. A compensating inflow of air from the adjacent region will occur, causing the surface pressure to rise. In order to further examine the effect of the thermal factor on the build-up of the Siberian high, Ding and Krishnamurti (1987) quantitatively estimated the heat budget of the Siberian high. Firstly, it was estimated via budget calculation. The apparent heat source (Q_1) and apparent moisture sink (Q_2) were calculated from large-scale variables according to the following equations (Yanai *et al.*, 1973):

$$Q_1 = C_p\left[\frac{\partial\theta}{\partial t} + V\cdot\nabla\theta + \omega\frac{\partial\theta}{\partial p}\right]\left[\frac{p}{p_o}\right]^{k}, \tag{2.3}$$

$$Q_2 = -L\left[\frac{\partial q}{\partial t} + V\cdot\nabla q + \omega\frac{\partial q}{\partial p}\right] \tag{2.4}$$

where θ is the potential temperature; q the specific humidity; ω the vertical velocity in p–coordinates; $\varkappa = R/C_p$ with R and C_p being the gas constant and the specific heat of dry air at constant perssure, respectively; L the latent heat of condensation for water; and $P_0 = 1000$ hPa.

The vertical P-velocity, ω, in (2.3) and (2.4) was calculated kinematically by integrating the mass continuity equation. The effect of terrain (as in Boville, 1985) at the Earth's surface, rather than at a pressure surface, was incorporated. The vertical velocity was adjusted using a thermodynamic constraint such that, at the top of the column, at 50 hPa, it was given by

$$\omega_T = \left[\frac{\partial\theta}{\partial t} + V\cdot\nabla\theta - \frac{(p_o/p)^{K}}{C_p}Q_R\right]\Big/\frac{\partial\theta}{\partial p} \tag{2.5}$$

where Q_R is the radiative cooling rate which was computed from a radiation scheme. As the second step, direct estimates of various terms of diabatiic heating with a physical parameterization approach was followed. Thus, the estimates of the heat source were both direct and indirect. The thermodynamic equation is written as

$$\frac{Q_1}{C_p} = \left[\frac{\partial\theta}{\partial t} + V\cdot\nabla\theta + \omega\frac{\partial\theta}{\partial t}\right]\left[\frac{p}{p_o}\right]^{K} = \frac{Q_R}{C_p} + \frac{Q_c}{C_p} + \frac{Q_s}{C_p}, \tag{2.6}$$

$$\frac{Q_2}{C_p} = \frac{-L}{C_p}\left[\frac{\partial q}{\partial p} + V\cdot\nabla q + \omega\frac{\partial q}{\partial p}\right] = \frac{Q_c}{C_p} + \frac{Q_L}{C_p}, \tag{2.7}$$

where Q_c is the net condensation heating rate, Q_s and Q_L are the sensible heating (or cooling) and drying (moistening) rates, respectively. This includes the surface sensible and latent heating, and the vertical diffusion (or mixing). Following Yanai *et al.* (1973) we express the heating by the relation

$$Q_s = \frac{\partial}{\partial p}\overline{s'\omega'}, \tag{2.8}$$

$$Q_L = L\frac{\partial}{\partial p}\overline{q'\omega'}, \tag{2.9}$$

where $s = C_pT + gz$ is the dry static energy. The overbars denote the Reynold's average over the grid scale and the primes denote the deviations from these averages. Small-scale eddies such as cumulus convection and turbulence, contribute to these eddy covariances. In the calculation of (2.6), Q_s is estimated as a residual. Thus, this term includes possible errors from the estimates of Q_1, Q_R and Q_c. Caution must be exercised in the interpretation of this residual. Figure 2.11 a, b and c illustrate the results of the computation of the apparent heat source Q_1 averaged over the regions indicated in Figure 2.6 for three stages of cold air outbreaks. The contribution of different components of Q_1, i.e, (a) the local tendency, (b) the horizontal advection, and (c) the vertical advection, are shown in these figures.

Over these very intense surface anticyclones, we find a weak apparent heat source (solid line) in the upper troposphere above 400 hPa, with an apparent heat sink of a greater intensity (around 2 °C day^{-1}) in the lower and middle troposphere. According to Eq. (2.6), under the condition of the winter Siberian high, Q_c (net condensation heating) is negligbly small; Q_s, the sensible heating is mainly significant

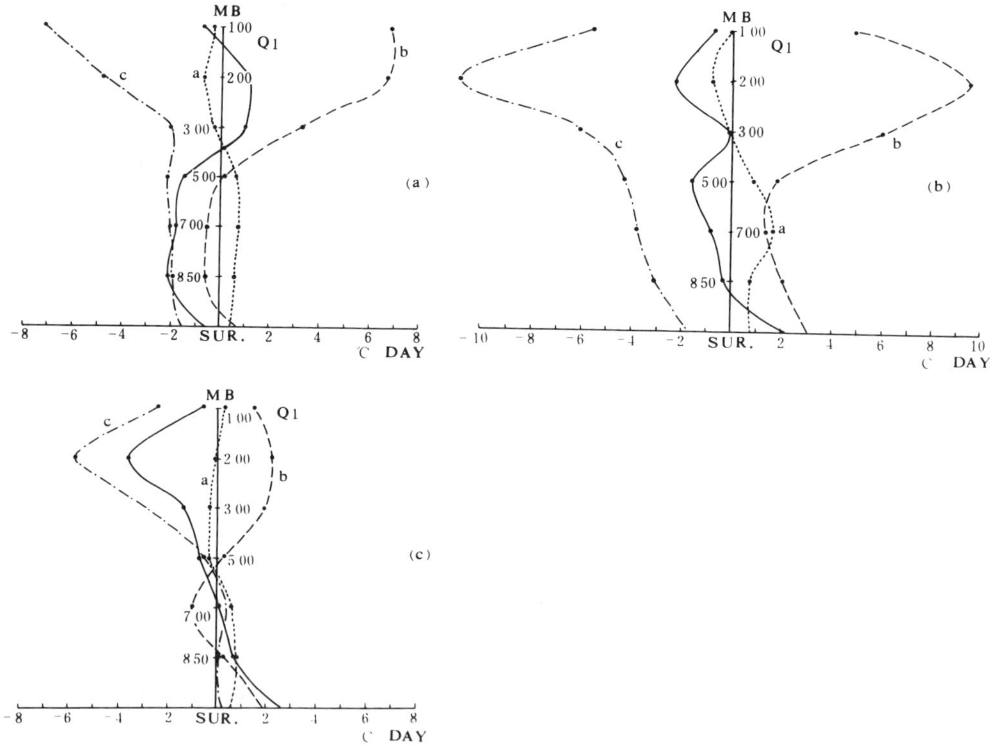

Figure 2.11. (a) Vertical profiles of area-averaged apparent heat source Q_1 and its three individual terms for region 1 (stage 1). The solid line denotes Q_1; curve a represents the local change term, curve b horizontal temperature advection term and curve c the vertical temperature advection term. Units are °C day^{-1}. (b) As in Fig. 2.11(a), except for region 2 (stage 2). (c) As in Fig. 2.11(a), except for region 3 (stage 3) (Ding and Krishnamurti, 1987).

at low level and near the tropopause. Therefore, for most of the tropospheric atmosphere, the component of the longwave radiation Q_R may make a major contribution to the cooling (in this case, solar radiation may be neglected because of late local time). This point will be further documented below. The local change of dry static energy is slightly positive in the lower troposphere, implying a weak cooling and weak heating, respectively. The major contributions to Q_1 arise from the horizontal and the vertical advective effects. Below 500 hPa the net heat sink Q_1 is largely balanced by the adiabatic warming. The sign convention for the interpretation of the components of the heat sources and sinks are the same as in Yanai *et al.* (1973). Because of the signs of Eqs. (2.3) and (2.6), the reader should note that the cold (warm) advection and adiabatic cooling (warming) are indicated by positive (negative) signs. Above 500 hPa cold advection (horizontal) and adiabatic warming are nearly in balance. The major question that might be raised is on the nature of the diabatic cooling below the 500 hPa surface that is balanced by the adiabatic warming. Since the intense Siberian high is essentially cloud-free, the profile of Q_1 must represent a net radiative cooling.

As the Siberian high moves towards central China, the local tendency shows a warming of the column below 300 hPa (Fig. 2.11b). The imbalance between the strong horizontal cold advection and the adiabatic warming is strongest in the upper troposphere and yields a weak net heat sink ($Q_1 < 0$) over most of the troposphere. Below 850 hPa, we see evidence for a net warming ($Q_1 > 0$). The processes are described later. Proceeding to the region of eastern China (Fig. 2.11c), the upper tropospheric adiabatic warming exceeds the cooling by horizontal advection. The net heat sink has an amplitude of nearly 3 °C day^{-1} at 200 hPa. In the lower troposphere, the salient feature is the net heating below 700 hPa. In this region, the time tendency is small. To offset the cooling by horizontal advection, a net heat source is required for balance. As the cold high moves sourthward, its static stability decreases progressively as a result of the cooling at high levels and the heating at low levels.

The vertical profiles of Q_2 for the three stages are shown in Figure 2.12. The small negative values of Q_2 are found below 850 hPa, suggesting a possible role of evaporation or sublimation of cloud and raindrops and ice crystals in this region of downward motion. Overall, regions 1 and 2 are very dry, suggesting insignificant condensation, except possibly for region 3 where stratiform clouds do exist.

We shall next address the geographical distribution of the heat sources. Figure 2.13a-d show the fields of vertically integrated Q_1 and its individual terms for region 1. Cooling dominates most of region 1 with a maximum located over the southeastern part of the high. Figure 2.6 shows the location of the high. The maximum cooling over the southeastern sector has a role in the development of very cold air behind the cold front. This belt of maximum cooling has the same NE-SW orientation as the leading cold front. The major contributions to Q_1 come from the horizontal advection and the adiabatic term, which are of opposite signs. The maximum values for both these terms are located in the southeastern sector of the high (see Figure 2.6 for the position of the high) with the adiabatic warming (negative sign due to Q_1) having a somewhat greater magnitude. This requires an overall stronger diabatic cooling for balance. We

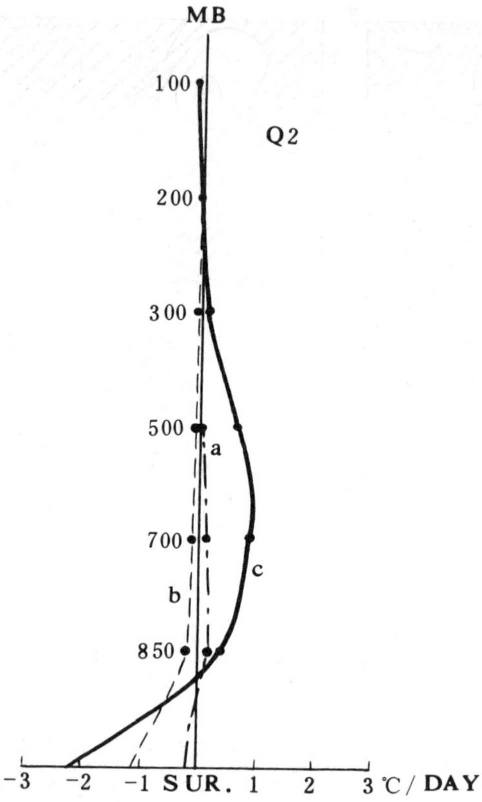

Figure 2.12. vertical profiles of area-averaged apparent moisture sink Q_2 for region 1 (stage 1) (curve a), region 2 (stage 2) (curve b) and region 3 (stage 3) (curve c). Units are °C day^{-1} (Ding and Krishnamurti, 1987).

also find that the maximum downward motion at 700 hPa is located in this same area. The pattern of ω at 300 hPa is also quite similar to that at 700 hPa.

Figure 2.13e-h show the horizontal distribution of Q_1 for region 2. A maximum cooling is found over the eastern part of the high, i.e., in northern and northeastern China (Fig. 2.13e), where strong cold air advection is present (Fig. 2.13f). In this region, just west of the major trough off the coast of East Asia, a strong downward motion exists, this contributes to a strong adiabatic warming (Fig. 2.13g), which dominates the horizontal advective term. Continuous and intense cooling over this part of the high contributes to an intensification of the cold front.

Region 3 (Fig. 2.13i-l) shows a region of significant heating to the east and south of the Siberian high, i.e., along the coastal areas of eastern and southern China. Its southwest to northeast orientation is parallel to the position of the cold front. This is usually the location of a quasi-stationary front over southern China. A significant upward motion is found over this belt of heating. The warm air ahead of the front flows northward with extensive precipitation, which often extends well behind the

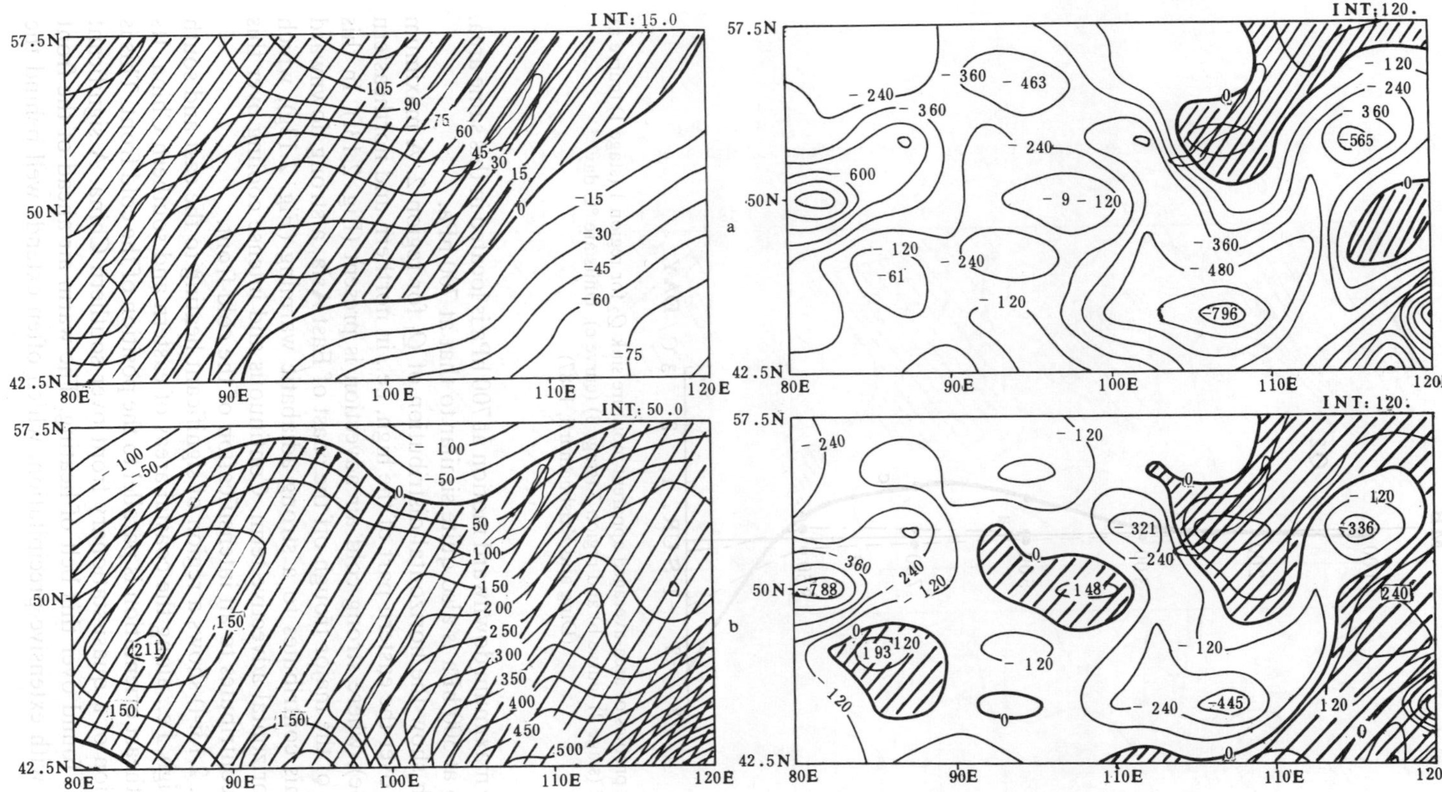

Figure 2.13 a-d. (a) Horizontal distribution of the vertically integrated local change term ($\partial T/\partial t$) for region 1. Positive area is shaded. Units are W m^{-2}. (b) Horizontal distribution of the vertically integrated horizontal temperature advection term ($\mathbf{V}\cdot\nabla T$) for region 1. Positive area (cold air advection) is shaded. (c) Horizontal distribution of the vertically integrated verical temperature advection term $\{\omega\partial\theta/\partial p[p/p_0]^{R/Cp}\}$ for region 1. Positive area is shaded. (d) Horizontal distribution of the vertically integrated apparent heat sources Q_1 for region 1. Positive area (heating) is shaded.

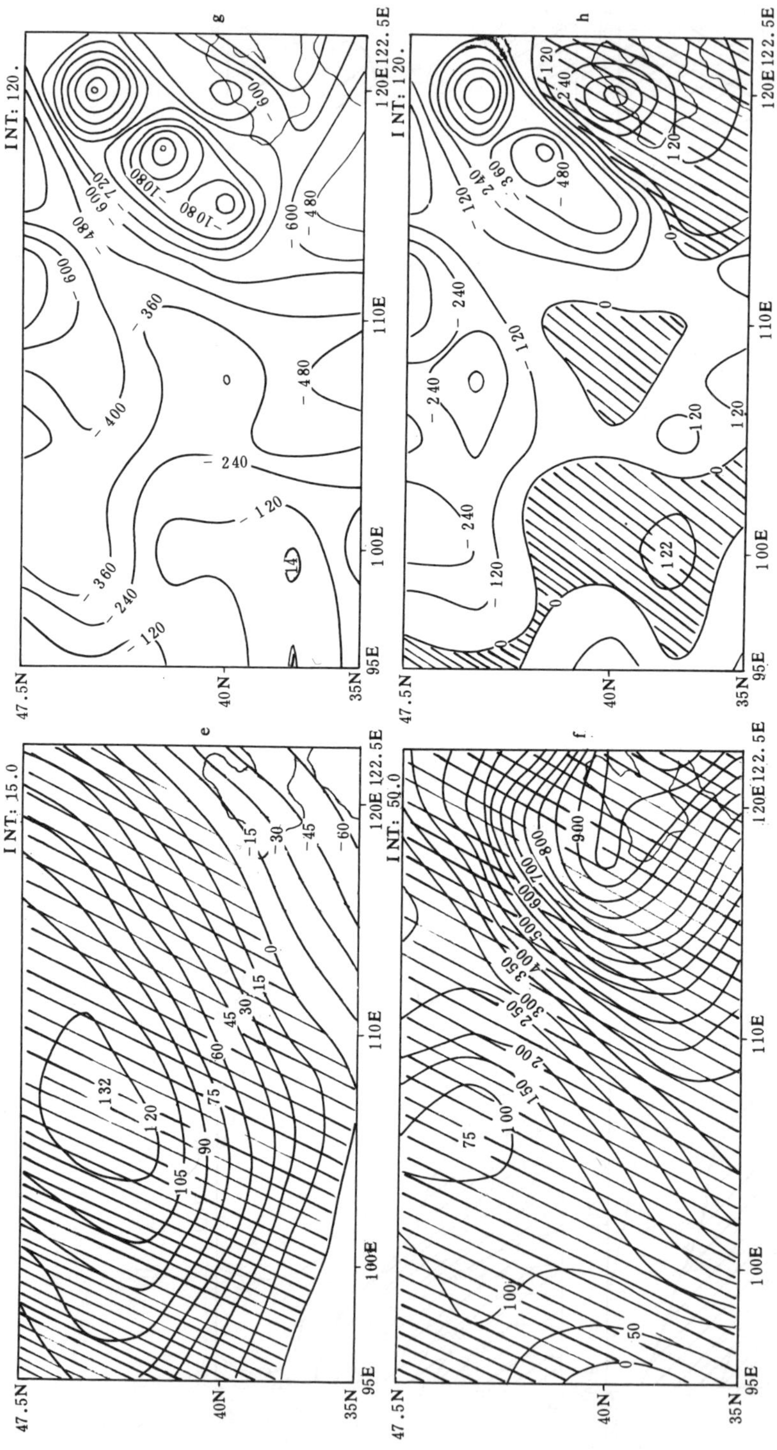

Figure 2.13. e-h (e) As Fig. 2.13a, except for region 2. (f) As Fig. 2.13b, except for region 2. (g) As Fig. 2.13c, except for region 2. (h) As Fig. 2.13d, except for region 2.

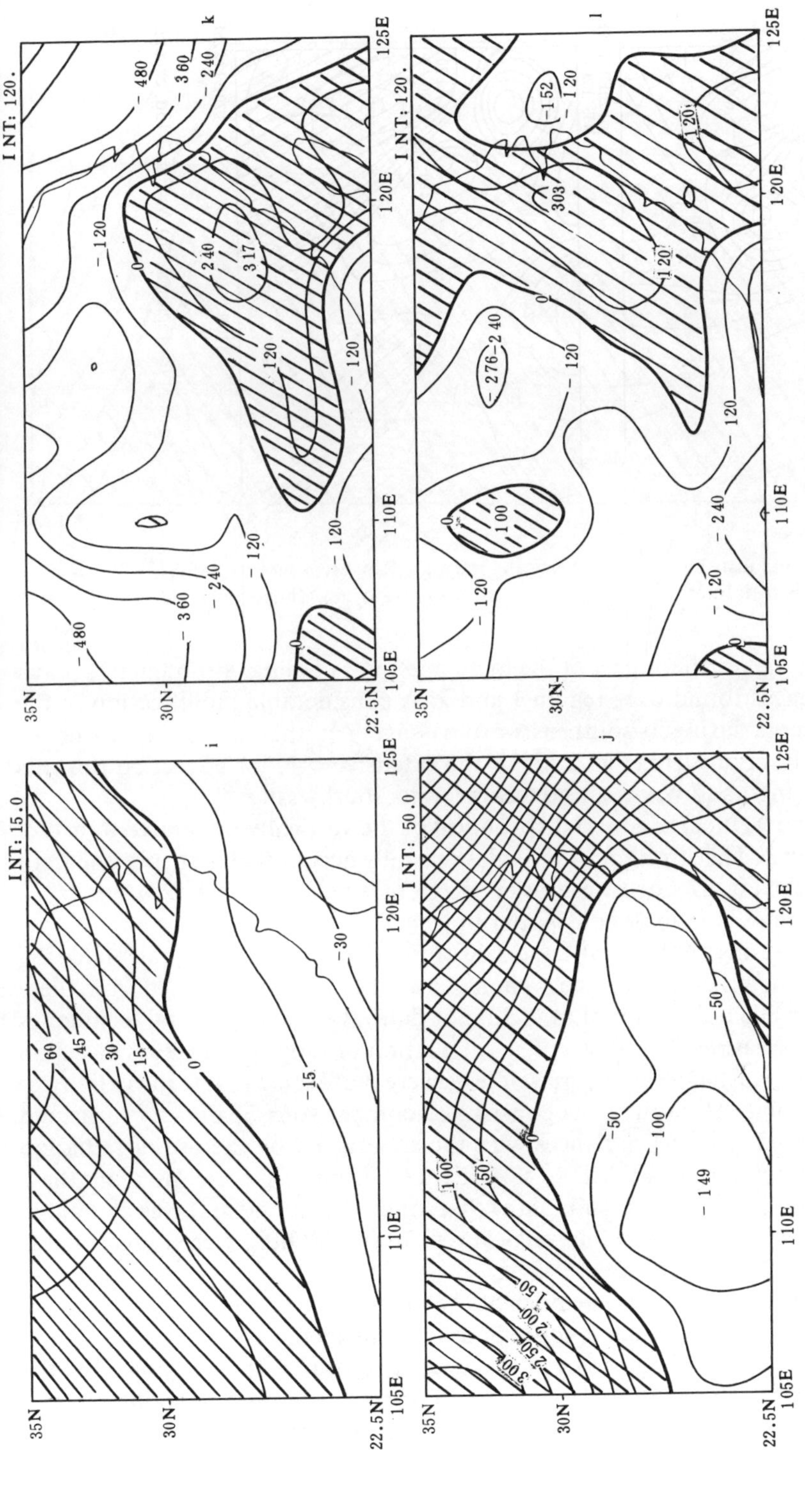

Figure 2.13. i-l (i) As Fig. 2.13a, except for region 3. (j) As Fig. 2.13b, except for region 3. (k) As Fig. 2.13c, except for region 3. (l) As Fig. 2.13d, except for region 3 (Ding and Krishnamurti, 1987).

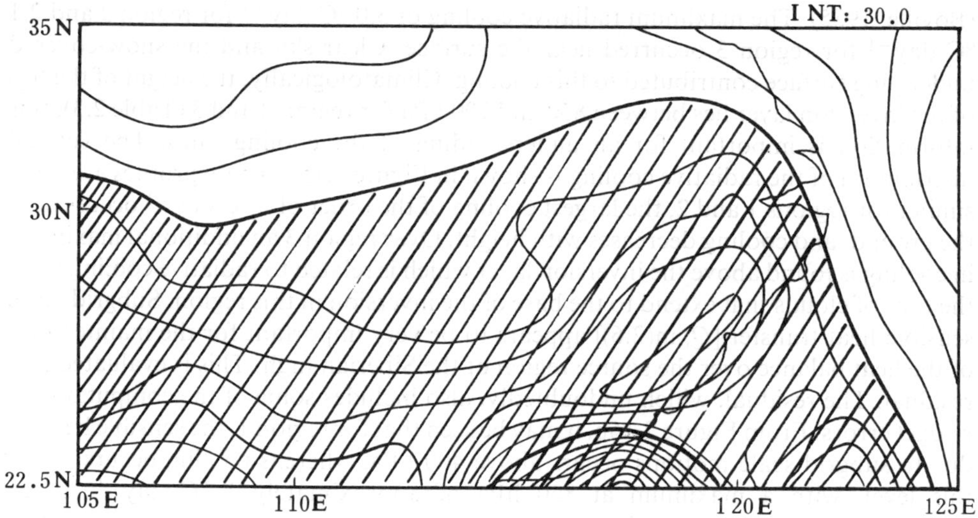

Figure 2.14. Horizontal distribution of the vertically integrated apparent moisture sink Q_2 for region 3. Positive values are shaded. Interval of isolines: 30 units. Units are W m^{-2} (Ding and Krishnamurti, 1987).

surface front. In the northern part of the high, one finds cooling, although it is much weaker than what is found over region 1 and 2. A considerable modification of the cold air mass occurs during its southeastward passage. Another interesting feature of the horizintal advection field (Fig. 2.13j) is the extensive cold air advection towards the north of the high and warm air advection to its southwest.

Figure 2.14 shows the horizontal distribution of the vertically integrated moisture sink Q_2 for region 3. The strong positive values usually accompany a strong apparent heating. A southwest to northwest orientation of the pattern of positive Q_2 (a moisture sink) follows an analogous pattern of Q_1.

The individual terms on the right-hand side of Eq. (2.6) were directly calculated via the physical parameterization schemes. The accuracy of the results derived from these schemes is very much a function of model input data, the computational techniques and the parameterization procedures. The calculation of radiative fluxes is based on Yang and Krishnamurti (1981). A series of tests were performed to test the validity of the radiation scheme. Here the tests constituted comparisons of directly measured (using instrumented aircraft) radiances with those obtained by the use of radiation algorithms. Datasets from BOMEX, GATE, and MONEX were used in these intercomparisons. Overall, the results show that longwave radiative fluxes computed by the emissivity method with a fine vertical resolution (50 hPa) were in reasonable agreement with the observations.

Figure 2.15 shows the vertical profiles of the area-averaged radiative cooling for the three stages (regions). The curve for region 1 represents the average for the region of origin. As expected, cooling occurred in a deep layer for all three regions, but the cooling rate varied from one region to another. For regions 1 and 2, a cooling rate of 1 to 2 °C day^{-1} was found below 300 hPa, in good agreement with other calculations

(Boville, 1985). The maximum radiative cooling of 3.0 °C day^{-1} for region 2 and 2.1 °C day^{-1} for region 3 occurred near the surface. Clear sky and the snow-covered underlying surface contributed to this cooling. Climatologically, the height of the top of the inversion layer lies between 850 and 700 hPa for region 2 and 3 (Table 2.3); this information is important for an understanding of the cooling rates. The vertical distribution of the radiative cooling is shown in Figure 2.15a. These profiles are quite similar for regions 1 and 2, the largest cooling of the order of 3 °C day^{-1} occurs near the ground, and cooling decreases with height. Over region 3 an additional maximum in cooling is found above the inversion layer which is related to radiative cooling from the tops of clouds that extend to the base of the inversion in this region. Subgrid-scale sensible heat transfer [Q_s in(2.6)] appears to be very important for the maintenance of the heat balance over the source region of the Siberian high. This is evaluated as a residual. The residual, which undoubtedly, also includes some of the computational budget errors. From Figure 2.15b, it may be seen that for region 1, the sensible heating due to subgrid-scale motions occurs above 400 hPa, whereas cooling occurs below this level, with a maximum at 850 hPa of approximately 1 °C day^{-1}. Large horizontal temperature gradients are present in the upper troposphere over this region with the result that the vertical wind shear is quite large. As a consequence, we believe that shear flow turbulence is possible in the upper troposphere (Keyser and Shapiro, 1986).

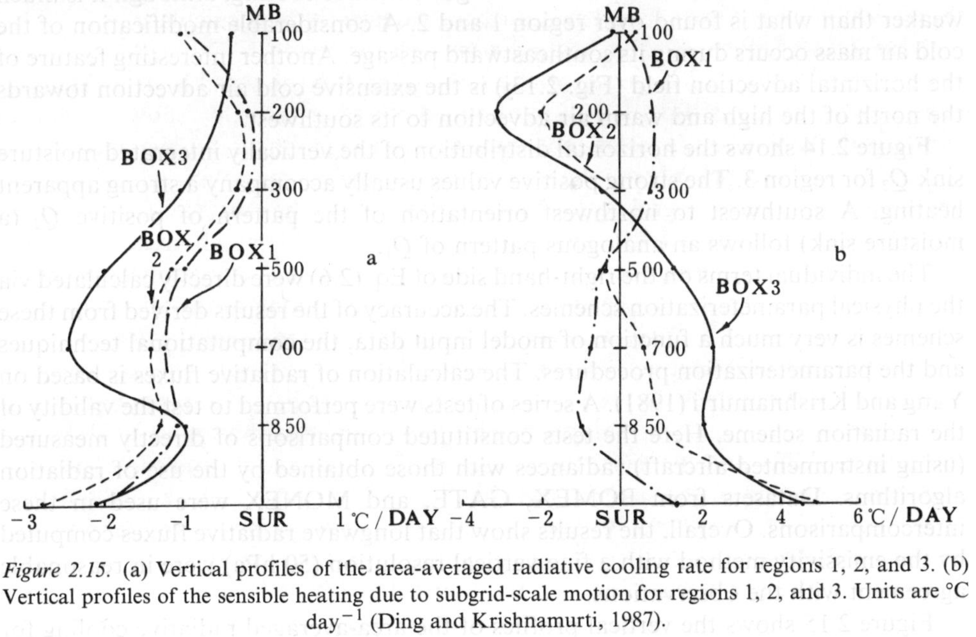

Figure 2.15. (a) Vertical profiles of the area-averaged radiative cooling rate for regions 1, 2, and 3. (b) Vertical profiles of the sensible heating due to subgrid-scale motion for regions 1, 2, and 3. Units are °C day^{-1} (Ding and Krishnamurti, 1987).

In order for systematic eddy heat fluxes (and convergence of eddy heat fluxes) to be present, a nonvanishing temperature-vertical velocity correlation should be present. Again, mechanical turbulence drives thermal turbulence, thus contributing

Table 2.3. Climatology of the inversion for regions 1, 2, and 3 (average for January and February of 1981–1983). T_d is temperature, T is dewpoint. Units are °C

Level (hPa)	Region 1				Region 2				Region 3			
	Ulan-hot 50°N, 92°E		Murren 40°N, 100°E		Eren 44°N, 112°E		Taiyun 38°N, 113°E		Hankou 31°N, 116°E		Shanghai 31°N, 122°E	
	T	$T-T_{dd}$	T	T_d	T	$T-T_d$	T	$T-T_d$	T	$T-T_d$	T	$T-T_d$
Sur.	−32.0	3.0	−25.3	2.8	−23.0	-	−6.6	9.9	19.0	1.0	3.2	2.6
850	−26.0	3.0	−19.8	3.3	−12.4	3.2	−5.7	11.7	14.0	4.2	0.0	7.6
700	−15.8	3.3	−18.8	5.5	−19.2	6.0	−11.9	12.1	6.0	10.7	−4.3	10.0
500	−309	5.5	−32.8	6.1	−33.4	6.1	−25.5	11.1	−5.9	20.6	−17.3	9.4
300	−55.1	4.0	−55.9	5.2	−54.4	5.7	−50.1	8.0	−31.4	5.9	−41.1	8.3
200	−61.5	4.6	−59.0	5.0	−54.3	6.1	−54.9	10.4	−52.1	2.8	−51.8	8.7
150	−58.5	5.7	−57.6	5.7	−53.5	6.9	−55.2	10.6	−65.2	-	−59.2	8.2
100	−58.5	6.9	−56.7	6.2	−53.5	7.1	−58.6	10.3	−77.5	-	−66.3	7.0

to the maintenance of the stratification of this region. Residuals in the order of 1 °C day^{-1} may, at first sight, seem within the range of observational and computational errors. However, we can rule out the former on the grounds that almost 20 cases were carefully composited to arrive at the results presented here. Furthermore, we do see a rather smooth transition (from region to region) in the results of the calculations of these sensible heat fluxes. This lends further credence to these calculations.

Near to the surface, sensible heating was weak for region 1. This appears to be a consequence of the convergence of the downward flux of sensible heat from the subgrid-scale motions. Boville (1985) obtained a similar result by showing that, in January, the subgrid-scale terms contribute to a heating in the mean over most of the globe, the only exception being the low levels over the winter polar regions north of about 60° N. His subgrid-scale heating rate near the surface for a region corresponding to region 1 was about 4 °C day^{-1}, which is much greater than the heating rate obtained here. His analysis includes oceanic areas in the zonal mean with the sensible heat flux at the ocean surface partly contributing to this discrepancy.

As the Siberian high moves southward, the layer of subgrid-scale heating becomes deeper. For region 2, heating occurs in a layer below 600 hPa, with a maximum of ≈ 5 °C day^{-1} at the surface. In the upper troposphere, a cooling of approximately 2 °C day^{-1} occurs over region 2. This change in the vertical distribution of Q_s decreases the static stability and increases the baroclinicity of the middle and high levels. In region 3, the height of the subgrid-scale heating layer ascends further to 400 hPa.

A significant heating (≈ 2 °C day^{-1}) occurs concurrently in the lower and middle troposphere. This result is similar to that obtained by Boville (1985).

Based on the above examination, the thermal structure of the Siberian high is mainly characterized by a deep layer of diabatic cooling, with the major contribution made from the radiational cooling. In response to the diabatic cooling, a downward motion in the deep layer of the troposphere can be induced, thus leading to mass convergence in the middle and upper troposphere and divergence in the lower troposphere. This condition is favorable for the development and the intensification of the Siberian high. The basic picture of anticyclogenesis resulting from radiative cooling is one of the velocity field adjusting to the mass field disturbances that result from the cooling. Therefore, with combined consideration of the result derived from the analysis of the budget of vorticity, one may suggest that the genesis and development of the Siberian high results from the combined effects of mass convergence at the middle and upper level and the advection of negative vorticity, as well as local thermal cooling. Note that heating in the upper troposphere due to subgrid-scale sensible heat transfer may be an important factor in the maintenance of mass convergence in the upper troposphere through enhancement of the warm upper-tropospheric anticyclone lying over the Siberian high. As compared with the previous studies of anticyclogenesis, the physical mechanism indicated here is to a certain degree, different from anticyclogenesis in Alaska (Bodurtha, 1952) that is caused by strong warm air advection prior to and during the period of anticyclogenesis in the lower and middle troposphere, west of the location of anticyclogensis, and the subsequent strong cold-air advection at 200 hPa over the region of anticyclogenesis in Alaska.

Recently, Curry (1987) investigated the contribution of radiative cooling to the formation of cold-core anticyclones with a nonlinear axisymmetric numerical model. The model results show that after 5 days of integration, the central surface pressure increase is 7 hPa for Wexler's cooling mechanism (internally determined radiative cooling rates with all condensed water is assumed to fall out immediately, unaffecting the radiative transfer), and 10 hPa for Curry's cooling mechanism (fully interactive calculation, thereby the radiative effects of the condensate that forms in the cooling air are included). A positive feedback loop is shown to exist between the formation of the condensate in the cooling air and anticyclogenesis. Radiative cooling from the condensate enhances anticyclogenesis and the large-scale meridional circulation associated with anticyclones replenishes the moisture in the condensate layer, thus enhancing the radiative cooling. This result shows that the presence of low-level clouds does not reduce the intensity of cold-core anticyclones. On the contrary, cloud-top cooling and cooling from ice crystals suspended in the air increases the lower tropospheric cooling, which results in stronger thermal forcing and therefore a more vigorous anticyclone. Results from the experiments employing externally specified radiative cooling rates show that the central surface pressure increase is largest for (i) increased amounts of cooling; (ii) increased horizontal extent of the cooling; (iii) decreased vertical extent of the cooling; (iv) decreased friction; and (v) increased latitude.

Lu *et al.* (1983) proposed a different mechanism responsible for the formation of a cold-core anticyclone. They demonstrated that the decrease in the temperature in the cold dome of the anticyclone is not only caused by the diabatic cooling at the surface, but also by ascending adiabatic expansion.

2.3. The Air Mass Transformation of the Siberian High and its Propagation

It is well known that when the Siberian high moves off to the East China Sea, it undergoes a significant transformation due to the strong, upward sensible and latent heat fluxes from the warm oceanic surface. A convectively mixed layer is observed (Nitta and So, 1980) below the inversion layer over the ocean. Subgrid-scale vertical transfer plays a major role in the modification of the cold high over the ocean. In this case, active cumulus convections develop in the moist and unstable stratification due to heating and evaporation from the underlying sea surface. They are considered to redistribute the heat and moisture in the free atmosphere.

The air-mass transformation over the ocean has been studied from various viewpoints. Several extensive analyses on the air-mass transformation over the tropical ocean were made (Yanai *et al.*, 1973) mainly for the purpose of evaluating the cumulus mass, heat, and moisture flux from the thermodynamic quantities of a large scale motions. From the viewpoint of large-scale synoptic situation, the air-mass transformation in the tropics takes place over the sea area of horizontally uniform temperature under the atmospheric situation of weak horizontal thermal advection and small local time change of temperature. The transformation there is characterized by the large amount of latent heat supply and small amount of sensible heat supply from the sea surface. The features of air-mass transformation in the high latitude are very different from those in the tropics. This transformation occurs mainly in relation to the cold air outbreak and cyclogenesis. The typical air-mass transformation in the polar region takes place under the situation of strong thermal advection over the sea where the air-sea temperature difference is extremely large. The amount of the sensible heat supply is large in the polar region in contrast to the tropical regions. The features of air-mass transformation over the middle latitude, such as the Kuroshio region, are more complex, because the polar and subtropical air-masses alternatively cover the sea area. In order to makes an observational and theoretical determination of cumulus mass, heat, and moisture flux over the Kuroshio area, the Air-Mass Transformation Experiments of 1974 and 1975 (AMTEX 74 and AMTEX 75) were carried out. In this sea region, the very warm Kuroshio current is predominant. The frequent air-mass transformation of cold air and the development of the extratropical cyclone in this area are considered to be closely related to the energy supply from the sea surface, especially over the Kuroshio area. Besides, the East China Sea is surrounded by relatively dense aerological and surface observation stations. This sea is, therefore, one of the most suitable places to study the air-mass transformation.

Table 2.4 presents the mean sensible and latent heat fluxes from the sea surface during AMTEX 74 and 75. It can be seen that the latent heat flux is significantly greater than the sensible heat flux. The total heat fluxes (385W m^{-2}) in AMTEX 75 is much greater than those in AMTEX 74 (260 W m^{-2}). This difference mainly results from the large-scale synoptic situation (Murti, 1976; Nitta, 1976; Ninomiya, 1975; Kung and Siegel, 1979). In the case of the cold air outbreak, the total heat flux from the Kuroshio warm current to the atmosphere amounted to about 800 W m^{-2}. Table 2.5 shows the vertical distributions of the total heat flux. The upward flux of the heat generally vanished at 500 hPa, with an obvious decrease in magnitude below 700 hPa. Therefore, the air-mass transformation of the Siberian high is mainly confined to the lower troposphere, with heating coming from the underlying sea surface. In spite of strong cold advection, the temperature fall ($\partial T/\partial t$) is small, because the cold advection is primarily balanced by the diabatic heating due to the energy supply from the sea. The results of the budget calculation also indicated that the diabatic heating is mainly confined in the lower layer below the inversion (Ninomiya, 1975; Ninomiya and Akiyama, 1976). Kung and Siegel (1979) also indicated that the lower 150–200 hPa layer from the sea surface serves as a sensible heat source. Cold air outbreaks foster a large heat sink in all layers except the lower troposphere, giving generally negative Q_1 values above the lower troposphere for the mean AMTEX 74 and 75 budgets. The apparent latent heat source exists throughout the mid-and lower troposphere for AMTEX 74 and 75. Nitta and So (1984) further demonstrated that large amounts of heat and moisture sources exist in the mixed layer. The moisture source has a maximum near the inversion base and the heat sinks occur in the inversion layer. These moisture sources and heat sinks in the inversion layer may be mainly caused by a moistening and cooling effect due to re-evaporation of cloud droplets detrained from the cloud top. Large radiational cooling occurs near the inversion base probably due to large upward terrestrial radiation fluxes above the cloud top. Small values of Q_1–Q_R and Q_2 above 700 hPa indicate that few clouds penetrate through this level. The sensible heat flux decreases linearly with height in the subcloud layer and becomes negative near the cloud base.

Table 2.4. Mean sensible (S) and latent heat flux (LE) from the sea surface during AMTEX 74 and 75. Units: W m^{-2}

	AMTEX 74			AMTEX 75		
Researchers	S	LE	S+LE	S	LE	S+LE
Kung & Siegel (1979)	41	168	209	58	256	314
Konda (1976)	75	210	285	105	313	418
Murty (1976)				185	237	422
Nitta (1976)	42	253	297			
Ninomiya (1975)	123	126	249			
Average	70	190	260	116	269	385

Table 2.5. A comparison of subgrid-scale total heat fluxes between the land and the ocean. Units: W m^{-2}

Researchers	Surface	850	700	500	Remarks
Ninomiya(1972)	442.8	232.8	81.5	0	Polar-air outbreak, 19, Jan 1968
Ninomiya (1975)	256.1	69.8	34.9	0	AMTEX 74
Ding & Krishnamurti (1987)	115	82	15	−35	Composite Siberian high over the subtropical land

Over the land in East Asia, the cold air actually also undergoes a considerable transformation process (Ding and Krishnamurti, 1987). The diabatic heating from the underlying surface commences shortly after it moves away from its source region and arrives in the Mongolian area and northern China. As the cold air flows further southward, this low-level heating becomes more significant and the depth of the heating layer also increases, which may extend upwards to 700 hPa over eastern China. When the high arrives at lower latitudes, the heating due to condensation in the lower troposphere further enhances the transformation process of the cold high. Figure 2.16 shows variations of area-averaged temperature profiles in the cold high as it arrives at different areas (Ding, 1990). In the source region, there is a temperature inversion layer below 800 hPa. As the cold high moves southward, the inversion layer gradually weakens and disappears. In northern China region, it cannot be detected, but, in eastern China, a weak inversion or isothermal layer is re-established in the middle troposphere that may be mainly caused by the strong dynamic descending motion in the high (Ding and Krishnamurti, 1987). Accompanying the evolution of the inversion layer, the temperature in the whole troposphere, especially in the lower troposphere, markedly increases, thus leading to a gradual transformation from a continental air mass into a warm oceanic air mass. The increase in the mean temperature near the surface and in the upper troposphere amounts to about 30 and 20 °C, respectively. The transformation process of the profiles of specific humidity is

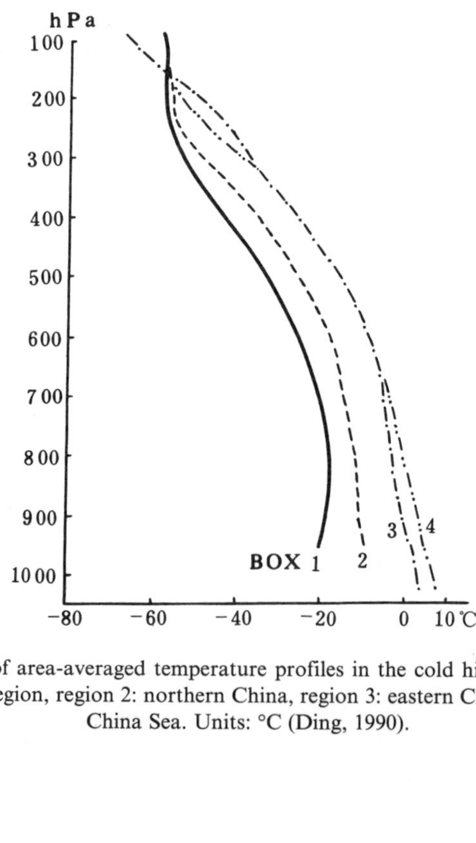

Figure 2.16. Variations of area-averaged temperature profiles in the cold high as it arrives at different areas. Region 1: source region, region 2: northern China, region 3: eastern China, and region 4: the East China Sea. Units: °C (Ding, 1990).

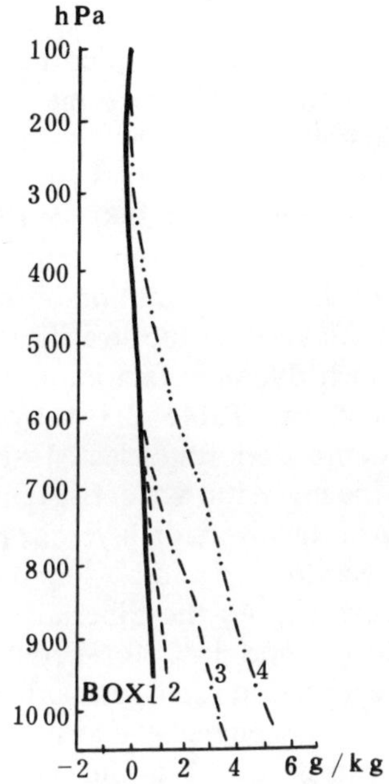

Figure 2.17. Same as Fig. 2.16, but for specific humidity. Units: g Kg^{-1} (Ding, 1990).

broadly similar to that of the temperature profiles (Fig. 2.17). The moisture in the cold high rapidly increases as it moves southward, especially in the lower and middle troposphere. It is worth noting the remarkable increase of moisture from region 2 to region 3 below 700 hPa which is almost equal to the magnitude of the increase of moisture over the ocean (from region 3 to region 4). This fact indicates the great importance of variations in the moisture field for the transformation of the cold high, even when the air mass remains situated over a land surface.

As previously indicated, the transformation of the cold high is mainly caused by the sensible heating and latent heating from the underlying surface. Based on the study of the heat budget of 19 strong Siberian highs, Ding and Krishnamurti (1987) pointed out that the heating rate from the land surface (114 W m^{-2}) is about half or one-third of that over the ocean (256–442 W m^{-2}) (see Table 2.5). But, their estimate was based on the budget calculation method, i.e., residual method. Some computational errors could be included in the results. Now, a direct estimate of the heat flux from the land surface for the different regions the Siberian high traverses has been attempted.

In order to take into account the effect of atmospheric stratification, the surface sensible heat (*fh*) and latent heat fluxes (*fq*) were estimated by use of the similarity theory by Zhao and Ding (1991). This computational procedure is, in many ways, similar to that suggested by Louis (1979). Table 2.6 summarizes the computational results for regions 2 and 3 where a major air mass transformation occurs. From the daily mean values, the total heat fluxes in regions 2 and 3 are 84 W m^{-2} and 75 W

m^{-2}, respectively, with the sensible heating making a major contribution (see Bowen ratio). These values of heat flux are slightly smaller than those obtained by Ding and Krishnamurti (1987), but they are of nearly the same order of magnitude. Therefore, their heating effect on the cold high from the land surface cannot be neglected. From Table 2.6 it is also seen that the heat flux assumes an obvious diurnal variation, with the maximum upward heat flux observed at 2:00 pm (Beijing Local Time) and very weak heat flux (generally downward) at the other observation times.

Another important aspect of the transformation of the Siberian high is the change in the static stability. The modification of the stratification not only depends on the diabatic effects but also on such dynamic factors as the vertical variations of the horizontal and vertical advection. Table 2.3 shows the vertical temperature distribution and dewpoint depression for selected stations in each region. The variations in the structure of the inversion are consistent with the changes in vertical velocity. The height of the top of the inversion layer for region 1 is near 700 hPa, with a temperature difference between the top and bottom of the inversion of 7 and 16 °C for Ulan and Murren, respectively. As the Siberian high moves to region 2, the inversion layer has weakened (11 and 1 °C for Erren and Taiyuan, respectively). Furthermore, the height of the inversion top is lower. The surface temperature as well as the temperature in the lower troposphere have by now significantly increased. Over region 3, no inversion layer can be detected and a normal temperature lapse rate prevails mainly due to the considerable surface heating and vertical mixing. Recently, Ding *et al.* (1991) made a diagnostic study of the static stability of the Siberian high, indicating that at the formative stage of the high, the stability tends to increase, due mainly to divergence term, vertical and horizontal flux terms, while the effect of radiative cooling is of minor importance. Therefore, the dynamic factor, especially dynamic descending motion, is a major one in determining the evolution of the stratification of the Siberian high.

The Siberian high and its associated winter monsoons usually show a marked low-frequency meridional movement (Lau and Lau, 1984; Pan and Zhou, 1985). These low-frequency perturbations with time scale between 10 and 30 days to migrate toward the subtropical Pacific, mostly accomplished by the successive development of

Table 2.6. The mean sensible heat and latent heat fluxes at the land surface in winter, averaged for five cases of cold waves. Beijing local time is used here. Units: W m^{-2} (Zhao and Ding, 1991)

Heat flux	Region	02	08	14	20	Daily mean
Sensible heat flux	region 2	−18	−6	266	−12	57
	region 3	−7	9	263	−8	49
Latent heat flux	region 2	−3	−1	78	−3	18
	region 3	−6	5	148	−7	35
Total heat flux	region 2	−21	−7	344	−15	75
	region 3	−13	14	351	−15	84
Bowen ratio	region 2	6.0	6.0	3.4	1.8	3.2
	region 3	1.2	1.8	1.4	0.5	1.4

new centers downstream of the primary disturbance located over the western Pacific. Here, further evidence is now presented to document the low-frequency mode of meridional propagation of the winter monsoon. Figures 2.18 (a) and (b) are $x-t$ diagrams of Q_1 showing the temporal evolution of heat sources and sinks along the trajectory of cold air outbreaks at 700 and 300 hPa, respectively. The selection of the time period from 16 January to 15 Februaty, 1981 covers a period of vigorous cold air outbreaks. First, the linear seasonal trend was removed from the once-daily Q_1 field. A time fitler was next applied to the above fields of Q_1 at 700 and 300 hPa, respectively, in order to obtain a smoothed distribution. From Figure 2.18, one notes that roughly, three cold outbreaks propagate from their source region to the coastal area of southern China during a period of one month. The strongest episode among these is 21–29 January, 1981. The propagation of heat sources and sinks from the northwest to the southeast along the trajectory at 300 hPa appears to precede a corresponding propagation at 700 hPa by almost 1–3 days. This southeastward propagation of Q_1 toward lower latitudes with the period of 10–20 days is closely related to the cold air outbreaks over eastern Asia. As a component of variations of the large-scale circulation features, the cold air outbreaks should be closely related to the low-frequency movement of troughs and ridges at middle and high latitudes. Therefore, the low-frequency mode of the southward propagation of the winter monsoon actually reflects the characteristics of low-frequency oscillation systems at

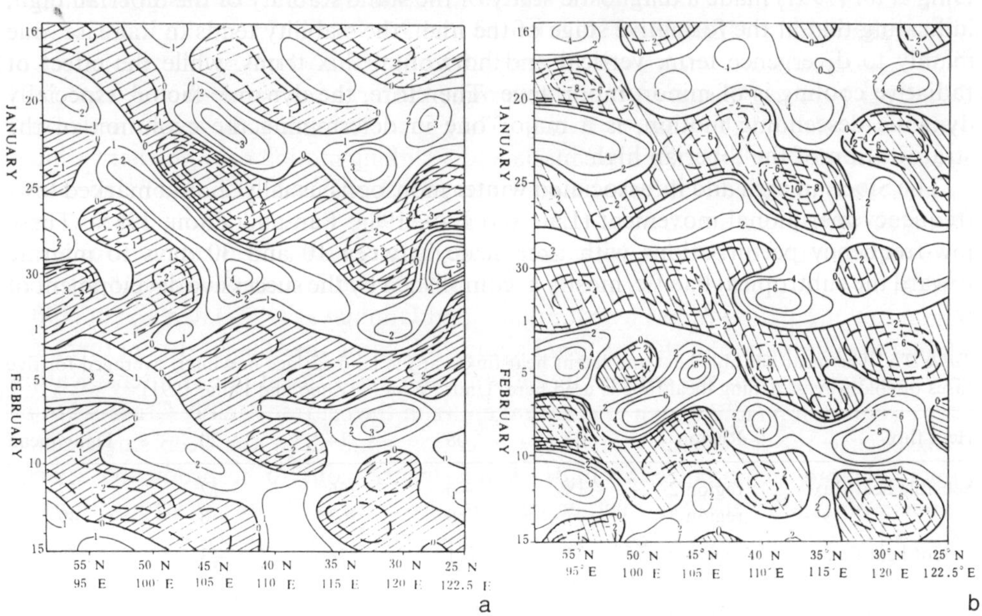

Figure 2.18. *x-t* charts of low-pass filtered apparent heat source Q_1 at (a) 700 hPa and (b) 300 hPa during the time period from 15 January to 15 February, 1981. The negative area is shaded. Solid lines denote positive (heating) Q_1 and dashed lines show negative (cooling) Q_1. Units: °C day^{-1}. The abscissa is taken along the trajectory of the cold air outbreaks (Ding and Krishnamurti, 1987).

middle and high latitudes and their effect on the circulation at low-latitudes (Krishnamurti and Gadgil, 1985).

2.4. The Winter Upper-Level Jet Stream in East Asia and the Western Pacific

The wintertime upper-level jet stream of East Asia and the western Pacific is the strongest jet stream in the world. Over 30 years ago, some investigators (Mohri, 1959; Academia Sinica, 1957, 1958) comprehensively described the large-scale characteristics of this jet stream, and documented its many important features, including the splitting of the jet stream by the large-scale topography of the Tibetan Plateau, the sudden seasonal northward jump, and the importance of the jet streams for the development of severe weather. But, the study of its smaller scale aspects, especially the vertical circulation of jet streaks, received less attention. In addition, the study of features of jet streams downstream over the western Pacific was limited by a severe lack of observations in this region, where the jet stream is at its strongest. However, during the last 15 year important nonstandard data sources have become available. These are winds observed by aircraft inertial navigation systems, winds derived from cloud motions observed by satellites, and indirect atmospheric temperature soundings by satellites. Therefore, the availability of data of these different sources provides the possibility of improving jet-stream analysis and a better understanding of the behavior of jet streams in East Asia and the western Pacific.

Murakami and Unninayar (1977) used the National Meteorological Center's (NMC) tropical wind analysis at 700 and 200 hPa for a 90-day winter period from December 1970 to February 1971 to determine the large-scale irrotational wind field over East Asia and the western Pacific. They found a large-scale mean meridional circulation with updrafts over Indonesia and a sinking motion over northern China. From a study of the eddy fluxes of momentum, Blackmon *et al.* (1977) concluded that the large-scale mean jet streams that appear over the western oceans develop as a result of thermally direct time-mean meridional circulations. Their conclusions are in agreement with those of Namias and Clapp (1949) and are also consistent with the time-mean circulation found by Murakami and Unninayar (1977). Cressman (1981) studied the circulation of the western Pacific, indicating that the jet stream there is a very steady feature of the winter circulation, with almost continuous intensification over eastern China and Japan and a weakening in the central Pacific. Cross-sections transverse to the jet stream in its intensification zone show the deep single-celled direct solenoidal circulation around a single frontal zone to be producing a large increase of kinetic energy. Further east, in the central Pacific, a strong indirect solenoidal circulation produces a strong decrease of kinetic energy.

Figure 2.19a presents zonal components of the average 200 hPa winds for East Asia and the western Pacific for the winter half-year (October through March). Inspection of these figures yield immediate information on the great intensity of the jet stream to the south of Japan, with a maximum zonal component of 62.4 m s^{-1}. This jet streak accelerates from the central Tibetan Plateau eastward down to the central and western Pacific with the entrance region to the east of 140°E and the exit

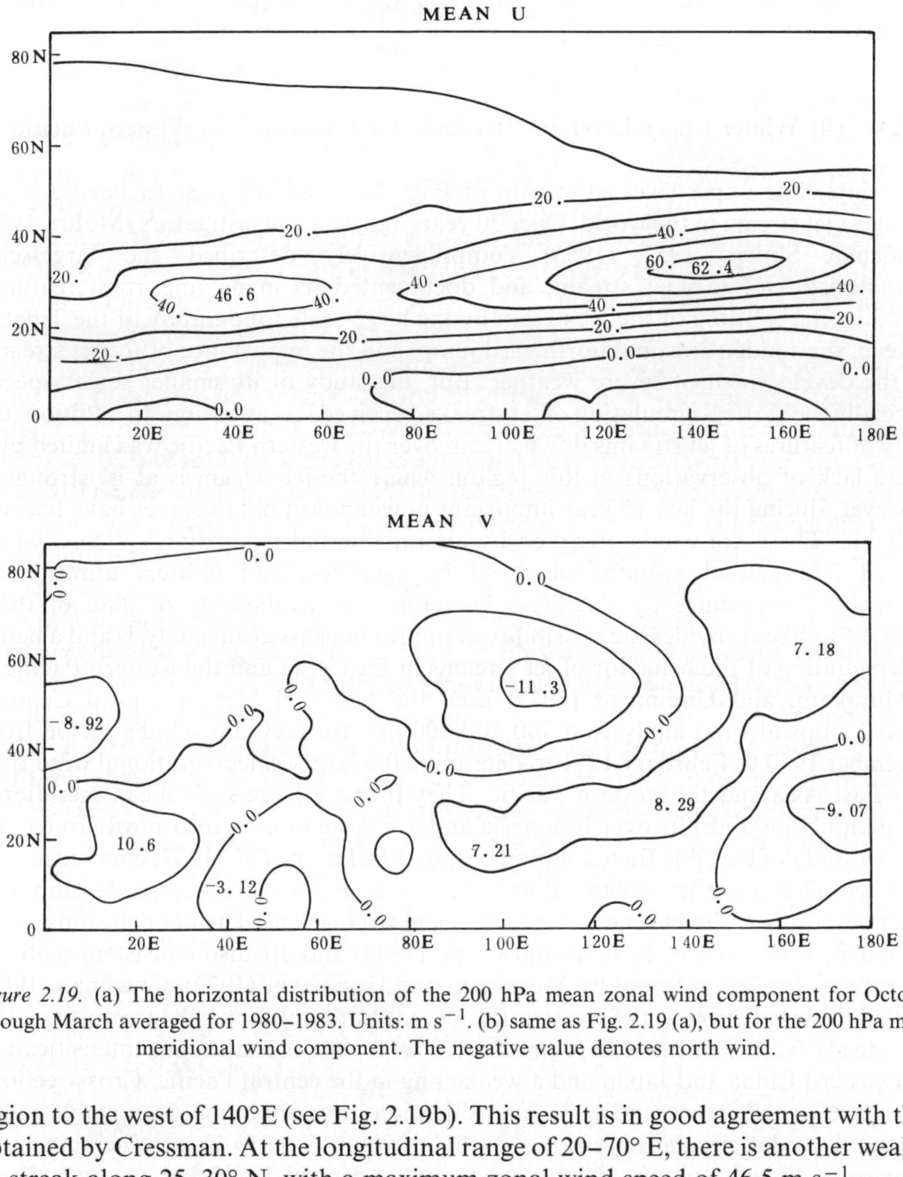

Figure 2.19. (a) The horizontal distribution of the 200 hPa mean zonal wind component for October through March averaged for 1980–1983. Units: m s^{-1}. (b) same as Fig. 2.19 (a), but for the 200 hPa mean meridional wind component. The negative value denotes north wind.

region to the west of 140°E (see Fig. 2.19b). This result is in good agreement with that obtained by Cressman. At the longitudinal range of 20–70° E, there is another weaker jet streak along 25–30° N, with a maximum zonal wind speed of 46.5 m s^{-1}.

This jet stream over western Asia is also a semi-permanent circulation feature in the winter half year that has a significant influence upon the weather over this region.

Murakami (1981) also noted the strong eastward acceleration of the jet stream at 200 hPa during FGGE (December 1978-December 1979) along 25° N. East of 70°E, this eastward acceleration is approximately 5 m s^{-1} per 1000 km. An analysis of the momentum over South Asia is primarily the result of local Hadley circulation. The time-averaged *u*-equation may be written as follows:

$$\frac{\partial \bar{u}}{\partial t}+\bar{u}\frac{\partial \bar{u}}{\partial x}+\bar{v}\frac{\partial \bar{u}\cos\phi}{\cos\phi\partial y}+\bar{\omega}\frac{\partial \bar{u}}{\partial p}+\frac{\partial \overline{u'u'}}{\partial x}+\frac{\partial \overline{u'v'}\cos^2\phi}{\cos^2\phi\partial y}+\frac{\partial \overline{u'\omega'}}{\partial p}=f(\bar{v}-\bar{v}_g) \tag{2.10}$$

~ 0 22 ~ 0 ~ 0 2 -3 21

Approximate values at 200 hPa along 27.5° N (jet axis) are shown in units of 10^{-2} m s^{-2}. First, note that the largest contribution (22 units) is due to the easterward acceleration of u. Conversely, momentum flux divergence due to transient eddies appears to contribute little to the maintenance of the upper-level jet stream. In fact, horizontal momentum flux divergence (2 units) is nearly compensated for by vertical momentum convergence (3 units). The $f\,(v-v_g)$ term, evaluated as a residual of the u-equation, amounts to about 21 units, indicating the existence of supergeostrophic v winds. In short, the jet stream and its substantial eastward acceleration are primarily maintained by southerly non-geostrophic v winds which correspond to an upper branch of the thermally direct, local Hadley circulation. Figure 2.20 is the correlation between the 200 hPa velocity potential average between the equator and 20° N and the zonal wind acceleration averaged over 25 to 40° N for different longitudes (10 degrees apart), based on twice-daily data during the 1983 to 1984 winter (Chang and Lim, 1985). Here, the 200 hPa velocity potential is used as a measure of the upper-level divergent outflow associated with deep convection. From left to right, the first extensive region of significant correlation is from 50 to 120°E, covering the longitude of the Indian Ocean and the maritime continent and overlapping with the entrance region for the East Asian jet. Chang and Lim suggested that the significant correlation is apparently consistent with the possibility that the jet acceleration is due to the Coriolis torque of the northward divergent flow induced by the tropical convection. A second area of large correlation is found just east of the dateline coinciding with the exit region of the East Asian jet. This correlation is likely due to presence of a thermally indirect circulation associated with the spatial deceleration of the jet. The remaining areas of large significant correlation are found over 100° W and 90 to 70° W, corresponding roughly to the entrance and exit regions of North American jet stream. For all the above regions, the largest correlation is found near zero lag.

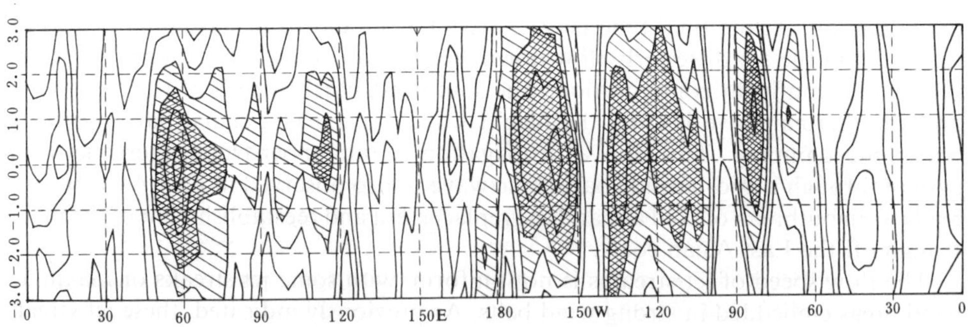

Figure 2.20. Time-lag correlation coefficient between velocity potential (averaged between 0 and 20° N) and mid-latitude jet accelaration (averaged between 25 and 40° N) at the same longitude for December to January 1983/1984. The lag interval is 12 h. Positive lag indicates mid-latitude leading. Thick lines indicates correlation coefficients of 0.3 and 0.4, significant at 96 and 99%, respectively (Chang and Lim, 1985).

By utilizing the zonal momentum equation, Chang and Lim (1985) made a crude estimate of the momentum budget as a function of the phases of the cold surges, as Murakami did. Near 100°E (the jet entrance region), the net Coriolis force term ($fv - \partial\phi/\partial\lambda$) is larger than the momentum flux convergence ($\partial uu/\partial\lambda + \partial uv \cos\phi/a \cos\phi\partial\phi + uv/a \tan\phi$). Near 180° (the exit region), the net Coriolis force term is small compared to the momemtum flux convergence term. The sign of the latter here indicates a removal of westerly momentum from the jet. The above analysis suggests that mid-latitude tropical interaction through the effect of the ageostrophic wind is most important at the jet entrance region over East Asia. This result is in good agreement with that obtained by Murakami and Unninayar (1977) and Murakami (1981).

The upper-level jet-frontal systems has a close relationship with the temporal evolution of baroclinic wave structure. Shapiro (1983) synthesized evidence into a hypothetical schematic depicting the migration of an upper-tropospheric jet-front system through a synoptic-scale baroclinic wave situated over North America (Fig. 2.21). The progression of events begins over western Canada with the confluence between a polar trough and a mid-latitude ridge (Fig. 2.21a), a flow configuration hypothesized by Namias and Clapp (1949) to be conducive to the formation and intensification of upper-level fronts and jets.

Roughly a day later (Fig. 2.21b), the jet and front have progressed to the inflection in the northwesterly flow upstream of a diffluent trough exhibiting a northwest-southeast tilt. The thermal wave lags the height wave by one-quarter wavelength, which places the front in a region of cold advection. The tilt in the height field is a sign of barotropic amplification provided that the background (zonally averaged) zonal wind becomes increasingly westerly with latitude in the vicinity of the developing shortwave disturbance, while the separation between the thermal and height waves is optimal for baroclinic amplification. The flow configuration in Figure 2.21b reflects the sense of the asymmetrical trough structure in the early stages of its development.

At the following time (Fig. 2.21c), the jet front system has reached the base of the longwave trough and assumed a curved orientation. The absence of asymmetries in terms of disappearance of the latitudinal tilt and phase separation between the thermal and height fields, indicates the cessation of barotropic and baroclinic amplification. At the final time (Fig. 2.21d), the jet and front have migrated to the inflection in the southwesterly flow downstream of the longwave trough, which has taken on a confluent configuration. The southwest-northeast tilt of the trough axis and the thermal wave leading the height wave, repectively, signal barotropic and baroclinic damping. This above propagation of an upper-tropospheric jet-front system through a mid-latitude baroclinic wave is also observed in East Asia. It often leads to an outbreak of cold air, and the replacement and reestablishment of the major trough off the East Asian coast.

The flow speed of jet streams is not uniform, with some jet streaks or maximum wind areas embedded in strong wind belts. As previously indicated, these jet streaks may induce tranverse ageostrosphic circulations across the jet axis. This is essential for the dynamics of the jet stream and has an important implication for synoptic development. A detailed discussion of this matter will be presented below. Figure 2.22 shows isotachs and winds direction vectors for the average 200 hPa winds for East

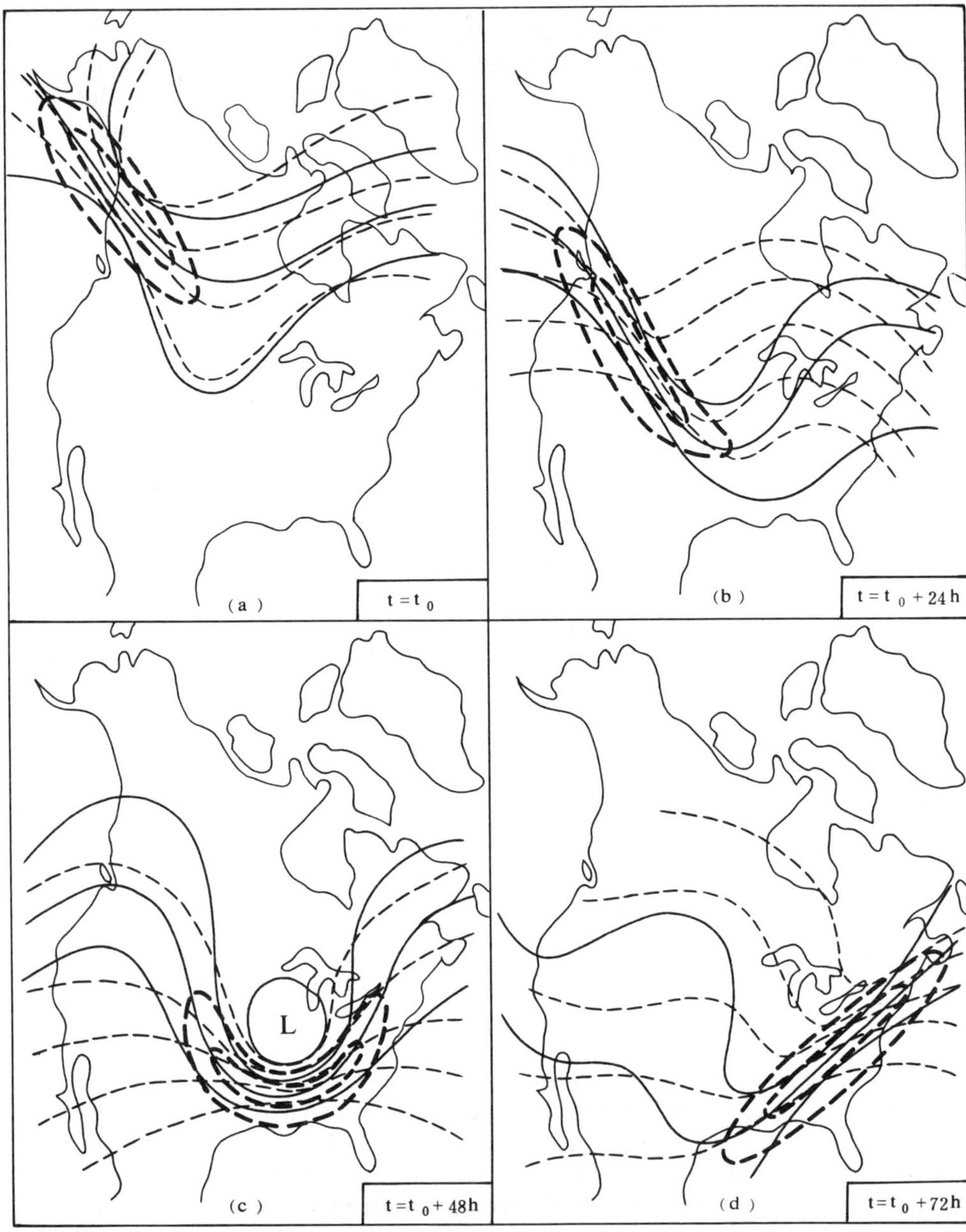

Figure 2.21. Idealized schematic depiction on a constant pressure surface of the propagation of an upper-tropospheric jet-front system through a midlatitude baroclinic wave over a 72 h period: (a) formation of a jet-front in the confluence between mid-and high-latitude currents; (b) a jet-front situated in the northwesterly flow inflection of an amplifying wave; (c) a jet front at the base of the trough of a fully developed wave; (d) a jet-front situated in a southwesterly flow inflection of a damping wave. Geopotential height contours, thick solid lines; isotachs, thick dashed lines; isentropes or isotherms, thin dashed lines (Shapiro, 1983).

Asia and the western Pacific for January (Cressman, 1981). The data for this map come from the average winds for the period 1968–1977, analyzed twice daily by NMC. The stationarity of the flow is assured by a steadiness factor exceeding 95%. Since the flow is so nearly zonal, we can make an interpretation from the equation of motion for du/dt, neglecting the viscous effects,

$$\frac{du}{dt}=f(\bar{v}-\bar{v}_g) \tag{2.11}$$

where the velocity components in the x and y directions are u and v, f is the Coriolics parameter, and v_g is the geostrophic wind in the y direction.

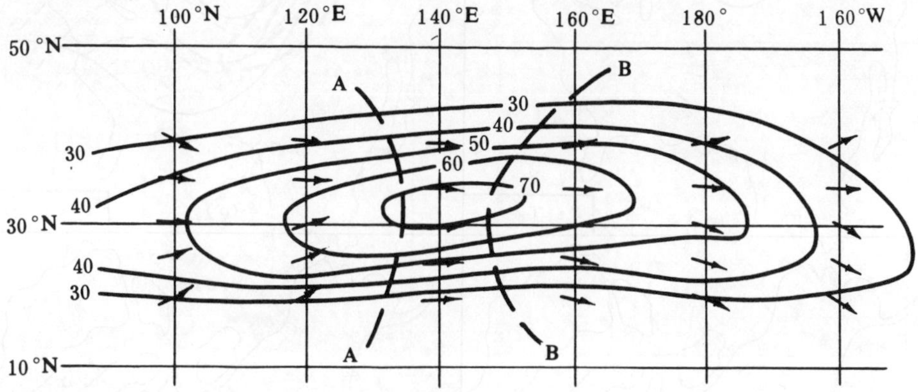

Figure 2.22. Speed (m s^{-1}) and direction of average January winds at 200 hPa for the period 1968–1977 (Cressman, 1981).

To left of the line AA, du/dt is clearly positive and $v > v_g$. Between the lines AA and BB, the wind vectors follow the isotachs, du/dt is near zero, and v is approximately equal to v_g. To the right of the line BB, du/dt is strongly negative and $v < v_g$. This distribution of v and v_g suggests two processes at work. The first suggestion is that there may be a direct solenoidal circulation in the jet stream entrance zone, left of AA, and an indirect circulation in the exit zone, right of BB. The second suggestion is that the air moving through a relatively stationary jet stream zone is following some form of inertial cycle (Newton, 1959). An example is presented below to describe the direct and indirect cell in the entrance and exit region of the jet stream over the western Pacific (Fig. 2.23). In this example, the jet stream speed zone is in its climatological position. It can be seen that, at the entrance region (Fig. 2.23a,b), a relatively simple direct solenoidal circulation extends through most of the depth of the atmosphere with cold air sinking and warm air rising. The kinetic energy generation takes place mainly above 500 hPa, indicating conversion from kinetic to potential energy. The maximum generation values coincide with the jet stream core. In the exit zone, a strong indirect circulation is shown by Figure 2.23 (c), (d), together with a very large conversion from kinetic to potential energy at levels above 600 hPa. Note that the

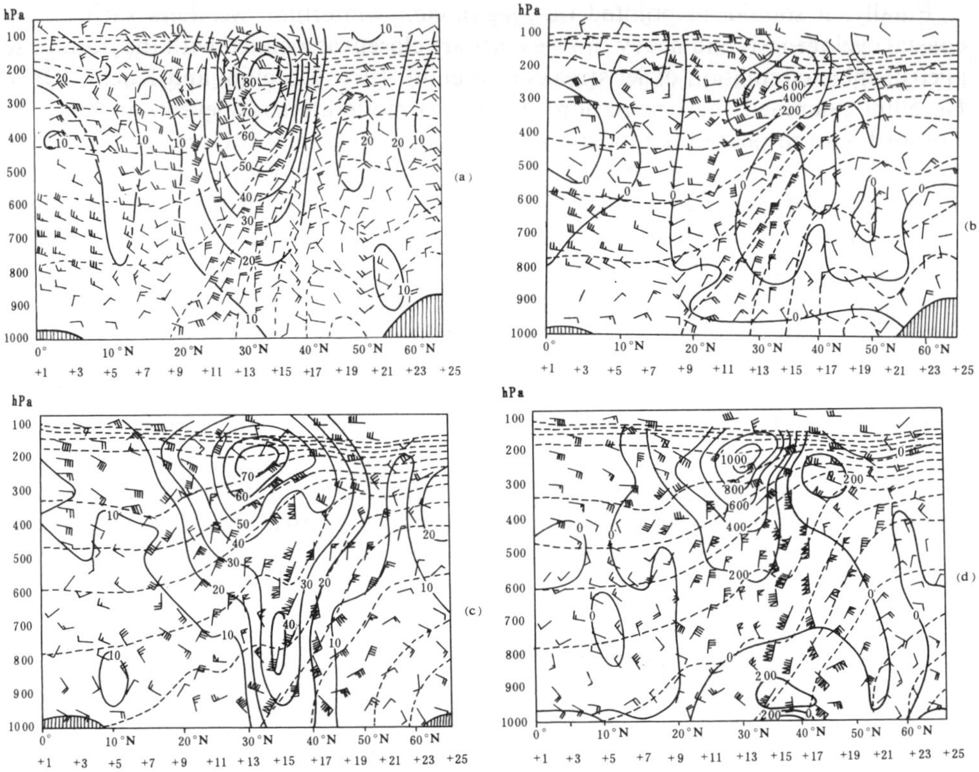

Figure 2.23. (a) Cross-section at the entrance region of the jet over the south of Japan, showing the vectors of the divergent component of the flow, isentropic surfaces, $\mathring{K}$ (dashed) and isotachs of total wind speed, m s^{-1}, (solid). All variables are 12 h forecast values, valid 00:00 GMT, 7 February 1980. (b) As Fig. 2.23 (a) except the solid lines are lines of $\mathring{K}$ (dk/dt) (J $\times$ kg^{-1}s^{-1} $\times$ 10^{-4}). (c) As Fig. 2.23 (a) except for the exit region. (d) As Fig. 2.23 (b) except for the exit region (Cressman, 1981).

irrotational flow in the jet core is toward the north in the accelerating portion and toward the south where the jet is in deceleration.

In summary, the wintertime jet stream of East Asia and the western Pacific gains energy as it crosses China, reaching maximum speeds in the vicinity of Japan. The zone of intensification (entrance region) is characterized by a great steadiness of flow and almost continuous conversion of potential to kinetic energy. The conversion is accomplished by a direct solenoidal circulation extending through practically the entire depth of the atmosphere. This circulation centers on a single deep frontal zone.

Farther east, over the central Pacific Ocean, the jet stream weakens (in the exit zone) as a deep indirect solenoidal circulation. In both entrance and exit zone, the maximum individual rate of increase or decrease of kinetic energy takes place in the vicinity of the jet stream core. The direct circulation in the entrance zone and the indirect circulation in the exit zone are in agreement with the mean circulation shown by Murakami and Unninayar (1977) (entrance zone) and by Blackmon *et al.* (1977) (entrance and exit zones).

Finally, it should be pointed out, significant wintertime weathers such as the coastal cyclogenesis and heavy snow events are closely associated with the transverse ageostrophic circulation of jet streaks (Uccellini, 1976; Uccellini and Kocin, 1987). Uccellini examined the orientation of light to moderate snow events in the U.S. Midwest with respect to the location of upper-level jet streaks. Uccellini and Kocin made an examination of eight heavy snow events along the East coast of the United States (Fig. 2.24), indicating that a particular configuration of two separate jet streaks can give rise to two interactive, transverse-ageostrophic circulations: (1) a direct circulation located within the confluent entrance region of a jet streak over the northeastern United States or southeastern Canada; and (2) an indirect circulation in the diffluent exit region of a jet streak associated with a trough in the southeastern United States. The interaction of the two circulations contributes to conditions suitable for heavy snowfalls along the East Coast. From Figure 2.24, it can be seen that indirect and direct transverse circulations that span the entire troposphere are located in the exit and entrance regions of southern and northern jet streaks, respectively. The rising branches of the transverse vertical circulations in exit and entrance regions of two jet streaks appear to merge, contributing to a widespread region of ascent that produces clouds and precipitation between the diffluent exit

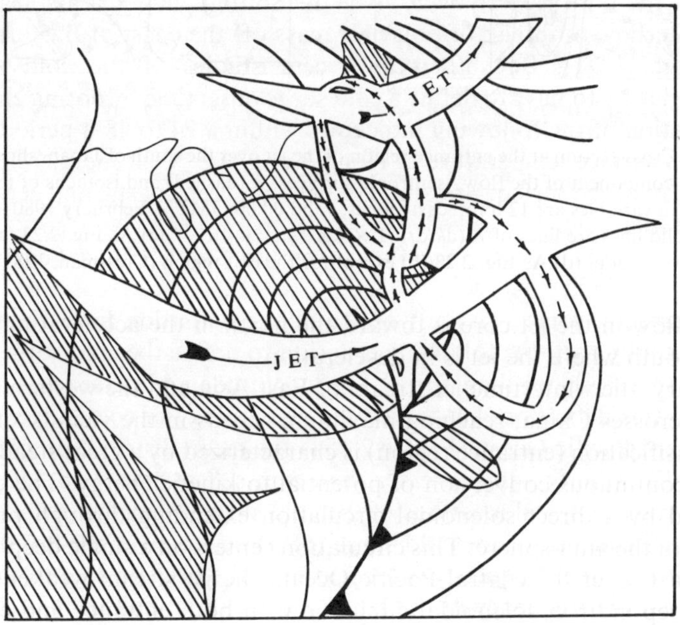

Figure 2.24. Three-dimensional schematic of jet-related circulation patterns during East Coast snowstorms. The transverse circulations are associated with the diffluent exit and confluent entrance regions of jet streaks embedded, respectively, at the base of troughs moving across the Ohio and Tennessee valleys and across southeastern Canada. Surface low and high pressure systems, isobars, and frontal positions are also indicated (Uccellini and Kocin, 1987).

region downwind of the trough near the East Coast and the confluent entrance region located over the northeastern United States.

Similar weather events may be observed in the East Asian coast in winter that have not been elaborated. Therefore, very little is known of the relationship between wintertime snow events and jet streaks. An inspection of the weather processes leading to snowstorm events in East Asia indicates that the synoptic model summarized by Uccellini and Kocin may be helpful to describe the production of heavy snow events in this region. A careful analysis of this matter still remains to be carried out.

2.5. The Development and Propagation of the Cold Surge in East Asia and the Western Pacific

When the winter monsoon arrives in southern China, the South China Sea and the tropical western Pacific, it can further intensify and usually becomes the prevailing northeasterly winds at low-level. These stronger low-level northerly or northeasterly winds are so-called cold surges, and their intensity is defined by an activity index of the cold surge (Chang and Lau, 1980). In a direct sense, the term cold surge was used to describe the rapid drop in surface temperatures over southern China and the freshening of the northeasterly wind over the South China Sea associated with the frequent wintertime outbreak of polar air mass off the coast of East and Southeast Asia (Ramage, 1971). But based on recent studies of the cold surge-related phenomena, it is customary to define a cold surge onset time by noting changes of one or a combination of the following indicators within a 24 to 48 h period:

(1) A drop in surface temperature at Hong Kong of 5 °C or more.
(2) An increase of the surface pressure gradient between coastal and central China of at least 5 hPa.
(3) A prevalent northerly surface flow over the South China Sea with a speed exceeding 5 m s^{-1}.

In the literature, some vaiations of the above definition have been used. Ding (1990) has recently used the following criterion to define the cold surge: a northerly wind with a speed greater than 5 m s^{-1} in magnitude and persistents for two days. According to this definition, he studied the climatology of cold surges in East Asia and the western Pacific Ocean. Figure 2.25 is the mean distribution of frequency of occurrence of northerly winds greater than 5 m s^{-1} at 1000 and 850 hPa, clearly showing the climatological locations of the cold surge. Near the surface (1000 hPa), the maximum frequency of the cold surge (65%) occurs over the South China Sea. Another region of high frequency is located over the oceanic area to the east of the Philippines, but the frequency of the latter is only about half of that of the former. Therefore, the strongest cold surges near the surface is observed over the South China Sea. This conclusion is consistent with that obtained by the study of cold surges for individual years. At 850 hPa, the frequency of occurrence of cold surge over the South China Sea significantly decreases, being half of the frequency near the surface, and the maximum of frequency shifts toward Malaysia and Indonesia. The cold surge activity

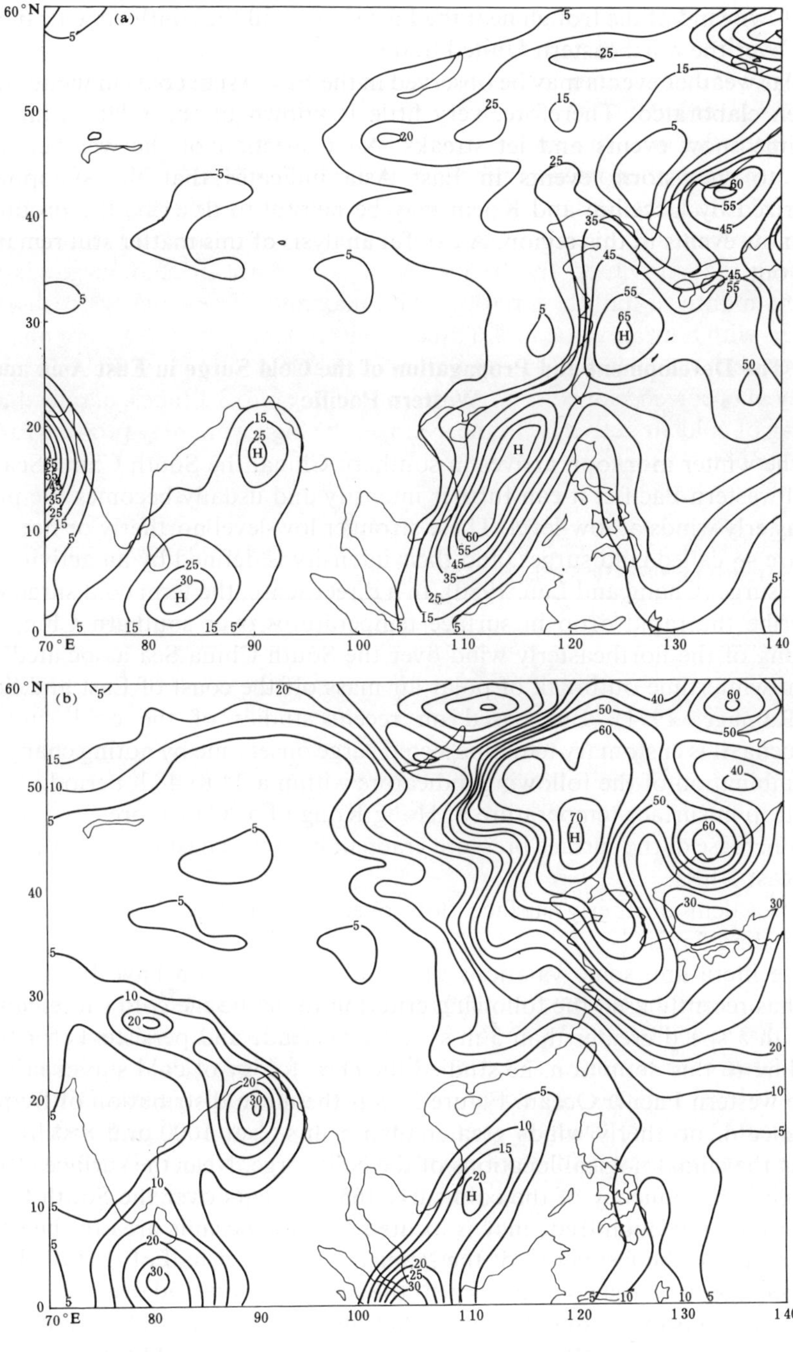

Figure 2.25. The mean distribution of the frequency of occurrence of north wind ($Vs < 0$) greater than 5 m s^{-1} in magnitude for five winters (December-February) of 1979–1984. (a) 1000 hPa and (b) 850 hPa (Ding, 1990).

over the western Pacific almost disappears (frequency 10%). Note that the cold surges over the South China Sea and the western Pacific nearly completely disappear above 700 hPa (not shown). From the above discussion, the cold surge is a phenomenon in the shallow layer below 700 hPa. In short, it is both frequent and strongest near the surface. Both the intensity and frequency of the cold surge decrease rapidly with height, so that the phenomenon nearly vanishes at 700 hPa.

Table 2.7 is the statistics of frequency of occurrence of cold surges over the South China Sea in winter (November-March) for 1978–1985 (Chen and Huang, 1988). It can be seen that in winter, the frequency of occurrence of cold surges assumes a significant month-to-month variation. The maximum frequency was observed in December, with a mean value of 4.6 times. The secondary maxima were observed in November and January, with mean values of 4 and 4.3 times. After February, the frequency of occurrence of cold surges decreased down to 3.1 times, mainly due to the weakening of cold air activity. On the average, the high wind may prolong for 4 days after bursts of cold surges.

Table 2.7. The statistics of frequency of occurrence of cold surges over the South China Sea in winter (November-March) for 1978–1985 (Chen and Huang, 1988)

1978–1979	Nov.	Dec.	Jan.	Feb.	Mar.	Total
1978–1979	3	5	6	4	1	19
1979–1980	6	2	3	2	1	14
1980–1981	3	8	4	4	2	21
1981–1982	4	4	7	4	2	21
1982–1983	6	5	2	3	2	18
1983–1984	3	4	4	3	0	14
1984–1985	3	4	4	1	2	14
Total	28	32	30	21	10	121
Average	4	4.6	4.3	3.1	1.4	17.3

It is very interesting to compare the patterns of the cold surge with the mean winter maps at 1000 and 700 hPa (Fig. 2.26). At 1000 hPa, there is a huge cold high, with its center located over central and eastern Siberia and Mongolia. Its associated ridge extends southward all the way down to the equatorial region (around 10° N). There is a significant pressure gradient over the East China Sea and the South China Sea where the most vigorous cold surge exists. At 850 hPa, one can only see the remnants of the cold high in the south, implying that the cold high that spreads down to the lower latitude is very shallow, mainly confined to the layer below 850 hPa. At 700 hPa, the synoptic situation in East Asia is quite different from that at the lower levels in that the basic air currents go around the Tibetan Plateau and split into two branches, with the warm air advected over southern China by the southwesterly airflow in advance of the semi-permanent trough in the southern branch of westerlies to the south of the Tibetan Plateau. This may be used to explain why the cold surge is mainly confined to the layer below 700 hPa.

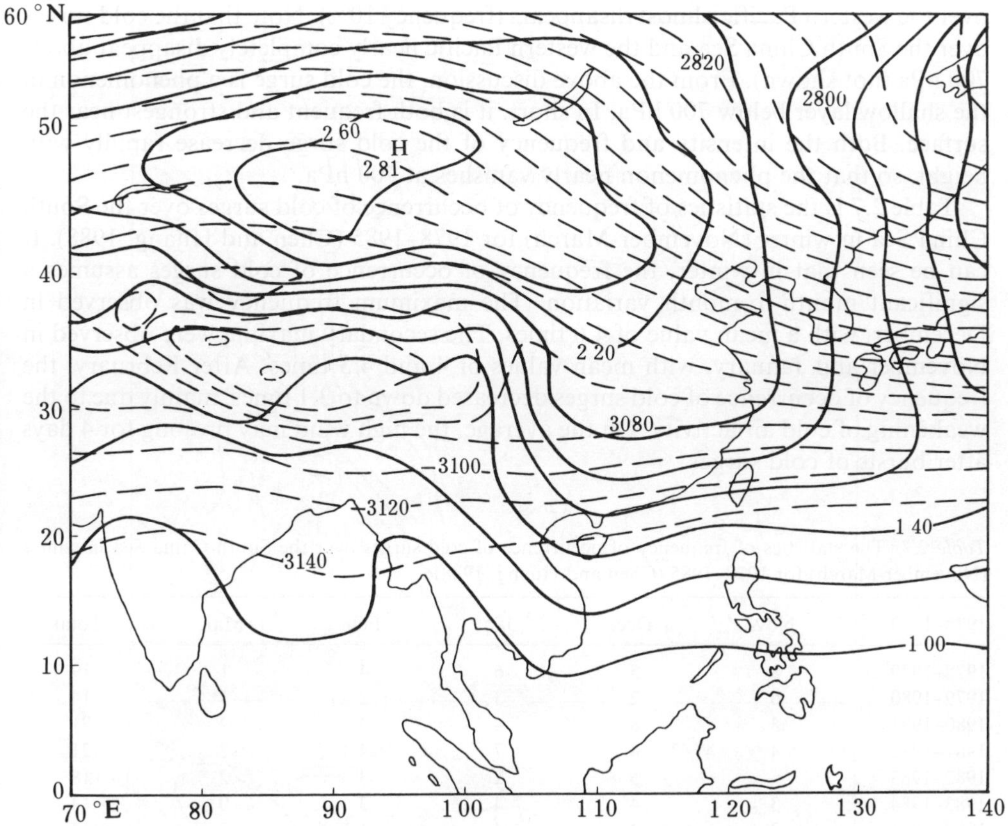

Figure 2.26. Five year (1979–1984) mean winter weather maps at 1000 hPa (solid lines) and 700 hPa (dashed lines). Unit of geopotential height: decameter (Ding, 1990).

The association between the behavior of the Siberian high and the cold surge may be more clearly seen through a contrasting study of winters of active cold surges and those of inactive cold surges. The winter of 1979–1980 was chosen as a year of inactive cold surges while the winter of 1983–1984 was selected to represent the active cold surges. A comparison of the frequency of occurrence of cold surges for these two winters (Fig. 2.27) shows that the most significant difference showed up in distributions over the South China Sea. At 1000 hPa, the frequency of occurrence of cold surges in that location for the winter of 1983–1984 was twice as high as that for the winter of 1979–1980 (80% as compared to 40%). At 850 hPa (not shown), the cold surges over the South China Sea for 1979–1980 winter were not detectable, and can be seen over the near-equatorial region of Indonesia and Malaysia. On the other hand, for the winter of 1983–1984, the cold surges over the central and southern parts of the South China Sea were still very significant. At 700 hPa (not shown), the cold surges for the winter of 1983–1984 still had a region of high frequency in Indonesia while the cold surges for the winter of 1979–1980 basically disappeared. Therefore,

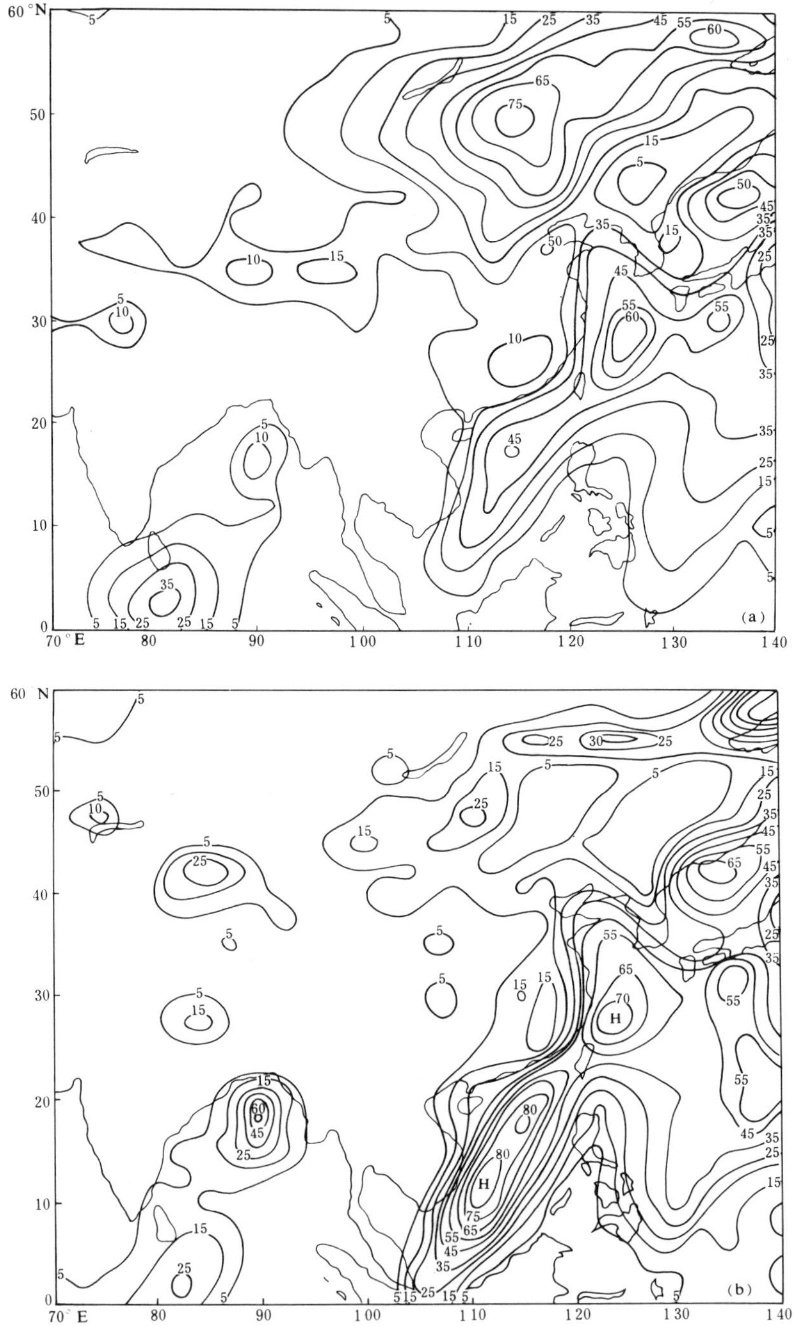

Figure 2.27. Distribution of occurrence frequencies of north wind greater than 5 m s^{-1} in magnitude at 1000 hPa for the winter of (a) 1979–1980 and (b) 1983–1984 (Ding, 1990).

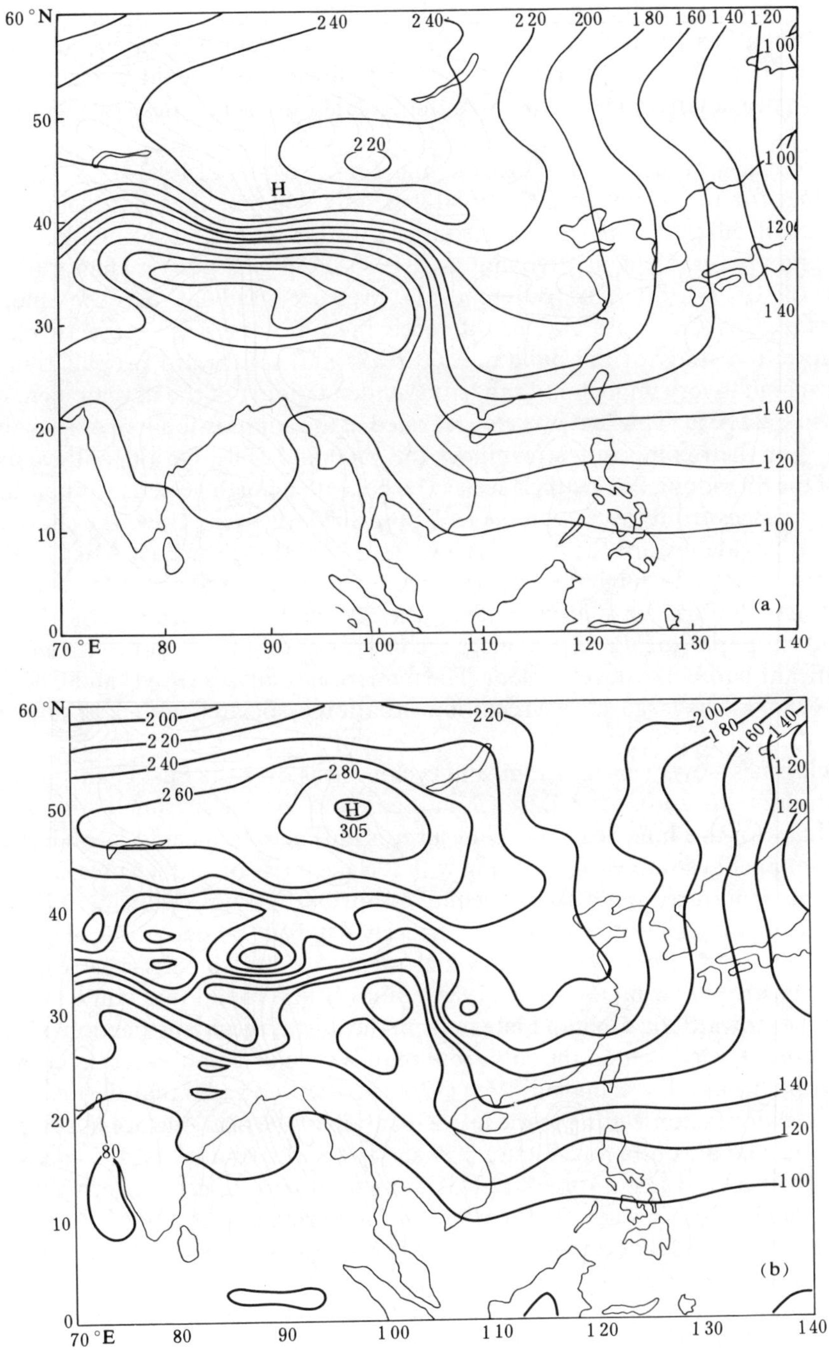

Figure 2.28. The mean weather charts at 1000 hPa (a) for the winter for 1979–1980, and (b) for the winter of 1983–1984. Units: gm (Ding, 1990).

the above comparison indicates that the frequency of activity of cold surges over the South China Sea for the winter of 1983–1984 was higher and extended upward to higher level (700 hPa). In relation to these difference in cold surges, it is very interesting to compare the intensity of the Siberian highs for these two winters (Fig. 2.28).

The intensity of the center of the cold high for the winter of 1983–1984 was 305 gm, while that for the winter of 1979–1980 was only 240 gm, thus producing pressure gradients of different intensity over southern China and the South China Sea. The mean pressure gradient for the winter of 1983–1984 was greater than that for the winter of 1979–1980. This difference in pressure gradient may be one factor responsible for the considerable difference in cold surges for these two winters. Therefore, the study of the buildup, outbreaks and southward propagation of the Siberian high is very important for a better understanding of the development of cold surges in East Asia. This fact was also revealed in the composite surge of Lau and Lau (1984). For their composite streamlines for 1000 and 500 hPa, the anticyclone just west of the Shandong Peninsula is seen to be beneath a northwesterly flow aloft at 500 hPa that comes from the Lake Baikal. Upon this large-scale flow, the shorter-scale waves propagate toward the mean longwave trough along the East Asian coast. The shortwave passing through the longwave trough amplifies and produces a circulation that thrusts the cold air southward from the reservoir of eastern Siberia. Therefore, the cold surge-producing disturbances are largely mid-latitude in origin and assume a significant baroclinic development. The numerous analyses (Boyle and Chen, 1987; Chen, 1988) of the large-scale circulation situation associated with cold surge events have indicated that intense baroclinic development occurs periodically, leading to anticyclogenesis over central China and cyclogenesis over the East China Sea. As the pressure tightens across the East China Sea, cold air bursts out of the continent toward the South China Sea and a cold surge is initiated. In the mid-latitude, a strong cold front pushes southward, bringing with it a steep rise of surface pressure, a sharp drop of temperature, and a strengthening of northerly winds (Ramage, 1971).

The strengthening of the low-level northerly winds is, by definition, an important indication of cold surge development which mainly results from the accelerating effect of the ageostrophic wind (Webster, 1981). Sumi (1985) found that while the cold air approaches toward the Tibetan Plateau from the north, the intensification of the low-level northerly wind behind the cold front mainly results from the acceleration of the ageostrophic wind. Lu, Ding and Wen (1988) further indicated that the ageostrophic wind is a major factor leading to the intensification of the near-surface northerly wind during the cold air outbreak. Figure 2.29 shows the distribution of the ageostrophic wind during a cold air outbreak (27–31 January 1980). It can be seen that strong ageostrophic winds flow out of the eastern and southern sectors of the cold high at 850 hPa. Near the cold front, the ageostrophic wind is particularly vigorous. Moreover, one may note that at the northeastern periphery of the Tibetan Plateau, there is very strong northerly ageostrophic wind, indicating that it may be associated with the effect of the flow going around the Plateau. It is this strong ageostrophic wind that stimulates the cold air to speed up the southward push of cold airs. For the examination of the effect of the large-scale topography of the Tibetan Plateau on cold

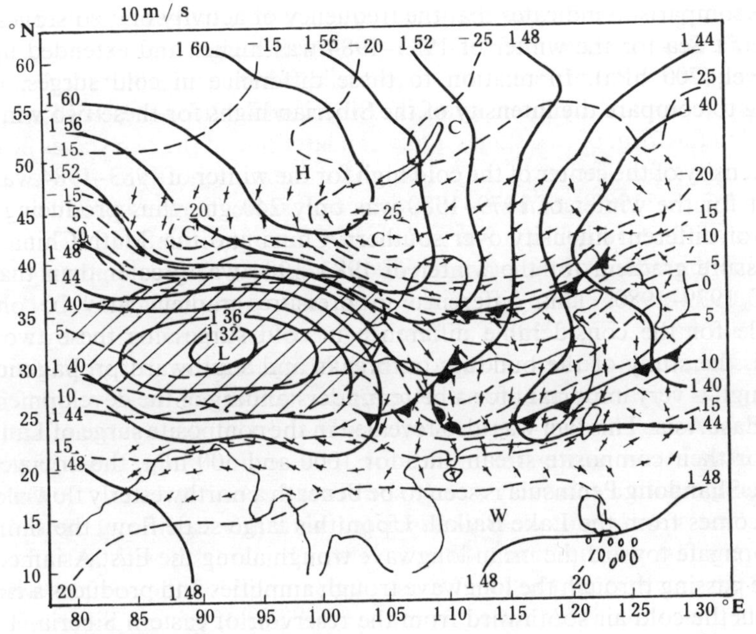

Figure 2.29. The distribution of ageostrophic wind (v-v_g) at 850 hPa at 12:00 GMT, 28 January, 1980. Solid lines represents the isopleths of geopotential height, with the interval of 4 decameters, dashed lines denote the isotherm lines, with the interval of 5 °C, and the short arrows the vectors of ageostrophic wind (m s^{-1}) (Lu, Ding and Wen, 1988).

surges, Zhu and Yang (1989) performed a numerical simulation, indicating that there is a belt of strong northerly wind at the eastern edge of the Plateau, with the northwesterly wind found to the north of 30° N and the northeasterly wind to the south of 30° N. Then the leading edge of the northeasterly wind spreads toward the South China Sea and finally arrives in the equatorial region, with a southward propagating speed of 50 m s^{-1}. This result is in good agreement with the observational evidence of cold surges over the South China Sea as indicated by Lau, Chang and Chan (1983) in their study of monsoon surge during winter MONEX.

Based on the above discussion, the development of a monsoon surge may be viewed as going through three distinct stages (Chang, 1988):

(1) The initiation stage which consists of the baroclinic processes triggered by the passage of shorter upper tropospheric troughs. Under this condition, the surge event is initiated through significant cold air outflow from an intense Siberian anticyclone. This strong cooling is mainly observed in the lower atmosphere southward of 40° N. Although cold surges occurs in other parts of the world, the East Asia winter cold surges appear to be unique in their intensity. As suggested previously, this seems to be a consequence of terrain distribution in the East Asia region (Zhu amd Yang, 1989; Xia and Zhu, 1989). The Tibetan Plateau restricts the movement of the polar air and leads to the establishment of the Siberian high and a very strong baroclinic zone off

the East Asian coast. Therefore, the intensity of the cold surges is probably due to the presence of a large reservoir of available potential energy.

(2) The equatorward propagation stage during which a mid-latitude frontal system evolves into a tropical dynamic process. In the mid-latitudes, a cold surge pushes southward in the form of two stages. Only can later one of the two stages be identified as a frontal passage. The early stage resembles the motion of gravity waves. Chang, Millard and Chen (1983) examined the surface pressure, temperature, dew point, and wind data over the South China Sea and vicinity during winter MONEX (1978–1979) to determine the timing of the passage of cold surges at various reporting stations. It was found that for more than half of the surge cases during winter MONEX, a surge occurs in two stages (Fig. 2.30). The first stage is often characterized by a significant rise in pressure, while the second stage by a sharp decrease in dew point temperature. Freshening of surface winds may accompany either or both stages. The time separation between the two stages is relatively short at the upstream stations, but generally long at the downstream stations. The second stage is associated with a frontal passage with a relatively steady southward speed of 11 m s^{-1}. On the other hand, the first stage is not clearly associated with any significant synoptic events. From its very fast propagation (40 m s^{-1}) which increases with an inferred depth scale, and an increase in the local cross-isobar angle of the surface wind during passage, the first stage is identified with a synoptic-scale gravity wave type motion.

Chen and Huang (1989) and Zhou (1989) have recently made further studies of the propagating characteristics of the gravity wave of cold surges based on station data and ship reports over the South China Sea. From Figure 2.31 it can be seen that when a cold surge and cold front simultaneously arrived in Hong Kong (14:00, 10

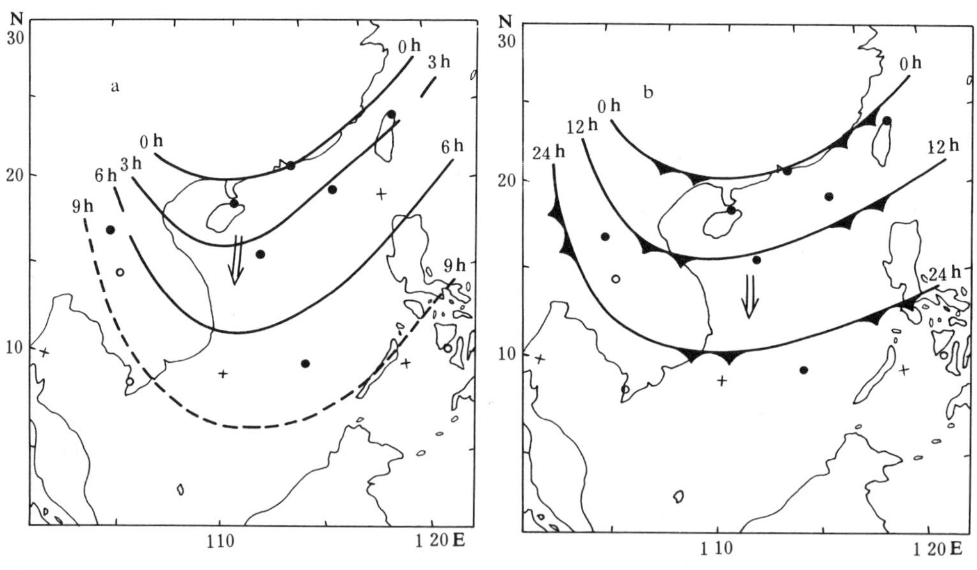

Figure 2.30. Schematic diagram for the progression of (a) unobstructed edge, and (b) unobstructed front (Chang, Millard and Chen, 1983).

December 1978) (a dual event), one may observe the pressure rise, and the rapid drop of temperature and dew point temperature. At 17:00, the cold surge passed over Dongshan station, and a similar change was still very evident. At 20:00, the cold surge arrived in Xishan, only causing sharp fall of the dew point temperature (a single event) and a rapid increase in wind speed. The thus estimated progressing speed of the cold surge is 27 m s^{-1}, smaller than the value of Chang, Millard and Chen (1983). Zhou (1989) also obtained a similar gravity-wave speed based on shipping reports from the South China Sea in 1986.

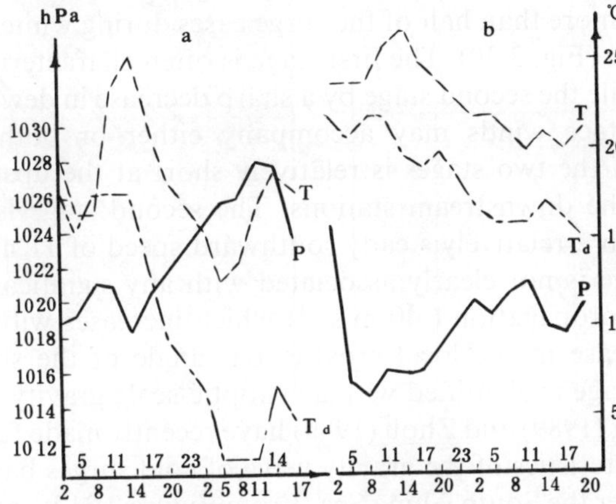

Figure 2.31. The temporal variations of surface elements at Hong Kong (a) and Dongshan (b) during the passage of the cold surge of 10–12 December 1978. Beijing local time is used here (Chen and Huang, 1989).

When an active surge front reaches about 20° N, northeastly winds freshen simultaneously over the South China Sea. An excellent case illustrating this situation was presented by Johnson and Zimmerman (1986). Figure 2.32 shows analyses at 900 hPa (1 km) because cold advection associated with the surge is strongest at low levels. On 10 December 1978 (Fig. 2.32 (a)) the surge is seen entering the northernmost portion of the region where strong cold advection is indicated and the wind speeds over the South China Sea are between 10 and 15 m s^{-1}. Isotherms are oriented northeast-southwest as a consequence of the influence of the Asian continent. Cold air has expanded southward and westward 11 December (Fig. 2.32 (b)) with an acceleration of the flow over most of the South China Sea south of 20° N to 15–20 m s^{-1}. Wind speeds near the China coast, on the other hand, have generally diminished. The analysis of 11 December suggests that low-level divergence is prevalent at that time over a significant portion of the northern South China Sea. By 12 December (Fig. 2.32 (c)) the cooling over the northern South China Sea has reverted to a warming, and the 850 hPa winds along the China coast have shifted to a generally westerly direction.

This cold surge extended down to the near-equatorial region. Figure 2.33 clearly demonstrates the deep southward penetration of the cold surge. In fact, the 10–12 December 1978 cold surge was the first of three moderate surge events during the

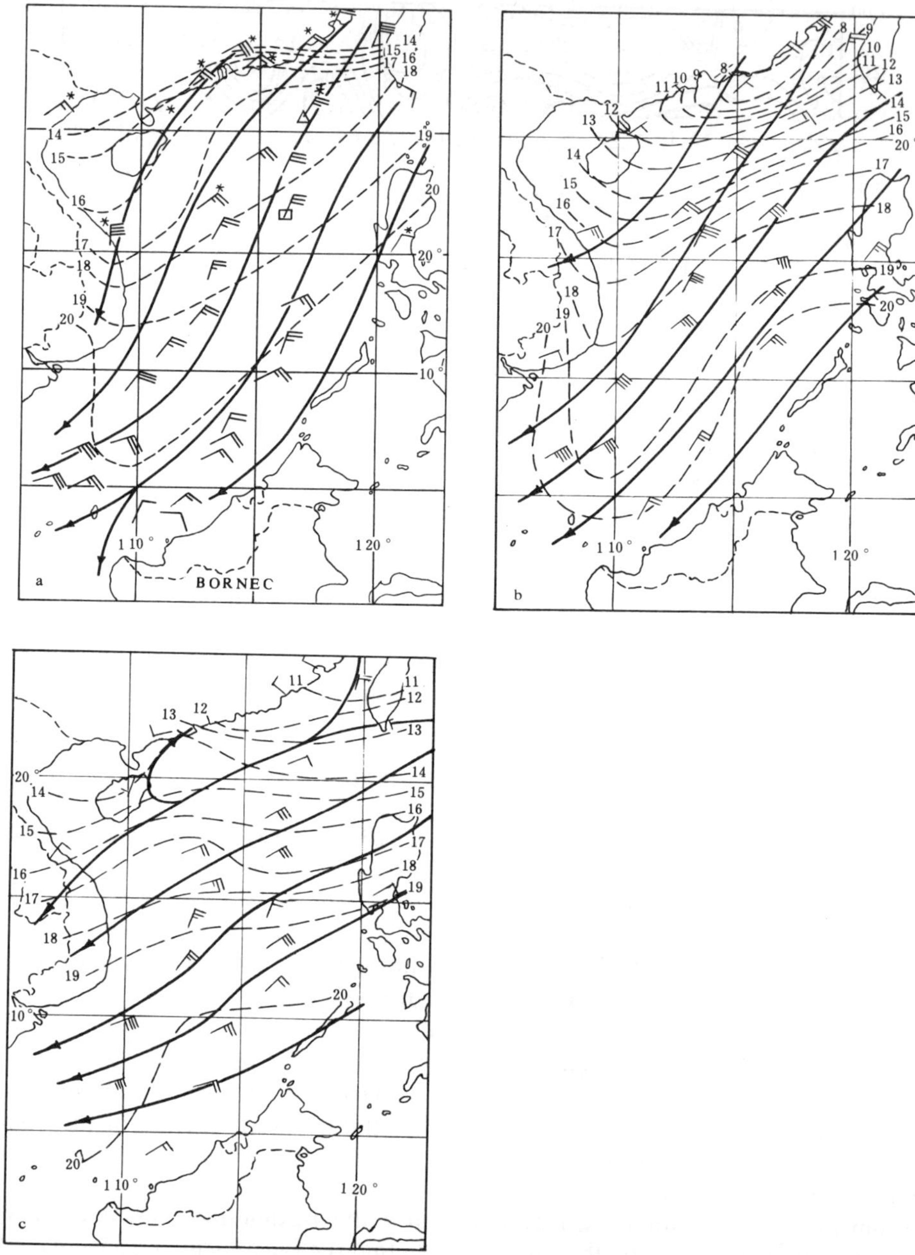

Figure 2.32. Streamlines and isotherms (°C) at 900 hPa at 12:00 GMT, 10 December (a), 12:00 GMT, 11 December (b), and 12:00 GMT, 12 December 1978 (c). One full wind barb = 5 m s^{-1}; one half barb = 2.5 m s^{-1}. The asterisk indicates that the wind is at 850 hPa. Observations over water (other than those from island stations) are based on aircraft dropwindsonde (WP-3D and Electra) and flight level (Hong Kong Islander) data (Johnson and Zimmerman, 1986).

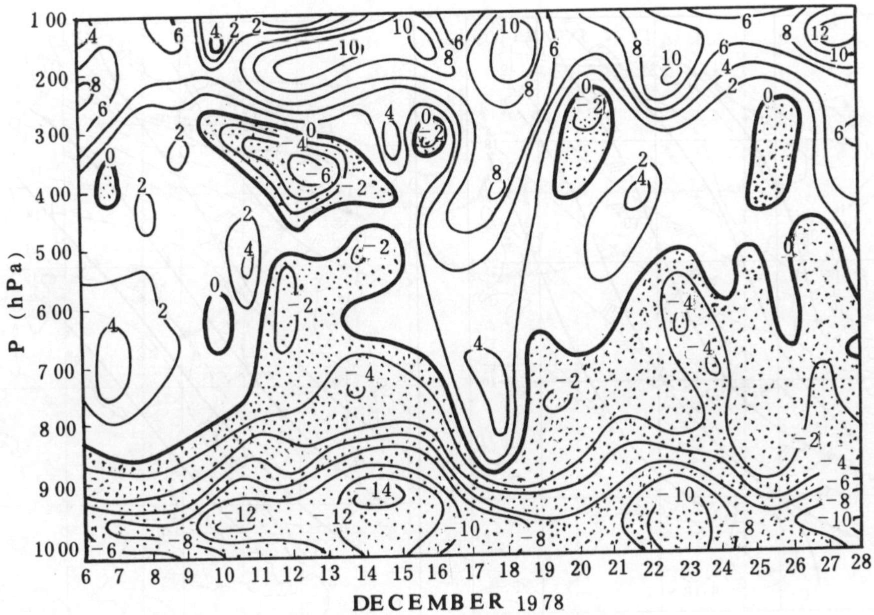

Figure 2.33. Time series of daily averaged values of the north-south component of the wind (m s^{-1}) at Priliv (7° N, 109° E) for the period 6–28 December 1978. Shading indicates northerly flow (Johnson and Zimmerman, 1986).

winter MONEX field phase. Modulations of the low-level northeasterly monsoon over the southern portion of the South China Sea by the surge are indicated in Figure 2.34. Surges on 10, 15, and 23 December are marked by an increase in the northerly flow at low-levels and , nearly concurrently, the southerly flow near 200 hPa, reflecting an intensification of the East Asian Hadley circulation at those times. The low-level flow is strongest ear 900 to 950 hPa, i.e., at or just above the top of the mixed layer. Because fronts usually stall and dissipate over the South China Sea, surges reaching very low latitudes should be of the edge type only.

Maximum 24 hour (12:00 to 12:00 GMT) cooling at 900 hPa for this event was about 8 °C near Hong Kong from 10 to 11 December (Fig. 2.34). Interestingly, some cooling (1 °C) extends south of 10° N over the western South China Sea, although the strong cooling by the surge is clearly limited to the northern portion of the South China Sea. From 11 to 12 December, there is a general warming (2 °C) over much of the northern South China Sea.

From the foregoing analysis, it can be seen that a freshening of low-level winds over the South China Sea clearly extends far beyond the region where the wind could have pushed the front. Based on the theoretical study by Lim and Chang (1981), such equatorial influences of cold surges may be viewed as transient motion under an adjustment of the pressure-wind imbalance. They used a linearized β-plane shallow-water equation model. The motion described by the shallow-water equation is to be interpreted as one vertical mode of the complete three-dimensional motion. The

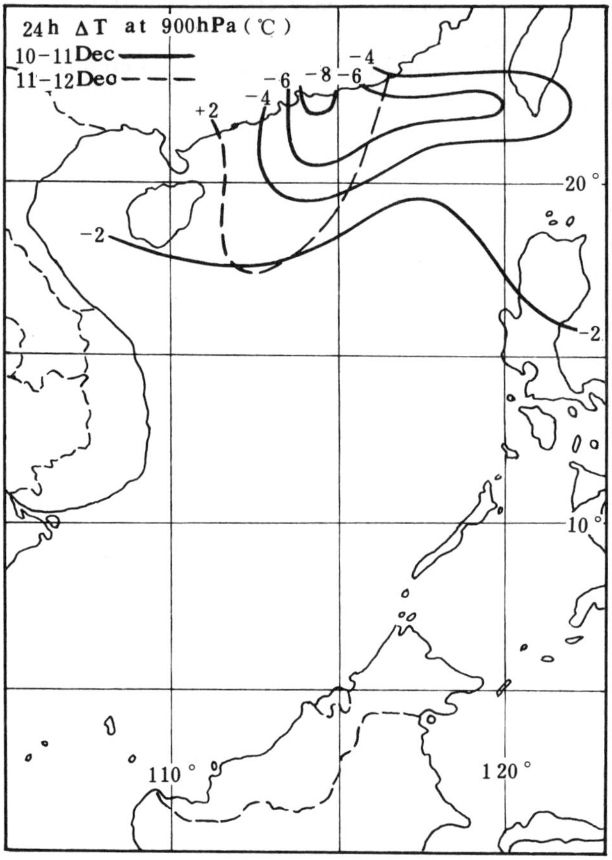

Figure 2.34. Twenty-four hour 900 hPa temperature changes (°C). Numbers in parentheses are for the 11–12 December period (Johnson and Zimmerman, 1986).

forcing term is specified to be a pulsed mass source with a Gaussian distribution centered at about 30° N, which simulates the effect of a sudden intense mid-tropospheric cooling. Figure 2.35 shows the solutions. On day 1, a belt of strong northerly and northeastly wind sweeps across the equator. Near the leading edge, the wind blows perpendicular to the isobars away from the high pressure areas. This reveals strong gravity-wave characteristics in the initial equatorward propagation of disturbances. More gravity-wave characteristics are exhibited on day 2 when a low center and southerly winds appear behind the leading surge winds, which have now penetrated deep into the Southern Hemisphere and are deflected eastward. From day 3 onward, the mid-latitude anticyclonic circulation around the region of forcing develops a marked northeast-southwest tilt. To its south, a belt of strong northeasterly winds sweeps from the subtropical latitudes to the equator. A southwesterly cross-equatorial countercurrent around the longitudes of the forcing becomes well-established after day 6. Sandwiched between the belt of strong northeasterly winds and the southwesterly cross-equatorial flow is an equatorial trough (shear line), south-southwestward from about 15° N to the equator. A cyclonic vortex also develops after day 7 in the Southern Hemisphere just to the south of the cross-equatorial current.

2.6. The Planetary-Scale Effect of Cold Surges

During cold surges, the planetary-scale circulation over the Asia- Pacific region appears to undergo significant short-term changes on a timescale of a few days. Previously, Ramage (1971) suggested that the occurrence of a cold surge off the southern China coast is related to an intensification of the local Hadley circulation over East Asia. Later, Chang and Lau (1980) investigated the changes in the planetary-scale feature, such as the Hadley type or Walker-type circulations in relation to the cold surges during December 1974, with a special focus on the sequential changes of the synoptic events associated with the cold surges. They showed that, in general, several mid-latitude and tropical circulation components vary in a coherent way during surge periods. Figure 2.36 is a schematic diagram describing the following series of events:

(1) Prior to the occurrence of a cold surge on the southern China coast, the cooling due to cold advection over northern China increases; this strengthens the East Asia local Hadley circulation through an increased sinking motion.

(2) Almost simultaneous with the strengthening of the heat sink in northern China, the East Asia jet stream centered near Japan intensifies as a result of upstream acceleration which is due to the increased upper level ageostrophic flow associated with enhanced local Hadley circulation.

(3) The variation of the West Asia jet stream centered over Afghanistan and Pakistan is out of phase with that of the East Asia jet stream, with the minimum of the later lagging nearly the maximum of the former. The apparent inverse relationship may be due to the Coriolis deceleration by the upper level equatorward meridional wind which is found upstream of the jet prior to and during the surge. This meridional wind also implies the possible occurrence of a reversed local Hadley-type circulation.

(4) The rapid eastward movement of an upper level trough which deepened over northern Japan appears to be a precursor to the outbreak of northeasterly cold air along the southern China coast.

(5) Immediately after a cold surge, the convection associated with pre-existing synoptic-scale disturbances in the equatorial South China Sea will intensify, thereby sustaining or increasing the already enhanced local Hadley circulation.

(6) However, the strengthening of the local Hadley cell does not continue for too long (beyond one day) after the surge, as the upper tropospheric outflow from the South China Sea region also accelerates along the equator toward the central Pacific and the east coast of Africa, thereby strengthening the east and west Walker cells.

The above evidence for the strengthening of the local East Asia Hadley circulation was consistent with the analysis of winter data by Boyle (1982), who showed that the 200 hPa southerly divergent wind is significantly correlated with the surface northerly

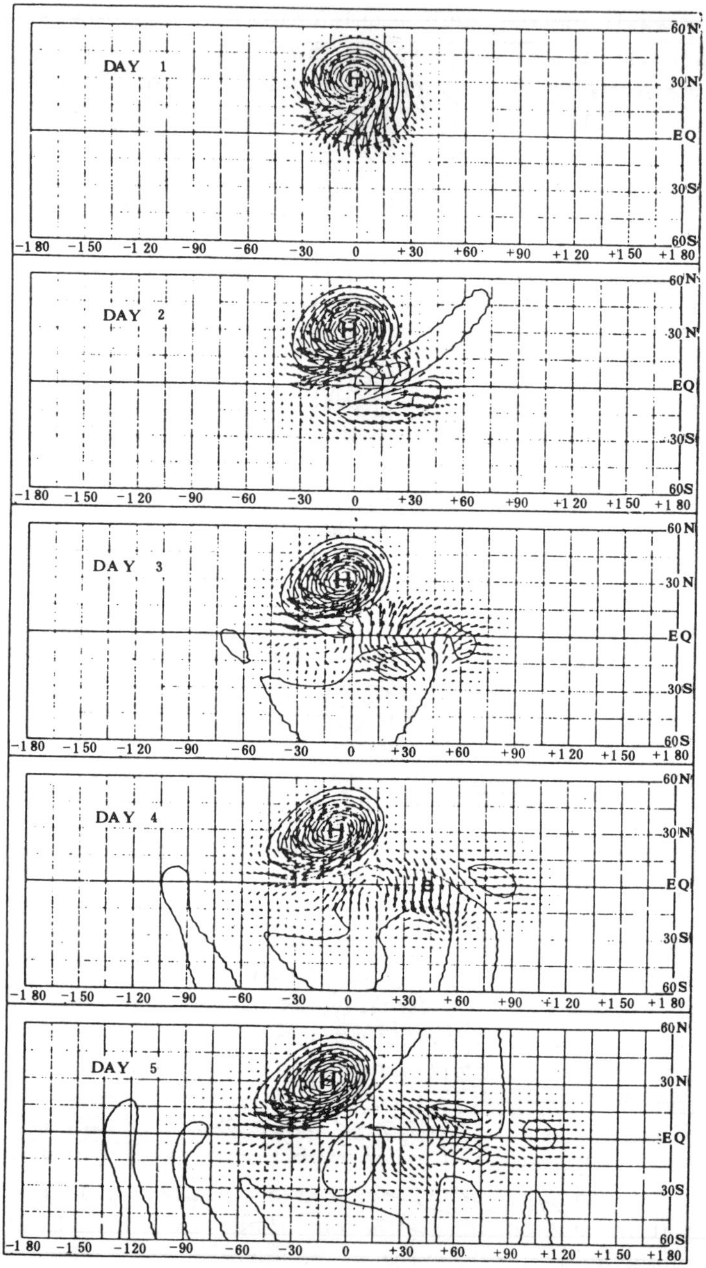

Figure 2.35. Transient local mass source at about 30° N turned on at time 0. The sequence of figure shows the spread of mid-latitude influence to the tropics. The Rossby mode response remains near the region of the forcing and exhibits a typical monsoon surge flow pattern. The ageostrophic flow component concentrates its energy into Kelvin waves and mixed Rossby-gravity waves which propagate eastward along the equator. This large area of complex motion, which constitutes the major response of the tropical

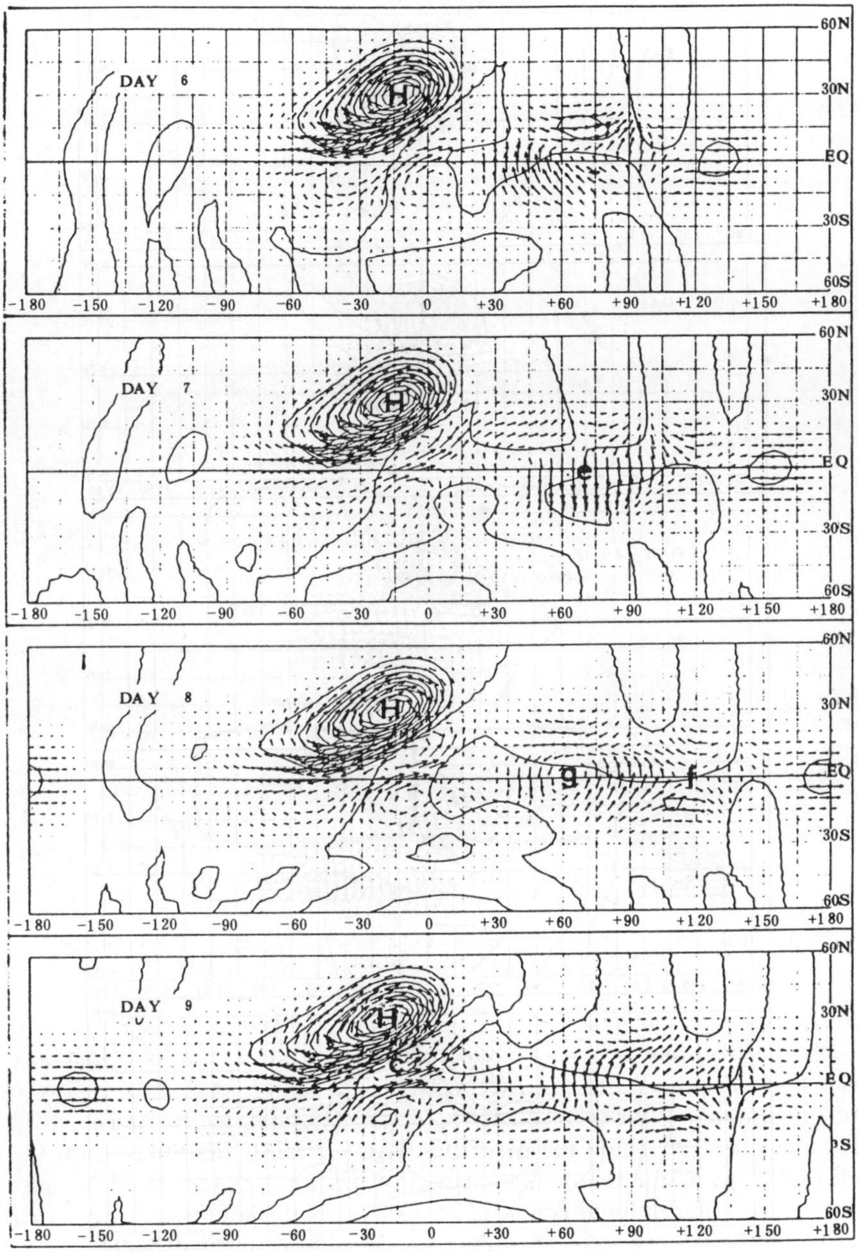

atmosphere to the mid-latitude forcing, bears a reasonable resemblance to the typical winter season flow pattern of the region. It suggests that the surge wind belt over the South China Sea and the quasi-permanent equatorial trough north of Borneo are basically features of a dynamical response of the tropical atmosphere to the intense anticyclonic development over China (Lim and Chang, 1981).

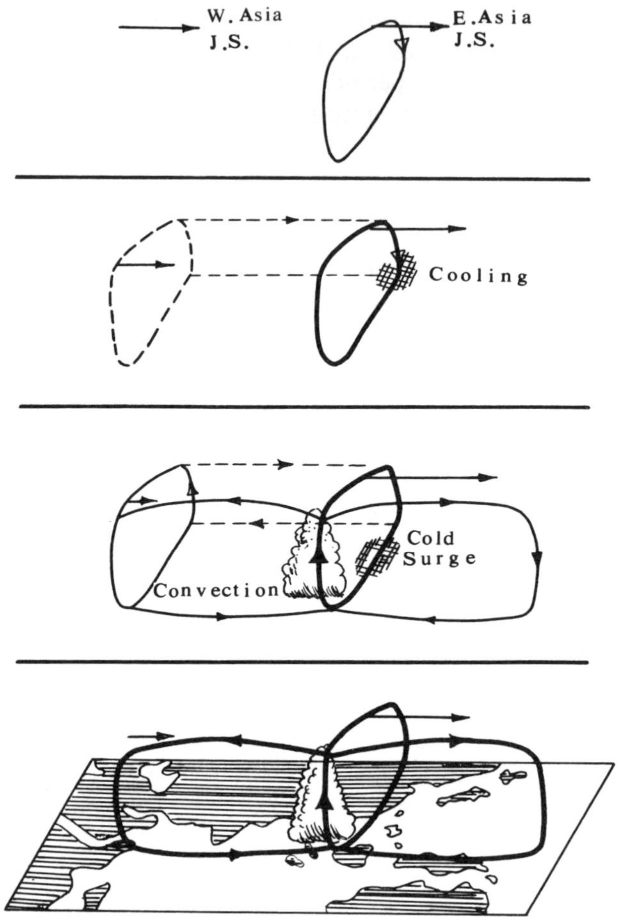

Figure 2.36. Schematic diagram showing the sequence of events before and after the cold surge in the South China Sea. See the text for details (Chang and Lau, 1980).

divergent wind within +48 h, with a maximum correlation at the 36 to 48 h lags. This suggests that the maximum enhancement of the upper level return flow occurs between 36 and 48 h after the occurrence of the maximum surge wind.

Yang and Zhu (1988) have recently made a numerical simulation to study the mid-latitude and tropical interaction during cold surges, and obtained a consistent result with the above observational analysis. This includes the southward propagation of cold surges as the gravity wave, the enhanced synoptic-scale disturbances in the near-equatorial region, and the enhancement of the local Hadley cell. However, they found that the Hadley cell has a southward migration during the cold surge and its intensity is associated with the effect of large-scale Tibetan Plateau.

The short-term planetary-scale mid-latitude and tropical interaction over East Asia demonstrates a significant difference between active and break periods of cold surges. It appears to be stronger during the active phases, while during break phases (inactive phases), the mid-latitude circulation components all exhibit reversed changes from those of the active periods and the variation in the tropics are less coherent, although still showing reversed tendencies. This contrast in the degree of organization of the tropical response between the active and inactive periods appears to suggest that during winter, the tropics are forced by the mid-latitudes rather than vice-versa, at least over East Asia and the Pacific. Table 2.8 shows the area-averaged

Table 2.8. Area-averaged parameters for active and inactive periods of cold surges (Chang and Lau, 1982)

Data and area	Circulation component represented	Remarks	
		Active period	Inactive period
200 hPa zonal wind over Japan and vicinity (U_J)	East Asia jet streak	From day of onset, U_J intensifies steadily by 18 m s^{-1} lasting over 6 days	At least 36 h prior to day of onset, U_J shows reversed tend (slackening of jet streak)
200 hPa zonal wind over India and Afghanistan (U_I)	West Asia jet streak	U_I weakens by 10 m s^{-1} over four days	U_I increases from day of onset to day 1.5, then decreases for three days resuming the increase
200 hPa zonal wind over equatorial Indian Ocean (U_W)	Western Walker circulation	An easterly acceleration of U_W with total speed increase of $\approx$ 4 m s^{-1}	U_W exhibits an out-of-phase relationship. Indian Ocean Walker cell weakens from -2 to day 1.5
200 hPa zonal wind over equatorial Pacific (U_E)	Western Walker circulation	A westerly acceleration U_E by 4 m s^{-1}	More noisy and less suggestive of trend
200 hPa meridional wind over East Asian coastal region and western Pacific (V_N)	Southerly flow south of the East Asia jet streak	V_N increases steadily by 15 m s^{-1}, but lags the onset of the surge by $\approx$ 1 day	V_N shows a reversed trend compared to the surge cases
200 hPa divergence over maritime continent or equatorial Pacific (D)	Large-scale deep convection in the major heat source region	D intensifies steadily, by 3×10^{-6} s^{-1} over 6 days	exhibits an out-of phase relationship with its decrease being half the increase during surge period
200 hPa zonal wind over Japan and vicinity (V_S)	Surge and break index	V_S shows an apparent increase in northerly wind	The relaxation of northerly wind actually starts 36 h prior to onset day. The composite surface wind remain southerly wind through day 8

parameters for active and inactive periods of cold surges (Chang and Lau, 1982). The results basically verify the short-term teleconnection model proposed by Chang and Lau (1980), but more complex processes in the acceleration of the mid-latitude jet streaks are indicated.

Using objectively analyzed wind data derived from winter MONEX, short-term planetary-scale interactions were further investigated (Lau *et al.*, 1983). During the winter MONEX, cold surges were known to be moderate at best. Results show that much of the planetary-scale interaction during that winter took place downstream of the Sea of Japan over the central Pacific in apparent connection with an eastward shift of the monsoon heat source from Indonesia to the vicinity of the dateline. The intensification of the 200 hPa jet streak over Japan is a precursor of cold surges and subsequent downstream development over the central Pacific. The mid-latitude-tropical interactions that took place over East Asia remain similar to that described above. A new feature observed during the winter MONEX is that the main convective heat source was found over the equatorial central Pacific and was enhanced 3–4 days after the surge onset. The enhancement coincided with the occurrence of a subtropical jet stream over Hawaii and a pronounced upper level trough-ridge system extending from the equatorial central Pacific to the west coast of North America. The eastward shift of the winter heat source is believed to be associated with the 1982–1983 ENSO event. Recent studies (Rasmussen and Carpenter, 1982) revealed that a weak local East Asia Hadley cell is associated with reduced surface easterlies over the central and western equatorial Pacific following the occurrence of an El Niño event, and the normal-type heat source over the Indonesian region moves eastward over the vicinity of the dateline.

Recently, Ding and Krishnamurti (1987) studied the planetary-scale effects during the passage of the Siberian high. They found that during cold air outbreaks, the mean major trough off the East Asian coast deepens and moves eastward (Fig. 2.37) by about 20° longitude (from 120 to 140°E). At the latter stage of a cold air outbreak, the upper-tropospheric Alaskan high amplifies significantly.

Figure 2.38 illustrates the velocity potential field obtained by averaging the three stages together at 200 hPa. This is the time-averaged picture of the divergent circulations associated with the windfield of Figure 2.37. The strongest center of divergent circulation (rising motion branch) is located over the equatorial central Pacific (150°E). This position is nearer or slightly farther to the east than those obtained in studies of cold surges (Chang and Lau, 1980; Lau and Chan, 1983).

Over the Eurasian continent and Africa, convergence (with associated tropospheric downward motion) predominates. If one inspects the changes in the divergent circulation on the planetary scale from stage 1 to stage 3 (not shown), a rapid eastward shift of the planetary-scale circulation ensues with an eastward shift of the strongest divergent circulation center by about 60° longitudes. This places the planetary-scale rising motion over the central equatorial Pacific, resulting in a weakening of the normal Walker circulation. An enhancement of the convection and precipitation over the equatorial eastern Pacific is also noted.

A clearer presentation of this feature is shown in Figure 2.39a, b, c where we have subtracted the mean velocity potential field at 200 hPa (Fig. 2.38) from those of the

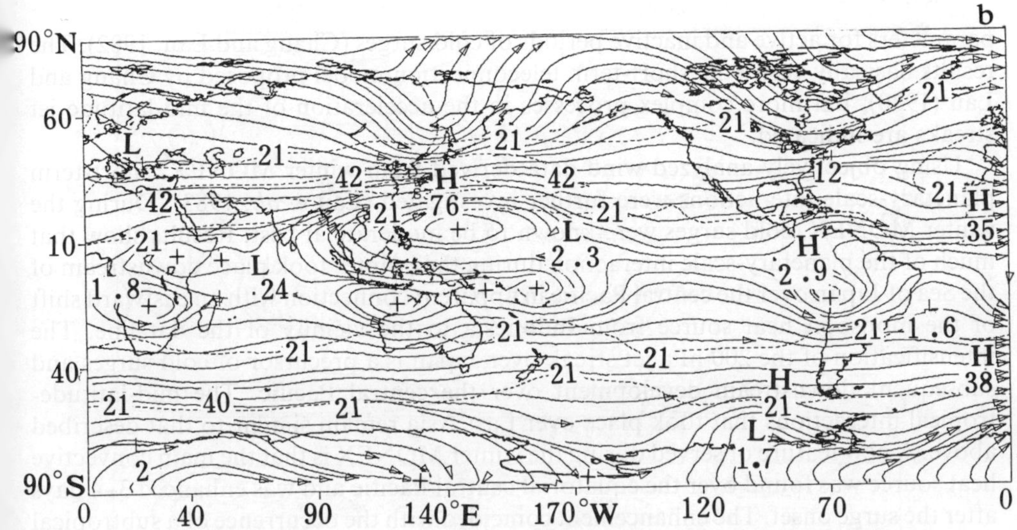

Figure 2.37. The mean flow pattern at 200 hPa composited for the 19 strong Siberian high cases. Solid lines denote the streamlines and dashed lines isotachs. Units: m s^{-1} (Ding and Krishnamurti, 1987).

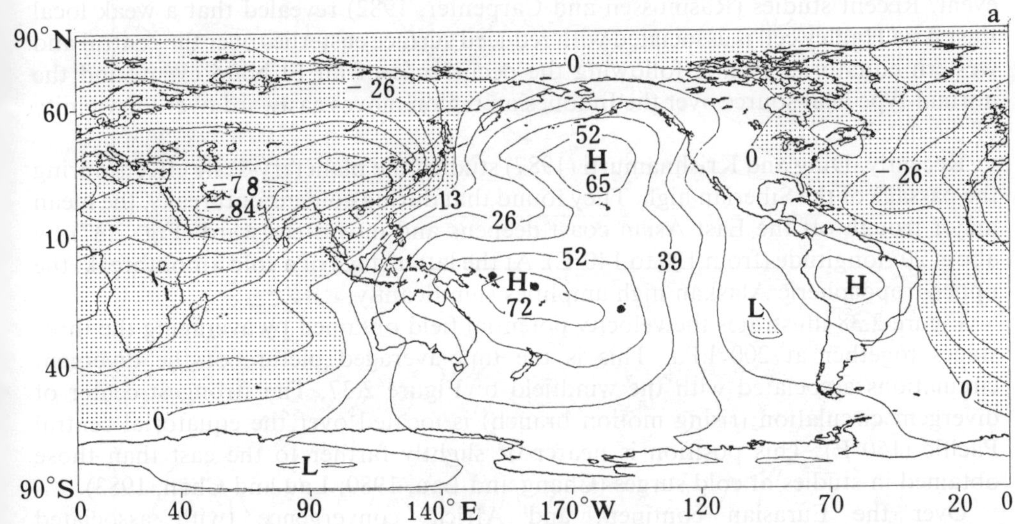

Figure 2.38. Mean velocity potential field for 200 hPa composited for the stages prior to, during, and after cold air outbreaks in East Asia. The positions of the divergent circulation centers for the three stages are indicated by black dots. The divergent region is shaded. Units: 10^5 m^2 s^{-1} (Ding and Krishnamurti, 1987).

respective stages. These fields of the divergent circulation anomaly clearly demonstrate the opposite patterns of divergent circulation prior to and following the cold air outbreaks. The anomalous strong divergent circulation at 200 hPa over the equatorial western Pacific, west of 170° W, and convergent circulation over the

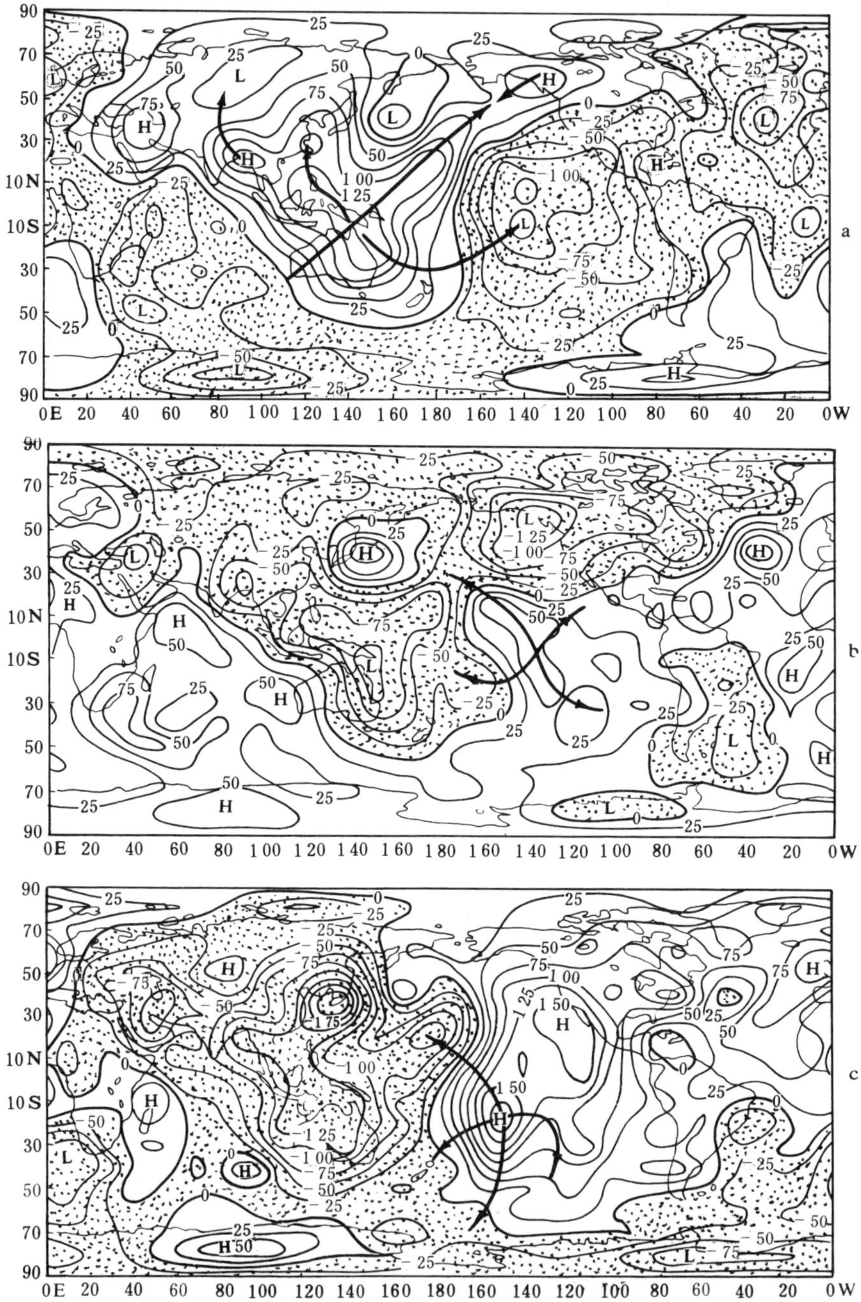

Figure 2.39. (a) Anomaly of velocity potential at 200 hPa for stage 1, obtained after subtracting the mean velocity potential field shown in Fig. 2.38 from the velocity potential field for stage 1. Negative anomaly (anomalous convergent region) is shaded. Units are 10^5 m^2 s^{-1}. (b) As in Fig. 2.39a, except for stage 2. (c) As in Fig. 2.39a, except for stage 3 (Ding and Krishnamurti, 1987).

equatorial eastern Pacific is indicative of the presence of vigorous normal-type Walker circulation for the early stage (Fig. 2.39c). However, as the Siberian high moves southward, the divergent circulation anomaly rapidly moves eastward and dominates the entire East Pacific (Fig. 2.39b).

The anomalous divergent circulation propagates and extends farther northeastward and southeastward and spreads its effect northeastward over North America (Fig. 2.39c) when stage 3 was the active center. The normal-type Walker circulation is now replaced by a reverse east-west circulation. Here, rising motions are found over the equatorial eastern Pacific. The sequence of events described before

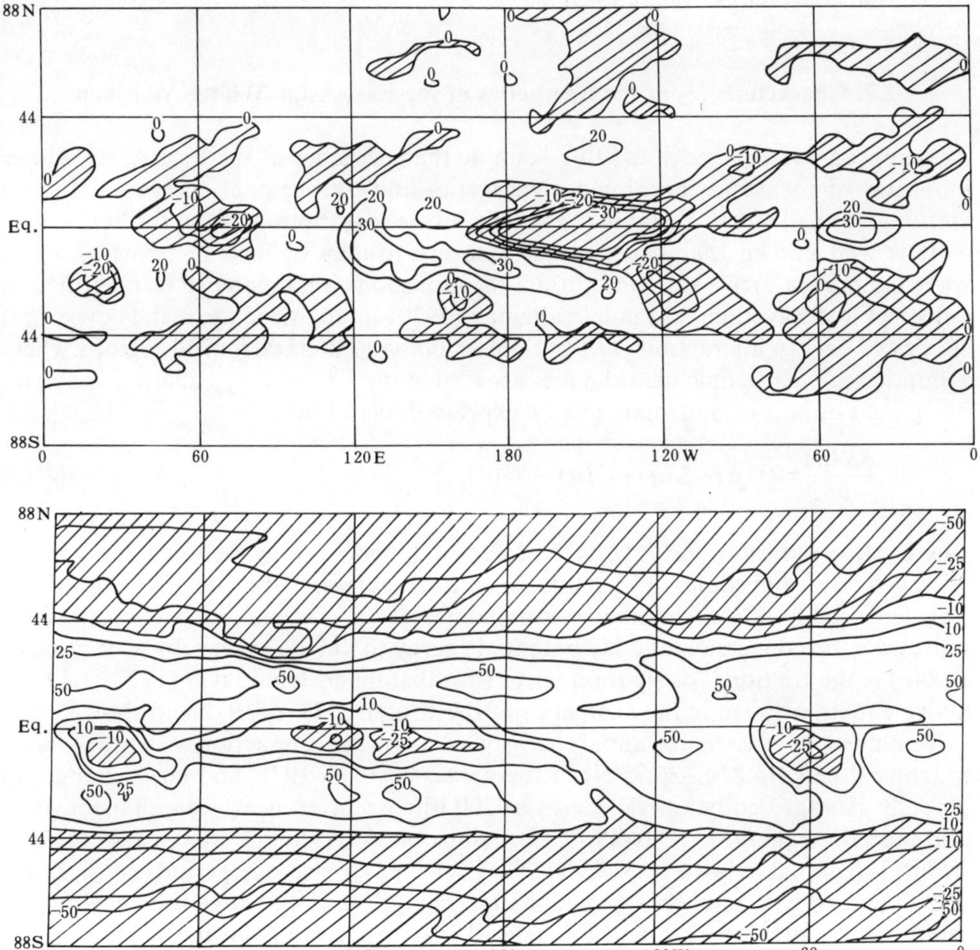

Figure 2.40. (a) Mean OLR (outgoing longwave radiation) distribution for the 4 winters (December-February) of 1980–1984. The vigorous convection areas are shaded. OLR values were obtained by subtracting 220 W m^{-2} from the original values. (b) Same as (a), but for the 14 cases of cold air outbreaks in East Asia.

seems to bear considerable resemblance to the atmospheric response in the tropics and mid-latitude regions associated with the El Nino phenomena, although they have a different timescale.

Based on the examination of the OLR data for the 14 cases out of the above 19 cold air outbreaks in East Asia, Ding (1990) provided an evidence to lend further support to the eastward shift of the heat source from its normal position to the central and eastern Pacific (Fig. 2.40a, b). Here, negative (positive) anomalies denote enhanced (reduced) convection or heating. For a normal winter condition, the vigorous convection is found over Indonesia (120–140°E). However, it can be seen that this center of convection shifted eastward to the central and eastern tropical Pacific during the cold air outbreaks.

2.7. Characteristics of the Energetics of the East Asian Winter Monsoon

Two approaches have been used to examine the energetics of the East Asian winter monsoon: the wavenumber domain energetics and the regional energetics. In this section, we shall stress the latter, with only a brief description of the former.

It is well known that scale interactions are usually defined in terms of zonal wavenumber. In such a scale resolution, the monsoon circulation can be identified as a distinct low wavenumber mode. Saltzman (1970) summarized the results of several studies of energy interactions in the Fourier domain in extratropical regions where synoptic-scale baroclinic disturbances are dominant. The energy equation for K(n), using the symbols of Saltzman, can be expressed as follows:

$$\frac{\partial K(n)}{\partial t} = M(n) + L(n) + C(n) - D(n), \tag{2.12}$$

where $M(n)$ indicates the rate of transfer of kinetic energy between the zonally averaged flows and eddies of wavenumber n. $L(n)$ represents the rate of kinetic energy transfer to eddy of wavenumber n from eddies of all wavenumbers. The $C(n)$ term represents the conversion of eddy potential energy to kinetic energy for wavenumber n. $D(n)$ is the frictional dissipation term. Murakami and Unninayar (1977) used the above equation to study the energetics of the winter of 1970–1971 and showed that wavenumber 8 receives substantial energy via wave-wave interaction with other waves in tropical belt (14.8° S–26.7° N) in the winter of 1970–1971, and large changes in globally averaged eddy kinetic energy at 200 hPa occurred nearly simultaneously at mid-latitude (region 1: 33.0–48.1° N), subtropical (region 2: 10.0–28.7° N) and tropical (region 3: 5.0–14.8° N) belts. They selected three 5 day periods (M periods) of 200 hPa eddy kinetic energy maxima ($k^* = ½[u^{*2} + v^{*2}]$), where the angle braces represent the time average and the asterisk the departure from it, and four 5-day periods (m periods) of 200 hPa k^* minima. During the M period, large-scale circulation features were characterized by an extremely strong jet stream near Japan and a deep upper trough over the central North Pacific. Large cloudiness over eastern Asia, associated with an above-normal local Hadley cell, possibly reflected enhanced northeast monsoon surges from the Siberian high. Similarly, *m* period synoptics showed a below-normal jet stream near Japan, weak oceanic trough over the central North Pacific, a depressed local Hadley cell, and less than season-mean clondiness over eastern Asia. From this study, it results that disturbance activity over the central North Pacific poleward of about 30° N is more pronounced during *M* period ($k^* > 800\ m^2\ s^{-2}$ at 35° N, 170° W), with a distinct band of large 200 hPa k^* extending from about 20° N, 150° W to 30° S, 170°E, crossing the equator around 170° W. This band nearly coincides with the axis of upper oceanic troughs embedded in equatorial westerlies over the North and South Pacific. Compared to *m* period, disturbance development in and around the upper trough appears to be more active during the *M* period. The 700 hPa k^* is quite large off the east coast of the Asiatic continent during the *M* period. A band of large k^* along about 130°E, from Taiwan to North Borneo, is probably associated with northeast monsoonal surges bursting out of the Siberian high and tropical disturbance activity over the South China Sea. Less frequent disturbance activity over the South China Sea in *m* is reflected by small 700 hPa k^*. In contrast, disturbance activity over the Timor Sea and the eastern South Indian Ocean becomes more pronounced during *m* as compared to the *M* period. Here, tropical disturbances developed in the Southern Hemisphere near the equatorial trough generally propagate west-southwestward along 10–15° S.

Later, Murakami (1978) further extended this study to three-year winters in 1970–1971, 1971–1972, and 1972–1973. They found that all transient waves behave as a source of kinetic energy for winter mean zonally-averaged flows with large contributions from waves in the wavenumber range of 3 to 8, but with regional differences. In region 1, wave-zonal flow interaction are substantially more pronounced than wave-wave nonlinear interactions. Regions 2 and 3 are characterized by the predominance of wave-wave interaction processes.

Recently, Chou (1985) estimated the energetics of the East Asian cold air outbreaks for the extended-range conditions and processes causing outbreaks of the cold air by use of Eq. (2.12). He indicated that the development of the ridges over the North Atlantic and North Pacific as a precusor of the beginning of the extended-range process of East Asian cold air outbreaks is often associated with the kinetic energy of wavenumber 2, and the establishment of the ridge over Ural mountains and development of the ridge over the Okhotsk Sea are often associated with the kinetic energy of wavenumber 3. In one or two days prior to the outbreak of cold airs, the kinetic energy of wavenumber 4 made the greatest contribution and the outbreak of cold air often corresponds to the development of wave 3. In addition, the mean kinetic energy ($K(0)$) and mean available potential energy ($P(0)$) at mid-latitude assume two cycle processes of increase and dissipation, the first increase starts with the development of the ridges over two oceanic basins, indicating the significant accumulation of a large amount of energy. Almost at the same time, the kinetic energy of wavenumber 2 ($K(2)$) first increases up to the maximum, mainly through the baroclinic conversion and then gradually decreases. In contrast, $K(3)$ undergoes an opposite change, with its rapid increase to maximum at the time of outbreak of cold airs, mainly through wave-to-wave interaction. Therefore, the development of the cold air outbreak over East Asia is characterized by unique global energetics, at least in the Northern Hemisphere mid-latitudes.

From the wavenumber domain energetics described above, it can be seen that the character of disturbance activity and its energetics differs significantly for different regions during the Northern winter. Therefore, it is very significant to investigate the regional energetics for the various limited regions relevent to the maintenance and development of circulation regimes. Kung and Chan (1981) have examined energetics characteristics of the Asian winter monsoon in the source region that is defined as a broad area including Siberia (40–60° N and 80–120°E, with an approximate area of 6.3×10^6 km^2), northeastern Asia (35–60° N and 120–140°E except for the area of the Sea of Japan region), the China Main (25–40° N and 105–120°E, with an approximate area of 2–3 $\times$ 10^6 km^2) and the Sea of Japan (35–40° N and 130–140°E, with an approximate area of 9.5 $\times$ 10^6 km^2). The source region thus designated is differentiated from adjacent regions in the south and east where the winter monsoon interacts with systems of southern origin and with the warm sea during the cold air outbreak. Further, the division of these four regions is decided by observing the mean flow pattern development of synoptic systems during the winter and the distribution of characteristic energy variables in the domain. The Siberia region is distinguished from other regions by the consecutive development of upper air ridges in association with the surface high, whereas the Northeast Asia region is under the influence of a major cyclone which is usually centered in the vicinity of Sakholin Island during the period of the Siberian ridge development. The development of a distinguishable upper air ridge in Siberia and an intense cyclone over the coast of northeastern Asia generally indicates a subsequent outbreak of the cold wave. During the outbreak period, the cold air flows from Siberia southward and southeastward to cover the the China Main and the Sea of Japan regions.

The kinetic energy equation (1.1) is used to estimate the kinetic energy balance of the atmospheric flow in the pressure layer over the area of analysis. In Eq. (1.1) $[\partial k/\partial t]$ is the local change of kinetic energy, $-[\nabla \cdot \boldsymbol{V}k]$ the horizontal flux convergence of kinetic energy, $-[\partial \omega/\partial p]$ the vertical flux convergence, $-[\boldsymbol{V}\cdot\nabla\Phi]$ the cross-isobaric generation, and $E]$ the dissipation of kinetic energy.

Table 2.9 shows vertical totals of the winter energy variables in four regions. Siberia, which is dominated by the anticyclonic type flow in winter, indicates a relatively low level of $[k]$, a general adiabatic destruction of $[k]$ in terms of negative values of cross-isobaric generation $-[\boldsymbol{V}\cdot\nabla\Phi]$ except in the lower troposphere, and the negative values of dissipation $[E]$ value above the lower troposphere. The negation $[E]$ obtained as the residual term of Eq. (2.13) should be considered as the net supply of kinetic energy to observed scales of motion from other scales unobservable (up-scale energy cascade) with the given upper air network. These energetic features in Siberia are very similar to those of the transient anticyclones over North America (Kung and Baker, 1975). The $-[\boldsymbol{V}\cdot\nabla k]$ term is a small positive value indicating the weak import of the kinetic energy. $-[\partial \omega k/\partial p]$ has generally insignificant vertical totals. It is readily obtained that the general adiabatic destruction of $[k]$ in this region is balanced out by the positive energy supply of sub-grid scale motion and the flux convergence through the boundaries. This energetic characteristic may reflect the unique dynamic behavior of the Siberian high in the source region. Later, Chan (1982) further documented this finding.

Table 2.9. Winter means of vertical totals of energy variables from the surface to 100 hPa. $[k]$ is in units of 10 Jm^{-2} and others in W m^{-2} (Kung and Chan, 1981)

Region	$[k]$	$-[\nabla \cdot Vk]$	$-[V \cdot \nabla\Phi]$	$-[\partial\omega k/\partial p]$	$[E]$
Siberia	19.61	1.47	− 7.88	−0.04	− 6.42
NE Asia	21.94	− 9.01	11.89	0.28	3.14
China Main	45.80	−43.29	60.25	0.25	17.09
Sea of Japan	44.47	− 1.98	36.39	0.57	35.06
Total area	27.18	− 9.28	13.74	0.06	4.52

The energy level of northeastern Asia region where the cyclonic flow-pattern dominates is also low. In contrast to the Siberia region, the generation of kinetic energy in this region is positive and the kinetic energy is dissipated (positive $[E]$ value). A small amount of the generated $[k]$ is exported outside this region. Therefore, this picture of the energetics of northeastern Asia, which is of normal-type, is basically opposite to that of the Siberia region. The resulting contrasting energetics of these two regions may be due to the difference in large-scale flow patterns and dynamics. The patterns of the vertical distribution of energy variables are similar to those of cyclones over North America (Kung and Baker, 1975) with the maximum of $-[V \cdot \nabla\Phi]$ and $[E]$ both in the lower boundary and at the jet stream level. The $-[\partial\omega k/\partial p]$ shows a supply of kinetic energy from the upper troposphere to the mid-and lower troposphere (not shown). The $[k]$ in the China Main and the Sea of Japan regions where the strong westerlies dominate, is high, more than twice the winter mean values in Sibera and northeastern Asia as well as North America (Kung, 1967), and the strength of the jet stream in this area is reflected on its vertical extent in the mid and upper troposphere of the extremely large $[k]$ values in these layers. The very strong cross-isobaric generation $-[V \cdot \nabla\Phi]$ in this area is observed and takes place in these layers with the maximum at the jet stream level. It well be pertinent here to note that, along with North America, this is one of the two general areas in the winter Northern Hemisphere where the poleward ageostrophic flow at the jet-stream level becomes significant. The generation is particularly strong in the China Main with the values at the jet-stream level about twice those over the the Sea of Japan (60.25: 36.39 W m^{-2}). This results has been further documented by the later works by Cressman (1984) who obtained almost similar estimates in this region (See Table 2.10)

Table 2.10. Kinetic energy for the Asian coast and eastern China for the period 12–16 December 1979. All numbers shown are volume integrals. $[k]$ is in units of J m^{-2} and others W m^{-2} (Cressman, 1984)

Area	k	$-[V \cdot \nabla\Phi]$	$-[\nabla \cdot Vk]$	$[E]$	$\partial[k]/\partial t$
Asian coast	52.6	46.6	−37.3	11.5	−1.2

The generation rate of $[k]$ is 46.6 W m^{-2}, equal to the mean value of the China Main and the Sea of Japan for Kung and Chan's case. The maximum of the generation rate is at jet stream levels. Cressman pointed out that the kinetic energy generation region of the China Main and coastal area is characterized by a large-scale direct solenoidal circulation or local Hadley cell, as well as by net strong sinking motions. The kinetic energy generation appeared as a result of a relatively small difference between the loss of potential energy by the sinking motion and a flux convergence of geopotential by upper tropospheric convergence in this region. The available potential energy of this region centered on the East Asian coast fuels the persistent acceleration of upper-level jet stream in this region (entrance region). On the other hand, production of kinetic energy over the nearly steady low-level cold flows from Siberia and northern China toward tropical oceans may well be of great importance.

There is some general net downward transport of generated kinetic energy from the mid- and lower troposphere in this area, as indicated by $-[\partial \omega k/\partial p]$. As presented in Table 2.9, 43.29 W m^{-2}, which is 72% of the generation 60.25 W m^{-2}, is exported out of the China Main through transport. Table 2.10 shows a similar percentage of the export rate (80%). This is in sharp contrast to the Sea of Japan, where only 1.89 W m^{-2}, that is 5% of the generation 36.39 W m^{-2},is exported out of the region. Because of the large export of generated kinetic energy from the China Main and negligibly small export from the Sea of Japan, the dissipaton $[E]$ over the China Main is only about half of the $[E]$ over the Sea of Japan despite a much stronger generation over China Main. This apparent role of the China Main region to supply kinetic energy to other regions and the self-supporting nature of the kinetic energy in the sense of the near balance of $-[V \cdot \nabla \Phi]$ and $[E]$ over the Sea of Japan, may be associated with the types of cyclonic flow dominant in these regions in winter.

Table 2.11 compares energy budgets over the China Main and the area of AMTEX based on energetics analysis by Kung (1977) during the 1974 AMTEX observation period, 14–27 February. The prevailing flow pattern during this period is approximately the same for the China Main and the East China Sea. It is readily noted that $[k]$ for both region is very high. However, the strong generation $-[V \cdot \nabla \Phi]$ over the AMTEX area is only about half of the generation over the China Main. A notable difference is also found in the large export of kinetic energy from the China Main and the import to the AMTEX area by $-[\nabla \cdot Vk]$. The dissipation $[E]$ is about three times larger over the AMTEX area than over China Main. The contrast between the China Main and AMTEX area is similar to that between the China Main and the Sea of Japan. It thus appears that the portion of the Asian winter monsoon over the general area of the China Main, East China Sea, and the Sea of Japan intensely generates kinetic energy in the upper troposphere. The strongest generation is over the China Main from which more than half of the generated kinetic energy is exported. The intense dissipation takes place in the eastern edge of this general area, such as the Sea of Japan and the AMTEX area. The generation in this region is not as strong as that over the China Main; however, very little kinetic energy generated is exported and most of it appears to be consumed in the dissipation process in the area.

During AMTEX 75 the upper air circulation was much more intense and a large trough system was visibly more dominant over the area than during AMTEX 74. The area was also dominated by a series of cold air outbreaks during AMTEX 75. Kung's (1977) computations of the vertical totals of $-[\nabla \cdot Vk]$, $-[V \cdot \nabla \Phi]$ and $[E]$ during this

Table 2.11. Comparison of the 14–27 February 1974 energy budgets over the China Main and the AMTEX 1974 area. $[k]$ is in units of 10 J m^{-2} and others in W m^{-2} (Kung, 1974).

Pressure layer (hPa)	$[k]$	$-[\nabla\cdot Vk]$	$-[V\cdot\nabla\Phi]$	$[E]$
China Main				
100–400	35.99	−31.24	34.31	2.33
400–700	9.37	−4.52	7.61	3.43
700–sfc	1.29	−0.22	3.86	3.02
Total	46.65	−35.98	45.78	
AMTEX 74				
100–400	30.78	10.65	13.88	18.72
400–700	10.20	−1.17	6.94	9.58
700–sfc	1.38	−0.19	1.89	2.07
Total	42.36	9.29	22.71	30.97

period are −25.07, 50.30, and 29.33 W m^{-2}, respectively. The values of generation and transport are closer to those over the China Main as shown in Table 2.11. The AMTEX 75 area showed an export of kinetic energy during AMTEX 75. Shen and Smith's (1977) kinetic energy budget analyses during a subperiod of AMTEX 75 also indicate a significant generation and export of kinetic energy from the AMTEX area during non-MCC (mesoscale cellular convection) periods.

It is very interesting to further discuss the energetics characteristics for cold air outbreaks in East Asia (Kung and Chan, 1981; Lu, Ding and Wen, 1988). Table 2.12 shows a comparison of the vertical totals of energy variables between the outbreak and non-outbreak periods. It is evident that there is no significant difference of energy variables in Siberia and northeastern Asia. The basic features of kinetic energy balance observed in the gross winter energy budgets of these regions remain essentially unchanged, despite the existence of an active anticyclone over Siberia and a strong cyclone over northestern Asia. However, a substantial increase in the energy transformations over the China Main and the Sea of Japan during the period of cold air ourbreaks was observed. Lu *et al.* obtained a similar result, with a particularly significant increase observed in the southern China region (See Table 2.13). Over the China Main, the kinetic energy level $[k]$ only increases by 19% in outbreak period from the non-outbreak period, but the generation $-[V\cdot\nabla\Phi]$ and dissipation $[E]$ increases by 61 and 171%. Over the Sea of Japan, $[k]$ increases by 17%, but the increase of $-[V\cdot\nabla\Phi]$ and $[E]$ are 122 and 152%, respectively.

It is interesting to note that there is a 13.37 W m^{-2} increase of horizontal transport $-[\nabla\cdot Vk]$ out of the China Main during the outbreak period; but there is no such increase over the Sea of Japan, despite the grossly intensified generation. It appears that the contrast of energetics between the outbreak and nonoutbreak periods over the China Main and the Sea of Japan area is that of the strength of energy transformation. During the outbreak period, energy transformations become extremely intense (78.17 W m^{-2}). Over North America, Chen and Bosart (1977) computed a composite quasi- Lagrangian energy budget of polar air penetration into the Caribbean Sea from off the North American continent, giving 64.6 W m^{-2} as the

Table 2.12. Comparison of vertical totals of winter energy variables from the surface to 100 hPa for the cold air outbreak and nonoutbreak periods. [k] is in units of 10 J m^{-2} and the others in W m^{-2} (Kung and Chan, 1981)

Region	Period	[E]	$-[\nabla \cdot Vk]$	$-[V \cdot \nabla\Phi]$	[E]
Siberia	outbreak	17.53	− 0.61	− 8.25	− 5.15
	nonoutbreak	20.57	2.84	− 7.70	2.07
NE Asia	outbreak	22.48	011.29	11.74	3.45
	nonoutbreak	21.60	− 8.03	11.94	27.36
China Main	outbreak	50.72	−51.56	78.17	10.06
	nonoutbreak	42.54	−38.19	48.43	57.25
Sea of Japan	outbreak	48.93	− 1.47	56.32	57.25
	nonoutbreak	41.83	− 2.12	25.40	−22.74

Table 2.13. The areal mean and vertically intergrated kinetic energy budgets for 5 cold air outbreaks of the 1979–1983 winters in the four regions. Units: W m^{-2} (Lu, Ding and Wen, 1988)

Region	$\partial[k]/\partial t$	$-[V \cdot \nabla\Phi]$	$-[\nabla \cdot Vk]$	$-[\partial \omega k]/\partial p$	[E]
A (80–120° E, 42.5–57.5° N, Siberia)	−2.90	6.16	−3.8	10.12	5.35
B (95–122° E, 35–47.5° N, Mongolia and northern China)	−3.87	37.12	−32.19	−0.01	8.78
C (105–125° E, 22.5–35° N, southern China)	2.21	48.89	−48.13	0.04	−0.96
D (105–130° E, South China Sea)	2.09	14.48	−11.57	−0.01	0.83

generation during the period of intensified cyclogenesis. Fuelberg and Scoggins (1980) obtained a generation of 70.5 W m^{-2} for a cold air outbreak over the eastern United States during a 36 h period. In view of the stronger jet stream in the area, a magnitude of 78.17 W m^{-2} for $-[V \cdot \nabla\Phi]$ over the China Main and 57.25 W m^{-2} for [E] over the Sea of Japan during the cold air outbreak seems consistent with these large transformations computed for major cyclones over North America. Finally, it should be pointed out that although there are significant differences in the magnitudes of the generation and the dissipation between the outbreak and nonoutbreak periods, the basic energetics features of each region remain unchanged throughout these two periods, indicating the basically stable energetics regimes of these regions in constituting the prevailing system of the Asian winter monsoon.

The above-described kinetic energy equation is applied to the general flows in a limited region. It is highly desirable to understand the characteristics of eddy motions in the atmosphere and their interaction with the mean flow. Thus, the eddy kinetic energy equation should be used for this purpose. Murakami (1977) used this type of kinetic energy equation to examine changes in regional energetics over the North Pacific (region 1), the South China Sea (region 3), tropical central North Pacific (region 2), and the Timor Sea (region 4) for active and inactive cold surge periods. The results have indicated that the upper-level zonal flow over region 1 is found to be barotropically stable, namely, zonally averaged flows gain energy from eddy motions through barotropic interaction processes. At 200 hPa, the zonal mean flows are more barotropically stable during the active cold surge period than the inactive cold surge period. Of interest is a large export of eddy kinetic energy across the southern boundary into the tropical North Pacific during the active period. In contrast to region 1, zonally averaged flows are barotropically unstable over region 2, where the oceanic trough dominates the upper troposphere. During the active period, eddy kinetic energy in the upper troposphere is maintained almost equally by barotropic instability and lateral coupling processes. The energy import across the northern boundary is substantially larger during the active period as compared to the inactive period. Over the South China Sea (region 3), the conversion of eddy available potential to kinetic energy was more pronounced than usual. An inverse relationship is found in the change of the baroclinic conversion processes over region 4.

Lau and Lau (1984) have also analysed the regional energetics of cold surges during the winter MONEX, but using a somewhat different approach (semi-Lagrangian method). Figure 2.41 shows the temporal variation for ensemble averages of $A_E + K_E$, C_A, C_K^E, and C_K^W for midlatitude disturbances accompanying cold surges. It is evident that the evolution of the individual quantities A_E *and* K_E is similar to that of their sum $A_E + K_E$. There is a weak indication that the occurrence of the maximum in A_E leads that of the maximum K_E by about half a day. The total eddy energy peaks on day +1. The development phase of the disturbances prior to that time is accompanied by active baroclinic processes (C_A) and, to a lesser extent, by barotropic processes in the wastern portion of the storm track (C_K^W). The magnitudes of A_E, K_E, C_A and C_K^W drop noticeably after day +1, whereas the dissipative barotropic processes in the eastern part of the storm track (C_K^E) remain active through day +2. These observations are consistent with other studies (Kung and Chan, 1981).

2.8. Explosive Cyclogenesis off the East Asian Coast

The climatology of extratopical cyclones has indicated that off the western and eastern coasts of the North Pacific and North Atlantic oceans, there often exist explosively deepening extratropical cyclones or "meteorological bomb" in the cold season which are defined as a development of the deepening rate of at least 12 hPa in 12 h at 45° N (Sanders and Gyakum, 1980). A notable meteorological bomb QEII assumed a surface pressure fall of about 60 hPa in 24 h (Gyakum, 1983 a, b). At the same time, the wind speed rapidly inreases up to about 30 m s^{-1} or greater. Sanders and Gyakum (1980) found that pronounced frequency maxima of explosively developing cyclones, predominantly a maritime, cold season event, occur in the westernmost portions of both the Atlantic and Pacific oceans within and just to the north of the warm waters of the Gulf Stream and Kuroshio current, respectively (Fig. 2.42). In fact, there are also a number of explosive events that occur over the North American landmass. Over the East Asian mainland continent, one may sometimes

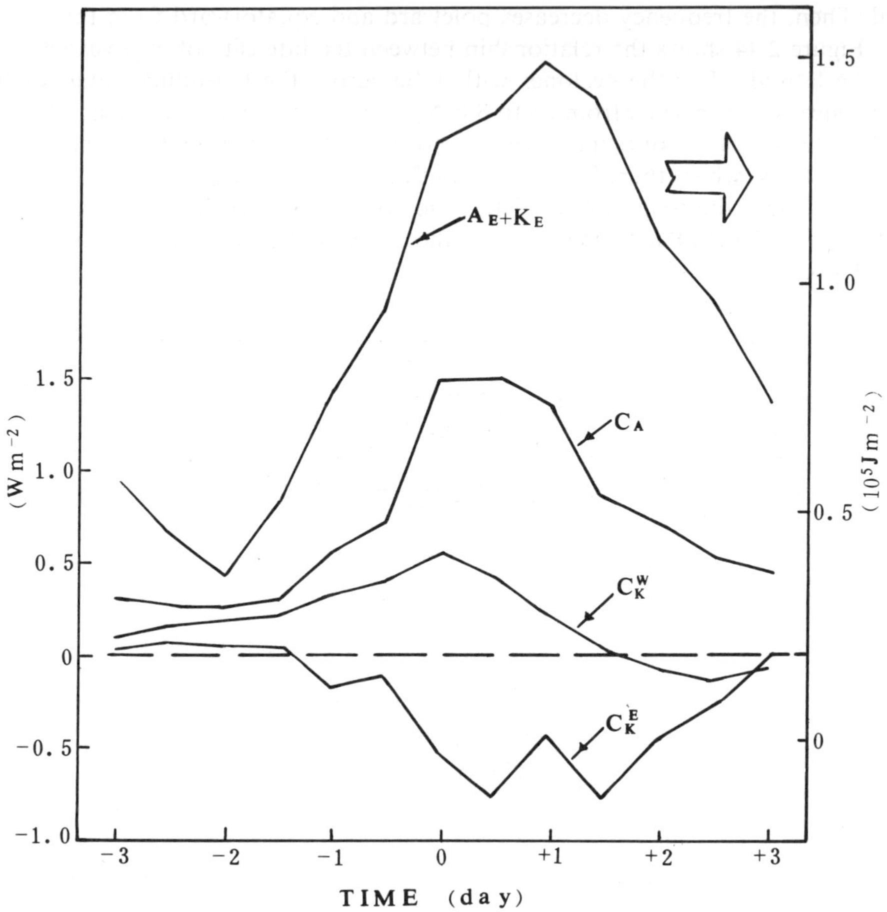

Figure 2.41. Temporal variation of ensemble averages of $A_E + K_E$, C_A, C_K^E and C_K^W for mid-latitude disturbances accompanying cold surges. A_E and K_E denote the available potential energy and kinetic energy of the high passed eddies; C_A and C_K are measures of the conversion of avialable potential energy and kinetic energy, respectively, from the quasi-stationary flow (represented by the low-passed data) to the high-passed eddies. Day 0 denotes the onset day of 11 composite cold surges, and days ± 1, ± 2, ± 3,... are the day numbers prior and post to the onset of cold surges. The abscissa runs along the storm track (Lau and Lau, 1984).

observe the explosive development of extratropical cyclones. Tian and Liu (1988) analysed a spring cyclone that rapidly developed in northern China, with a surface pressure fall of 22.5 hPa in 18 h, exceeding the mean deepening rate of 1 hPa in 1 h. It can be seen that in this region (Figure 2.43); the explosive marine cyclones occur to the north of 25° N, predominantly in the midlatitude between 30° N and 55° N. Of 388 explosive cyclones for 20 years, 372 cases occurred in the region of 30–55° N, accounting for 95.9% of the total. A particular spatial cluster of occurrence frequency (299 cyclones) was found to be in the region of 35–50° N, accounting for 77.1% of the total. Then, the frequency decreases poleward and equatorward from this latitude belt. Figure 2.44 shows the relationship between the intensity of explosive cyclones and the latitude. For the cyclones with 1 Bergeron, the latitudinal range of their occurrence is rather wide, from 25 to 80° N, with the maximum found at 45–50° N. As the intensity of cyclones increases, the position of occurrence of cyclones tends to move to 40° N, where there is a strong confluent upper level jet streams.

Figure 2.43 is the geographical distribution of explosive cyclones over the western portion of the North Pacific in winter for the 20 years of 1968–1987 (Ouyang, Lu and Hu, 1990).

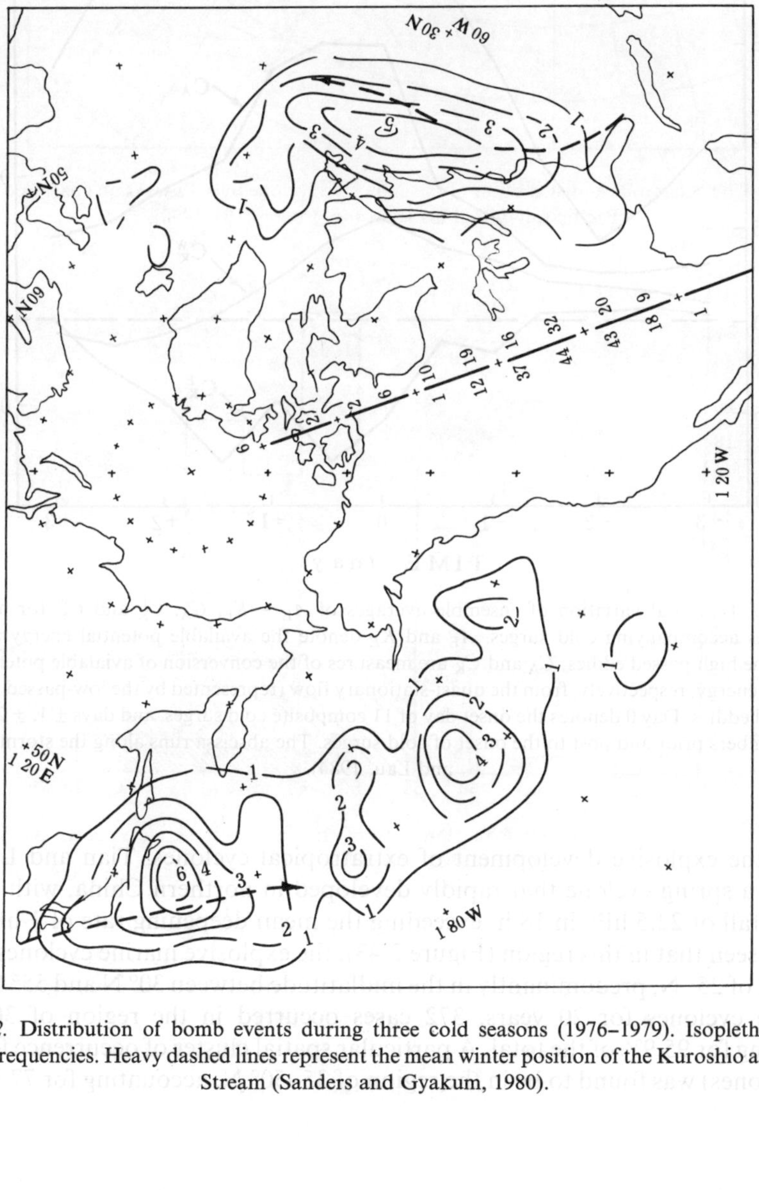

Figure 2.42. Distribution of bomb events during three cold seasons (1976–1979). Isopleths represent smoothed frequencies. Heavy dashed lines represent the mean winter position of the Kuroshio and the Gulf Stream (Sanders and Gyakum, 1980).

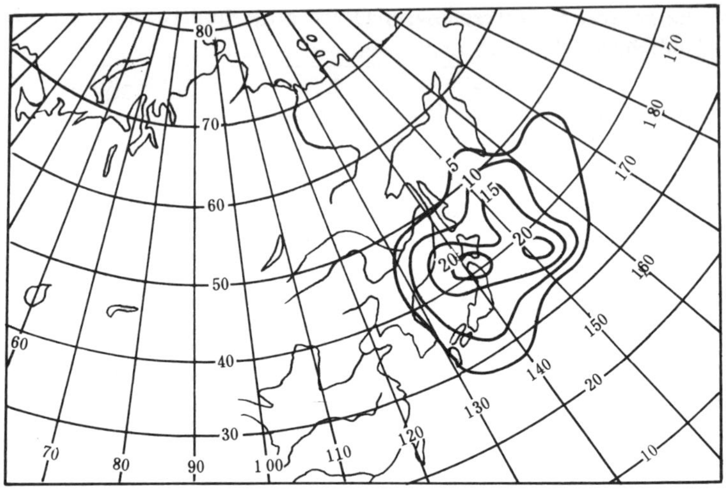

Figure 2.43. The geographical distribution of the explosive cyclone over East Asian coast and Northwest Pacific for 1968–1987 (Ouyang, Lu and Hu, 1990).

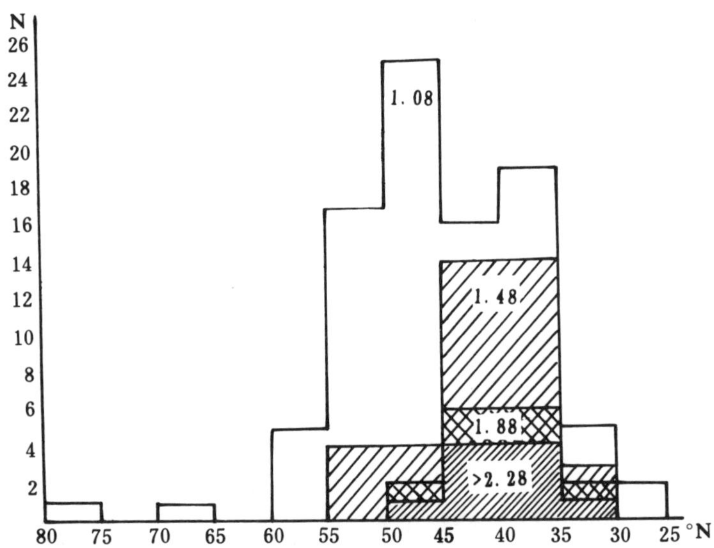

Figure 2.44. The relationship between the intensity of the explosive marine cyclone and the latitude (Ouyang, Lu and Hu, 1990).

The annual variation of occurrence frequency of explosive marine cyclones is very marked, with the frequencies for January-April and for September-December considerably higher than those for other months (Table 2.14) (Murty *et al.*, 1983). The maximum occurrence frequency is observed in October. Li and Ding (1989) indicated that the peak frequency of cyclones over the North Pacific occurred in December-February, in which, on an average, one cyclone may form every two days.

Table 2.14. Number of occurrences of explosive cyclones over the North Pacific during January-December of 1955–1977 (Murty *et al.*, 1983)

Month	Jan.	Feb.	Mar.	Apr.	May	June	July	Aug.	Sep.	Oc.t	Nov.	Dec.
No. of occurences	43	35	36	32	16	12	5	3	27	67	55	51
Monthly mean	1.9	1.5	1.6	1.4	0.7	0.5	0.2	0.1	1.2	2.9	2.4	2.2

Ouyang et al. have studied the distribution of the intensity of explosive marine cyclones. The number of cyclones smaller than 1.4 Bergeron (general intensity) is 294 in total, accounting for 76% of 388 cases for 20 years. The number of cyclones of the moderate intensity (1.5–1.9 Bergeron) is 69, accounting for 18%. The number of intense cyclones (> 2.0 Bergeron) is 25, accounting for 6%. During the period of 20 years, an extremely intense case of 3 Bergeron occurred during the period of 00:00 GMT, 19–21 December, 1981 over the oceanic area from the southern Korean Peninsula to Kamtchaka. The central surface pressure of the cyclone dropped from 1016 hPa at 00:00 GMT, 19 December to 989 hPa at 00:00 GMT, 20 December, a deepening rate of 27 hPa in 24 h (intensity of 1.5 Bergeron). At 00:00 GMT, 21 December, the surface pressure further fell down to 932 hPa, with a deepening rate of 57 hPa in 24 h (intensity of 3 Bergeron). The total surface pressure fell in 48 h amounted to 84 hPa.

Figure 2.45 shows the frequency distribution of the maximum pressure fall in 24 h. For 388 cases, the mean pressure fall in 24 h at the explosive developmental stage is -23.8 hPa day^{-1}, slightly greater than Roebber's value of -22.3 hPa day^{-1}. The cyclones have the deepening rate of 21–25 hPa day^{-1}, accounting for 38.1% of the total (Ouyang, Lu and Hu, 1990).

Figure 2.46 is the central surface pressure at the explosive developmental stage, showing that most of the cyclones have a central surface pressure of 975–980 hPa, with a mean value of 970.3 hPa, which is equivalent to that of tropical cyclones of moderate intensity.

From the viewpoint of the horizontal scale, the marine extratropical cyclone may fall into two categories: synoptic and subsynoptic-scale cyclones that can undergo explosive development if the conditions are adequate. Observations indicated that there are two kinds of subsynoptic-scale marine cyclone; one of them is the so-called 'intermediate-scale' cyclone that may develop into an intense extratropical cyclone. This problem will be addressed later.

The other kind of marine cyclone is known to be the polar low that mainly occurs in the polar cold air mass behind the cold front over the North Pacific and North Atlantic oceans. Then this polar low often develops into an intense cyclone through the instant occluded process.

Sanders and Gyakum (1980) report that the explosive cyclogenesis was generally found about 400 nautical miles downstream from a mobile trough at 500 hPa which had typically migrated for several days across the upwind continent before triggering the explosive surface event. The event almost always occurs poleward of the main belt

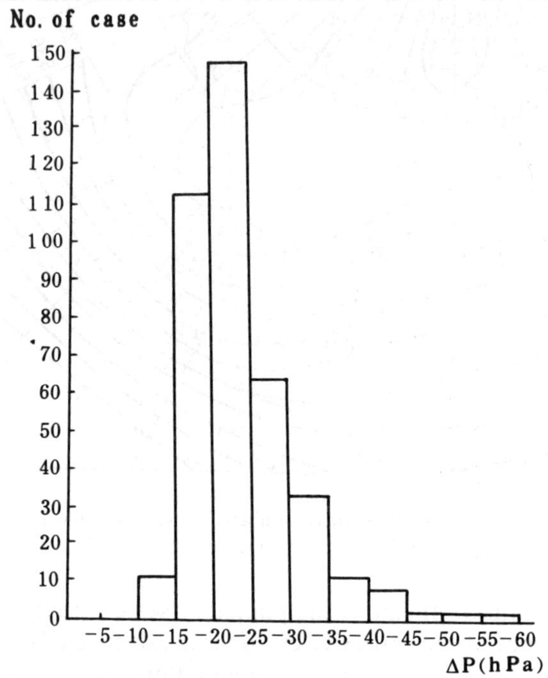

Figure 2.45. The frequency distribution of the maximum pressure fall in 24 h (Quyang, Lu and Hu, 1990).

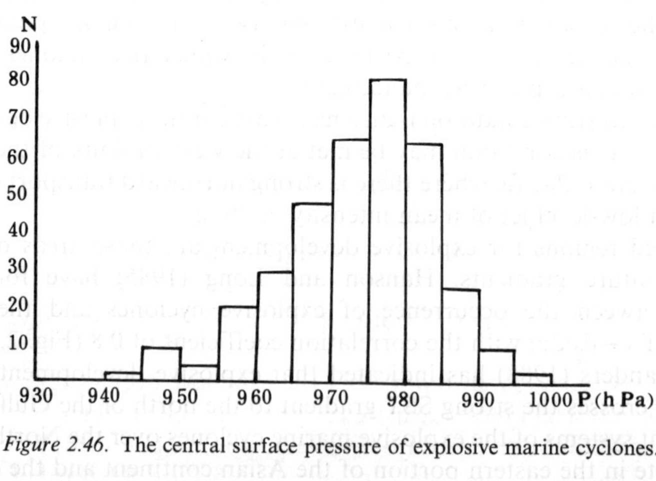

Figure 2.46. The central surface pressure of explosive marine cyclones.

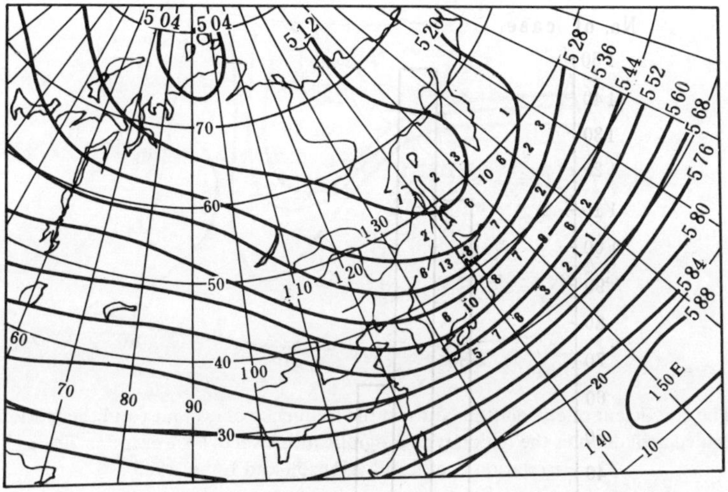

Figure 2.47. The explosive marine cyclones for December-February of 1968–1987 and the mean 500 hPa topographic field for January (Ouyang, Lu and Hu, 1990).

of westerlies and within or ahead of the planetary scale trough. Figure 2.47 clearly suggests the occurrence of the explosive cyclones in advance of the East Asian major trough and to the north of the westerly jet stream (Ouyang, Lu and Hu, 1990; Li and Ding, 1988).

The intense baroclinic zone is also favorable for the explosive development of marine cyclones. The East Asian region is highly baroclinic, marked by a large vertical wind shear crowned by the strong wintertime jet maximum and frequent southeastward outbreaks of cold air. As the upper level, the shortwave trough moves eastward, the cold air spreads behind the trough. At low-level, this event reflects the southeastward push of the Siberian high; at the same time, there is the vigorous warm air advection at the western flank of the subtropical high over the western Pacific. Therefore, in the coastal area and neighboring oceanic region an apparent low and middle-level baroclinic zone may be created, in which the cyclones are easy to maintain and gain an explosive enhancement.

An adequate moisture condition is also necessary for the explosive development of marine cyclones. This condition may be met at the western flank of the subtropical high over the western Pacific where there is strong northward transport of warm and moist air and a low-level jet of mean intensity of 20 m s^{-1}.

The preferred regions for explosive development are those areas of strong sea surface temperature gradients. Hanson and Long (1985) have found a close relationship between the occurrence of explosive cyclones and the meridional gradient of SST ($-\partial s/\partial y$) with the correlation coefficient of 0.8 (Fig. 2.48).

Recently, Sanders (1988) has indicated that explosive development of cyclones occurs when it crosses the strong SST gradient to the north of the Gulf Stream.

The incipient systems of the explosive marine cyclones over the Northwest Pacific mostly originate in the eastern portion of the Asian continent and the coastal area.

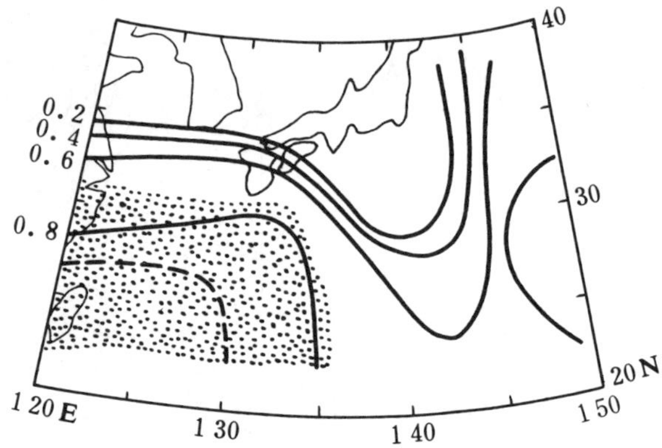

Figure 2.48. The coefficient of correlation between the frequency of cyclones and the meridional gradient of SST. The shaded area denotes the region with a significance level of 95% or more. The isolines represent the correlation coefficient (Hanson and Long, 1985).

After the cyclones form, they often drift northeastward or eastward and rapidly develop into an explosive cyclone. Finally, the cyclones move over the Kamtchaka Peninsula and Aleutian Islands. In recent years, numerous investigators have examined the physical factors leading to the genesis and explosive development of extratropical marine cyclones, including the advection of vorticity and thermal advection, sensible and latent heating, air and sea interaction (sensible and latent heat fluxes at the interface), the ageostrophic acceleration effect, the baroclinic instability and so on. However, no definite conclusions have been drawn concerning primary factors or mechanisms that work for the explosive deepening of cyclones. Recently, Li and Ding (1989) have made a diagnostic study of an explosively deepening cyclone over the Northwest Pacific ocean that occurred on 6–9 February, 1983 (Fig. 2.49). An explosive development occurred on 8 January, with the surface minimum pressure dropping by 33.9 hPa in 24 h, equivalent to 1.5 Bergeron, and the rapid increase in low-level wind speeds greater than 20 m s^{-1}. Large-scale conditions created the marked coastal frontogenesis that was favorable for the explosive development of an incipient cyclone. An eastward moving shortwave trough (Fig. 2.49 (c)) brought the cold air from Mongolia, while the thermal advection at the western flank of the subtropical high that was carried by a significant low-level jet brought the warm and moist air in juxtaposition with the cold air, thus producing a strong barolinic zone off the coastal zone in which the cyclone was embedded. The considerable potential instability was observed in the eastern and northeastern sectors of the cyclone, and this favors the growth of cumulus convection there. The principal results of using the seven-level nonlinear balance ω-equation and the Sawyer-Eliassen equation for the secondary circulation, have indicated that large-scale latent heat release is very crucial to explosive development of cyclones. If there is sufficient moisture transport, positive feedback process between the ascent of air and large-scale heating would work. The cumulus heating and transport of momentum and heat in the area of the cumulus played an important role during the explosively deepening stage. Thermal advection

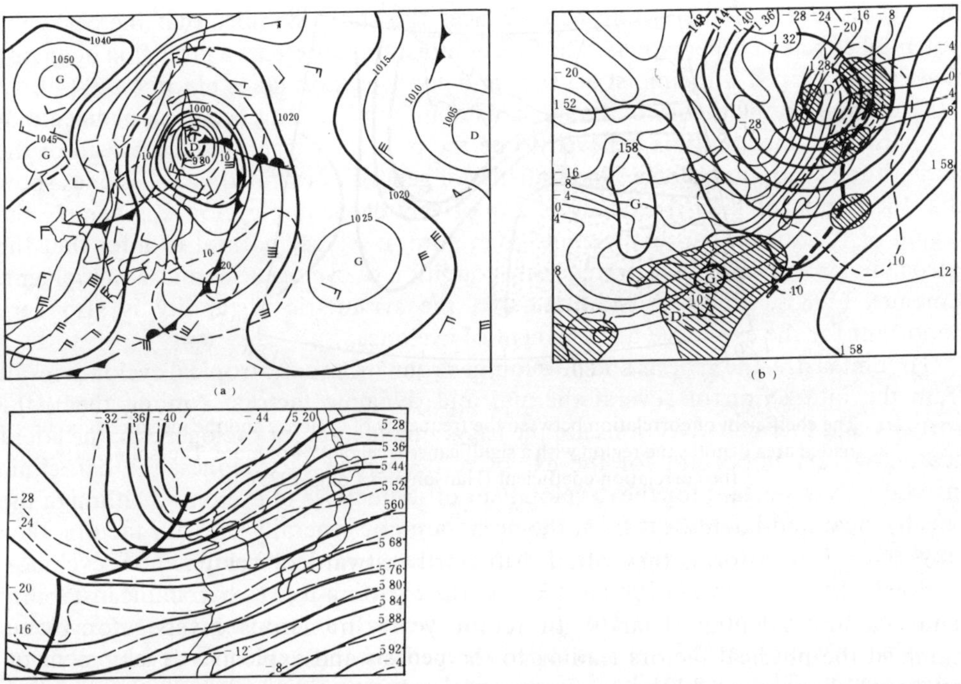

Figure 2.49. (a) The surface weather map at 12:00 GMT, 8 January 1983. Solid lines represent isobars and dashed lines isotachs (m s^{-1}). (b) The 850 hPa weather map at 00:00 GMT, 8 January 1983. The thin dashed lines denote isotachs with intervals of 10 m s^{-1}, and the bold dashed lines, the axis of the low-level jet; the shaded area denotes the region of $(T-T_d) \leq 1$ °C. The solid lines represent the isolines of geopotential height, with the intervals in decameters. Thin lines, isotherm lines. (c) Same as (b), but for the 500 hPa weather map at 00:00 GMT, 7 January 1983. (Li and Ding, 1989).

is the initial triggering condition for large-scale heating and the conditional instability for cumulus convection.

A recent case study of East Asian coastal cyclogenesis was performed by Chen and Dell'Osso (1985) using the ECMWF numerical model. They have also indicated the overwhelming importance of latent heat release in the proper simulation of the cyclone. The baroclinic upper-level wave acted as a trigger to initiate the cyclonenesis. Full development was critically dependent upon the release of latent heat in the cyclone circulation.

Chen *et al.* (1983) demonstrated that the release of latent heat and moisture and heat fluxes from the ocean were very important in producing the observed development of the AMTEX 75 cyclone. Their work indicated that, in this case, a substantial depression could develop even in the absence of significant upper-level support if latent heat were available.

Relative to the latent heat, fluxes of moisture and heat from the warm oceanic surface produce what could be termed as secondary effects. The heat introduced into the lowest levels of the atmosphere reduces the static stability that increases the potential effects of other effects. The flux of moisture represents a large flux of energy that can be tapped if the moist air rises and condensation takes place.

According to the theory of dynamic instability, it has long been recongnized that the baroclinic instability is believed to be the primary triggering mechanism of the midlatitude synoptic systems. The statistics of genesis of extratropical cyclones show that the region of high frequency of explosively deepening cyclones is namely the region of the strong baroclinic zone. Therefore, it is very natural to infer that the baroclinic process may be a prerequisite condition of cyclogenesis and development. Emanuel (1983), however, indicates that the symmetric instability is also very important for the explosive development of cyclones.

To summarize, the genesis and development of marine extratropical cyclones result from the interaction of several thermal and dynamic factors. Among them, the vorticity and thermal advections are the triggering factors for cyclogenesis; the latent heating is very important for the explosive deepening of cyclones; the baroclinic instability is important for the development of some types of cyclones. Although the sensible heat and latent heat from the ocean are not generally primary factors, they may create the environmental fields favorable for the rapid development of cyclones.

Besides the typical baroclinic wave in which growth is due to baroclinic instability triggered by an upper-level trough in the westerlies, another type of cyclonic disturbance was observed which is called the medium-scale cyclones or intermediate-scale cyclone. This cyclone has a horizontal wavelength of 1000–2000 km and is mostly confined to the lower half of the troposphere. Cyclones of this type are usually fast-developing systems which produce large amounts of precipitation and generate strong low-level winds. They are considered one of most severe weather events affecting the heavily populated coastal areas.

Table 2.15 shows the seasonal variations of the cyclogenesis frequency of these two types (Nitta and Yamamoto, 1973). The interannual variability of the frequency is

Table 2.15. Frequency of cyclogenesis during the period 1966–1970 (Nitta and Yamamoto, 1973)

Intermediate-scale disturbance		Upper-level through passage	
Month	Frequency	Month	Frequency
Sept.	8.2	Sept.	9.2
Oct.	10.0	Oct.	10.0
Nov.	9.8	Nov.	13.2
Dec.	8.8	Dec.	15.2
Jan.	14.4	Jan.	13.4
Feb.	11.6	Feb.	−4.6
Mar.	11.8	Mar.	16.8
Apr.	13.8	Apr.	18.0
May	14.0	May	17.2
June	15.2	June	14.4
July	9.2	July	10.2
Aug.	10.0	Aug.	9.6

about 12% of the annual total for both types, and the frequency reaches a peak in the presummer months. A close relationship was found between the cyclogenesis of intermediate-scale disturbances and the monthly mean 500 hPa flow pattern. The frequency of the cyclogensis increases when the mean pattern shows a zonal or flat flow. The frequency of cyclogenesis decreases when the upper level troughs anchor to the west of Japan and the cyclogenesis of the baroclinic waves becomes active.

Chen *et al.* (1985) observed that a major winter cyclone developed from a medium-scale cyclone over the ocean northeast of Taiwan during the initial phase of the AMTEX 75 (Figs. 2.50 and 2.51). On 13 February 1975, an anticyclonic system

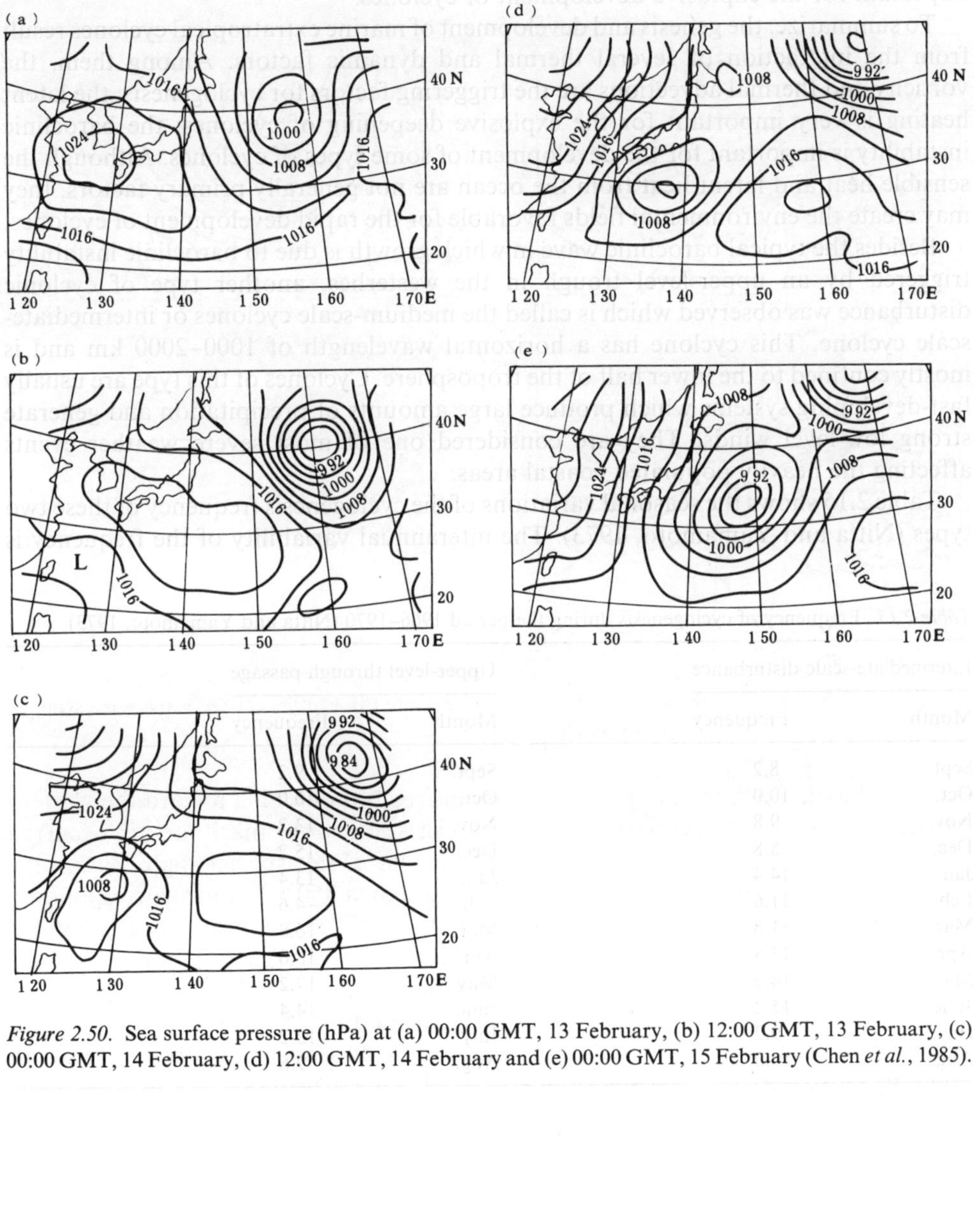

Figure 2.50. Sea surface pressure (hPa) at (a) 00:00 GMT, 13 February, (b) 12:00 GMT, 13 February, (c) 00:00 GMT, 14 February, (d) 12:00 GMT, 14 February and (e) 00:00 GMT, 15 February (Chen *et al.*, 1985).

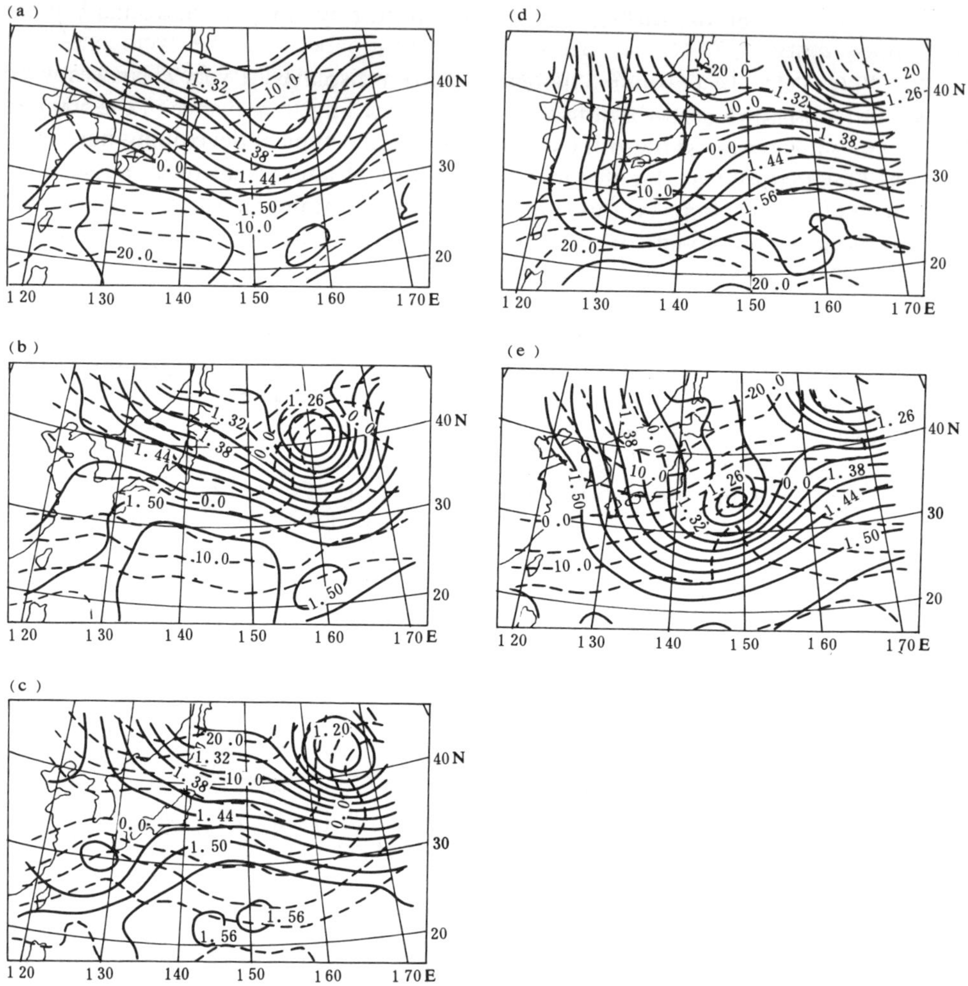

Figure 2.51. 850 hPa height (solid lines) (km) and temperature (dashed lines) (°C) at the same synoptic times as Fig. 2.50 (Chen *et al.*, 1985).

associated with a continental polar air mass moved out of China toward the western Pacific. A northward transport of warm moist subtropical maritime air occurred along the course of the Kuroshio current at the rear of this surface high. Within the region of the warm and moist air, a shortwave trough at 850 hPa and a corresponding surface and pressure trough developed at 12:00 GMT, 13 February (Fig. 2.50 (b)). The disturbance was a very shallow system. During the initial period of this system's development, there was no evidence that forcing by an approaching synoptic-scale system was involved. The distribution of zonal wind did not suggest the existence of barotropic or baroclinic instability in the lower troposphere. By 00:00 GMT, 14 February (Fig. 2.50 (c)), a closed circulation had formed at the surface. At this stage, the coupling between the surface disturbance and the synoptic-scale trough became more discernible.

By 12:00 GMT, 14 February, the entire system displayed a very clear westward tilting of the pressure perturbations with the thermal ridge in advance of the 500 hPa trough, i.e., the system showed the characteristics of a typical developing baroclinic system. This feature was due to the diabatic effect of latent heat release associated with the low-level cyclone. At this stage, the cyclone exhibited a warm-core structure aligned with the maximum concentration of moisture. It seems clear that the CISK mechanism enhanced the medium-scale cyclogenesis.

The combined effects of several physical processes may be involved in the genesis and subsequent development of this cyclone into a large-scale system. These include the heat flux from the Kuroshio current that rapidly destabilized the lower layer of the polar air mass, mild lifting due to the northward warm advection that caused latent heat release and the formation of a low-level pressure trough, further release of latent heat in the region of the low-level disturbance resulting in the development of the medium-scale cyclone, and the vertical coupling of the cyclone and incoming upper-level synoptic system leading to the development of a severe extratropical cyclone.

The later study of this cyclone (Chang, Perkey and Chen, 1987) further documented the importance of latent heat release which seemed to contribute to the vertical decoupling and the decrease in zonal baroclinicity, thus then to strengthen the ascent aloft and convergence below in the vicinity of the cyclone. Strong ascent enhanced the thermally direct east-west circulation across the upper trough, while convergence resulted in further latent heating and, consequently, larger ageostrophic components and convergence near the surface. Thus, kinetic energy was generated to support the system's development despite the decoupling in the vertical and the weakening of the horizontal baroclinicity.

The above analyses suggested that the observed explosive oceanic cyclogenesis may be a result of latent heat enhanced regional baroclinic processes in which several physical factors cooperatively work. The numerical study of this cyclone using a fine-mesh numerical model, provided the quantitative assessment of the physical factors described above (Chen *et al.*, 1983), stressing the importance of the diabatic effects of the latent heat release, boundary layer moisture and heat flux from the ocean surface.

The East China Sea has been identified as a major source region of oceanic cyclones observed over the Pacific. Among them, some may gain an explosive deepening under favorable environmental conditions and combined effects of several physical processes. Among the physical factors, the latent heat release has a profound impact upon the development of cyclones.

References

Blackmon, M. L., Wallace, J. M., Lau, N. C., Mullen, S. L., 1977: "An observation study of the Northern Hemisphere wintertime circulation". *J. Atmos. Sci.*, **34**, 1040–1053.

Bodurtha, F. T., 1952: "An investigation of anticyclogenesis in Alaska". *J. Appl. Meteorol.*, **7**, 118–125.

Boville, B. A., 1985: "The thermal balance of the NCAR community climate model". *J. Atmos. Sci.*, **42**, 695–709.

Boyle, J. S., Chen, T.-J., 1987: "Synoptic aspects of the wintertime East Asian monsoon". pp. 125–160 in Chang, C.-P., and Krishnamurti, T. N., eds.: *Monsoon Meteorology*, Oxford Univ. Press, New York.

Boyle, J. S., 1982: "Synoptic study of the initiation of surges". Paper presented at Winter MONEX Workshop, Monterey, CA, 14–15 June, 1982.

Chan, P. H., 1982: "Energetics of winter synoptic features over East Asia". *J. Meteorol. Soc. Japan*, **61**, 426–438.

Chang, C. B., Perkey, D. J., Chen, W.-D., 1987: "Observed dynamic structure of an intense oceanic cyclone". *Mon. Wea. Rev.*, **115**, 1127–1139.

Chang, C. P., Lau, K. M., 1980: "Northeasterly cold surges and near equatorial disturbances over the winter MONEX area during December 1974. Part II: Planetary-scale aspect". *Mon. Wea. Rev.*, **108**, 298–312.

Chang, C. P., Lau, K. M., 1982: "Short-term planetary-scale interactions over the tropics and mid-latitudes during northern winter. Part I: Contrasts between active and inactive periods". *Mon. Wea. Rev.*, **110**, 933–946.

Chang, C. P., 1988: "Dynamics of winter monsoon surges". Lecture notes of the Second WMO/MOSTE Regional Workshop on Asian Winter Monsoon, Kuala Lumpur, 27 June–1 July, 1988, 1–8.

Chang, C. P., Millard, J. E., Chen, G. T. J., 1983: "Gravitational character of cold surges during winter MONEX". *Mon. Wea. Rev.*, **111**, 293–307.

Chang, C. P., Lim, K. G., 1985: "Tropical mid-latitude interactions over Asia and the western Pacific ocean during the 1983–84 northern winter". *Mon. Wea. Rev.*, **113**, 1345–1358.

Chen Qianjin, Huang Aifen, 1989: "Some features of the South China Sea cold surge during winter season". *Meteorol. Monthly*, **15**, 49–53.

Chen Shoujun, Dell'osso, L., 1987: "A numerical case study of East Asia coastal cyclogenesis". *Mon. Wea. Rev.*, **115**, 477–487.

Chen T. C., Chang, C. B., Perkey, D. J., 1983: "Numerical study of an AMTEX 75 oceanic cyclone". *Mon. Wea. Rev.*, **111**, 1818–1829.

Chen T. C., Chang, C. B., Perkey, D. J., 1985: "Synoptic study of a medium-scale oceanic cyclone during AMTEX 75". *Mon. Wea. Rev.*, **113**, 349–361.

Chen, T. G., Bosart, L. F., 1977: "Quasi-Lagrangian kinetic energy budgets of composite cyclone-anticyclone couplets". *J. Atmos. Sci.*, **34**, 452–470.

Chen T. S., 1988: "Synoptic patterns leading to cold surges over Southern China". Lecture notes for the Second WMO/MOSTE Regional Workshop on Asian Winter Monsoon. Kuala Lumpur, 27 June–1 July, 1988, 45–51.

Chou, Y. Y., 1985: *Medium-Range Weather Prediction*, Science Press, Beijing, 420 pp.

Cressman, G. P., 1981: "Circulation of the western Pacific jet stream". *Mon. Wea. Rev.*, **109**, 2450–2463.

Cressman, G. P., 1984: "Energy transformation in the East Asia-western Pacific jet stream". *Mon. Wea. Rev.*, **112**, 563–574.

Curry, J., 1987: "The contribution of radiative cooling of the formation of cold-core anticyclones". *J. Atmos. Sci.*, **44**, 2527–2592.

Dallavalle, J. P., Bosart, L. F., 1975: "A synoptic investigation of anticyclones accompanying North American polar air outbreaks". *Mon. Wea. Rev.*, **103**, 941–957.

Ding, Y. H., Krishnamurti, T. N., 1987: "Heat budget of Siberian high and the winter monsoon". *Mon. Wea. Rev.*, **115**, 2428–2428.

Ding, Y. H., 1990: "A statistical study of winter monsoons in East Asia". *J. Tropical Meteorol.*, **6**, 119–128.

Ding, Y. H., 1990: "Build-up, air mass transformation and propagation of Siberian high and its relation to cold surge in East Asia". *Meteorol. Atmos. Phys.*, **44**, 281–292.

Ding, Y. H., 1991: "The planetary-scale effect of outbreaks of cold airs over East Asia". *J. Appl. Meteorol.*, **2**, 124–131.

Emanuel, K. A., 1983: "On assessing local conditional symmetric instability from atmospheric soundings". *Mon. Wea. Rev.*, **111**, 2016–2033.

Fuelberg, H. E., Scoggins, J. R., 1980: "Kinetic energy budget during strong jet stream activity over the eastern United States". *Mon. Wea. Rev.*, **108**, 69–77.

Hanson, H. P., Long, B., 1985: "Climatology of cyclogenesis over the East China Sea". *Mon. Wea. Rev.*, **113**, 697–707.

Johnson, R. H., Zimmerman, J. R., 1986: "Modification of the boundary layer over the South China Sea during a winter MONEX cold surge event". *Mon. Wea. Rev.*, **114**, 2004–2015.

Joung, H. J., Hitchman, M. H., 1982: "On the role of successive downstream development in East Asian polar outbreaks". *Mon. Wea. Rev.*, **110**, 1224–1237.

Keyser, D., Shapiro, M. A., 1986: "A review of the structure and dynamics of upper-level frontal zones". *Mon. Wea. Rev.*, **114**, 452–499.

Krishnamurti, T. N., Gadgil, S., 1985: "On the structure of the 30 to 50 day mode over the globe during FGGE". *J. Atmos. Sci.*, **36**, 1880–1894.

Kung. E. C., Chan, P. H., 1981: "Energetics characteristics of the Asian winter monsoon in the source region". *Mon. Wea. Rev.*, **109**, 854–870.

Kung, E. C., Baker, W. E., 1975: "Energy transformations in mid-latitude disturbances". *Q. J. Roy. Meteorol. Soc.*, **101**, 793–815.

Kung, E. C., 1962: "Diurnal and long-term variation of the kinetic energy generation and dissipation of a five year period". *Mon. Wea. Rev.*, **95**, 593–606.

Kung, E. C., 1977: "Large-scale energy transformations in the intense winter monsoon over the Kuroshio region". *J. Meteorol. Soc. Japan*, **55**, 498–510.

Lau, K. M., Chan, P. H., 1983: "Short-term climate variability and atmospheric teleconnections from satellite-observed outgoing longwave radiation". Part I: Simultaneous relationships. *J. Atmos. Sci.*, **40**, 2735–2750.

Lau, K. M., Chang, C. P., Chan, P. H., 1983: "Short-term planetary-scale interactions over the tropics and midlatitudes". Part II: Winter period. *Mon. Wea. Rev.*, **111**, 1327–1388.

Lau, N. C., Lau, K. M., 1984: "The structure and energetics of midlatitude disturbances accompanying cold air outbreaks over East Asia". *Mon. Wea. Rev.*, **112**, 1309–1327.

Li, Changqing, Ding, Y. H., 1989: "A diagnostic study of an explosively deepening oceanic cyclone over the Northwest Pacific Ocean". *Acta Meteorol. Sinica*, **47**, 180–190.

Lim, H., Chang, C. P., 1981: "A theory of midlatitude forcing of tropical motions during winter monsoons". *J. Atmos. Sci.*, **38**, 2377–2392.

Lu Wentong, Ding Yihui, Wen Shigeng, 1988: "A study on the ageostropic wind during winter monsoon over East Asia". *J. Acad. Meteorol. Sci.*, **3**, 138–150.

Mohri, K., 1959: "Jet streams and upper fronts in the general circulation and their characteristics over the Far East (Part II)". *Geophys. Mag.*, **29**, 334–412.

Murakami, T., Unninayar, M., 1977: "Atmospheric circulation during December 1970 through February 1971". *Mon. Wea. Rev.*, **105**, 1024–1038.

Murakami, T., 1977: "Changes in regional energetics over the North Pacific, South China Sea and the Indonesian Sea during winter". *Mon. Wea. Rev.*, **105**, 1508–1520.

Murakami, T., 1978: "Winter circulations and wavenumber domain energetics at 200 hPa". *J. Meteorol. Soc. Japan*, **56**, 215–231.

Murakami, T., 1981: "Orographic influence of the Tibetan Plateau on the Asiatic winter monsoon circulation". Part I. Large-scale aspects. *J. Meteorol. Soc. Japan*, **59**, 40–66.

Murti, L. K., 1976: "Heat and moisture budgets over AMTEX area during AMTEX 75". *J. Meteorol. Soc. Japan*, **54**, 370–381.

Murty, T. S., McBean, G. A., McKee, B., 1983: "Explosive cyclogenesis over the Northeast Pacific Ocean". *Mon. Wea. Rev.*, **111**, 1131–1135.

Namias, J., Clapp, P. F., 1949: "Confluence theory of the high tropospheric jet stream". *J. Meteorol.*, **16**, 638–645.

Newton, C. W., 1959: "Axial velocity streaks in the jet stream: Ageostrophic "inertial" oscillations". *J. Meteorol.*, **16**, 638–645.

Ninomiya, K., 1972: "Heat and water-vapor budget over the East China Sea in the winter season". *J. Meteorol. Soc. Japan*, **50**, 1–17.

Ninomiya, K., 1975: "Large-scale aspects of air-mass transformation over the East China Sea during AMTEX 74". *J. Meteorol. Soc. Japan*, **53**, 285–303.

Ninomiya, K., Akiyama, T., 1976: "Structure and heat energy budget of mixed layer capped by inversion during the period of polar outbreak over Kuroshio region". *J. Meteorol. Soc. Japan*, **54**, 160–174.

Nitta, Ts., Yamamoto, J., 1973: "A diagnosis of the formation of intermediate-scale disturbances near Japan, the western Pacific and Southeast Asia". *Pap. Meteorol. Geophys.*, **24**, 289–309.

Nitta, Ts., 1976: "Large-scale heat and moisture budgets during the air-mass tranformation experiment". *J. Meteorol. Soc. Japan*, **54**, 1–14.

Nitta, Ts., So, S. S., 1980: "Structure and heat, moisture and momentum budgets of a convective mixed layer during AMTEX 75". *J. Meteorol. Soc. Japan*, **58**, 378–393.

Ouyang Ziji, Lu Meizhong, Hu Zhiming, 1990: "Some statistic characteristics of the explosive cyclones in Asia and Northwest Pacific". *J. Inst. Meteorol. Air Force*, **11**, 23–30.

Pan Hanming, Jia Xiue, Yang Xiuzhen, 1985: "The climatological features of outbreaks of cold air in China". Paper collection of Beijing Meteorol. Center, SMA, 120–131.

Pan, H.-L., Zhou, F. X., 1985: "The 10–20 day tropical-midlatitude interaction during winter monsoon season". *J. Meteorol. Soc. Japan*, **63**, 829–843.

Peterssen, S., 1956: *Weather Analysis and Forecasting*, 2nd edn., McGraw-Hill, New York.

Ramage, C. S., 1977: *Monsoon Meteorology*, Academic Press, New York, 296 pp.

Rasmussen, E. M., Carpenter, T. H., 1983: "Variation in the tropical sea surface temperature and surface wind field associated with the Southern Oscillation/El Ñino". *Mon. Wea. Rev.*, **110**, 345–384.

Saltzman, B., 1970: "Large-scale atmospheric energetics in the wave-number domain". *Rev. Geophys. Space Phys.*, **8**, 289–302.

Sanders, F., Gyakum, J. R., 1980: "Synoptic dynamic climatology of the bomb". *Mon. Wea. Rev.*, **108**, 1589–1606.

Shapiro, M. A., 1983: "Mesoscale weather systems of central United States", in: Anthes, R. A., ed, *The National STORM program: Scientific and Technological Bases and Major Objectives.* , UCAR, P. O. Box 3000, Boulder, CO 80307, 3.1–3.77.

Sheu, Y. T., Smith, P. J., 1977: "Kinetic energy budget analyses during a subperiod of the Air Mass Transformation Experiment 1975". *Mon. Wea. Rev.*, **105**, 1501–1507.

Staff Members, Academia Sinica, 1957–58: "On the general circulation over Eastern Asia. Part I, Part II, Part III". *Tellus*, **9**, 432–446; **10**, 58–75; 299–312.

Sumi, A., 1985: "A study on cold surges around the Tibetan Plateau by using numerical models". *J. Meteorol. Soc. Japan*, **63**, 377–395.

Tao, S. Y., 1955: "A summary of activities of cold airs in East Asia for winter half year". *The Central Meteorol. Observatory of China.*

Tao, S. Y., 1957: "A process of outbreak of cold air in East Asia during the breakdown of blocking situation". *Acta Meteorol. Sinica*, **28**, 63–74.

Uccellini, L. W., 1976: "Operational diagnostic applications of isentropic analysis". *Natl. Wea. Dig.*, **1, 4–12.**

Uccellini, L. W., Kocin, P. J., 1987: "The interaction of jet streak circulations during heavy snow events along the east coast of the United States". *Weather and Forecasting*, **2**, 289–308.

Webster, P. J., 1981: "Mechanism determing the mean and transient structure of the large-scale winter monsoon: cold surges". Report of the International Conference on Early Result of FGGE and Large-scale Aspects of its Monsoon Experiment. Tallahassee. FL. WMO. Geneva, Section 4, 20–29.

Wexler, H., 1937: "Formation of polar anticyclones". *Mon. Wea. Rev.*, **65**, 229–236.

Wexler, H., 1951: "Anticyclones", in: Malone, T. F., ed., *Compendium of Meteorology*, *Amer. Meteorol. Soc.*, 621–628.

Xia Youlong, Zhu Baoahen, Huang Boyin, 1991: "The numerical simulation on the formation and evolution process of the Asian continental high in winter". *Quart. J. Appl. Meteorol*, **2**, 131–139.

Yanai, M., Esbensen, S., Chu, J. H., 1973: "Determination of bulk properties of tropical cloud clusters from large-scale heat and moisture budgets". *J. Atmos. Sci.*, **30**, 611–627.

Yang Song, Zhu Qiangeng, 1989: "The Numerical study of cold surge structure and circulation interaction between mid-latitudes and tropics during cold surge". *J. Tropical Meteorol.*, **5**, 227–234.

Yang, D., Krishnamurti, T. N., 1981: "Potential vorticity of monsoonal low level flows". *J. Atmos. Sci.*, **38**, 2675–2695.

Zhao, Q., Ding, Y. H., 1991: "A study of physical processes of air-mass transformation of the Siberian high over land". *Acta Meteorol. Sinica*, **9**, 170–180.

Zhou Xuequn, 1989: "An analysis of two cold surges". *J. Tropical Meteorol.*, **5**, 57–63.

Zhu Qiangen, Yang Song, 1989: "A numerical experiment on the effect of the Tibetan Plateau on cold surges". *Acta Meteorol. Sinica*, **3**, 448–457.

CHAPTER 3

THE SHORT-RANGE FLUCTUATIONS OF MONSOONS AND THEIR ASSOCIATON WITH THE MAJOR WEATHER EVENTS IN CHINA

3.1. Climatological Aspect of Heavy Rains in China

China, as are other countries, is greatly affected by summer monsoons and their associated heavy rainfalls. A notable example for severity and duration of rainfalls is Meiyu (Plum Rains) occurring over the Yangtze River Valley. Almost every year, much damage is caused in China by flash floods or persistent floods caused by excessively heavy rains. In some years, the regional loss of property and lives reaches disaster proportions, exemplified by the extensive and devastating flood of the Huaihe River (in the central part of China) in August 1975. Much emphasis has been placed in recent years in China on research dealing with heavy rain with the aim of improving the prediction of heavy rainfall. The summer monsoon rainfalls in China mainly include the pre-summer heavy rainfall in South China, the Meiyu (Plum Rains) over the Yangtze River and Huaihe River Valleys and the mid-summer heavy rainfall in northern China. For years, Chinese meteorologists have paid more attention to these problems. In the 1950's, the research into this aspect was focused on large-scale circulation conditions and rain-bearing synoptic systems (Xie *et al.*, 1956; Tao *et al.*, 1957). During the past 20 years, however, emphasis has been placed on the diagnostic analysis and theoretical study of meso- and small-scale systems.

It is believed that three large-scale aspects are responsible for the occurrence of heavy rains in China. First, the summer monsoon coming from India and the western Pacific plays an important role. As indicated in Chapter 1, the rainy season in China generally begins with the onset of the summer monsoon and ends with its withdrawal. Rainfall intensity and its variation are closely related to the fluctuations of the summer monsoon. Chinese forecasters, through experience, expect heavy rains to occur during active monsoon periods in East Asia, if other conditions are favorable. Under such conditions, the southwest monsoon may reach as far as the Yangtze River, and even northern China. The interannual variability in the frequency of heavy rainfall occurrence in China is large and is also closely connected with the interannual variability of the summer monsoon regime.

Second, the positions of the subtropical highs over the western Pacific determine the seasonal shift of the main rain belt in China, as indicated in Chapter 1.

Finally, the interannual variability of heavy rains is closely related to anomalies of the general circulation in the middle and high latitudes of the Northern Hemisphere, especially in the East Asia sector. Blocking highs, located over the Ural Mountains, Lake Baikal, the Sea of Okhotsk, or the Caspian Sea, and the longwave troughs over the Ural Mountains and Lake Baikal, are believed to be key circulation features determining conditions, favorable or unfavorable, for heavy rains. For example, in the summer of 1972, when many unusual weather events occurred on a global scale, an anomalous longwave trough persisted over the eastern Seaboard of Asia (130–140° E), replacing the mean high-pressure ridge for normal years.

The following aspects of the general circulation are of significant importance in the formation of heavy rains: (1) Large-scale flow controls the synoptic or subsynoptic-scale weather systems, influencing their movement and intensity, and causing such rain-bearing systems to repeatedly occur in the same region, thus producing persistent heavy rains. (2) Large-scale flow patterns, to a certain extent, determine the location and intensity of heavy rainfall. (3) These flow patterns also determine the moisture sources and transport into the heavy rain regions. Major moisture sources available for the heavy rains in China are the Bay of Bengal, the South China Sea, and the East China Sea. Low-level southerly jets and easterly jets emanating from these source regions play a very important role in initiating and maintaining heavy rainfalls.

Table 3.1 gives an overview of excessively heavy rainfalls in China during the period 1931–1977 (Ding and Reiter, 1982). Most of these rain episodes exceeded 400 mm in 24 hours. Some records from the early 1930's were not very accurate, but are given for reference and comparison. Several significant features of a general climatological character for heavy rinas in China become evident from this table.

Table 3.1. Major excessively heavy rains in China during the past 50 years in decending order of precipitation amount. Data in the table are compiled based on various sources, but mainly on Tao *et al.* (1980)

Province or City	Location	Time of occurence	Maximum rainfall (mm) 24 h	3days	Entire period	Main rain producing synoptic system
Taiwan	Xinliao	17–19 Oct. 1967	1672	2749	2749	An easterly wave disturbance in the northern section of the Carla typhoon
Many provinces in Central China	Yangtze River and Huaihe River	June and July 1954	not available		2000	8–9 low pressure disturbances
Taiwan	Baixin	10–12 Sept. 1963	1248	1684	1684	Gloria typhoon
Henan	Linzhuang	5–7 Aug. 1975	1060	1629	1631	Weakening Nina
Hebei	Zhangmao	1–10 Aug. 1963	950	1458	1600* (or 2050)	Low-level SW vortices
Guang-	Taishan	24–30 May 1973	850	not available	1268	Shearlines
Hubei and Hunan	Wufeng	3–8 July 1935	not available		1200	Low-level vortices

Table 3.1. (continued)

Province or City	Location	Time of Occurence	Maximum rainfall (mm)			Main rainproducing synoptic system
			24 h	3days	Entire period	
Shanxi and Inner Mongolia	Maousu Desert	1–2 Aug. 1977	1050	1050	1050	Shearline and upper cold dront
Guangxi	Laohutan	11 July 1960	634	1036	1036	Vortices and upper trough
Jiangsu	Chaoqiao	3–4 Aug. 1960	653	934	934	Shirley typhoon
Fujian	Fengchao	18–19 Sept. 1956	593	871	871	Freda typhoon
Anhui	Zhilaichao	15–17 Aug. 1975	539	809	809	Ora typhoon
Shanxi	Yangquan	22–24 Aug. 1966	not available	800	800	Remnants of tess typhoon
Liaoning	Heigou	7–8 Aug. 1962	657	794	794	Extratropical cyclone transformed from Opal typhoon
Hubei	Duzhenwan	9 Aug. 1975	629	629	629	Nina typhoon
Shanxi	Middle Yellow River	6–9 Aug. 1933	400	not available	600-800*	Two upper troughs
Many provinces in Central China	Middle and lower Yangtze	July 1931	not available		800*	Six low pressure disturbances
Zhejiang	Shiling	1–2 Aug.	564	688	688	Wanda typhoon
Shanghai	Tangqiao	21 Aug. 1977	581	592	592	Mesoscale cloud clusters north of Amy typhoon
Shandong	Shifuzhi	12 Aug. 1974	499	572	572	Shearline in the northern section of Lucy Typhoon
Jiangxi	Dongxiang	17 Aug. 1953	499	533	533	Nina typhoon
Beijing	Zaoshulin	26–28 July	479	518	518	Rita typhoon
Henan	Yuanqu	14–19 July 1958	367	not available	500	Upper trough
Shanxi	Ansai	5–6 July 1977	400	400	400	Two upper troughs
Beijing	Beijing	1–2 July 1973	169	200	200	Low-level SW vortices

* The cumulative rainfall given here is not very accurate and should be taken as a rough estimate.

(1) Heavy rains in China stand out by their severity and duration when compared with those in other countries of the world in similar climatic regions. In China, the maximum of rainfall intensity for 1 hour reached 198.5 mm and was recorded at Linzhuang, Henan Province, on 5 August 1975. The maximum rainfall amount recorded for 24 hours is 1672 mm and occurred at Xinliao, Taiwan Province, on 17 October 1967 (Paulhus, 1965). The second highest 24 h rainfall with 1248 mm was also recorded in Taiwan on 10 September 1963. Although these amounts do not break the world record of 3240 mm for a 24 h rainfall recorded at Réunion in the tropical southwest Indian Ocean, they are much greater than the corresponding records established in similar climatic zones of the United States. Here, according to the statistics of maximum 24 h rainfall, the most extreme, with 983 mm, was recorded in Florida (Cram and Tatum, 1979). The most extreme cumulative rainfall amount for three days in China is 2749 mm, somewhat less than the world record of 3240 mm. In China the occurrence of 24 h rainfalls greater than 1000 mm is not unusual. This has been observed four times during the last 20 years. A unique case occurred in the desert area of northwestern China on 1–2 August 1977. In about eight hours, from midnight to early morning, one huge thunderstorm complex dumped 1050 mm of rainfall over a small area of less than 900 km^2 (Cai and Zhang, 1978). The duration of heavy rains in China ranges from a few hours to 63 days, with an apparent preference for time spans of two days to one week. Out of the 26 cases listed in Table 3.1, 14 cases of heavy rains persisted for three days or longer. Prolonged periods of heavy rains are seen to be more characteristic for China than for other countries in similar climatic regions.

(2) If one plots the locations of excessively heavy rains for the past 50 years, with 24–hour amounts greater than 200 mm (Tao and Ding, 1981), one finds that almost all these occurrences are contained in three bands: South China, the Yangtze River Valley, and northern China (Fig. 3.1). In addition, a few heavy rains were recorded in coastal regions and were caused by typhoons. There are two gaps with a dearth of heavy rains between these major rain disaster regions. These gaps are caused by the less frequent visitations by frontal zones and by the sudden, stepwise northward jumps of major circulation systems. More importantly, this characteristic distribution of heavy rains is closely related to the frequency of synoptic systems originating, or passing, over the Tibetan Plateau. For example, pre-summer heavy rains in southern China are mainly produced by disturbances in the southern branch of the westerlies (the subtropical jet stream) passing to the south of the Plateau. Heavy rains in the Yangtze River Valley are frequently caused by low-level vortices and shearlines coming from the Plateau and are not the sole consequence of typhoons. When the upper frontal zone shifts northward as the season progresses, these low-level vortices tend to change their tracks from an eastward to northward or northeastward movement, thus leading to the occurrence of heavy rains in northern China,or even in northwestern China.

(3) Based on the properties of rain-producing synoptic systems, heavy rains in China may be classified into four types. The first type is caused by typhoons, or their remnants, and by extratropical cyclones transformed from typhoons. It has been a popular belief that typhoons are the most important and most severe rain-producing synoptic systems in China. Statistics in 15 coastal provinces (Tao *et al.*, 1980) show

that the maximum rainfalls recorded in 12 of these provinces were caused by typhoons. Also, from Table 3.1 it can be seen that 14 cases, or 56% out of a total of 26 cases, were caused by typhoons. In particular, the three most severe 24 h rainfalls were due to them. Therefore, typhoon-producing heavy rains are of primary importance for the precipitation climate of eastern China.

The second type of heavy rain is caused by low-level vortices, or wind shearlines, associated with these vortices. From Table 3.1 it can be seen that eight cases out of the 26 were caused by such vortices and shearlines. One often observes, especially during the Meiyu season, that a succession of low-level vortices, sometimes five or six of them, move eastward from the Tibetan Plateau along more or less northeastward-oriented shearlines, leading to prolonged heavy precipitaion. For example, the heavy rainfalls in June and July of 1954 were caused by eight or nine low-pressure disturbances, most of them having originated over the Plateau and passing over the Yangtze River Valley under the steering effect of wave troughs. Similarly, the major floods of early July 1931 in the Yangtze River Valley were caused by five to six such low-pressure disturbances and the floods of July 1935 in the same region were due to two pressure disturbances.

The third type of heavy rain is caused by upper troughs and attendant cold fronts. The rain-producing disturbance tends to slow down as it approaches a blocking anticyclonic region, leading to prolonged rainy period. In addition to these typical rainfall patterns, a fourth type which entails flash floods may occur with relatively

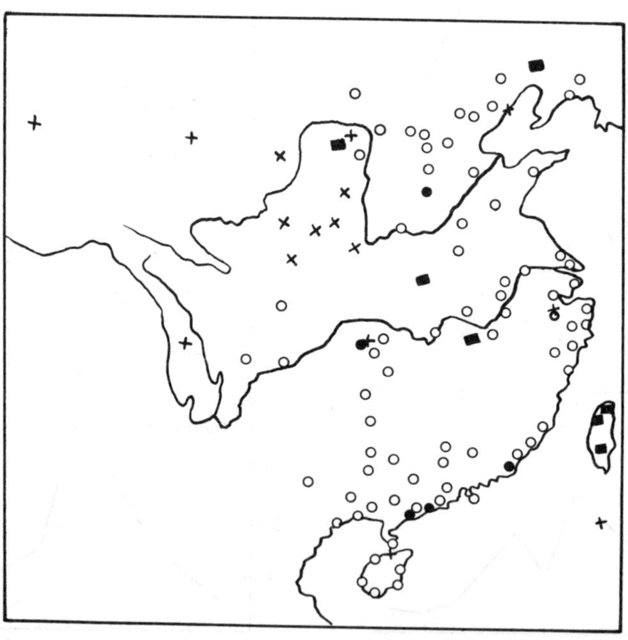

Figure 3.1. The distribution of heavy rains for 1953–1977 with the rainfall amount for 24 h exceeding 1000 mm (black square), 800 mm (black dot), 400 mm (white circle) and 200 mm (cross) (Tao and Ding, 1981).

localized thunderstorm complexes that repeatedly develop under favorable large-scale synoptic conditions.

3.2. Pre-Summer Heavy Rainfalls over Southern China

As the season progresses, the long-term mean position of the rain belt in East Asia moves northward in steps, from spring to summer. During April to June, the rain belt swings over southern China, thus causing the pre-summer rainy season in this region. Figure 3.2 is the annual range of the long-term mean (1960–1979) rainfall amount at several selected stations of southern China (Huang, 1986). It can be seen that the distribution of rainfall amount for most of southern China assumes a marked double-peak pattern, with the peaks observed in May to June (the pre-summer rainy season) and August (the typhoon rainy season caused by typhoons and other tropical weather systems), respectively, and even with the former peak of rainfall amount being greater than the latter. Thus, the pre-summer rainy season is the major rainy season for southern China.

Figures 3.3a, b further indicate this feature. Over most of southern China, the total precipitation amount (500–1000 mm) for April-June accounts for 40–50% of the

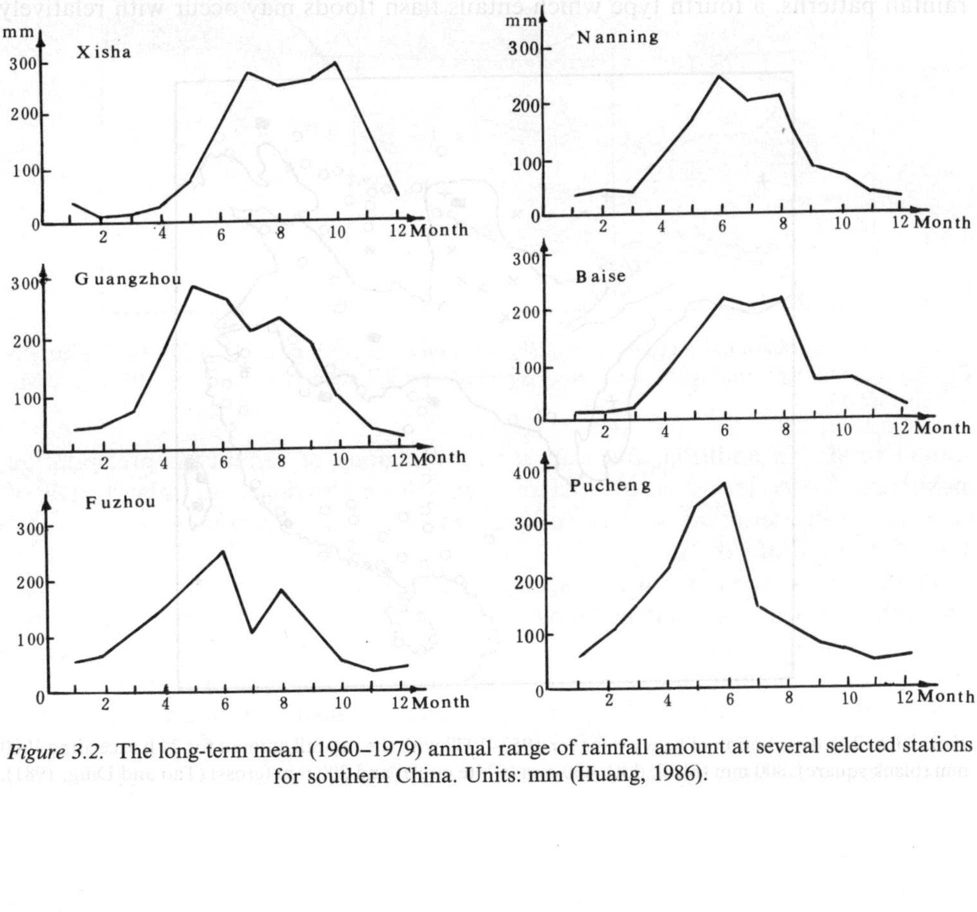

Figure 3.2. The long-term mean (1960–1979) annual range of rainfall amount at several selected stations for southern China. Units: mm (Huang, 1986).

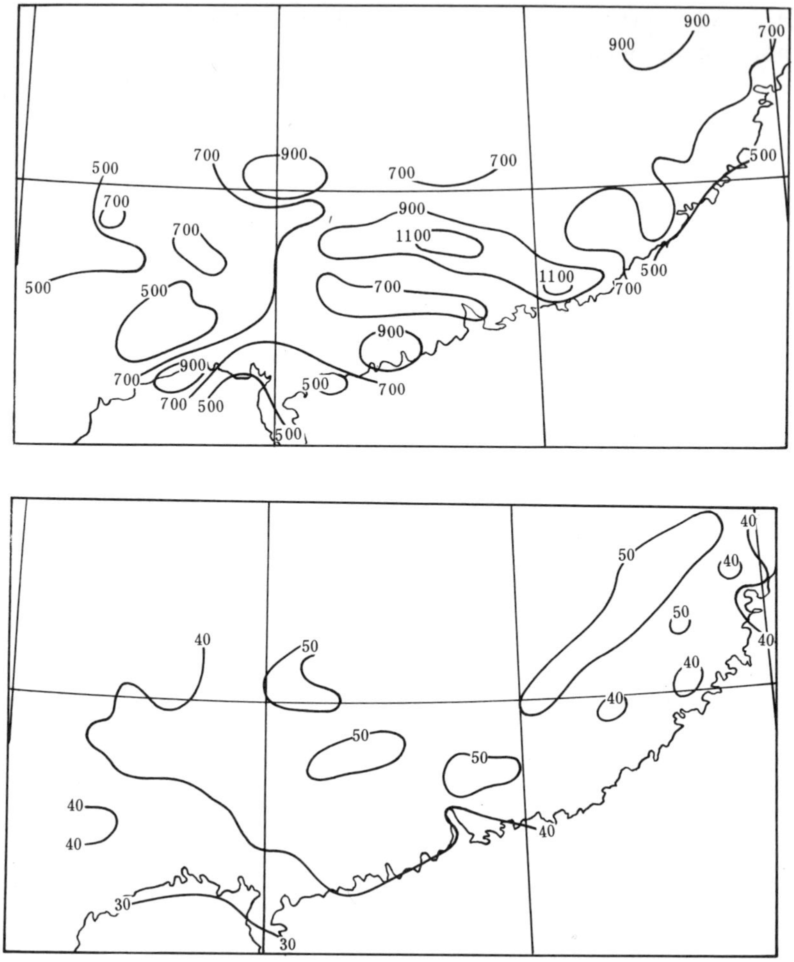

Figure 3.3. (a) Distribution of mean total precipitation amount for April-June 1959–1979. (b) The same as Fig. 3.3 (a), but for percentage (%) accounting for annual precipitation total. Units: mm (Huang, 1986).

annual totals. In addition, one can see four regions of significant precipitation maximum: the eastern coastal area of Guangdong Province, the inland area of Guangdong Province, the western coastal area of Guangxi Province and the coastal area of the Gulf of Tonkin.

The inhomogeneous feature of spring and summer precipitation over southern China shows up in short- and medium term fluctuations. Figure 3.4 is indicative of the periodic variation of precipitation for March-November 1976–1980, with the period of 40–60 days dominating. Note that the pre-summer rainy season is really a major part of this medium-term fluctuation of precipitation, lasting from the early part of May to the middle part of June. Analysis of the power spectrum for the precipitation at Guangzhou further shows that there is also a 12–day fluctuction besides a 40–60 day fluctuation. The above two fluctuations of precipitation over southern China clearly reflect the corresponding oscillations of the summer monsoon with the same periods that will be particularly discussed later. When the easterly wind weakens, the precipitation significantly increases and the peak of precipitation occurs shortly before the transition from the easterly to westerly wind.

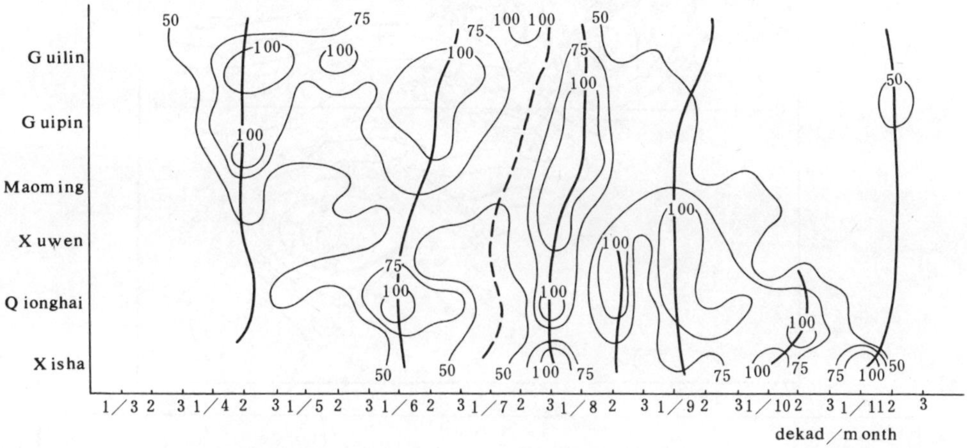

Figure 3.4. Temporal cross-section of 10-day total precipitation over southern China along 110° E for March-November 1976–1980. Units: mm (Huang, 1986).

The pre-summer rainy season over southern China occurs under special large-scale synoptic and circulation conditions that are characterized by the following events: the first northward jump of the subtropical high over the western Pacific, the onset of the East Asian summer monsoon over the South China Sea, and a gradual weakening of activities of cold air in East Asia, but still with frequent intrusions into the region of southern China. So, this seasonal rain belt is caused by the combined effect of the synoptic systems in westerlies and the summer monsoon flow located in southern China.

The inspection of daily wind fields or flow patterns has indicated that a prevalence of the southeasterly wind at 850 hPa at Xisan Island station leads to the beginning of the rainy season over southern China and the change from the southeasterly wind into a southwesterly direction in the middle part of May heralds the peak spell of the pre-summer rainy season. This event is also an indication of the onset of the summer monsoon over the South China Sea, and at the same time the beginning of the seasonal change in circulation pattern at low latitudes. To illustrate this relationship, Figure 3.5 shows the composite flow patterns for 5 days before and after the abrupt change in the circulation pattern at 850 hPa, averaged for 1974–1980 (Shen *et al.*, 1983). Prior to the onset of the summer monsoon, the subtropical high dominates the coastal area of southern China and the South China Sea. The relatively weak westerly wind at the northern flank of the subtropical high prevails over southern China and frequently meets with eastward moving troughs in the southern branch in the westerlies or cold airs coming from the north over the northern part of southern

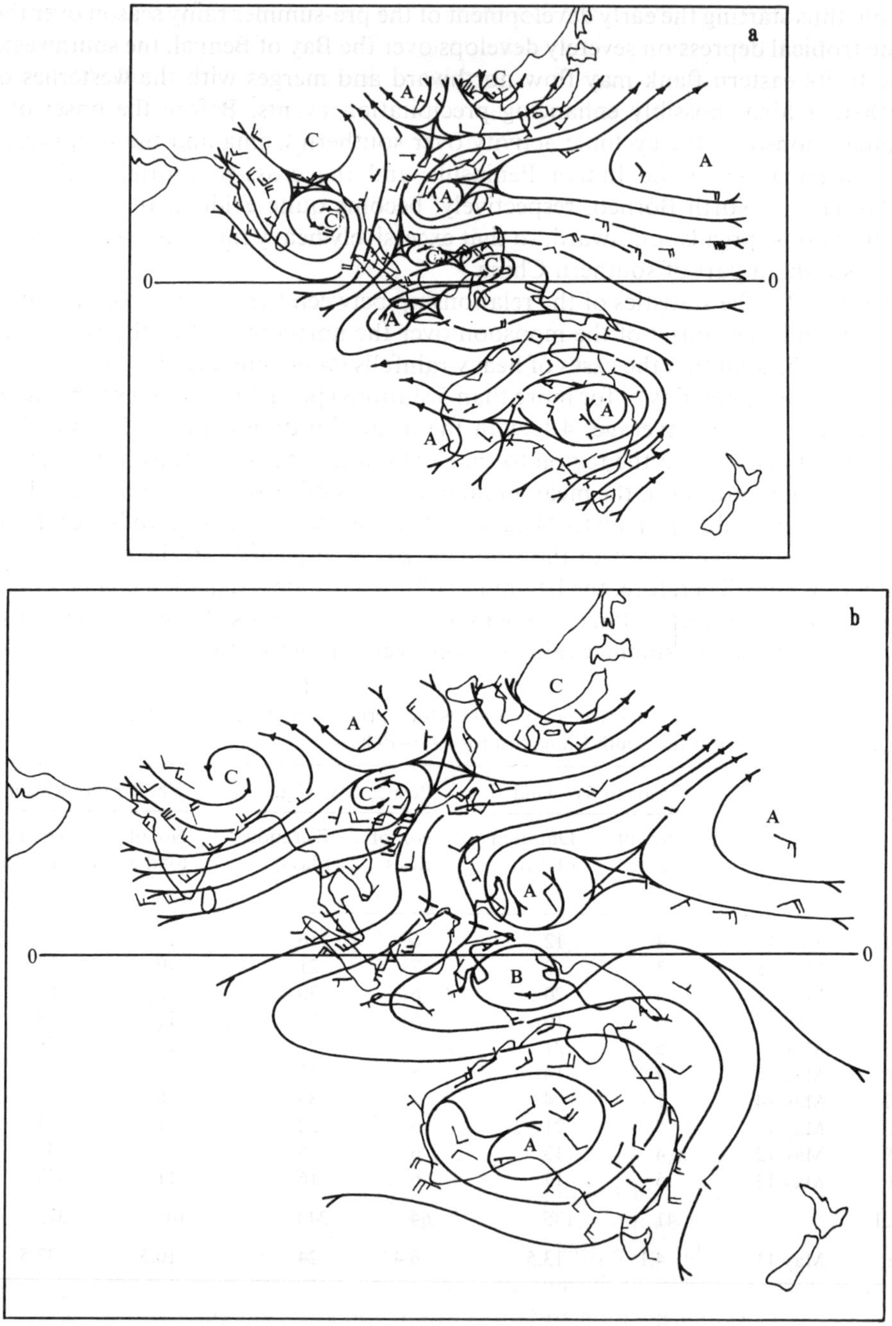

Figure 3.5. (a) The 5-day mean flow pattern prior to the onset of the summer monsoon over the South China Sea, averaged for 1974–1980.(b) The same as (a), but for the 5-day mean flow pattern post to the onset (Shen *et al.*, 1983).

China, thus starting the early development of the pre-summer rainy season over there. If the tropical depression severely develops over the Bay of Bengal, the southwesterly flow to its eastern flank may flow northward and merges with the westerlies over southern China, possibly enhancing precipitation events. Before the onset of the summer monsoon, the cyclonic activity over southern China and the southwesterly flows originating in the Indian Peninsula and the cross-equatorial airflow over Malaysia and North Borneo, respectively, become quite evident. In this case, pre-summer rains spreads southward, so that excessively heavy rainfalls may be observed in the southern part of southern China.

Table 3.2 is the statistics of the relationship between the processes of pre-summer rains and dates of onset of the monsoon over the northern part of the South China Sea. It can be seen that the cases of heavy rainfalls (> 40 mm day^{-1} for more than 10 stations or > 80 mm day^{-1} for more than 5 stations) prior to the onset of the summer monsoon are, on an average, 4.1, with the mean duration being 13.5 days. Of 135 cases, most occurred in the region to the north of 23° N. In contrast, after the onset of the summer monsoon, the mean number of cases of heavy rainfall increased up to 6.4 and the mean duration up to 24 days. At the same time, the events of severe rainfall moved down to the region to the south of 20° N, especially in the coastal area of southern China. Therefore, the beginning of low-latitude circulation changes in Asia in summer and the onset of the summer monsoon over the South China Sea marks the peak stage of the pre-summer rainy season over southern China.

Table 3.2. The statistics of the cases of heavy rainfalls for April-June and the onset of the summer monsoon over the northern part of the South China Sea for 1971–1980

Year	Date of onset of the monsoon	Prior to SW wind		Post to SW wind		Whole rainy season	
		No. of cases	Duration (days)	No. of cases	Duration (days)	No. of total cases	Total duration (days)
1971	May 16	4	12	6	16	10	28
1972	May 5	3	3	7	21	10	24
1973	May 11	6	20	6	30	12	50
1974	May 5	4	10	10	39	14	49
1975	May 14	5	21	6	27	11	48
1976	May 5	4	14	5	21	9	35
1977	May 14	1	4	5	30	6	34
1978	May 18	5	21	6	22	11	43
1979	May 12	4	13	6	18	10	31
1980	May 13	5	17	6	16	11	33
Total		41	135	64	240	105	375
Ave.	May 11	4.1	13.5	6.4	24	10.5	37.5

As the summer monsoon sets in, the cross-equatorial airflow may flow northward up to southern China, this causing the air temperature and humidity at low-level to increase, especially in layer of 500–1000 m. Figure 3.6 shows an example for 1978. In

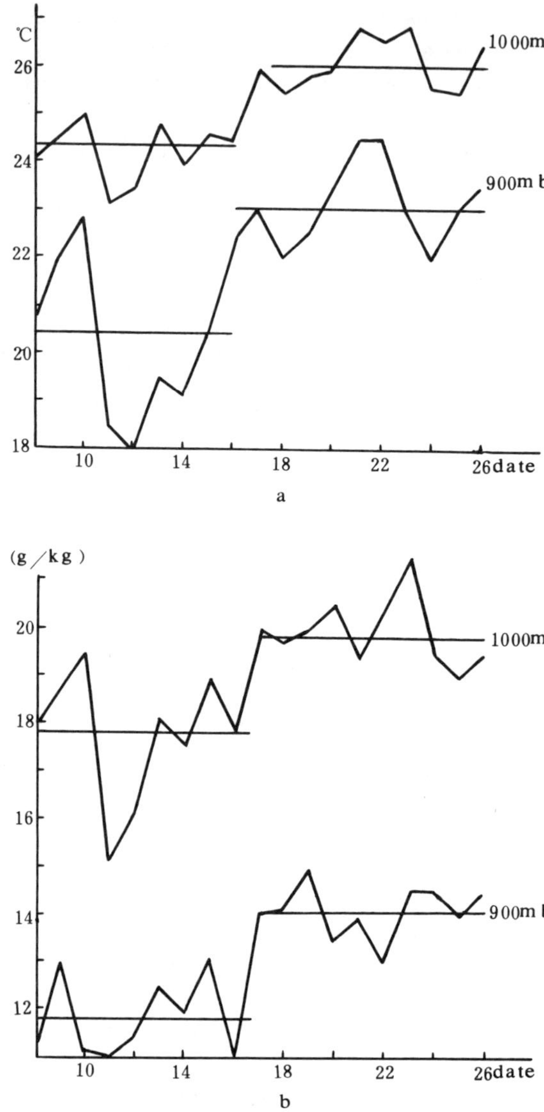

Figure 3.6. The day to day variations of temperature (a) and specific humidity (b) at 900 hPa at Haikou for 8–26 May 1978 (Huang, 1986).

this year, the cross-equatorial aircurrent reached the northern part of Hainan Island on 15 May. On the next day, the wind speed near the surface increased up to 12 m s^{-1}. As a result of the intrusion of this strong southerly wind, the air temperature at 1000 and 900 hPa abruptly rose by 2–8 °C and the specific humidity by 2–3 g kg^{-1}. Concurrently, extensive heavy rainfalls occurred, over Hainan Island, the western part of Guangdong Province, and the southern part of Guangxi Province. Some of

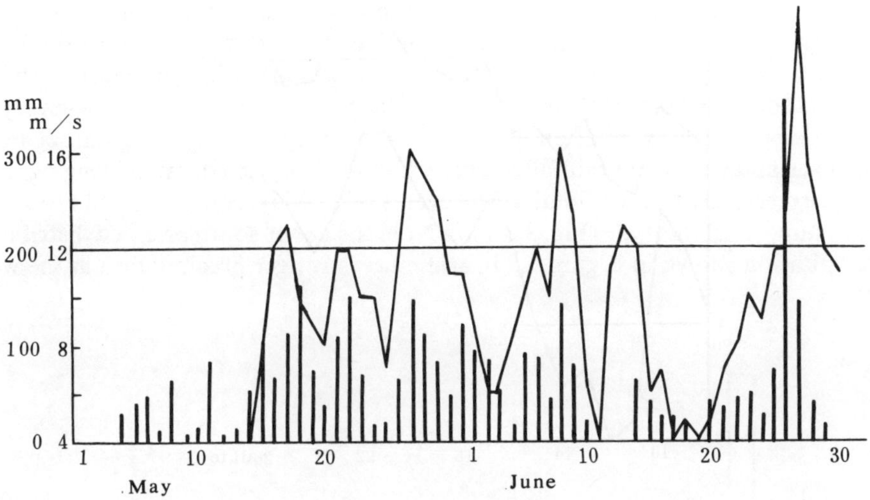

Figure 3.7. The temporal variations in the maximum wind speed below 1000 m at Tongfan station (curve) and the daily rainfall amount over the coastal area to the west of the Pearl River delta (vertical barb) (Huang, 1986).

these regions had daily rainfall amounts greater than 100 mm. Figure 3.7 indicates the relationship between the short-term fluctuation of the monsoon and the precipitation variability along the coastal area of the eastern Guangdong Province from mid-May to the end of June 1978. For 10 days before 15 May, the precipitation amount recorded in this region totaled 318 mm. But for 10 days after this day the precipitation amount was drastically augmented, totaling 574 mm, almost twice the amount of the previous one. In 1977 and 1979, the almost same event happened. Thus, this provided further evidence for the statement that the peak stage of the pre-summer rainy season over southern China starts with the establishment of the cross-equatorial airflow over the coastal area of southern China. From Figure 3.7 one can notice that a process or event of heavy rainfall is usually accompanied by the occurrence of southerly monsoon airflow (> 12 m s^{-1}). In particular, after 10 June the four monsoon surges occurred with wind speeds exceeding 12 m s^{-1}; each surge caused heavy rainfall with the amount nearly equal to and greater than 150 mm day^{-1}. To a certain extent, the short-term variability of precipitation of the pre-summer rainy season is modulated by the activity of monsoons in East Asia, especially monsoon surges. Based on the recent statistics of monsoon surges from April to August made by Yong *et al.* (1987), 37 of the 47 cases studied (79%) were accompanied by precipitation episodes.

During the pre-summer season, most heavy rains result from the appropriate interaction and horizontal juxtaposition of cold air and moist air. Based on synoptic and weather radar observations, a complete precipitation process often consists of four types of precipitation of different characters: the shearline in the warm section, low-level jet in coastal area, the prefrontal low-level jet and cold front. But not all the precipitation episodes necessarily assumes these four types of precipitation. The heavy rains associated with the shearline in the warm sector ahead of the cold front are mainly caused by the meso-scale north-south oriented bands of heavy rainfall that are often triggered by convergence between southwesterly and southeasterly winds. Sometimes they have the same orientation as the meso-scale southerly low-level jet in the boundary layer. Precipitation of this type mostly occurs in the coastal regions of Guangdong and with great rainfall intensity (100–200 mm h^{-1}), often bringing about localized excessively heavy rainfalls. The events of 27 May 1978 and 13 May 1979 are typical examples of this type (Fig. 3.8 a, b). Note the north-south oriented distribution of precipitation shown in Figure 3.8 b, and much stronger precipitation in the warm sector.

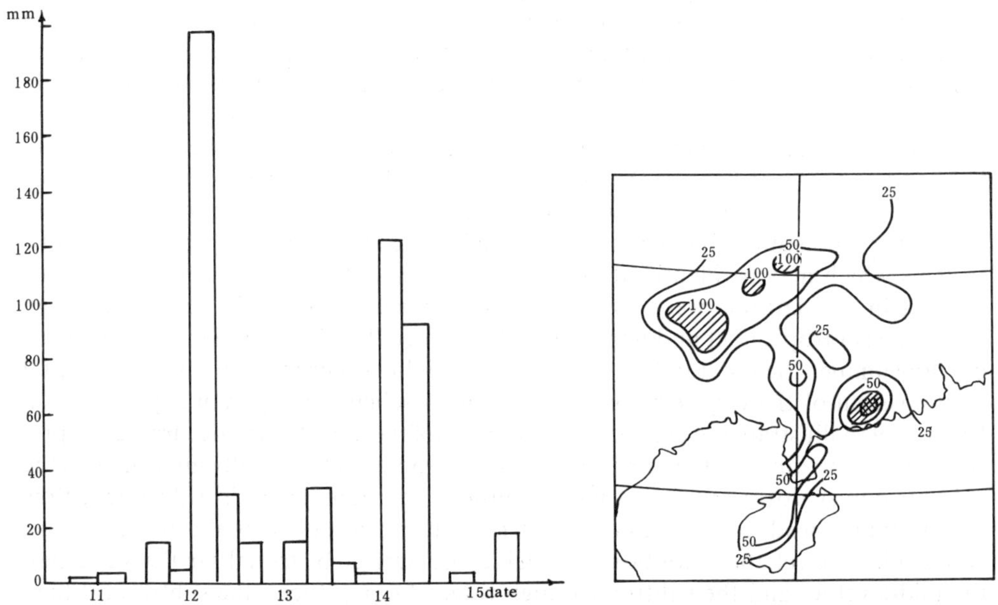

Figure 3.8. (a) The distribution of 6 hour precipitation at Yangjian station for 11–13 May 1979; (b) the distribution of daily precipitation on 27 May 1978. Units: mm (Huang, 1986).

When the southwesterly airflow consistently prevails over the coastal areas of southern China, the region of heavy rains caused by the shearline in the warm sector disappears. If the low-level southwesterly wind rapidly enhances and reaches the intensity of a low-level jet, a corresponding region of heavy rains, (west-east oriented), will be observed which may last for several days and produce unprecedented events of heavy rain. In particuar after the last part of May when the onset of the southwest monsoon surge induces the creation of a low-level jet in the northern part of the South China Sea, a region of persistent heavy rains may occur along the coastal area. This problem will be addressed in detail later.

The heavy rainfalls associated with the prefrontal low-level jet occur during the process of the southward intrusion of cold air. In this case, the prefrontal low-level jet is usually observed. When the frontal system and its attendant low-level jet move southward into Guangdong Province, they may produce an extensive area of heavy rains. However, this type of heavy rain generally has a weaker intensity than the first two types.

The heavy rainfall associated with a cold front or stationary front occurs within the frontal zone and behind of the cold front. In general, this type of region of heavy rains is relatively narrow, with a shorter duration.

The dominating type of heavy rainfall during the pre-summer rainy season over southern China is the first type of heavy rains which is characterized by convective precipitation and extensive thunderstorms. It has a greater rainfall intensity, shorter duration, and marked fluctuation of rainfall amount.

The major rain-bearing synoptic systems for the pre-summer heavy rains have been identified as three kinds. The first one is the surface baroclinic zone (southern China cold front and stationary front), synoptic-scale convergence zones (monsoon trough, frontal low-pressure trough), and warm and moist zone. The second type of synoptic-scale rain-bearing system includes a low-level jet, low vortices and shearlines in wind field and convergent lines in the boundary layer. These synoptic-scale systems may cause the occurrence of heavy rainfalls, mainly as the triggering mechanism and suppliers of moisture. The third type is related to the middle and upper-level synoptic systems which may be identified as troughs on the southern branch in westerlies or troughs in mid-latitude westerlies, mid-tropospheric cyclones, the subtropical high ridge, the subtropical jet and upper-level divergence flow field. A significant episode of pre-summer heavy rainfall requires the right superposition and horizontal juxtaposition of these systems or part thereof. A schematic diagram is presented in Figure 3.9 to illustrate the characteristic situation favorable for the occurrence of pre-summer heavy rains. At the upper-level, the subtropical jet generally intensifies as the cold air comes down and has a more southern position. There is a divergent flow pattern that is produced by the northwesterly or northeasterly flow ahead of the ridge. At the middle level, the synoptic situation is characterized by troughs in the westerlies, the mid-level jet and the subtropical high ridge. Among them, the subtropical high ridge is a major component. At low level and at the surface, there are low vortex, shearline, cold front, low-level jet, convergence line and low pressure trough.

As previously indicated, the precipitation in the warm sector is most significant for the pre-summer rainy season. The physical conditions for this type of heavy rainfall have been estimated based on the data obtained during the experiment on pre-summer heavy rainfall over southern China (1977–1978). The salient feature is the presence of a deep moist layer, with the near-saturation layer extending upward to 500 hPa, and even higher to 200 hPa. The strong baroclinic layer is mainly observed below 850 hPa, indicating that the activity of the cold air associated with heavy rainfalls is concentrated in the boundary layer. The atmospheric stratification in the warm sector is usually unstable below 600 hPa. There is significant convergence below 850 hPa, with the maximum found at 950 hPa. In the middle and upper troposphere, the divergence and anticyclonic vorticity often lies over regions of heavy rainfall.

The transport of moisture into the region of heavy rainfall is crucial for the development and maintenance of heavy rain. Under the case of warm sector

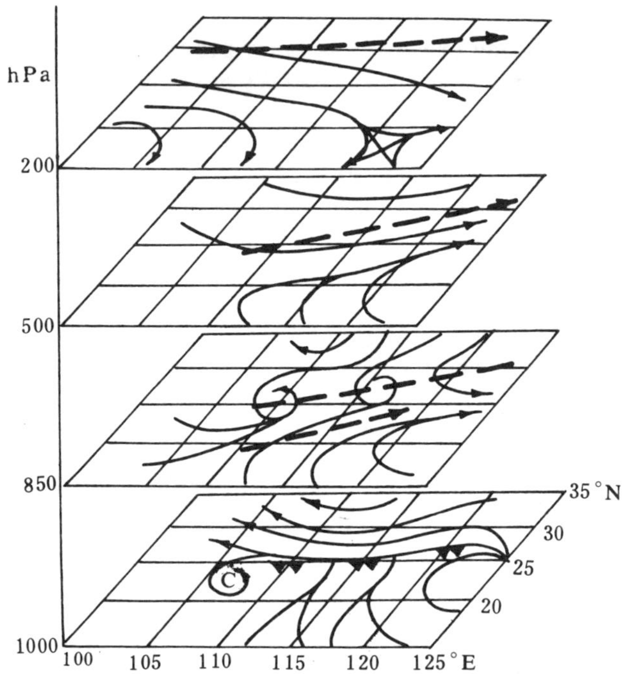

Figure 3.9. Schematic diagram of the synoptic situation characteristic of the occurrence of the pre-summer heavy rainfalls. The thin solid line denotes the streamline, the bold broken arrow, the jet axis, and the bold broken line the shearlines (Huang, 1986).

precipitation, a greater portion of moisture is transported in from the southern boundary in the layer below 850 hPa, especially below 900 hPa. At the western boundary, the inflow of moisture at 850–700 hPa is most significant. Above 500 hPa, the moisture flux is negligibly small. The horizontal convergence of moisture also has its maximum at 900 hPa and then decreases upward and downward.

The moisture available for the pre-summer heavy rainfall is believed to have two regions of origin: the Bay of Bengal and the northern part of the South China Sea. The estimates of trajectories of water vapor based on the data obtained during the field experiments have indicated the importance of the latter region of origin of moisture for the pre-summer heavy rainfalls which will be elaborated in Chapter 6.

The activity of the low-level jet is closely related to the heavy rainfalls in southern China. Many events of pre-summer heavy rainfalls were accompanied by the activity of low-level jets. The low-level jet is usually defined as the maximum wind or jet streak with wind speeds greater than 12 m s^{-1} in the layer of 500–3000 m (about 900–700 hPa). Sometimes, the low-level jet with wind speeds less than 16 m s^{-1} is classified as the weak jet, the low-level jet with the wind speed between 16 and 20 m s^{-1} the moderate jet, and the low-level jet with the wind speeds greater than 20 m s^{-1}, the strong jet. Figure 3.10 is the wind profiles for Guangzhou for May of 1973 and 00:00 GMT, 7 May 1973, respectively. It can be seen that the wind speed in the lower

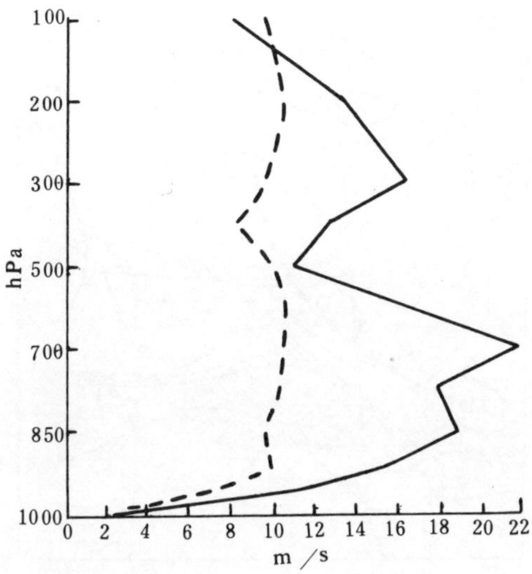

Figure 3.10. The wind profiles for Guangzhou, (a) for the 00:00 GMT, mean wind speed of May 1973 (dashed line); (b) for the 00:00 GMT, 7 May 1973 (solid line). Units: m s^{-1} (Guangdong Observatory, 1976).

troposphere for the individual day, 7 May, is considerably greater than the monthly mean condition, with two peaks found at 700 and (22 m s^{-1}) 850 hPa (19 m s^{-1}), respectively. The low-level jet on 7 May 1973 was associated with the process of heavy rainfall over the eastern part of Guangdong Province.

Based on the statistics of the relationship between the low-level jet (> 16 m s^{-1}) and the heavy rainfalls in Guangdong Province made from data for May-June 1970–1973, it has been found that out of 19 processes of low-level jets, 16 were accompanied by heavy rainfalls, with 1–5 days lagging. Therefore, the probability of the occurrence of heavy rainfalls is as high as 84.2% once there is a low-level jet. The statistics for Guangxi Province (1957–1974) have revealed a similar result.

The heavy rainfalls related to the low-level jet are generally located in the left front sector of the jet where there is maximum convergence. The statistics has indicated that about 80% of heavy rainfalls occurred in this manner.

The genesis and development of low-level jets are attributable to two types of synoptic process. The first type is associated with the intensification of the subtropical high and development of low-pressure systems to the west of the jet. When the trough in the southern branch of the westerlies moves eastward to the east of the Tibetan Plateau, it may often induce the genesis of a low vortex or low pressure trough at low level. As these systems move farther eastward and develop, the pressure gradient would enhance. If, at this time, the system can acquire a development, the ageostropic wind would increase, thus augmenting the wind speed to generate low-level jet. Figure 3.11 is an obvious example. The generation of a low-level jet at 850 hPa on 16 May 1977 is closely associated with the eastward movement of a SW vortex from Sichuan Province.

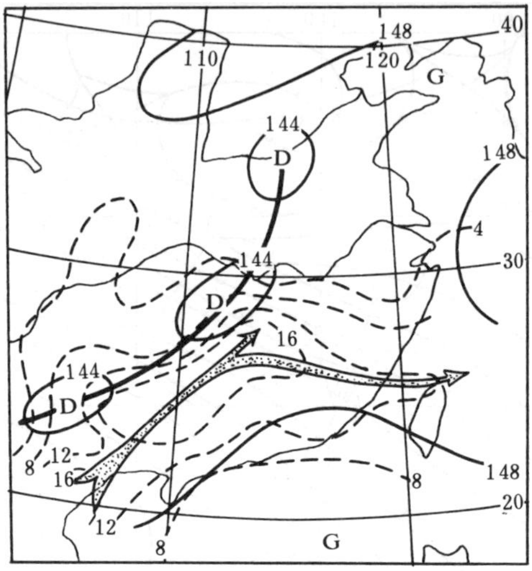

Figure 3.11. The weather map at 850 hPa at 00:00 GMT, 16 May 1977. The solid line represents the isopleth of geopotential height (decameter), the dashed lines isotachs, and the double arrow the jet axis (Huang, 1986).

There is a part of a low-level jet whose formation is related to the enhancement of the southwest monsoon over the Indo-China Peninsula and the northern part of the South China Sea. The wind maximum is first observed in the prevailing southwest monsoon airflow over the Indo-China Peninsula, then it gradually moves northward and eventually influences the region of southern China (Fig. 3.12). This type of low-level jet is usually relatively strong and has a larger scale and more persistent duration (Zhong and Liang, 1984). Sometimes, the low-level jet may also result from the enhancement of cross-equatorial airflows. In these cases, the low-level jet over southern China, as a matter of fact, is part of a large-scale monsoonal airflow.

The vertical circulation associated with a low-level jet is characterized by the upward motion ahead of the jet streak and the downward motion behind the jet streak. Figure 3.13 is divergence field around the low-level jet at 12:00 GMT, 20 June 1978. The left front sector has a strong convergence with a minimum of -7.4×10^{-5} s^{-1}, where the heavy rainfall usually occurs.

The low-level jet is a very warm and moist high-speed airflow. The dewpoint temperature has a marked positive anomaly along and to the left of the jet axis, with 700 hPa being the most moist (Li *et al.*, 1988). The warm tongue of the temperature always runs with the jet zone. The low-level jet may transport abundant moisture northward.

The wind field associated with low-level jet is super-geostrophic (Li, 1980) (Fig. 3.14). The strong ageostrophic component would excite the inertial gravity wave, thus causing the genesis of meso-scale systems. This contention has been documented with the relationship between wind field and precipitation during the pre-summer rainy season.

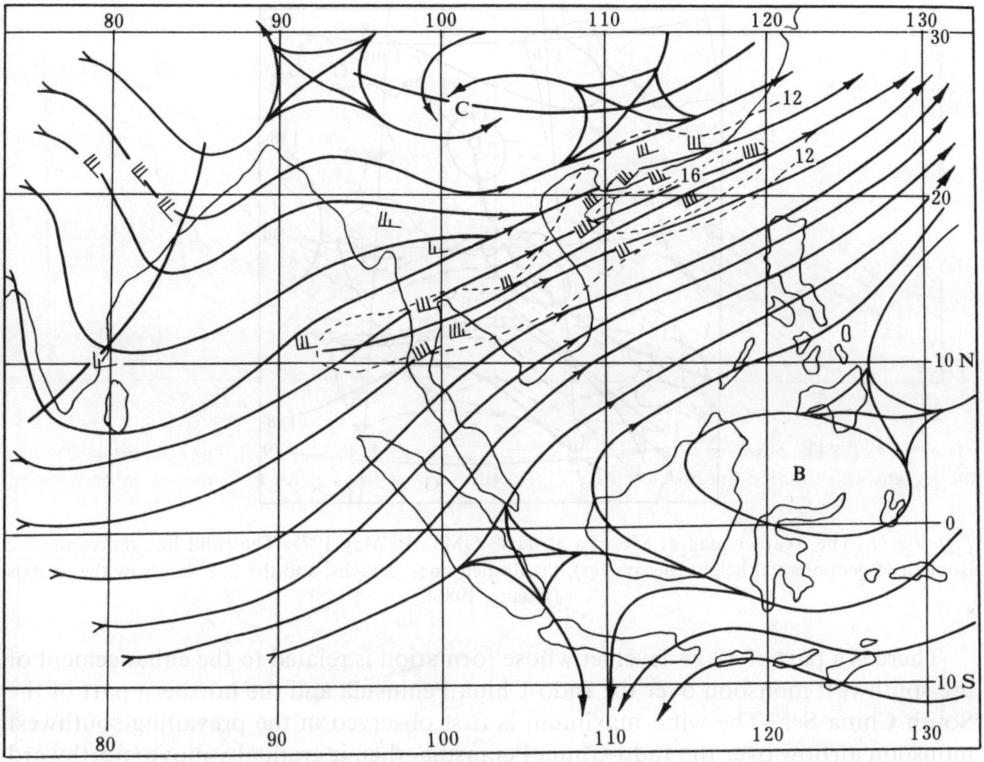

Figure 3.12. The streamline chart at 00:00 GMT, 30 May 1977. The dashed line denotes isotach. Units: m s^{-1} (Zhong and Liang, 1984).

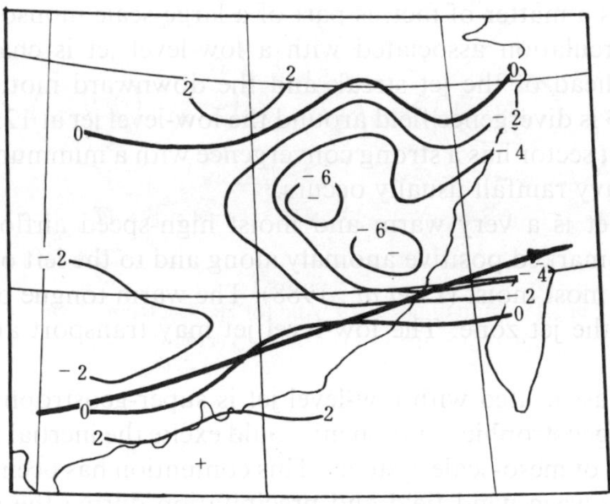

Figure 3.13. The horizontal distribution of divergence around the low-level jet axis at 550 hPa at 12:00 GMT, 6 June 1978. The negative sign denotes convergence. Units: 10^{-5} s^{-1}.

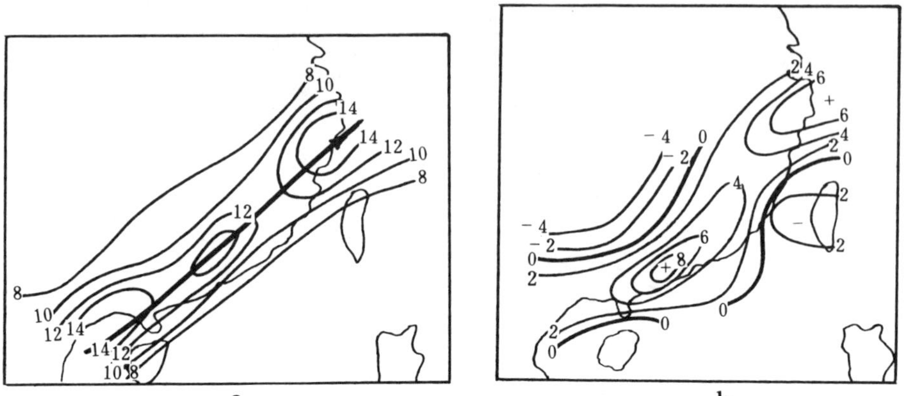

Figure 3.14. (a) The observed wind field at 550 hPa at 12:00 GMT, 17 May 1978. The bold arrow denotes the jet axis and the thin line isotach (m s^{-1}); (b) the same as (a), but for the super-geostrophic wind (Li, 1980).

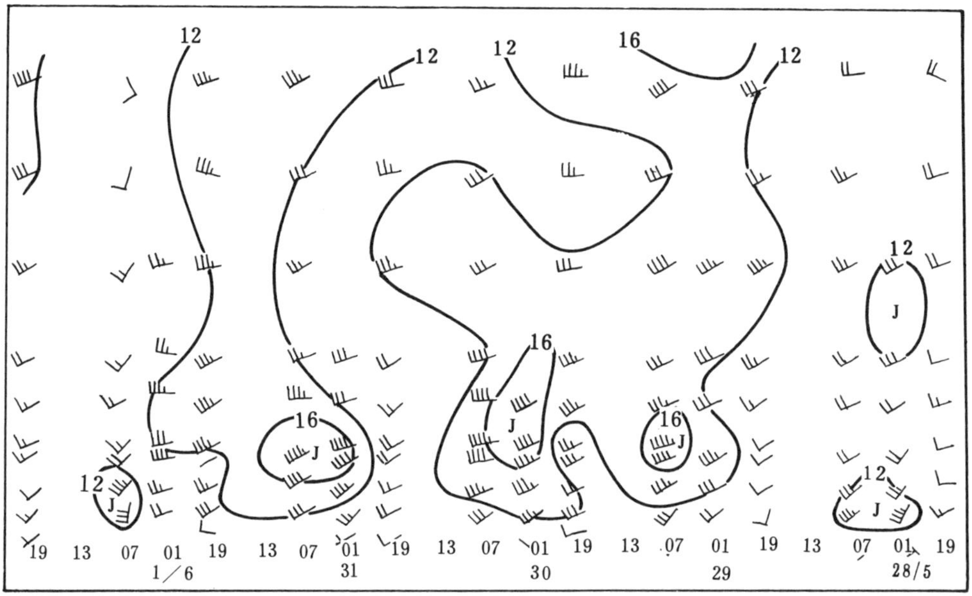

Figure 3.15. Height-time cross-section of upper air wind for Haikou from 29 May to 1 June 1977. The solid line represents isotach (m s^{-1}) (Huang, 1986).

The low-level jet has a marked diurnal variation in wind speed (Fig. 3.15). This condition is, in many ways, similar to that of the Somali Jet or the Mid-West of the United States.

It has been recognized that the heavy rainfalls are mainly caused by meso-scale synoptic systems. The meso-scale analysis of several dozen pre-summer heavy rainfalls has indicated that every process of the pre-summer heavy rainfall in southern China can find the activity of meso-scale systems, generally including over 10 rain masses (rainfall intensity > 5 mm h^{-1}). These rain masses have a horizontal scale of 30–40 km, mean life span of 5 h and mean movement speed of 25 km h^{-1}, steered by the mean wind at 500 or 700 hPa. The formation of rain masses are greatly controlled by local terrains, and, to a lesser extent, sea breeze, with the preferred region of origin observed in the valley area, of rivers, lake area and horn-shaped terrains.

During the pre-summer heavy rains, the meso-scale systems may be classified into the follwing types: the meso-scale cyclone, the meso-scale anticyclone, the meso-scale shearline, the convergence line in the southerlies, the dewpoint front and energy front, the meso-scale convective systems, and the squall (Jiang *et al.*, 1990). A brief description of some of them will be given (Lian and Bao, 1980). The meso-scale cyclones are generated along the shearline, especially at the crossing point of two shearlines. They may produce excessively heavy rainfalls. For example, 25 meso-scale cyclones were observed during the process of the case of the heavy rainfall "77.5" (27 May-1 June 1977, the eastern part of Guangdong Province). That produced an acumulative rainfall amount of over 500 mm over the extensive area (Fig. 3.16).

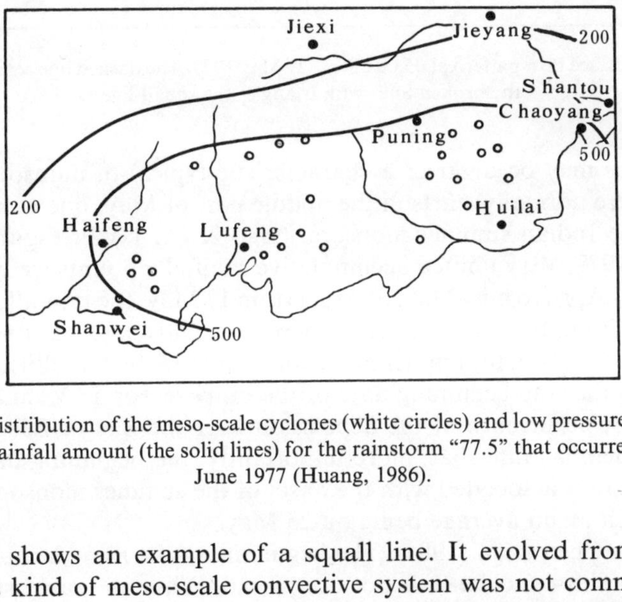

Figure 3.16. The distribution of the meso-scale cyclones (white circles) and low pressure (black circles) and the accumulative rainfall amount (the solid lines) for the rainstorm "77.5" that occurred during 27 May-1 June 1977 (Huang, 1986).

Figure 3.17 shows an example of a squall line. It evolved from a meso-scale shearline. This kind of meso-scale convective system was not commonly observed during the experiment on the pre-summer heavy rains over southern China. When this squall line passed through Sihui County, it brought about high winds of over 16 m s^{-1}, a rainfall of 200 mm in 18 minutes, and a temperature drop of 6–7 °C per hour. Its mean motion speed was 50 km h^{-1}.

The rainy season in Yunnan Province in the southwest of China is quite different from the pre-summer rainy season over southern China in that it does not result from the process of seasonal changes of monsoons. Near the surface in Yunnan Province, a perpeptual southwesterly wind prevails almost all the year round, and only the dry

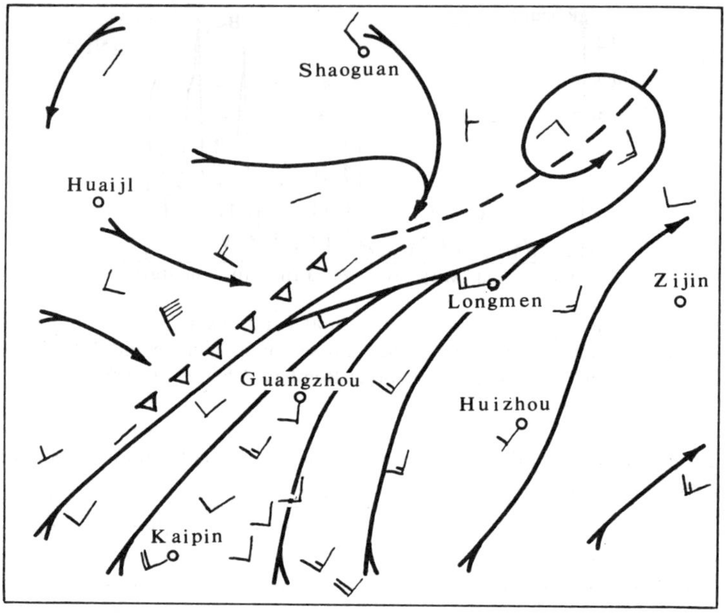

Figure 3.17. The surface flow pattern at 05:00 GMT, 29 May 1977. The dashed line represent the shearline; the broken lines with triangles the squall line.

and wet seasons may be divided, a characteristic typical of monsoon climate. The rainy season here normally starts in the middle part of May, one month earlier than the onset of the Indian summer monsoon (Gao *et al.*, 1962). Figure 3.18 gives an example from 1979. Very limited accumulative rainfall amounts were recorded from the beginning of April to mid-May. Starting from 17 May, the rainfall amount sharply increased, indicating the onset of the rainy season. This date was one month earlier than the onset of the Indian summer monsoon. Table 3.3 further illustrates this point. It also suggests that the beginning date of the rainy season in Yunnan is nearly the same as the first occurrence of the depression over the Bay of Bengal. Chen *et al.* (1984) clearly demonstrate that, more importantly, the beginning date of the rainy season in Yunnan is associated with the onset of the summer monsoon over the Bay of Bengal, which on an average being on 25 May.

According to Chen *et al.* (1984), The beginning of the rainy season in Yunnan is about 23 May. This concidence is not occasional. While the rainy season sets in, a significant change in some other major circulation features also may be observed, including the occurrence of a low-level southwest monsoon jet and the southern branch of a tropical easterly jet, and the axis of the anticyclone at 100 hPa over Southeast Asia suddenly moves northward from 20° N to about 25° N, along with a shift of the major trough, and the transformation of the Kunming quasi-stationary front into the movable cold front which becomes the major rain-producing systems in this region.

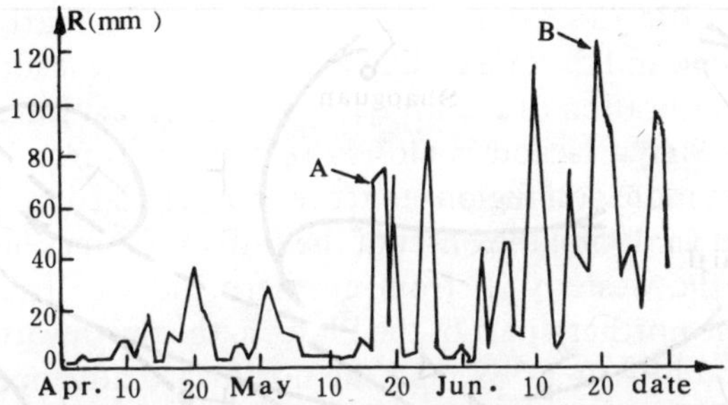

Figure 3.18. The curve of accumulative rainfall amount at 7 stations in Yunnan Province for April-June 1979. The letter A indicates the starting date of the rainy season in Yunnan and B the date of the onset of the Indian summer monsoon at Bombay (Chen *et al.*, 1984).

Table 3.3. The beginning dates of the rainy season in Yunnan Province and the dates of the onset of the Indian summer monsoon (Chen *et al.*, 1984)

	1975	1976	1977	1978	1979
Dates of onset of rainy season in Yunnan	May 7	May 2	May 14	May 12	May 17
Onset of monsoon at Bombay	June 18	June 5	June 16	June 9	June 19
The first occurrence of depression over the Bay of Bengal	May 4	April 30	May 10	May 14	May 6

3.3. The Meiyu (Plum Rains) over the Yangtze River Valley

In the Yangtze River Valley, the rainy period from mid-June to mid-July is the noted Meiyu or Plum Rains. This rainy season is characterized by the abundant precipitation, high relative humidity, cloudy or overcast, short daily duration of sunshine, weak surface wind, and the frequent occurrence of heavy rainfalls. The precipitation is mostly of a continuous and stable nature, but sometimes with thunderstorms and showers. During the Meiyu season, the Yangtze River Valley is vulnerable to flooding and droughts, due to the relative concentration and severity of rainfall amount. Based on the statistics of May-September 1966–1975, about 40% of heavy rainfalls occur during one month from mid-June to mid-July, which corresponds to the Meiyu season.

Different definitions of the Meiyu have been used in various observatories. Thus, the starting and ending dates of the Meiyu period vary from place to place. The Meiyu season in the middle and lower Yangtze River Valleys is usually defined by the duration of the rainy period at 5–6 representative stations. In general, if the duration of a rainy period persists for at least six days at 5–6 stations, this may be defined as the Meiyu season, with other circulation conditions being satisfied. Based on the statistics made by Nanjing Observatory for 27 years from 1954 to 1980 (Xian and Bao, 1986), the starting date of the Meiyu is, on an average, 20 June and the ending date

10 July. Therefore, the Meiyu season spans about 20 days. The early onset of the Meiyu may occur early in June and the late ending may occur in the last part or at the end of July, thus the longest duration of the Meiyu season may last for about 2 months.

The onset of the Meiyu period is closely related to a prominent change in the circulation features in monsoon regions that are characterized by the following major events (Yeh, Tao and Li, 1958): the onset of the Indian summer monsoon, the abrupt northward jump of the westerly jet from northern India or from the south of the Tibetan Plateau to the northern part of the Plateau, the northward shift of the axis of the anticyclone at 100 hPa to 30° N, and the northward advance of the subtropical high over the western Pacific from 20° N to 25° N. The recent study (Wang and Li, 1982) has indicated that the beginning and maintenance of the Meiyu is also related to the cross- equatorial airflow at about 110° E. Therefore, the Meiyu in East Asia is directly developed under the influence of the Indian southwest monsoon or cross-equatorial airflow coming from the Southern Hemisphere. Based on recent statistics of the relationship between the onset date of the southwestern monsoon over Bombay and the beginning of the Meiyu made by Chen *et al.* (1984), the onset of the Indian monsoon is, on an average, three days earlier than the beginning of the Meiyu, which is similar to previous findings.

Figure 3.19 is a schematic map of large-scale circulation features characterizing the Meiyu in East Asia. It can be seen that the Meiyu front is formed through the meeting of monsoon airflows and the polar cold airmass or transformed polar continental airmass. The monsoon airflows come partly from the Indian Ocean and partly from the cross-equatorial airflow over the South China Sea and the western Pacific (Si, 1989). They eventually merge into a strong southwest monsoonal flow over southern China and the Yangtze and the Huaihe River Valleys. The onset of the Meiyu is almost simultaneous with the occurrence of the stable and persistent southwest monsoon in this region (Lin *et al.*, 1983). Concurrently, the isopleth of θ_{se} = 340 K also abruply moves over this region, indicating that the inflow of a large amount of warm and moist air into the Meiyu region.

The ending of the Meiyu in the Yangtze River Valley is characterized by the replacement of the southwest monsoonal airflow by the southeast wind. This condition is created by the westward extension of the subtropical high over the western Pacific which is accompanied by the advance from the south to the north and from the region off the coast inland of the southeasterly wind at the southwestern flank of the subtropical high. This kind of southeasterly wind, prevailing after the Meiyu season that starts the mid-summer of the year, is called the Bo-Zhao wind by Su Dongpo, a noted poet during the Song Dynasty.

During the Meiyu season, a persistent flow pattern is observed at the mid- and high latitudes over Eurasia, characterized by a blocking situation upstream and/or downstream of the Meiyu region at 50–70° N and a zonal flow prevailing in the latitude band of 35–50° N. The shortwave troughs often move southeastward down to the Yangtze River Valley, bringing cold air over this region, thus forming and maintaining the stationary Meiyu front. Figure 3.20 is the compositing 850 hPa flow pattern derived from 10 cases of heavy rainfall with thunderstorm activity during the Meiyu season for 1973–1983 (Zhang *et al.*, 1986).

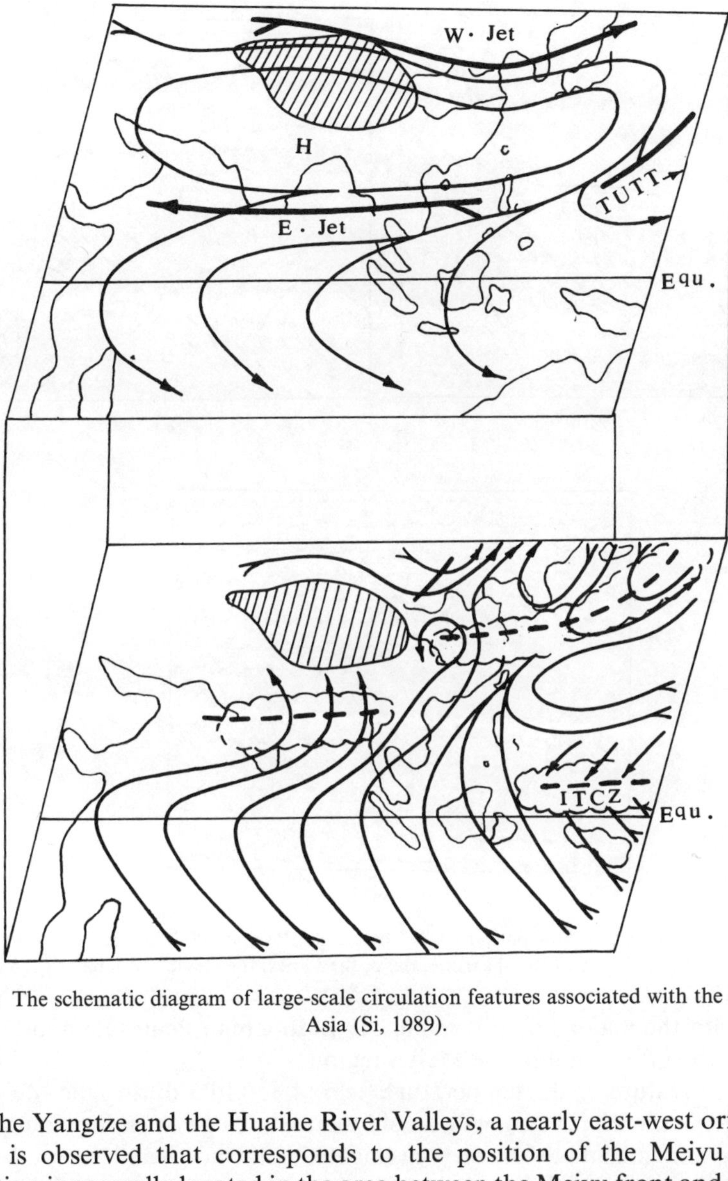

Figure 3.19. The schematic diagram of large-scale circulation features associated with the Meiyu in East Asia (Si, 1989).

Over the Yangtze and the Huaihe River Valleys, a nearly east-west oriented wind shearline is observed that corresponds to the position of the Meiyu front. The precipitation is generally located in the area between the Meiyu front and the 700 hPa shearline, about 5 latitude wide (Qiu and Ding, 1979). The southwest airflow is found to the south of the shearline, while the easterlies of the rear part of the modified continental cold high are observed to the north of the shearline. There are two vortices along the shearline (mainly near the surface), located in the Sichuan Basin and the eastern part of the Dabie Mountain, respectively. Note the strong wind axis in the southwest monsoonal airflow originating in the Indian Ocean that reaches the intensity of a low-level jet (12 m s^{-1}). It splits into two branches while propagating

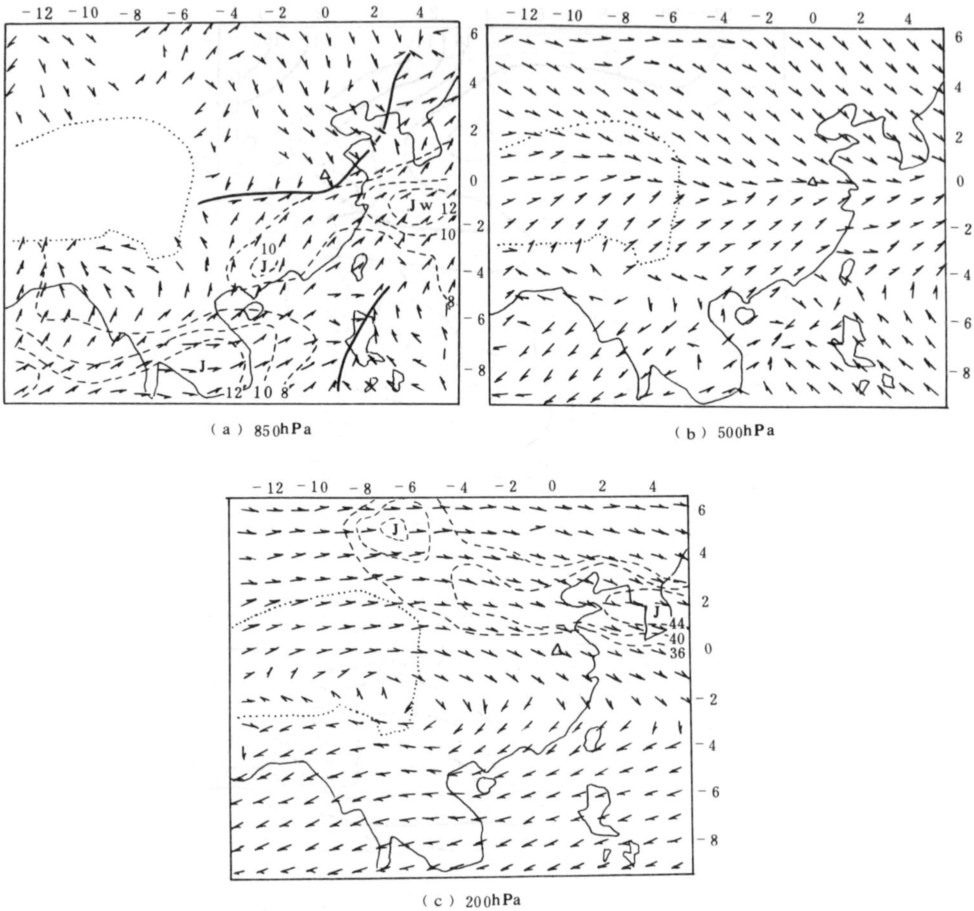

Figure 3.20. The compositing flow pattern at 850 hPa, 500 hPa and 200 hPa (see the text). The full line represents the position of the shearline; the dashed lines, the isotach (Zhang *et al.*, 1986).

eastward, with the major one traversing the Indo-China Peninsula and the Gulf of Tonkin and then flowing into the Meiyu region.

The salient feature of the temperature field at 850 hPa during the Meiyu season shows up in a sandwich pattern, with a cold thermal trough extending southwestward into the Meiyu region and the warm areas located to the north and the south of it (Fig. 3.21). The temperature of the northern warm area may be sometimes higher than that of the southern warm area, due mainly to strong sensible heating from the underlying surface in the early summer (Sato, 1985). The shearline is located within the cold thermal trough. Above 700 hPa, the temperature field is characterized by an east-west oriented warm belt capping the low-level cold trough.

The moisture field is shown in Figure 3.22, which indicates an extensive region of high moisture from the South China Sea to the Yangtze River Valley. This high moisture loading region almost corresponds to the location of low-level jet and

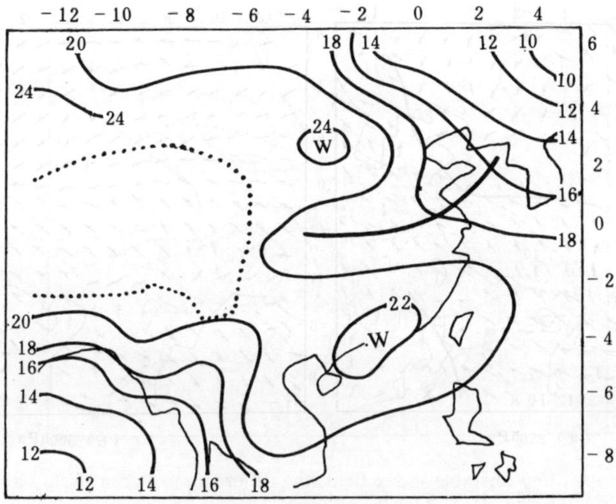

Figure 3.21. The compositing temperature field at 850 hPa. Units: °C (Zhang *et al.*, 1986).

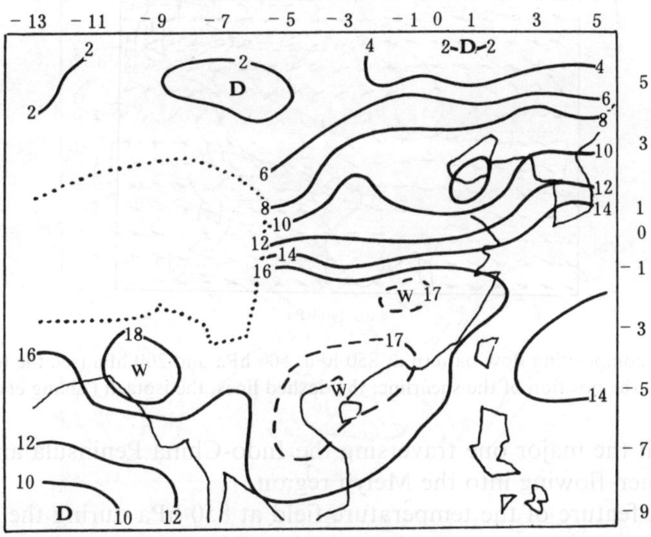

Figure 3.22. The compositing dewpoint field at 850 hPa. Units: °C (Zhang *et al.*, 1986).

constitutes the prominent dewpoint front or the energy front with the dry cold air on the northern side, with the moisture (dewpoint) gradient of 4–7 °C/5 latitudes, significantly greater than the temperature gradient (1.5 °C/5 latitudes). Therefore, the Meiyu front is mainly characterized by the moisture contrast across the front rather than the temperature contrast.

It is very interesting to note the thermal stability distribition field (Fig. 3.23). In the region of heavy rainfall during the Meiyu, there is a weakly unstable condition.

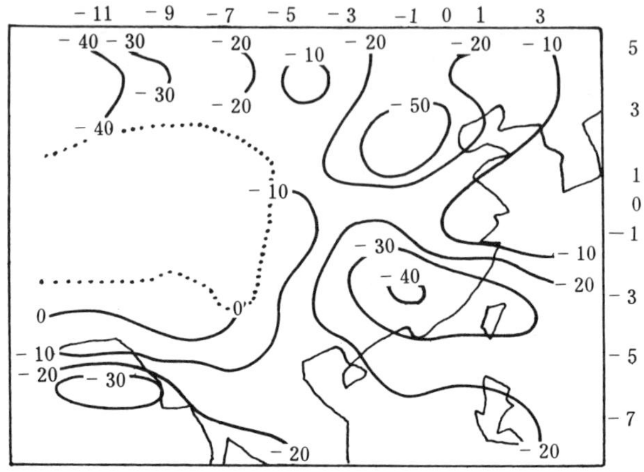

Figure 3.23. The compositing unstable index field, the difference between $(\theta_{se})_{500}$ and $(\theta_{se})_{sur}$. Units: °C (Zhang *et al.*, 1986).

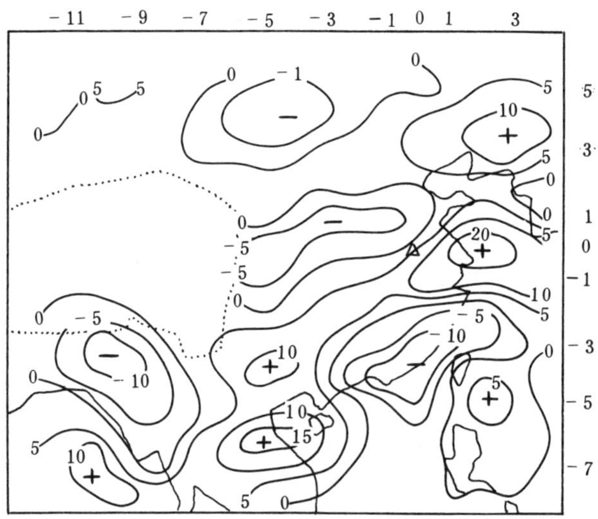

Figure 3.24. The compositing vorticity field at 850 hPa. Units: 10^{-5} s^{-1} (Zhang *et al.*, 1986).

However, there is the strongly unstable stratification to the north or south of this region.

The pattern of the vorticity during the Meiyu period distributes zonally. The belt of positive vorticity at low level is in coincidence with the belt of heavy rainfall, with the belts of negative vorticity located to the north and the south of it, respectively (Fig. 3.24). At high level, a belt of negative vorticity lies over the belt of positive vorticity, thus creating a self-maintaining mechanism for the heavy Meiyu rainfalls. The pattern of divergence also assumes the feature of zonal distribution, with the belt

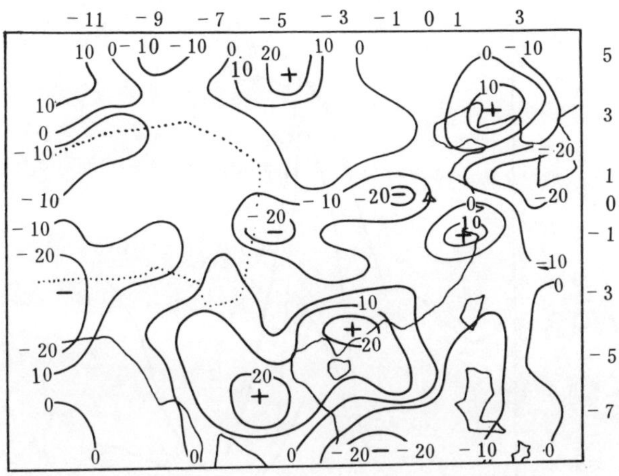

Figure 3.25. The compositing field of *p*–vertical velocity at 500 hPa. Units: 10^{-4} hPa s^{-1}.

of convergence in the layer of 850–500 hPa being coincident with the belt of heavy rainfall. At 200 hPa, a significant belt of divergence lies over the region of heavy rainfall (not shown). Figure 3.25 is the pattern of *p*–vertical velocity ($\omega = dp/dt$) for the Meiyu period, indicating a wide area of ascending motion in the Meiyu region. To the south and the north of this region, are the regions of descending motion. A similar result was obtained by Qiu and Ding (1979) in the study of the circulation structure during the Meiyu period of 1973.

From satellite images, one may clearly see a continuous cloud band over the region of the Meiyu, generally located to the south of the shearline at 700 hPa. The convective cloud masses or ensembles are often observed to be embedded in the extensive cloud band. Figure 3.26 show this relationship (Qian *et al.*, 1986). It can be seen that the maxima of total cloudiness, thunderstorm activity, precipitation amount, and convective cloudiness are basically located at the same latitude, about 30° N. The convective cloudiness accounts for about 30% of the total cloudiness. Note that the thunderstorm activity, precipitation, and convective cloudiness have the nearly same width (5° latitude wide). Therefore, the heavy rainfalls during the Meiyu period are of the nature of convective precipitation.

The Meiyu is mainly produced by the Meiyu front which has been addressed by numerous investigators (Xie, 1956; Tao, 1957; Qiu and Ding, 1979). Figures 3.27–3.29 represent compositing structures of the Meiyu front (Qiao *et al.*, 1986 a). The front has an apparent structure of frontal zone, mainly above 500 hPa. The thermal gradient is directed from the cold air in the north to the warm air in the south. Above the frontal zone, there is an upper-level westerly jet at 200 hPa. At 700 hPa, the structure of the frontal zone cannot be detected, only with an observed shearline that forms between low-level easterly and westerly winds. Note that the low-level jet at about 700 hPa seems to be a downward extension of the upper-level jet. This problem will be discussed later. Below 850 hPa, there is a significant temperature gredient,

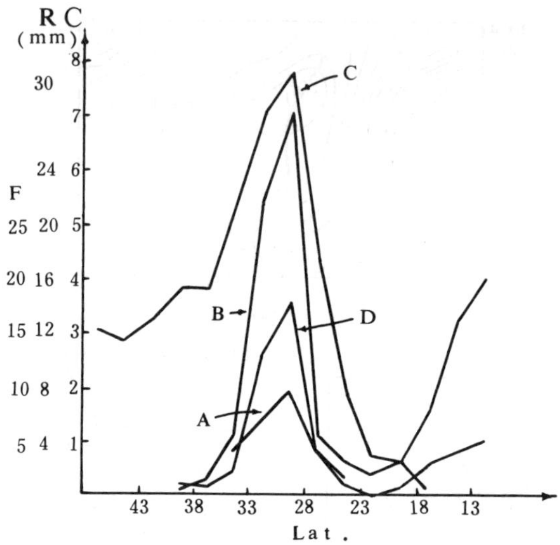

Figure 3.26. The latitudinal distributions of total cloudiness (C), precipitation (B), thunderstorm activity (D), and convective cloudiness (A) along 110–120° E averaged for the four Meiyu processes of 1980–1983. F represents frequency and C octas of cloudiness (Qian *et al.*, 1986).

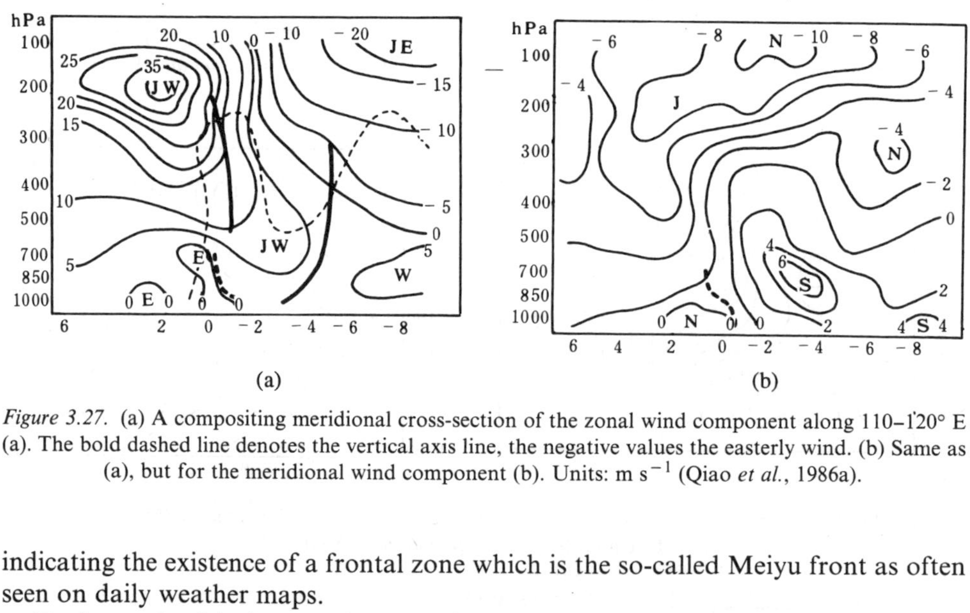

Figure 3.27. (a) A compositing meridional cross-section of the zonal wind component along 110–120° E (a). The bold dashed line denotes the vertical axis line, the negative values the easterly wind. (b) Same as (a), but for the meridional wind component (b). Units: m s^{-1} (Qiao *et al.*, 1986a).

indicating the existence of a frontal zone which is the so-called Meiyu front as often seen on daily weather maps.

To the north of the heavy rains, there is a deep layer of dewpoint front, with a mean gradient of 7–9 °C/5 latitude, while to the south there is a vertical deep layer of high moisture. Therefore, the Meiyu front is essentially a dewpoint front or energy front, with the thermal contrast or gradient being less distinctive.

The θ_{se} cross-section also indicates an obvious frontal zone above 500 hPa and its nonexistence below 700 hPa.

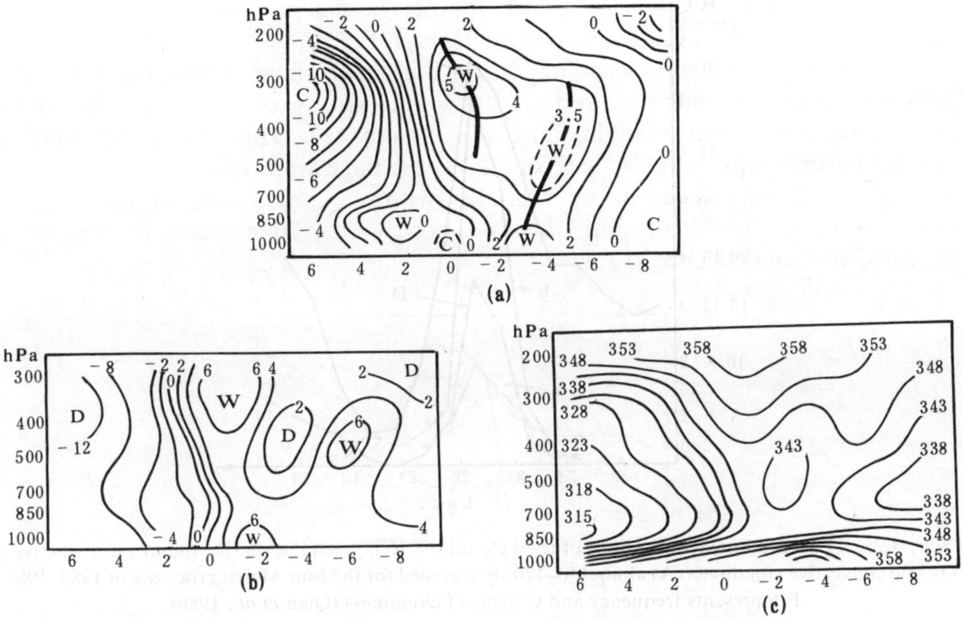

Figure 3.28. The compositing meridional cross-sections of temperature deviation (a), the dewpoint temperature deviations (b) and θ_{se}(c) along 110–120° E. Units: °C (Qiao *et al.*, 1986a).

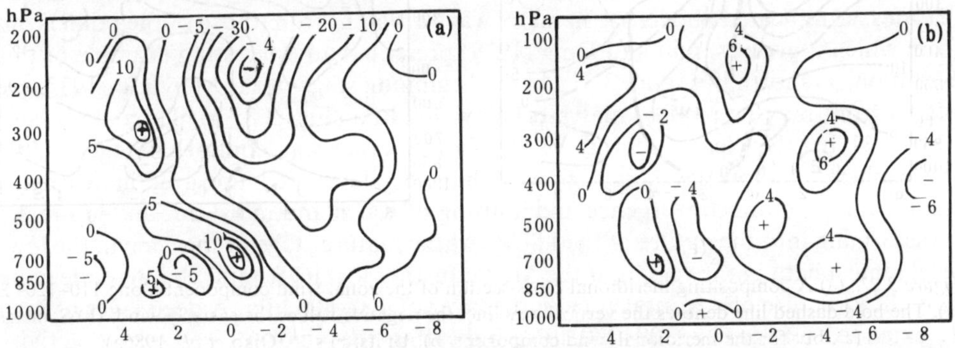

Figure 3.29. The compositing meridional corss-sections of vorticity (a) and divergence (b) along 110–120° E. Units: 10^{-5} s^{-1} (Qiao *et al.*, 1986a).

A layer of high positive vorticity below 500 hPa corresponds to the wind shearline, with the maximum in the layer of 850–500 hPa. Above 500 hPa, there is a deep layer of negative vorticity. The convergence layer associated with the Meiyu shearline is deep, extending upward to 300 hPa. Below 850 hPa, however, the strong convergence is located to the south of the shearline, while in the layer of 850–300 hPa, it is found to the north of the shearline. Above 300 hPa, a strong divergence exists.

The meridional cell associated with the Meiyu region is unique, and broadly similar to the condition of the ITCZ in the tropics (Fig. 3.30). Over the heavy rains of the Meiyu, there is a deep ascending airflow that results from a merging of the cold airflow coming from the north and the monsoonal airflow coming from the tropics. The vertical axis of the ascending airflow is almost erect. This belt of confluencing airflows corresponds to the Meiyu front. Note the huge monsoon cell to the south of the region of the Meiyu heavy rainfalls, with the descending branch located equatorward of 10° N. Overall, this is a positive meridional cell, as previously indicated by Qiu and Ding (1979).

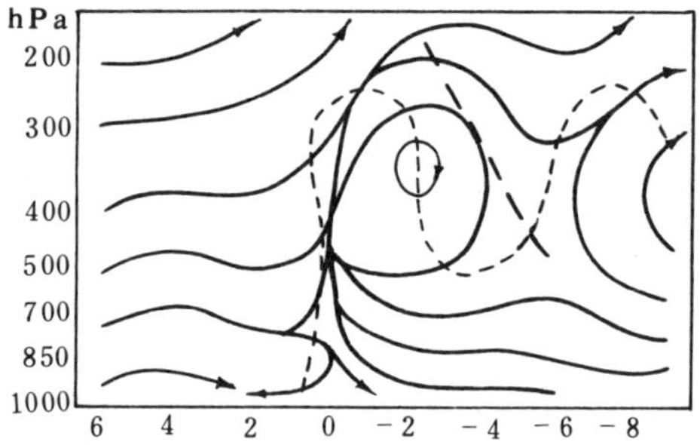

Figure 3.30. The compositing meridional cell associated with Meiyu heavy rainfalls. The dashed line denotes the isopleth of relative humidity greater than 80% (Qiao *et al.*, 1986).

The Meiyu front is a major synoptic-scale rain-producing system. Its unique features, as indicated above, are in many ways different from those of the polar front from which it evolves. Qiu and Ding (1979) presented an interesting example of this evolution, as shown in Figure 3.31. At the beginning stage of the Meiyu for 1973, the Meiyu front may be readily analysed below 400 hPa (Fig. 3.31 a). Three days later (Fig. 3.31 b), the clear structure of the frontal zone was confined in the middle troposphere. At low level, a region of high θ_{se} developed upward, indicating a significant moisture convergence and a lifting of warm and moist air, and gradually transforming into the nature of the ITCZ. At the mature stage of the Meiyu, the low-level atmosphere in the region of the Meiyu further lost its frontal character, leaving only the structure of a frontal zone in the layer of 450–600 hPa. A deep layer of high θ_{se} extended upward to 300–200 hPa, indicating the vigorous convective activity involved. Note that the slope of θ_{se} isolines tilts southward, very much similar to the structure of the equatorial front analysed by Xie (1956). The above evolutive process of the Meiyu front demonstrates the crucial importance of convective activity that forced the Meiyu front to be transformed into a belt of strong convection and to lose its frontal character in the lower troposphere. Therefore, the more prominent features of the Meiyu front are the contrasts of airflows (shear line) and moisture (dewpoint front) rather than temperature.

A study on the structure and the vorticity budget of a Meiyu system for the period 10–15 June, 1974 also revealed similar results (Chen and Chang, 1980). They pointed out that during the mature and decaying stages of the monsoon trough (nearly

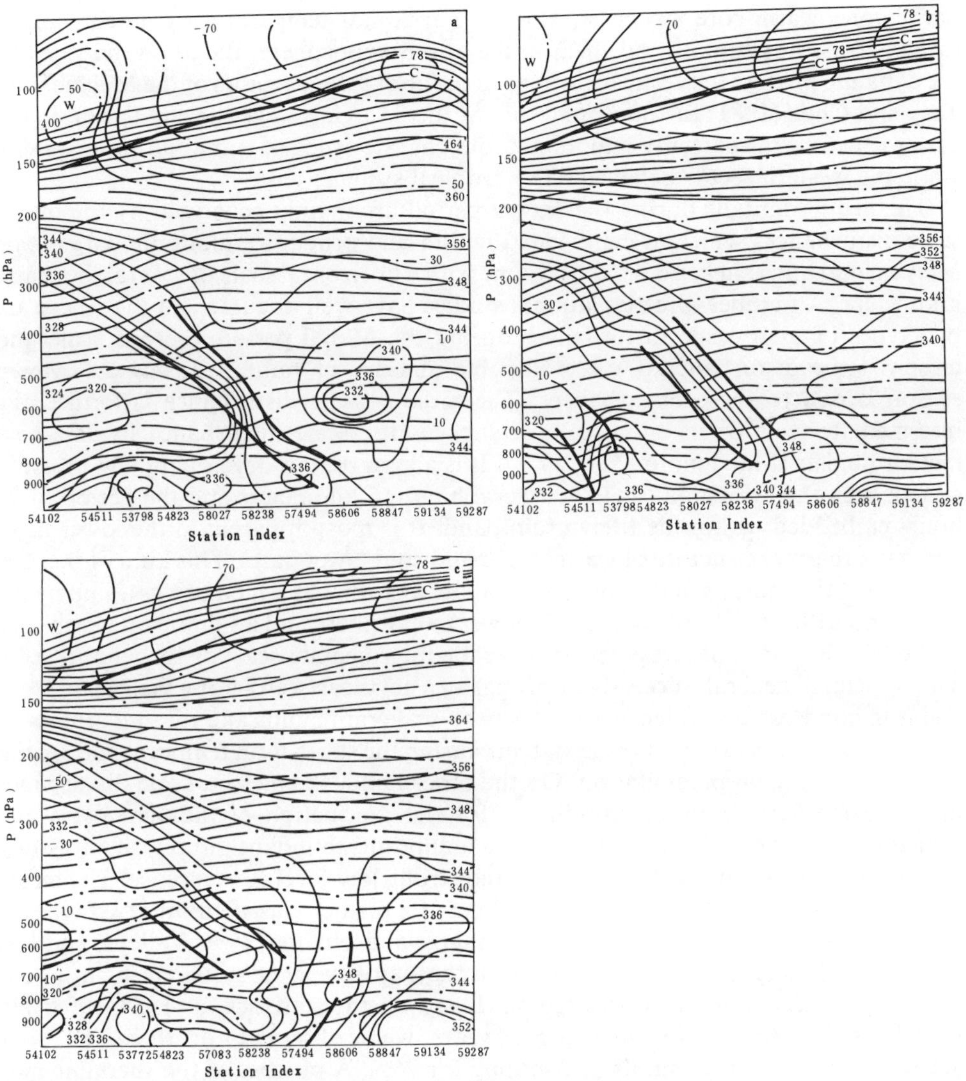

Figure 3.31. The meridional cross-sections across the Meiyu front along 116° E for 17 June 1973 (a), 20 June 1973 (b) and 22 June 1973 (c). The full lines represent θ_{se} isolines (K) and the dotted-full lines, the isotherms (°C) (Qiu and Ding, 1979).

correrponding to the Meiyu front), the structure of the eastern sector (southern Japan) and the central section (East China Sea) is different from that of the western section (southeastern China). In the former two sectors, the structure resembles a typical mid-latitude baroclinic front (like a polar front) with a strong vertical tilting toward an upper-level, cold core low and a strong temperature gradient. On the other hand, the western sector assumes a semi-tropical disturbance with an equivalent barotropic, warm-core structure, a weak horizontal temperature gradient, and a rather strong horizontal wind shear in the lower troposphere. Based on the study of the dynamic process for vertical motions in the above three sectors of the Meiyu front, Tsay and Chen (1981) further found that at the decaying stage the circulation in the eastern sector still partially maintained the charateristics of a mid-latitude system, while the western sector behaved like a tropical system.

The heavy rainfalls during the Meiyu period are mainly generated by the meso-scale disturbances which are embedded within and propagate along the large-scale Meiyu cloud and rain band or frontal zone with a horizontal scale of several thousand kilometers. A number of observational studies have well documented that there are two types of meso-scale disturbances during the Meiyu period: meso-α scale and meso-β scale disturbances. A considerable amount of rainfall during the Meiyu season is due to these disturbances. The meso-α scale disturbance is termed the intermediate-scale or medium-scale or subsynoptic scale disturbance by Japanese meteorologists (Matsumoto *et al.*, 1971). It is a kind of non-developing low-pressure system with the horizontal scale of several hundred kilometers and the period of 20 hours embedded within the Meiyu rain band. It is most apparent in the lower layer and has a thermal structure of warm upper and cold lower parts. This kind of thermal structure is the same as that of the intermediate disturbance which brings about heavy rains around Japan. A cold core in the lower troposphere and a significant warm core in the middle and upper troposphere over the intermediate-scale disturbance is seen. This system, in general, successively propagates northeastward along the Meiyu front and is in good correspondence with the meso-scale rain bands and rainfall centers. It is the major rain-bearing synoptic system causing the short-term temporal and spatial variations of Meiyu precipitation. On the other hand, the meso-α scale disturbance may provide very favorable conditions for the production of thunderstorms and rainstorms during the Meiyu season, including an abundant supply of moisture, warm and moist unstable stratification, and strong low-level convergence. The meso-α scale systems during the Meiyu period may be further classified into two types: a Changjiang-Huaihe shearline over the Yangtze and Huaihe River Valleys, and low vortex. The Changjiang-Huaihe sheraline is the major synoptic system bringing about the Meiyu heavy rainfalls in this region. Based on the statistics for 1954–1958, 110 shearlines for April-September for 5 years were observed, of them, 84 were accompanied by heavy raifalls, accounting for 76%. A process of the sheraline may cause the rainstorms for 5–7 days. A recent study has indicated that the Changjiang-Huaihe shearline is, in fact, a reflection of the airflow pattern of the meso-scale cloud and rain bands during the Meiyu period. It has relatively shallow systems with the vertical extent confined in the layer below 500 hPa, and being especially obvious below 850 hPa. The sheraline is also of baroclinic structure, tilting northward. There is a marked belt of airflow divergence aloft lying over the low-level sheraline. The temperature and pressure fields corresponding to the meso-scale rain band are mainly characterized by the apparent thermal and dew-point frontal zone, but the frontal zone does not extend very high, generally, only up to about 850 hPa. There is a belt of convective heating in the meso-α scale rainband.

So far no acceptable viewpoint has been put forward concening the dynamic mechanism of the formation of this kind of intermediate-scale rainband or shearline. Some investigators argue that a baroclinic wave of Ri <1 is likely because Ri, the Richardson number of the Meiyu frontal zone, is very small under the mean condition and it is very easy to satisfy the criteria of the unstable growth of the intermediate-scale distrubance, while other investigators suggest that this disturbance should be an unstable wave incorporating the effect of cumulus convection. Lu and Wang (1988) also have a similar argument, indicating that the development of the intermediate-scale disturbance results from the combined baroclinic-CISK condition. Recently, Zhang (1988) discussed the meso-scale stability of the baroclinic airflow, showing that in the case of a thin sheared layer and small Ri, the Eady mode may extend down to the regime of the meso-α scale wave. The characters thus obtained still have the quasi-geostrophic property. This mechanism seems to be used for explaining the propagating-type meso-scale disturbance in the Meiyu front.

There are at least two kinds of meso-scale low vortices that may bring about heavy rains during the Meiyu season. One kind is the southwestern vortex which will be dealt with in Chapter 5. This low vortex is one kind of meso-scale synoptic system generated on the leeside of the Tibetan Plateau. It is most evident at the 850–700 hPa level, while at the surface and at 500 hPa, it cannot easily be detected. After the southwestern vortex forms, it tends to be stationary locally if there is no upper-level trough to steer it out of the Sichuan Basin. Howerer, if the vortex is steered out, it may move eastward along the Meiyu shearline in most cases, and, in some cases, moves northeast or southeastward. Meanwhile, it can cause heavy rainfalls along its track. When an extended trough is connected with the southwest vortex to build up the flow pattern of a northern trough and southern vortex, this synoptic condition may often cause extensive heavy rainfalls. Figure 3.32 is an example of a southwest vortex that moved eastward and developed, and then caused an east-west elongated belt of precipitation.

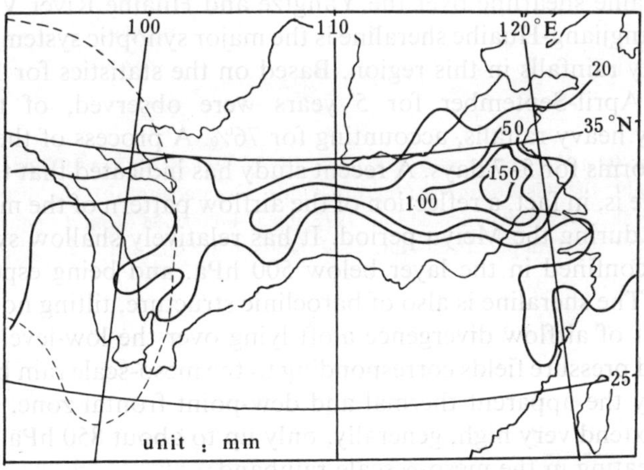

Figure 3.32. Rainfall pattern for the period 1–3 July 1972. Units: mm (Tao *et al.*, 1980).

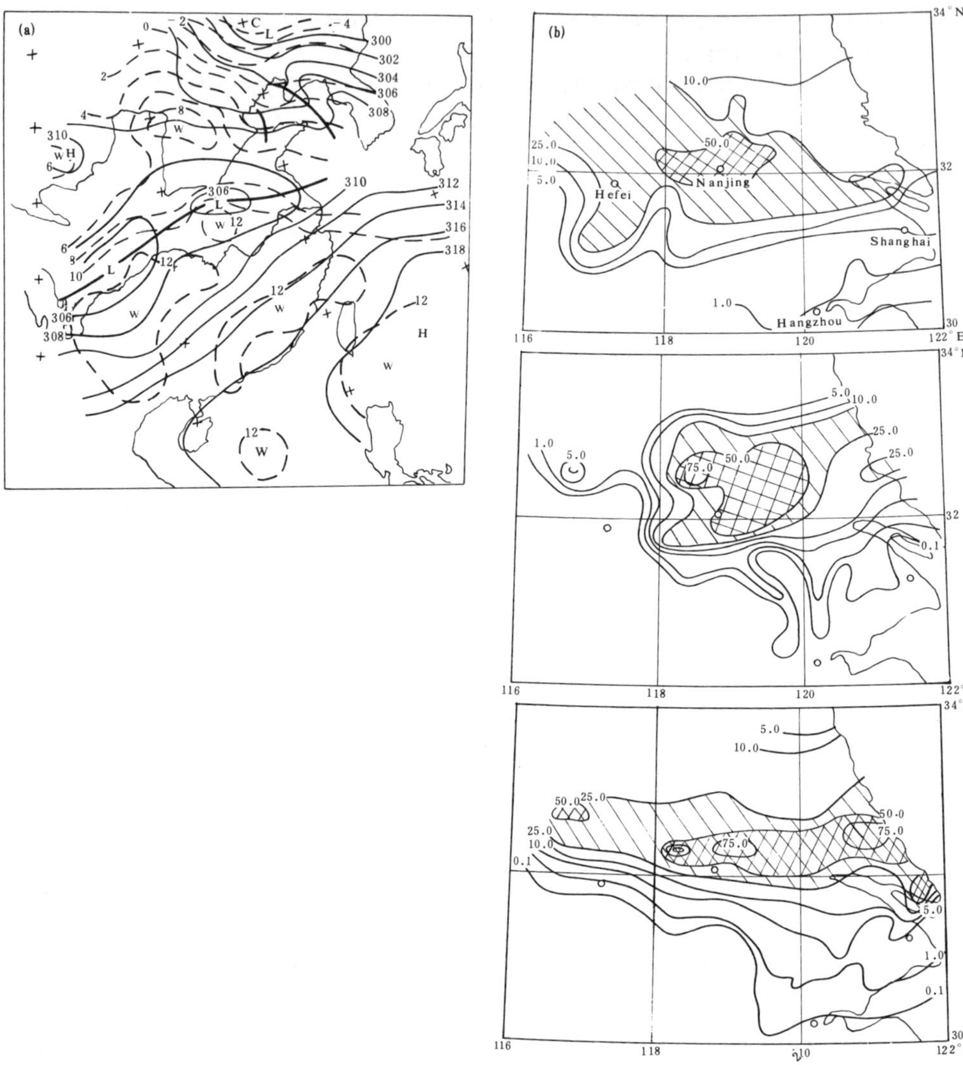

Figure 3.33. (a) The 700 hPa synoptic map at 00:00 GMT, 21 June 1972. The full lines represent isopleths of geopotential height; the dashed lines isotherms. Units: Decameters. (b) The distribution of rainfall amount for 18:00–24:00 GMT (top panel) 20 June 1972, 00:00–06:00 GMT (middle panel), and 06:00–12:00 GMT (bottom panel), 21 June 1972. Units: mm (Zhang and Li, 1979).

An other kind of low vortex is the intermediate-scale cyclone with a horizonatal scale of 1000–3000 km. Figure 3.33 is an example of the heavy rainfall caused by an intermediate-scale low vortex that was initially generated in Henan Province and then moved eastward. The starting time of the heavy rainfall was recorded after 18:00 GMT, 20 June 1972. During its genesis and development, alltogether three heavy rainfalls occurred that were located in the western part of the Tabie Mountain, i.e. the southeastern sector of the Henan low vortex. Note the three episodes of heavy rainfall shown in Figure 3.33(b).

The intermediate scale cyclone is different from the synoptic-scale cyclone in that the latter is closely related to the trough in the upper troposphere, while the former is laid aloft at 500 hPa, by flat and straight westerlies. The cyclonic disturbance is mainly confined in the layer below 500 hPa and in the upper troposphere there is a high-pressure ridge lying over it. Therefore, this kind of low vortex is also characterized by the negative vorticity at the middle or upper levels. The thermal structure of the intermediate cyclone shows the warm core in the upper troposphere. The differences in the structure between the synoptic-scale cyclone and the intermediate-scale cyclone suggest a possibly different dynamic mechanism of their formation. In most cases, the former is initiated by the baroclinic mechanism, while the latter is greatly dependent upon the release of condensational latent heat, as indicated by Si (1976) and Chen *et al.* (1980). That is to say, the intermediate-scale cyclone is generated through the CISK mechanism. But, some of the intermediate-scale cyclones may also gain their growth from the baroclinic energy source. Jin (1982) analysed the structure of an intermediate-scale disturbance over the Valley of the Changjiang and Huaihe Rivers, indicating that this disturbance formed in the low-level cold frontal zone. The genesis and development of the low vortex has a significant relationship to the release of the available potential energy of the baroclinicity of tropospheric frontal zone. Zhao (1988) also emphasized a similar result based on his study of the enegetics of the cyclogenesis on the Meiyu front. He pointed out that at the developmental stage, the baroclinic energy conversion was very significant, especially in the middle and upper troposphere. Zhao *et al.* (1982) simulated the genesis and development of a meso-α scale low vortex and found that the meso-α scale vortex may be generated in the large-scale synoptic system. The dynamic factors play a major role at the genesis stage, while at the developmental stage the condensational latent heating produced by precipitation plays an important role in the maintenance and intensification of the meso-α scale low vortex.

Utilizing a 10–layer moist meso-scale model, Huang *et al.* (1987) simulated a heavy rainfall process during the Meiyu season, with a special emphasis on the structure and evolution of the intermediate-scale system. They found that the low-level vorticity field, especially the low-level positive vorticity center, is closely related to the center of heavy rains. The maxima of 3 hours precipitation developed and moved along with the vorticity maximum center. While the low-level vorticity maximum moved near or under the center of the negative vorticity at the upper level, it intensified and, concurrently, the upward motion and precipitation also intensified. These results are broadly consistent with the observational evidence.

When the intermediate-scale cyclone is connected with the upper-level trough in the westerlies, it is likely to develop into a baroclinic synoptic-scale cyclone. Many cases of this evolutive event have been studied, among which is the work by Xu and Jiao (1984). They analysed the structure of two cyclones over the Valley of the Changjiang and Huaihe Rivers by using conventional meteorological data during the period of 23–25 June and 23–25 August 1980. The results show that the mean wavelength of the cyclones is about 2000–2500 km and the vertical extent may reach up to 7–8 km high. The axis of the cyclones slopes pronouncedly northward with height (Fig. 3.34). The strongest cyclonic circulation and vorticity are found at about 700 hPa. The frontal zone is stronger in the upper troposphere, but weaker in the lower troposphere, only with cold air being observed near the center and to the north of the cyclone at and below 700 hPa. The heavy rainfall is found in the front sector of the cyclone, which is consistent with the regions of strong upward motion and low-level convergence.

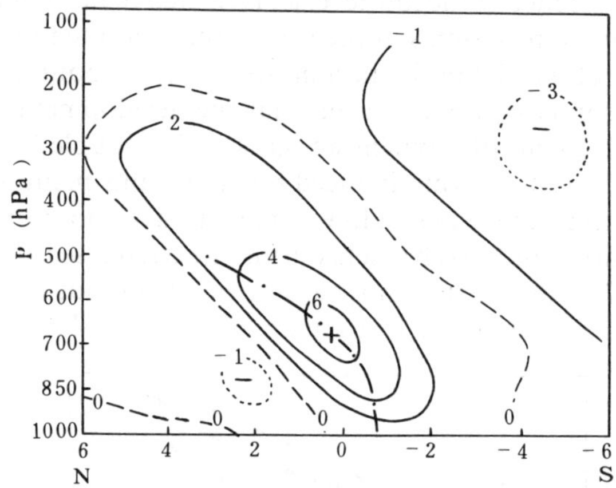

Figure 3.34. The cross-section of the relative vorticity for the average cyclone for 1980 (Units: 10^{-5} s^{-1}). The dotted-dashed line denotes the axis of maximum vorticity (Xu and Jiao, 1984).

Previously, due to the lack of data, very little work was done on meso-β scale systems and, consequently, little was known of their structures and dynamic characters. The Meiyu experiment over the middle and lower Yangtze River Valley (1980–1984) provided the first opportunity of studying this system on the horizontal scale of 25–250 km, with a denser network of upper-air and surface observations. It has been found that the meso-β scale system occurs in advance of the forward-tilting minor wave trough which is located near the Meiyu cloud and rain bands, the right side of the upper-level jet, and the left side of the lower-level jet. In general, this system is associated with the meso-β scale shearline. In the south of the meso-β scale shearline, there is a belt of strong wind or a conveying belt of warm and moist air. To the left of this belt, rain masses of rainstorms occur. At the time of development of the rain masses of rainstorms, the shearline has the marked temperature gradient, which possibly indicates that the meso-β scale system at this time has the character of significant baroclinicity at the low level. At the middle and upper levels, the meso-β scale system is of the nature of significantly convective systems, with convergence found at low and middle levels, and divergence at upper levels due to the release of condensational latent heat. In the boundary layer, there is the dry and cold descending air produced by the drag effect of precipitation particles. The rainstorm is generally located somewhere between the meso-high and meso-low.

As observed during the pre-summer rainy season over southern China (Section 3.2), the low-level jet is also a major component of the Meiyu season that is closely related to the occurrence of heavy rainfalls. During heavy Meiyu rainfalls, the low level jet is a super-geostrophic southwesterly jet in the lower troposphere (Sun, 1986). But, at present, the argument about the cause of the formation of this jet is controversial. Japanese meteorologists suggest that the low level jet in the Meiyu regions results from the up-and down exchange of momentum by cumulus convection in the heavy rainfall region (Ninomiya and Akiyama, 1974); some investigators stressed that the occurrence of the low level jet is related to development and eastward movement of low pressure systems in the west; others pointed out that the low-level jet streak is a result of inertial oscillation in the planetary boundary layer. Recently, Uccellini and Johnson (1979) pointed out that the development of the low level jet results from an adjustment of momentum-mass and is coupled with the upper-level jet. In general, the isallobaric wind induced by mass adjustment in the exit region of the upper-level jet streak may cause a low level jet to develop. But, statistics and case studies of the formation of low-level jets during the Meiyu season (Huang and Gao, 1982; Si *et al.*, 1982) indicate that most low level jets (about 76%) occur below the entrance region rather than at the exit region of the upper-level jet streak. This is a

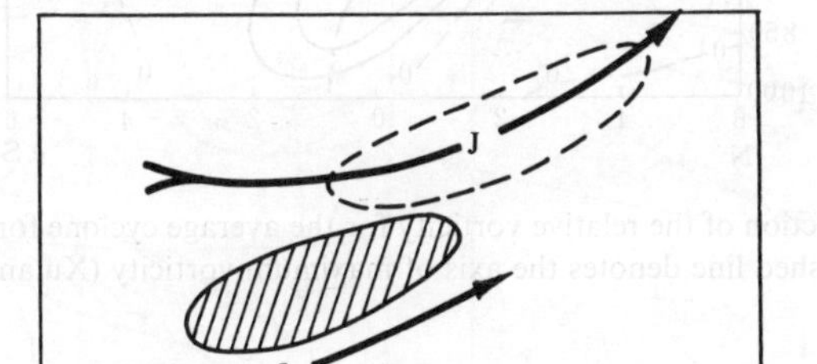

Figure 3.35. Total rainfall amount (solid lines) for 00:00 GMT, 11 July–00:00 GMT, 12 July 1969 and its relation to the upper-level jet (J) and low level jet(J_L). Units: mm (Si, 1988).

feature of the low-level jet during the Meiyu rainstorm period. Figure 3.35 shows this relationship (Si, 1988). The developmental condition for this kind of low-level jet in the entrance region is that, first, there is an extensive area of southwesterly wind and, then, as the upper-level jet moves eastward, the ageostrophic equilibrium near the streak of upper- level jet causes a region of upper level divergence in the right sector of the exit region of the jet, thus leading to convergence through mass adjustment and production of isallobaric winds in the lower troposphere.

This kind of isallobaric wind may cause the development of a low level jet while acting upon the low-level southerly winds. Xiao and Chen (1984) also discovered similar evidence, i.e., the low-level jet occurred at a distance of 550 km to the south of the upper-level jet entrance region at 700 hPa (Fig. 3.36). Extensive precipitation occurred over the region between the axis of the upper-level jet and that of the low level jet (see also Fig. 3.35). This region corresponds to the area of high temperature and moisture to the north of the subtropical high. Xiao and Chen discussed the acceleration of a low-level jet using the interaction of advective and adjustment

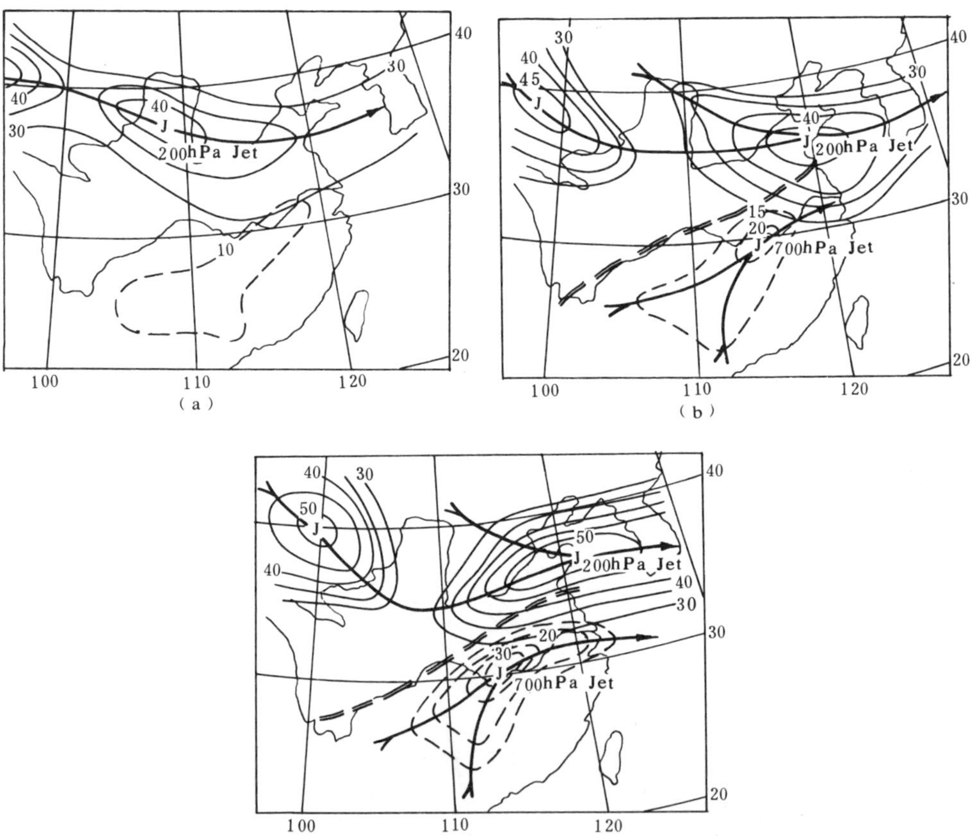

Figure 3.36. The patterns of jet streams at 200 (solid line) and 700 hPa (dashed line) (a) for 12:00 GMT, 10 July, (b) 00:00 GMT, 11 July, and (c) 12:00 GMT, 11 July 1969 (Xiao and Chen, 1984).

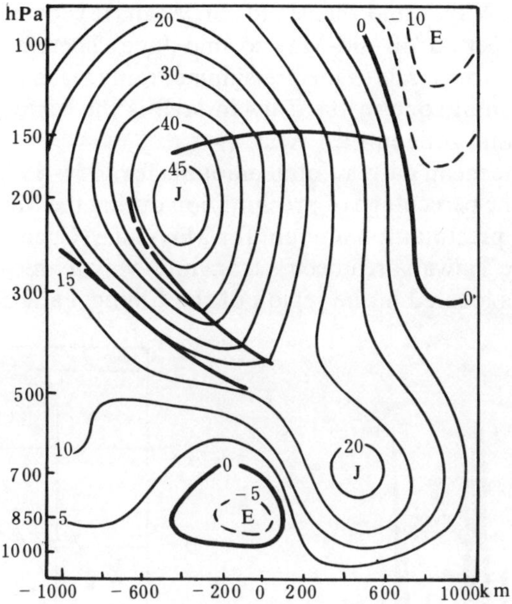

Figure 3.37. A composite cross-section of wind speeds across the shearline, averaged for eight cases of 1968–1971. Units: m s^{-1} (Xiao and Chen, 1984).

processes and the instability of the inertial gravitational wave during the adjustment process. They further made a composite picture of the cross-section of jet streams across the shearline by using eight cases for 1968–1971, and well documented the coupling relationship between the the upper-level jet and low-level jet (Fig. 3.37). It can be seen that the wind maximum of the upper-level jet is located at 200 hPa, with a wind speed of 48m s^{-1}, while the low-level jet has its maximum of about 20 m s^{-1} at 700 hPa, with the shearline in between.

In addition, Chen (1989) pointed out that this coupling also may occur between the low-level jet and the tropical easterly jet in South Asia. This vertical coupling is quite essential for the occurrence of heavy precipitation and the latent heat release acts as a positive feedback process for such a vertical coupling.

In the Taiwan area, the Meiyu rainfall is identified as one of the most destructive and disastrous natural hazards (Wu, 1990). Heavy regional precipitation (in excess of 100 mm daily), especially flash floods, may cause serious local hazards. For instance, a flash flood that occurred in central Taiwan on 7 August 1959 led to great property losses, 699 deaths and 377 missing persons. The flash floods that occurred on 28 May and 3 September 1981, as well as on 3 and 10 June 1984, led to great property losses resulting from heavy rainfall. But, it should be pointed out that the Meiyu (or Baiu) in the Taiwan area generally occurs during the period of late spring to early summer when the transition circulation regime from the northeast monsoon in winter to the southwest monsoon in summer occupies the area. As indicated previously, the mean position of this phenomenon migrates northward with time in the area to the South of the Yangtze River Valley and the islands of Japan. It is over southern China and Taiwan area in the period of mid-May to mid-June. Therefore, the Meiyu in the Taiwan area in fact corresponds to the pre-summer rainy season over southern China. The rain-bearing frontal zone in the Taiwan area is the eastward extension of the major planetary frontal zone in East Asia.

Figure 3.38 are the mean daily rainfall amounts for 1956–1975 at Teizhong (Chen and Wu, 1978) and the percentage of precipitation during the Meiyu period (15 May–15 June) of the total precipitation amount for May-June (Chen, 1983). It can be seen that the Meiyu in the Taiwan area occurs from mid-May to mid-June, and the major part of the rainfall is located in the region of the Chung Yang Shan Mo.

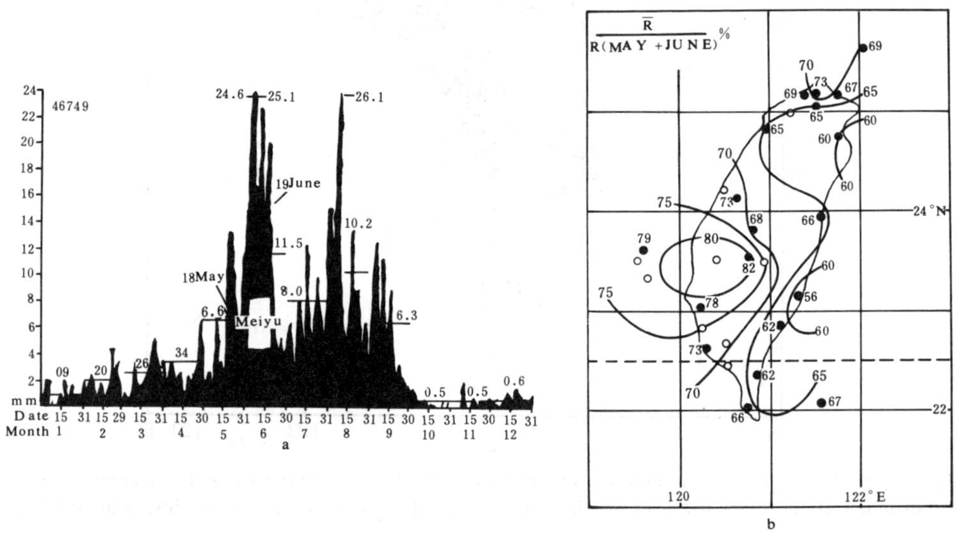

Figure 3.38. (a) The mean daily precipitation amount at Taizhong averaged for 1956–1975. (Units: mm); (b) the percentage of Meiyu precipitation (15 May-15 June) for the total precipitation amount for May-June (Chen and Wu, 1978).

The meso-α scale systems producing the Meiyu precipitation in the Taiwan area are the Meiyu front, meso-scale lows (stationary or moving), the low-level jet, and the outflow boundary. These systems have been extensively studied by meteorologists in the Taiwan area (Chen, 1989). Because of the importance of the meso-scale system for heavy rains during the Meiyu, the TAMEX (Taiwan Area Mesoscale Experiment) was carried out during May-June 1987. The three key scientific objectives of this program were to study:

(1) the mesoscale circulation of the Meiyu front, (2) the evolution of meso-scale convective systems in the vicinity of Taiwan, and (3) the effects of the orography of the Meiyu front on meso-scale convective systems. The field phase of TAMEX was carried out from 1 May to 29 June 1987, with a total of 13 operational missions. The observing network for TAMEX is shown in Figure 3.39.

The network consists of a conventional meteorological radar, seven rawinsonde

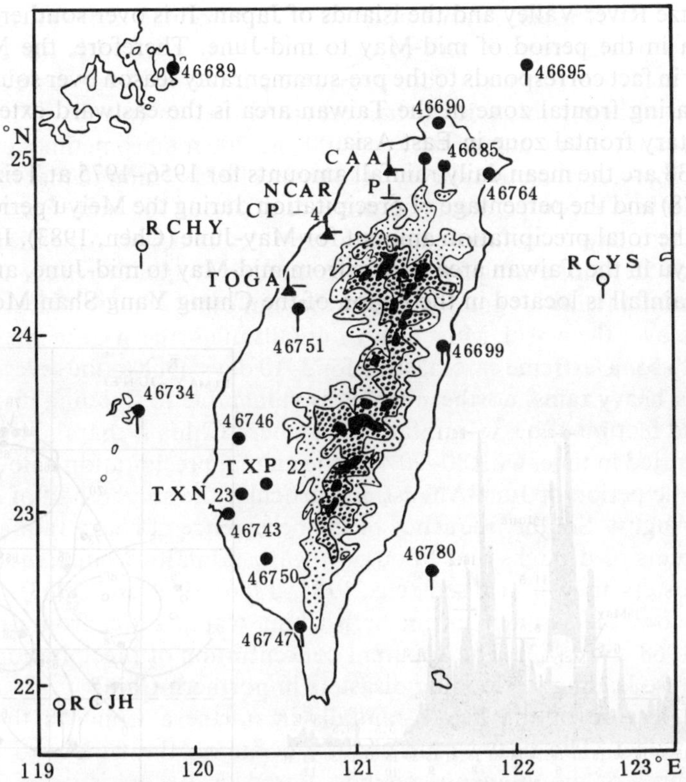

Figure 3.39. Locations of data-gathering sites during TAMEX in and near Taiwan. Black circles represent radiosonde stations, black triangles weather radar sites and white circles, research ships. (Kuo and Chen, 1990).

stations, four pibal stations, and a low-level minisonde, three research ships, a U.S. NOAA P–3 aircraft, and two Doppler radars, in addition to existing observation equipment. The main results obtained from this project have been summarized in the paper by Kuo and Chen (1990) and the proceedings of the Workshop on TAMEX Scientific Results (1990). The reader is recommended to refer to them for more details.

3.4. Heavy Rainfalls over Northern China During the Summer Monsoon

The rainy season in northern China starts in the middle or last part of July (July and August are two months with frequent occurrences of heavy rainfall in northern China). The excessively heavy rainfalls during the first part of August 1963 ("63.8" heavy rainfall) in the Haihe River Valley, in the middle part of July 1958 ("58.7" heavy rainfall) in the middle Yellow River Valley, and in the first part of August 1977, in the northern part of Shanxi Province, all occurred during such a time period. Climatologically, the first salient feature of the heavy rainfalls in northern China is their severity and long persistence. Northern China is situated in the temperate zone at 40–45° N. In summer, warm and moist monsoonal airflows may advance northward up to this region, and at the same time, meet with the cold air that frequently comes from the north, thus providing a favorable large scale condition for the formation of heavy rains. It is often recorded that an episode of heavy rainfall brings about rainfall amounts of between 100 and 200 mm, sometimes even as much as 400–500 mm. According to a survey of 23 cases of heavy rainfall that have occurred since 1959, 10 case recorded a rainfall amount of about 500 mm. For "63.8" heavy rainfall, 865 mm day^{-1} was recorded at a very localized site. An extremely intense rainfall rate of 53.1 mm during 5 minutes occurred in Shanxi Province. These seldom observed rainfall intensities greatly exceed those in other climatological zones at the same latitude over the world. Most heavy rainfalls in northern China may persist for 2–3 days, with some extreme cases lasting for 3–10 days. Due to the severity and long-persistence of heavy rains, northern China is vulnerable to flooding disasters.

The second feature of heavy rainfalls in northern China is that their occurrence is very concentrated in time, with 80–90% of the annual precipitation amount recorded during the time period of June-August, in particular the second half of July and the first half of August. So, the frequency of the occurrence of heavy rainfalls is closely related to events of droughs and floods. If heavy rainfalls occur frequently, severe flooding disasters may be caused (e.g. 1954, 1956, 1959 and 1963), while if the frequency is low, the summer in northern China would suffer from droughts (e.g. 1960, 1965, 1968 and 1972). The apparent concentration of precipitation spells is an important cause leading to flooding disasters in northern China

The third feature of the heavy rainfalls in northern China is the significant topographic effect. The heavy rains mainly occur on the windward sides of the mountain ranges or mountaneous regions. Based on the statistics of 190 cases of rainstorms in Hebei Province, 60.4% of them occurred on the windward slope, 34.2% on the plain region, and only 5.4% on the leeside and the Plateau region. Therefore, the terrain feature is a significant factor for heavy rainfalls in northern China.

The heavy rainfalls in northern China occur when the summer monsoon reaches its peak stage and the northernmost position. Figure 3.40 is the 10-yr mean wind field at 850 hPa in July (Liang, 1987). It can be seen that the southwesterly monsoon consistently prevails over southern, northern and northeastern China. To the south of the Yangtze River Valley, the frequency of occurrence of the southwest airflow has its peak value of the year, generally with 60% or greater. To the north of the Yangtze River Valley, the frequency may reach 40–50%. At 700 and 500 hPa, the southwest monsoon airflow advances northward up to the eastern part of northern China and southeastern part of northeastern China. As the southwest monsoon penetrates deeply into northern China and the southern part of northeastern China (40–45° N), the monsoon rainy belt moves northward up to the same regions which are generally located in the southwest monsoon airflow to the south side of the shearline, or to the northwestern sector (left side) of the axis line of maximum southerly winds. The rainy belt runs nearly parallel to this axis line.

The fluctuation of intensity of summer monsoon is closely related to the precipitation variation in northern China (Shi and Chao, 1983). Based on the statistics for 1958–1979, the precipitation is generally above-normal when the

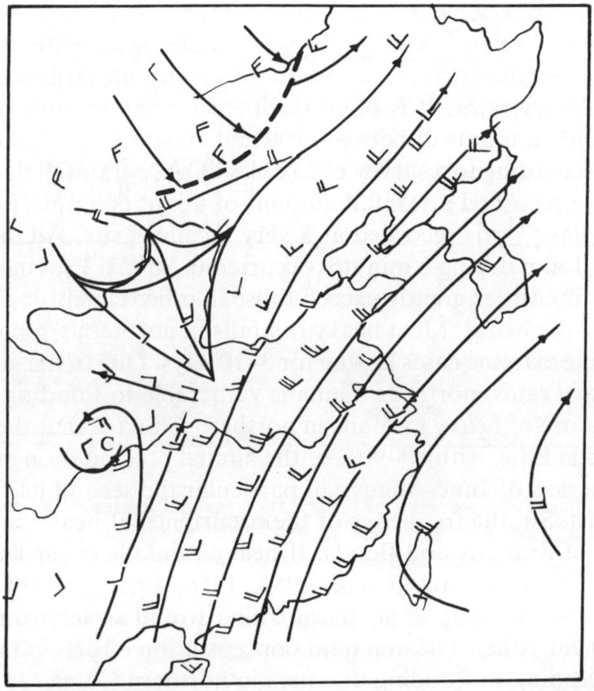

Figure 3.40. The 10-yr (1960–1969) mean wind field at 850 hPa for July. The thick dashed line denotes the shearline (Liang, 1987).

summer monsoon is relatively active in northern China (for example, 1973, 1977, 1979) and vice versa (for example, 1972, 1974, 1976, 1979, 1980), with the exception of some individual years. Therefore, the activity of summer monsoon in northern China sets the stage for the occurrence of the rainy season in this region. Its short term fluctuation is, to some extent, associated with the corresponding fluctuation of precipitation in northern China, although not closely due to the complexity of the rainfall process.

One interesting finding by Liang (1988) is the good correlative relationship of the rainfalls between northern China and the Indian monsoon rainfall in summer. By use of the long-time series of precipitation data in India and Tianjin, which is representative of the northern China region, Liang has found that the summer rainfall between India and northern China has a stable and significant positive correlative relationship, especially with the fairly consistent occurrence of droughts and flooding events in these two regions (Fig. 3.41)

The statistics have shown a marked similarity of the annual range of the precipitation amount with the summer rainfall amount accounting for nearly the same percentages of the total annual amount. A possible link of precipitation between these two regions is associated with the development of the monsoon troughs in India. When this condition is true, the monsoonal airflow in India and the significant low-

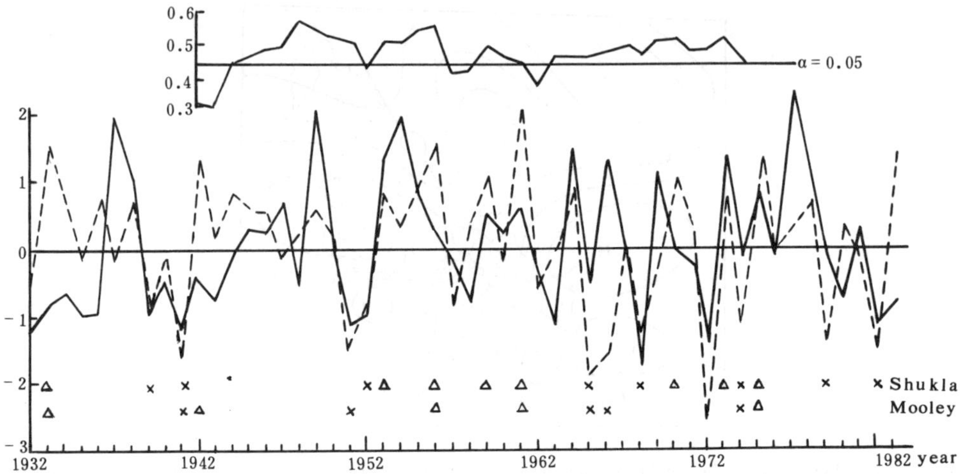

Figure 3.41. The standardized time series of precipitation amount for the whole Indian region (dashed curve) and Tianjin (full line) for 1932–1983. The top panel represents the running mean correlative coefficient, triangles flooding years and crosses dry years (Liang, 1988).

level jet (southerly) in China will enhances, thus causing an increase in precipitation in these two regions. Some possible teleconnections have been suggested by several authors (Huang, 1985).

Based on the analysis of 33 cases of heavy rainfalls in northern China for 1958–1976 (Ding *et al.*, 1980), the circulation patterns and synoptic situation favorable for heavy rainfalls in northern China have been identified. The heavy rainfalls mainly occur under the circulation situation of the blockage of the low-pressure trough coming from the west by the stationary high downstream (Fig. 3.42). When the longwave trough is located at 100–110° E, it is most favorable for the occurrence of heavy rainfalls, because in such a case, the northern China region is located in

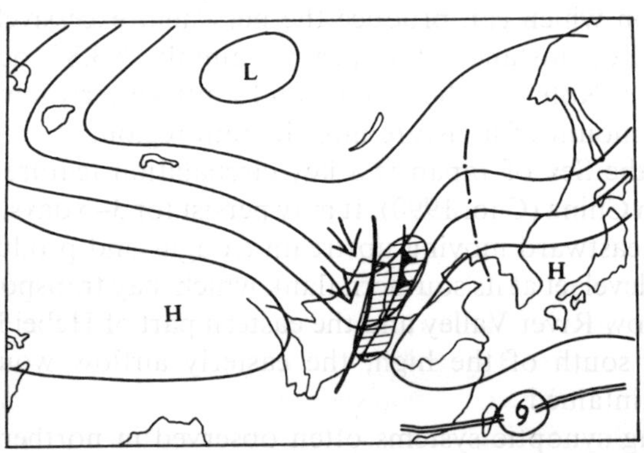

Figure 3.42. The major circulation types favorable for the occurrence of heavy rainfalls in northern China. The shaded area represents the region of heavy rains and the arrows the tracks of the cold air (Ding, Li *et al.*, 1980).

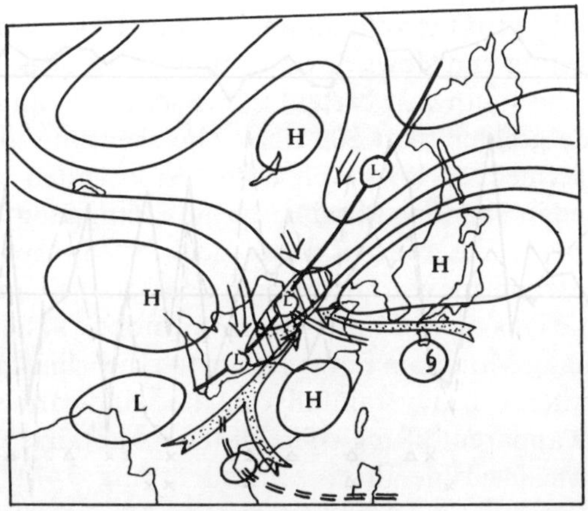

Figure 3.43. Type of circulation favorable for persistent heavy rainfall in northern China. The shaded arrows denote the warm and moist airflows associated with the low-level jets; double arrows the cold air (Ding, Li *et al.*, 1980).

advance of this longwave trough. Therefore, one key problem with the prediction of heavy rainfalls in northern China is to determine if there will be an upper-air trough to build up in the longitudinal range of 100–110° E. On the other hand, the presence of a stable high ridge (or blocking high) or the subtropical high downstream is also an important circulation condition for the occurrence of heavy rainfalls in northern China. When the high ridge anchors in the region of 120–140° E, it can produce a marked blocking situation, retarding or hampering the eastward movement of the upstream trough, thus causing heavy rainfalls along the surface frontal zone in advance of the retarded upper air trough. If the high behind the upper air trough amplifies, the thus-developed circulation situation of two highs facing each other would be more favorable for the heavy rainfall in northern China (Fig. 3.43). A north-south extended low-pressure trough or shearline is located in between. This is a circulation situation which can produce the persistent excessively heavy rainfall in northern China. For instance, the heavy rainfall "63.8" occurred under this circulation situation. Note that the easterly and southwesterly low-level jets both may transport a large amount of moisture into the rain region.

The high over the Sea of Japan is a key circualation feature for creating heavy rainfall in northern China (Cao, 1990). It may persist for 3–5 days, thus playing a dual role: blocking the eastward moving upstream troughs and producing an easterly or southeasterly low-level jet at its southern flank which may transport moisture into the regions of the Yellow River Valley and the eastern part of Hebei Province. If there is a typhoon to the south of the high, the easterly airflow would become further intensified and maintained.

The rain-bearing synoptic systems often observed in northern China are upper troughs with accompanying cold fronts, the wind shearline in the warm sector, the extratropical cyclones in the Yellow River Valley and the upper cold vortex. In general, the eastward moving upper trough with accompanying cold front can seldom produce heavy rainfalls with the rainfall amounts of 100–200 mm. However, if the upper trough is deepening and slows down its movement or is connected with a low vortex (for exmaple, the southwest vortex) in the south, it can cause heavy rainfalls with rainfall amounts greater than 200 mm. The shearline in the warm sector in northern China is defined as low-level, east-west oriented wind shearlines that generally are generated in Shadong Province or the Changjiang-Huaihe region. Then, as the southerly airflow to the south of it enhances, often reaching the intensity of a low-level jet, the shearline moves northward, and, at the same time, so does the associated rain region. This type of rain – bearing synoptic system is mainly observed to the east of the Taihang Mountain range. When the shearline moves over the Beijing area, it may cause intense heavy rainfalls on the windward side of the Yanshan Mountain, due to the apparent lifting effect of the mountain for the southerly wind. For example, a warm shearline that occurred during 9–14 July 1958 poured a torrential rainfall of 315 mm on Zhunhua Country. The heavy rainfall is, in general, located in the region in advance of the shearline, or between the shearline at the surface and 700 hPa, with the axis of the shearline tilting northward. Therefore, this pattern of rainfall is in many ways similar to that caused by the warm front (Fig. 3.44). During the analysis of 12 cases of warm shearlines, nine of them were found to have low-level jets with wind speeds of 20–30 m s^{-1}.

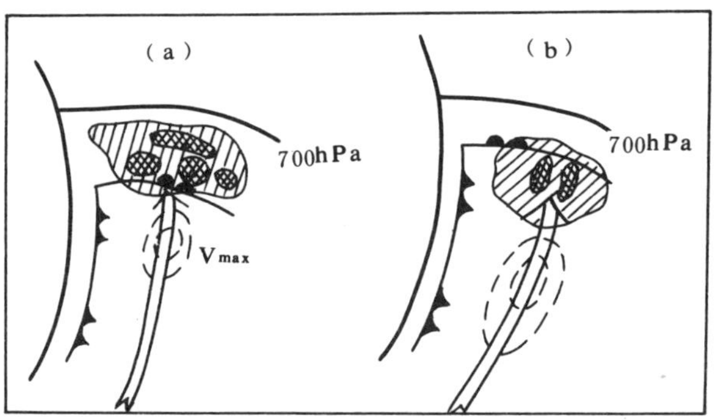

Figure 3.44. Schematic diagram of the relationship between the warm shearline and the region of heavy rainfalls (Ding, Li *et al.*, 1980).

The extratropical cyclone over the Yellow River Valley may generally cause rainfall events of 50–150 mm, or over 200 mm in some limited areas. These heavy rainfalls often last for 1 day or so, sometimes as long as 2–3 days. Under stable circulation conditions, the cyclone may be successively generated, thus producing persistent heavy rainfall episodes. An extratropical cyclone occurring during 29–31 July 1975 dumped a rainfall amount greater than 200 mm in the Tangshan area, Hebei Province, with a maximum of 531 mm at Baigezhuang. During this event, there were two conveyor belts of moisture, one coming from the Bay of Bengal, across southestern region of China and the middle Yangtze River Valley and reaching the northern China plains, and the other coming from the oceanic area to the south of

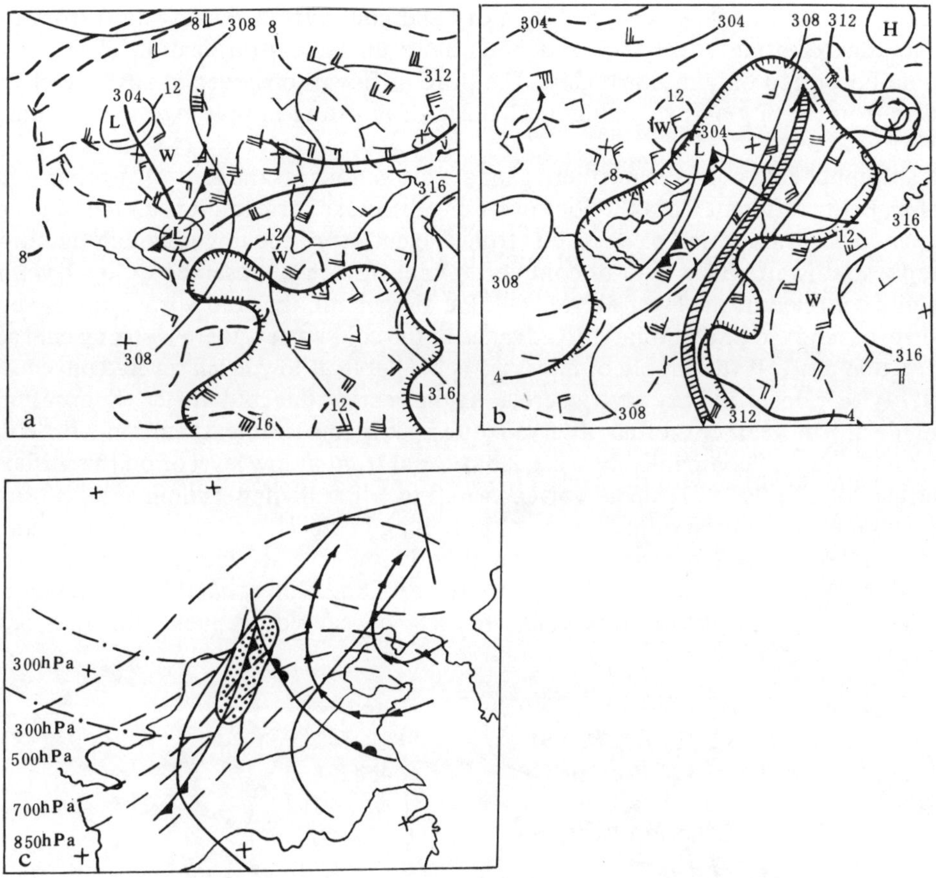

Figure 3.45. (a) 700 hPa weather map at 00:00 GMT, 27 July 27 1976, (b) 700 hPa weather map at 00:00 GMT, 28 July 1976, and (c) the three-dimensional airflow structure on the isentropic surface. The arrow denotes the low-level jet; the region enclosed by the comb-shaped line represents the region with $T-T_d<4°C$ (Ding, Chen *et al.*, 1980).

Japan, across the Yellow Sea and Bohai Bay, and reaching the eastern part of Hebei Province, in the form of a low-level easterly jet.

The extratropical cyclone sometimes originates in northwestern China as a low vortex (northwest vortex). After connecting with an upper trough or a shearline, it may rapidly develop into an intense extratropical cyclone around the Big Bend of the Yellow River. Figure 3.45 shows such an example (Ding *et al.*, 1980). The low-level jet is very marked, stretching from southern China to northern China (Fig. 3.45 a,b). Figure 3.45 (c) is the pattern of the three-dimensional airflow structure. There were three branches of major airflows important for the occurrence of heavy rainfall. The warm and moist southeasterly airflow ascended around 100 km in front of the cold front and reached up to 700–500 hPa over the Big Bend of the Yellow River. From the slope of isentropic surface, it was estimated that the vertical velocity here was very great. Here, this airflow is capped by a dry and cold airflow coming aloft from the west, thus creating a very marked potentially unstable stratification. The heavy rainfall occurred only in this region. The third airflow is observed in advance of the warm front, acting as a recycling mechanism of moisture in that region (Browning, 1973).

The upper cold vortex in northern China is also known as the East Mongolian cold votex. Being different from the southwest or northwest vortexes, it is a kind of deep, upper cold vortex which often results from the cut-off of the upper trough that dips deeply southward. This kind of cold vortex, not obviously observed at low level, is mainly confined in the layer above 700 hPa. In general, the cold vortex moves very slowly or tends to be stationary. In advance of the cold vortex, there is strong cold air advection aloft. If the supply of moist air is available at low level, severe convective activity or rainstorm may be triggered. At the rear of the cold vortex, intermittent thunderstorm weather is often observed, due mainly to a southeastward moving cold thermal trough accompanied by a secondary cold front at low level or on the surface. During 20–20 July 1977, a cold vortex moved over the Beijing-Tianjin area, causing heavy rainfall, with the rainfall amount in Beijing being 71 mm and Daxing County 146 mm (Fig. 3.46). Note that a strong southerly low-level jet with a maximum wind speed of 10 m s^{-1} was located in the warm sector. Ding, Sun *et al.*(1980) analysed the meso-scale system within this cold vortex that caused frequent thunderstorm activities.

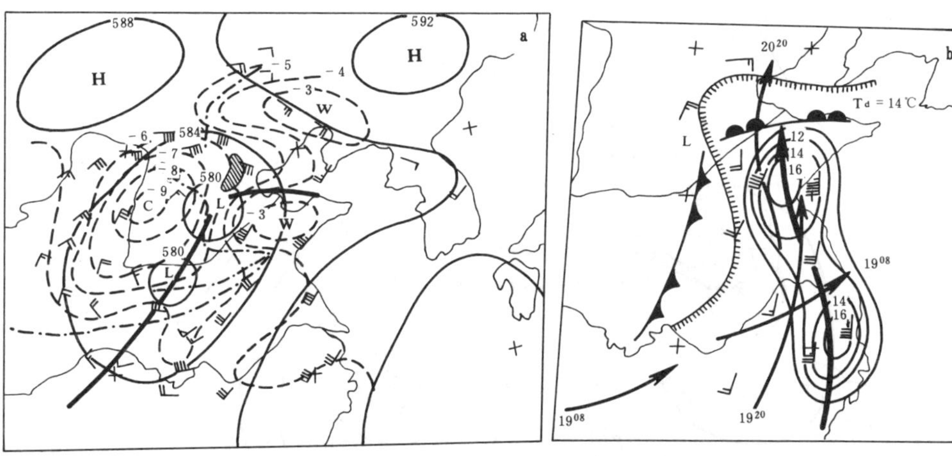

Figure 3.46. (a) The 500 hPa weather map at 00:00 GMT, 20 July 1977. The full lines represent the isolines of geopotential heights (Units: decameter), the dashed lines isotherm lines, and the shaded area the region of heavy rain; (b) the wind field at 850 hPa at 00:00 GMT, 20 July 1977. The thin lines represent isotachs and the thick solid line with an arrow denotes the axis of the low-level jet (Ding, Sun *et al.*, 1980).

Another case of an upper cut-off low, characterized by prolonged thunderstorm activity over northern China with local hailstorm, occurred between 11 and 17 July 1974. During the five days between 12 and 16 July, six thunderstorm systems were reported. The characteristic features of the large-scale circulation revealed an unusual excursion of the high-pressure ridge towards high latitudes and the development of a cut-off upper cold vortex over Mongolia. As the cut-off low became quasi-stationary, moving only very slowly southward, the flow to the rear of the cold trough provided frequent incursions of potential instability into the vicinity of the trough line.

As a matter of fact, the episode of heavy rainfall in northern China is not caused only by the single rain-bearing system mentioned above. It often involves the interaction of two or more rain-bearing systems. In particular, tropical synoptic systems may play an important role in causing heavy rainfalls in northern China. For example, the remnants of typhoons or weakening depressions may bring about excessively heavy rainfalls that will be examplified later, when they merge with the mid-latitude system . Sometimes, the meridional connection of the upper trough and the southwest vortex may produce excessively heavy rainfalls. The low-level jet plays a very important role in the formation of the heavy rainfall in northern China, as it does for the Meiyu or the pre-summer rains in southern China. Based on the analysis of 33 cases of rainstorms in northern China, 21 of them (with a maximum wind axis greater than 12 m s^{-1}) found the existence of low-level jets, accounting for 64% of the total cases. Among 21 cases of low-level jets, 15 are southerly or southwesterly and 6 are easterly. The latter developed either over the sea or on the land when there was typhoon activity. Figure 3.47 illustrates this point (Zhao and Sun, 1980). On 21 and

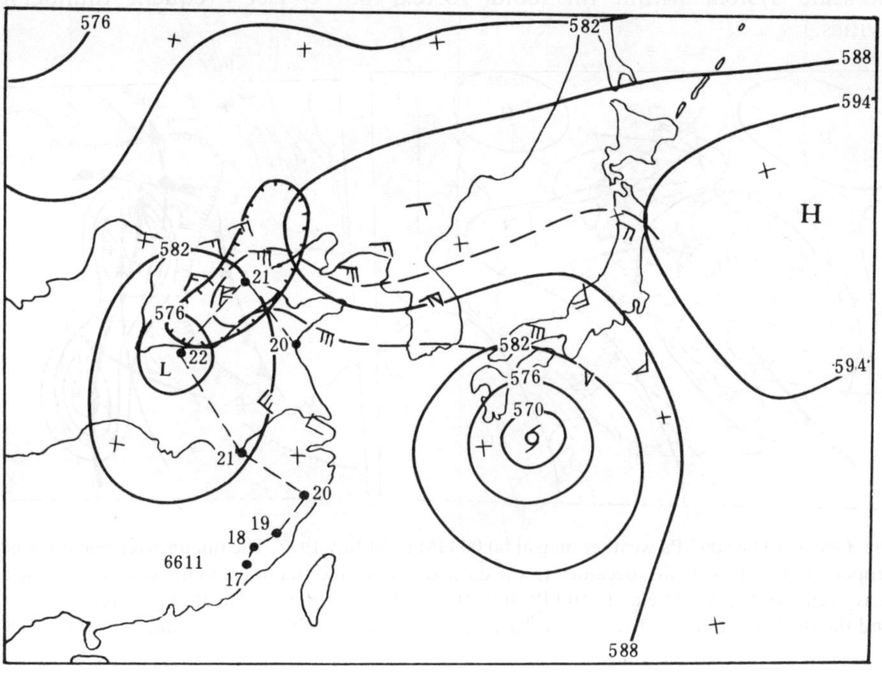

Figure 3.47. The 500 hPa weather map at 12:00 GMT, 22 August 1966. The shaded area denotes the rain region and the region enclosed with dashed line the area of the easterly low-level jet with wind speeds greater than 12m s^{-1}. The track of the weakened depression is indicated by a dashed line with numbers. Units of geopotential height: decameter (Zhao and Sun, 1980).

22 August 1966, the easterly low-level jet ran from Japan and Korea to northern and northeastern China, thus contributing much to the heavy rainfall of 800 mm in Pingding County, Shanxi Province, through the transport of a large amount of moisture from the sea. Note that a typhoon was located to the south of Japan.

Under the influence of a large-scale southwesterly airflow, the wind streak of the southwesterly low-level jet tends to propagate northeastward. Whenever the wind streak arrives in Henan Province, heavy rains occur downstream in Hebei Province, if other conditions are favorable. Once the wind streak propagates up to Hebei Province, the heavy rainfall moves out of this region. For the easterly low-level jet, a similar situation is true. Whenever its wind streak arrives at the Korean Peninsula, heavy rainfall is often observed to occur. An example of a markedly propagating downstream low-level jet is shown in Figure 3.48. The wind streak along the jet axis was located in Guangxi Province on 12:00 GMT, 27 July 1966. As the subtropical high moved northward and extended westward, the wind streak in this southerly low-

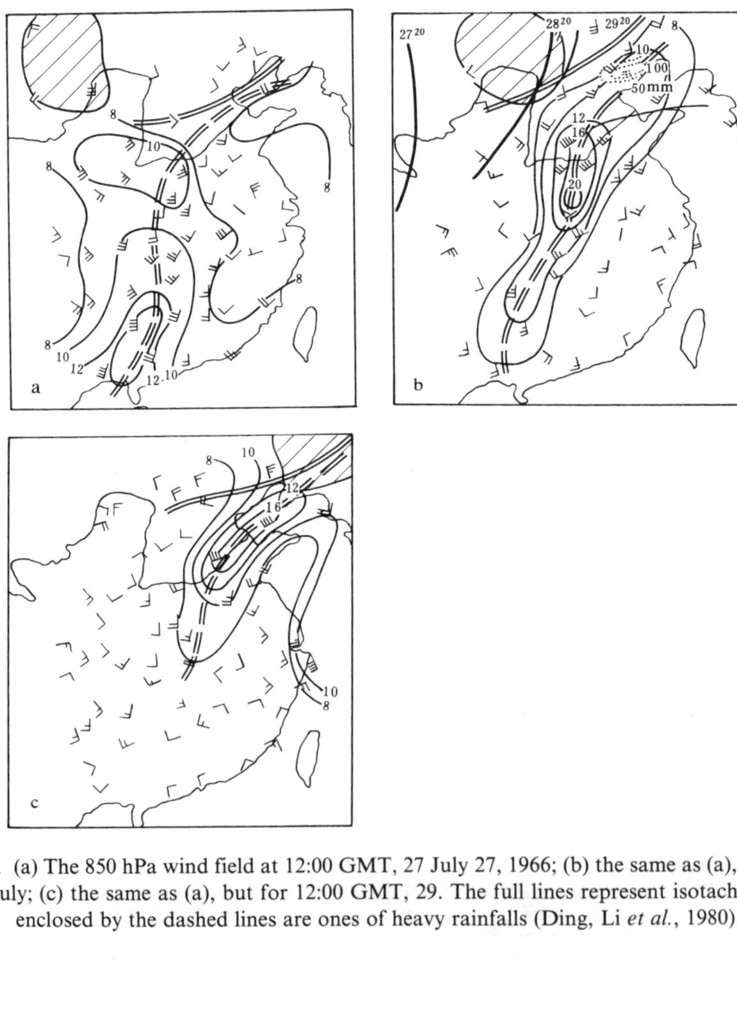

Figure 3.48. (a) The 850 hPa wind field at 12:00 GMT, 27 July 27, 1966; (b) the same as (a), but for 12:00 GMT, 28 July; (c) the same as (a), but for 12:00 GMT, 29. The full lines represent isotachs; the regions enclosed by the dashed lines are ones of heavy rainfalls (Ding, Li *et al.*, 1980).

level jet propagated northward. The next day, it arrived in the Huaihe River-Yellow River regions (Fig. 3.48 (b)). At this time, it caused a heavy rianfall in the eastern part of Hebei Province, with a maximum rainfall amount of about 200 mm. On the third day, the low-level jet further moved over the Bohai Bay, ending the process of heavy rain in Hebei Province.

A number of investigators have studied the environmental conditions for the occurrence of heavy rainfalls and severe convective weather (Ding *et al.*, 1980; Lei and Wu, 1980; Tian and Du, 1983). They include: (1) the potential instability of the air layer, or temperature inversion capping the unstable air; (2) the low-level tongue of moist air or convergence of moisture; (3) the mechanisms needed to release the instability, such as an area of low-level convergence, gravity wave, density current, orographic lifting, etc.; (4) low-level and upper-level jets; (5) strong vertical wind shear for a thunderstorm system; and (6) the entrainment of dry cold air at the middle level for severe weather. The physical conditions between heavy rains and local severe storms show significant differences (Ding *et al.*, 1980), as indicated in Table 3.4. In the case of severe storms, there is a temperature inversion or isothermal layer near the surface, whereas no inversion exists for the heavy rainfall event. Above 700 hPa, the temperature is much lower than in the case of heavy rains, with a difference of 8 °C at 400 hPa. Consequently, the lapse rate of temperature below 7 km for severe storms is 1–3 °C km^{-1} greater than for heavy rain, mainly due to the strong cold air advection in the mid- and upper-troposphere over severe storms. The layer of

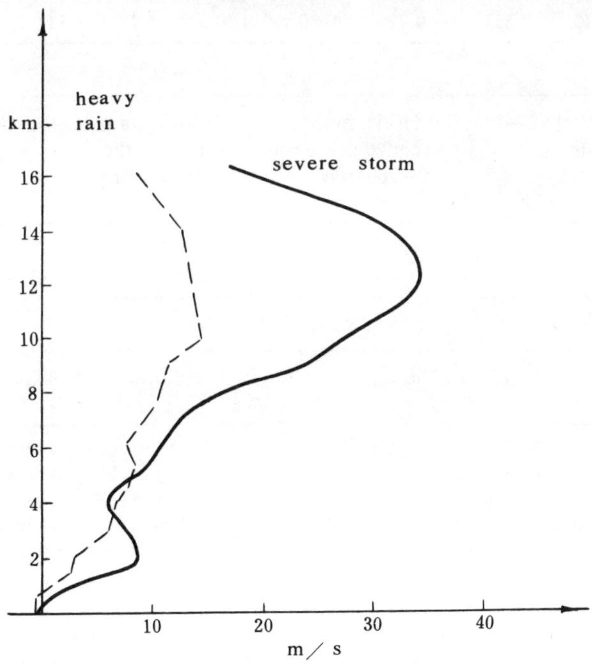

Figure 3.49. Comparison of environmental zonal wind associated with heavy rainfall (solid line) and severe storms (dashed line) (Ding *et al.*, 1980).

potential instability for severe storms is deeper and the level of free convection is higher than that for heavy rains. This suggests that stronger triggering mechanisms are needed for severe storms, but once the outbreak of severe storms occurs, their development will be more vigorous and rapid than heavy rains. There are also significant differences in several physical quantities characteristic of moisture content and transport. For example, the horizontal convergence of moisture from the surface up to the 300 hPa level for heavy rains is three times as large as for severe storms. Heavy rains occur and develop under the condition of weak vertical shear, whereas severe storms develop under strong vertical shear (Fig. 3.49). This large difference in vertical shear mainly results from the difference in the wind speed at upper levels. The severe storm often occurs below the axis of the upper-level jet in contrast to heavy rains which occur in the region 200–500 km to the south of the jet axis.

Table 3.4. The differences in the physical conditions between heavy rains and severe convective weathers (Ding *et al.*, 1980)

Weather type	CCL (hPa)	FCL (hPa)	Height of tropo-pause (hPa)	Mean lapse rate 1–9 km	Mean lapse rate 10–12 km	Height of 0 °C level (hPa)	Lift index (°C)	K index	Max. T below 900 hPa	Max. specific humidi-ty 900 hPa
H.R.	935	820	119	0.63	0.67	600	3.5	35.1	27.3	17.5
L.S.	835	670	227	0.72	0.40	630	5.2	34.8	24.5	13.5

Weather type	PMP of 927–275 hPa (cm)	Horiz. moist. convergence sur. -300 hPa 10^{-4} g cm^{-2} s^{-1}	Moist. trans. through the top of PBL 10^{-4} g cm^{-2} s^{-1}	Vert. shear of u 10^{-3} s^{-1}
H.R.	6.8	1.90	2.00	1.0
L.S.	3.1	0.85	0.55	3.5

Weather type	θ_{se500}	θ_{se850}	$\theta_{se500-850}$	T_{d500}	T_{d850}	T_{500}	T_{850}
H.R.	75.0	78.3	-4.9	-4.4	17.0	-2.2	19.4
L.S.	56.4	63.4	-6.8	-17.3	12.1	-8.4	18.1

Remarks: H.R. = Heavy rainfall. L.S. = Local strom. FCL = Free convective level. PBL = Planetary boundary layer. CCL = Convective condensational level. PMP = Probale maximum precipitation.

Based on the sounding data of 45 rainstorms in Beijing, Tian and Du (1983) have analysed temperature, wind, and humidity conditions. It was found that during the rainstorms, the air is abnormally moist and unstable. For 850, 700, and 500 hPa, the departure of relative humidity all exceeds 23%, while for 200 hPa, the departure

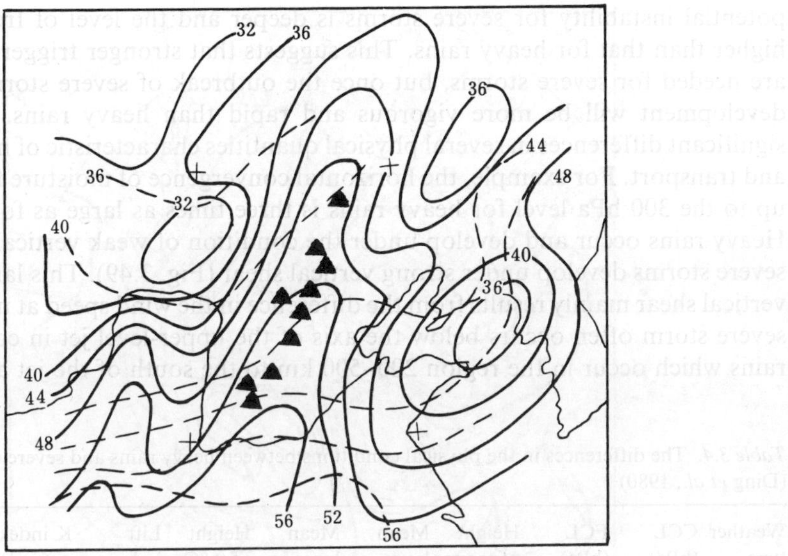

Figure 3.50. The distribution of T_σ (the moist static total temperature) at 850 hPa (solid line) and 500 hPa (dashed line) on 5 June 1975. The black triangles show the hail zones (Wu and Bei, 1982).

exceeds 16%. The height of the condensational level is relatively low, so that it is very easy to reach the free convection level (FCL). For a rainstorm to occur, the warm air advection at lower and middle levels should be strong, whereas there is cold air advection at high level (above 400 hPa). At the same time, a considerable degree of vertical wind shear is needed. The Ri number becomes smaller during the rainstorms.

Some investigators have emphasized the importance of the so-called Ω-shaped energy systems for the occurrence of rainstorms and severe convective systems (Ma, 1980; Wu and Bei, 1982). During the 12–36 hr prior to the occurrence of rainstorms and severe convective weather, it is often observed that there is a kind of pre-existing subsynoptic scale Ω-shaped energy system, i.e., region of high moist static energy with a region of low moist static energy at either side (to the east and to the west) (Fig. 3.50). This system is mainly created by the deep layer of warm and moist upward airflows and the convergence of moisture, especially for rainstorms.

Finally, some cases of excessively heavy rainfalls in northern China will be described below to further elucidate the role of large-scale circulation for the development of heavy rainfalls:

Under certain circulation conditions, low-level vortices will move northward and can generate rainstorms and severe local storms in northern China. The heaviest and most extensive rainstorm in the history of weather records of northern China occurred under such circumstances during the period 1 to 10 August 1963, with rainfall of 2005 mm during approximately one week recorded at the Zhangmao station on the eastern slope of the Taihang Mountains (see the dashed line in Fig. 3.51). Moist and warm easterly to southerly to northeasterly flow guided southward by the blocking high pressure ridge, converged towards that shearline. High pressure

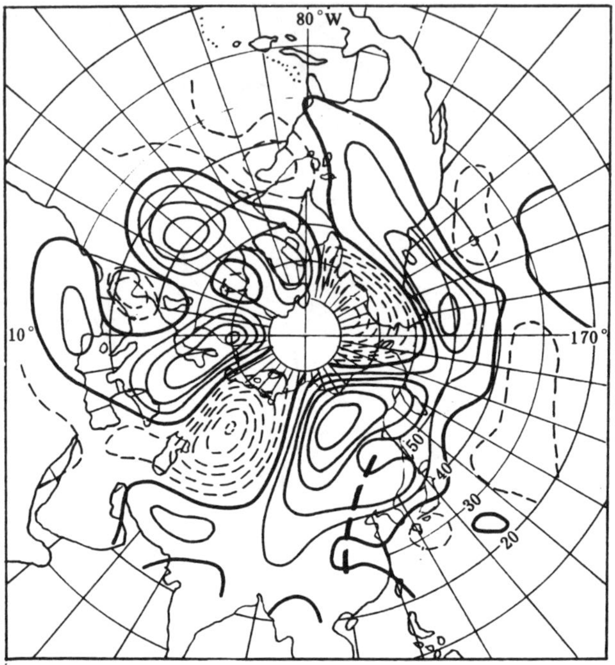

Figure 3.51. 500 hPa anomalies, truncated with wave number 4, for the period 1 to 10 August, 1966. The heavy dashed line indicates the mean position of shearline. Thin solid line denotes positive anomalies and the thin dashed lines negative anomalies. Analysis interval: 25 geopotential meters. Heavy solid lines: zero anomaly (Ding and Reiter, 1982).

anomalies also dominated Tibet and the region southeast of China. The most intensive rainfall episode occurred during the passage of low-level "southwest " vortices which brought about an increase of wind speed in the easterly flow.

In a similar synoptic development, heavy rainstorms in Beijing on 1 and 2 July 1973 were caused by a northward moving "southwest" vortex under the steering influence of a very strong south-north oriented upper-level jet ahead of a deep polar trough. A south-north oriented high-pressure ridge prevailed over the eastern Seaboard of China. These unusually well-developed meridional circulation patterns provided the steering current for the low-level "southwest" vortex which had originated over the Tibetan Plateau. A total rainfall amount of about 200 mm during slightly longer than 24 hours was produced in Beijing, which was second to the record established on 26 June 1893.

Occasionally, northward or northeastward moving typhoons suddenly change their tracks, turn towards the west or west-northwest, and penetrate deeply inland if a blocking high to the north of the typhoons builds up rapidly, amplifies and retrogrades. Under such conditions, the typhoon, though usually greatly weakened, drifts in an easterly current and finally becomes nearly stationary because of the blocking effect of the high-pressure ridge to the north of it. If other conditions are favorable, prolonged heavy rainfalls may ensue. An unprecedented case appeared early in August 1975 when the typhoon Nina dumped torrential rainfall with a maximum of 1692 mm during three days over a narrow, elongated band of 120 km long and 50 km wide along the windward side of the Funiu Mountain Range (Fig. 3.52). Figure 3.52 shows the 500 hPa circulation pattern for 12:00 GMT, 7 August 1975. Its most prominent feature is a large blocking high located to the north of the tropical depression. Due to the blocking action of this high-pressure region, typhoon Nina showed considerably slow movement two days after landfall, and eventually changed its direction of propagation from northeastward into westward. Finally, the depression became nearly stationary and probably made a small loop over the hilly region in the southern part of Henan Province. The erratic movement lasted for more than 20 h and was influenced by an extratropical weather system (cold front) as well as by terrain features. Torrential rains fell mainly in the northeast quadrant of the depression. The nearly stationary circulation pattern allowed organized mesoscale thunderstorm systems to develop repeatedly and produce exceptionally large rainfall amounts over a limited mountainous area. The severity of this rainfall resulted in a devastating flood within the basin of the middle and upper Huaihe River. From a statistical view point, an event of this magnitude might occur about once in 400 years.

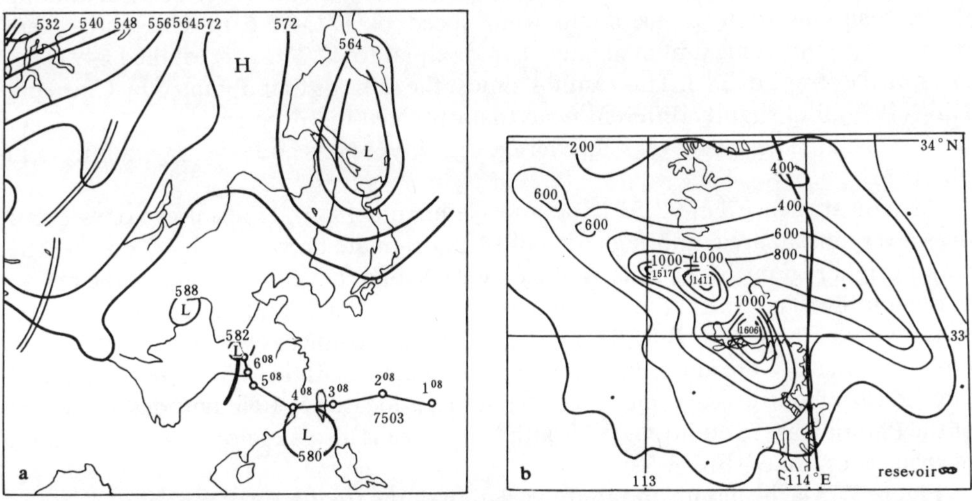

Figure 3.52. (a) 500 hPa analysis for 12:00 GMT, 7 August 1975. Height contours are drawn for every 80 m, 580 represents 5800 m and the typhoon track is given by a solid line. (b) Cumulative ranifall isopleths for 5–7 August 1975 (black lines).The 100 m terrain contour is dashed (Ding and Reiter, 1983).

The occurrence and persistence of the heavy rain was closely linked to the strong low-level easterly flow established between the blocking high and typhoon Nina. The easterlies emanated from the East China Sea and reached low-level jet strength. They transported a great amount of moisture into the heavy rain region. Many analyses of heavy rainfalls in northern and central China indicate that the easterly low-level jet is of vital importance.

3.5. Typhoon Activity over the Western North Pacific Ocean and its Relationship to Monsoons

The Western North Pacific basin is a region where the tropical cyclone most frequently occur from the global point of view, accounting for about 33% of a total number of approximately 80 tropical cyclonces each year. An average of about 26 tropical cyclones each year may reach at least tropical storm intensity (Elsberry *et al.*, 1987). Roughly two-thirds of these become typhoons. A comprehensive book on typhoons over the western Pacific by Chen and Ding (1979) is devoted to describing the climatology, environmental conditions of the genesis and development, motion, weathers and oceanic phenomena associated with typhoons. Recently, Wang and Fei (1987) published a book that was written specifically for typhoon forecasters. The reader is referred to both these books to obtain a complete picture of typhoon activity. In this section, only the updated regional climatology of typhoons over the western North Pacific Ocean basin and the relationship between typhoon activity and monsoons will be discussed.

Table 3.5 shows the statistics of annual frequency of tropical cyclones over the western North Pacific for 1951–1980 (Wang and Fei, 1987). During this 30-yr period, 1104 tropical cyclones have been observed, with an annual average of 36.8. Among them, typhoons (with a maximum wind speed of 17.2–32.6 m s^{-1}) and severe typhoons (with the maximum greater than or equal to 32.7 m s^{-1}) totalled 849, with an annual average of 28.3. This result is almost the same as that obtained by Chen and Ding (1979), but slightly different from that obtained by Elsberry *et al.* (1987). Over the South China Sea, each year, an average of 3.7 typhoons and severe typhoons were observed.

As indicated in Figure 3.53 (Xue and Neumann, 1984), tropical cyclones occur over a very broad area, although 91% of them originate between 5 and 22.5° N. The major origin regions of the genesis of tropical cyclones have three regions: the region to the southwest of Guam, the central part of the Caroline Islands and the central part of the South China Sea. But, the regions where the tropical cyclones intensify into typhoon intensity are different from their regions of origin. The major region of intensification of tropical cyclone into typhoons is located 5–10 longitudes to the east of the Philippines. In addition, the South China Sea is also a region of intensification of tropical cyclones (Fig. 3.54).

Figure 3.55 is the month-to-month variation of the frequency of typhoons over the western North-Pacific and the South China Sea. Although highly seasonal, this is the only region where tropical cyclogenesis has been observed in all months of the year. The typhoon season is usually between July and October, accounting for 69% of annual total.

Gray (1968) estimated that about 80% of all tropical cyclones originate in or just poleward of the Inter-tropical Convergence Zone (ITCZ) or monsoon trough. Therefore, preferred regions of genesis vary seasonally with the annual migrations of the ITCZ. Enhancement of the flow on either side of this monsoon trough increases the low-level relative vorticity and makes conditions more favorable for genesis. The westerlies on the equatorial side of the ITCZ may increase due to cold surges from the

Table 3.5. Annual frequency of tropical cyclones over northwestern Pacific for 1951–1980 (Wang and Fei, 1987)

Year	Western Pacific (including the South China Sea)					South China Sea				
	D	T	Ts	T+Ts	D+T+Ts	D	T	Ts	T+Ts	D+T+Ts
1951	10	5	15	20	30	3	1	0	1	4
1952	5	10	21	31	36	1	2	1	3	4
1953	8	9	17	26	34	3	4	0	4	7
1954	10	8	15	23	33	4	4	0	4	8
1955	9	8	20	28	37	4	1	2	3	7
1956	14	5	18	23	37	9	2	0	2	11
1957	5	4	18	22	27	2	1	1	2	4
1958	3	11	22	33	36	2	5	1	6	8
1959	8	7	17	24	32	5	2	0	2	7
1960	10	8	22	30	40	6	6	2	4	10
1961	10	13	20	33	43	2	5	2	7	9
1962	5	8	24	32	37	3	3	1	4	7
1963	8	5	20	25	33	5	2	0	2	7
1964	4	13	23	36	40	0	3	3	6	6
1965	9	13	19	32	41	4	5	1	6	10
1966	9	11	24	35	44	3	1	4	5	8
1967	13	15	25	40	53	5	0	1	1	6
1968	9	6	23	29	38	2	2	1	3	5
1969	10	8	14	22	32	3	2	0	2	5
1970	21	15	12	27	48	6	5	1	6	12
1971	15	11	25	36	51	5	2	0	2	7
1972	6	8	23	31	37	1	2	5	7	8
1973	5	11	13	24	29	3	5	4	9	12
1974	4	21	16	37	41	0	5	0	5	5
1975	7	8	15	23	30	2	2	1	3	5
1976	4	8	17	25	29	1	0	3	3	4
1977	6	10	12	22	28	0	1	0	1	1
1978	10	15	15	30	40	2	2	1	3	5
1979	11	11	13	24	35	9	2	1	3	9
1980	7	12	14	26	33	3	2	0	2	5
Total	255	297	552	849	1104	95	75	36	111	206
Aver.	8.5	9.9	18.4	28.3	36.8	3.2	2.5	1.2	3.7	6.9

Remarks: D = Tropical depression (<17.2 m s^{-1}); T = Typhoon (17.2–32.6 m s^{-1}); Ts = Severe typhoon (> 32.7 m s^{-1})

winter hemisphere, which raise pressures at the equator and lead to stronger pressure gradients and the westerly flow along the equator, as shown in Figure 3.56 (Love, 1985 a, b). Examples of interactions between longwaves in the Southern Hemisphere and equatorial westerlies in the western North Pacific during the Northern Hemisphere summer were given by He and Yang (1981).

The easterly flow on the poleward side of the monsoon trough can increase as a result of the intensification of the subtropical ridge. Cyclones seldom form until this

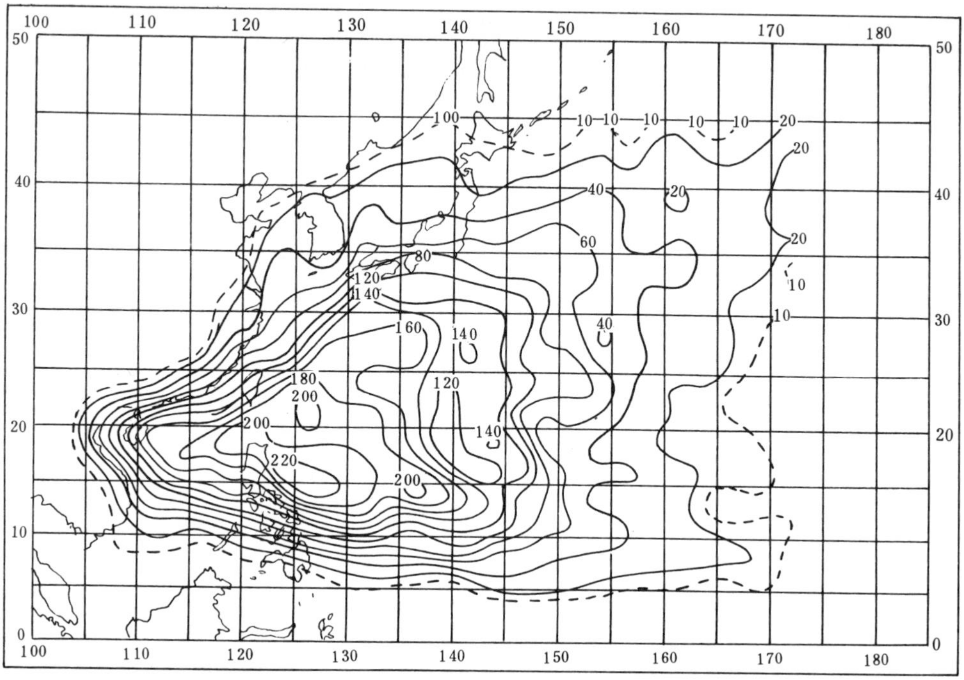

Figure 3.53. Number of storms passing within 75 n.mi. of grid point in the western North Pacific for the period 1948–1982 (Xue and Neumann, 1984).

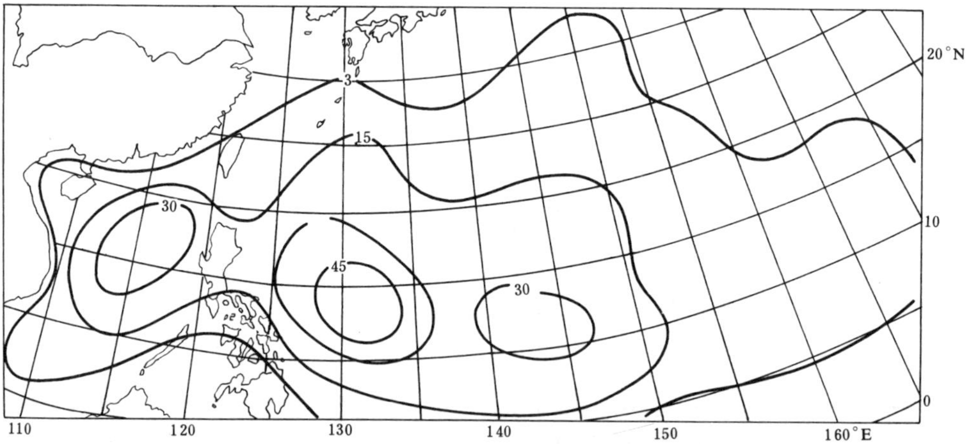

Figure 3.54. The distribution of the frequency of initial position of typoons in the area 5° × 5° lat./lon. over the western North Pacific for 1951–1980 (Wang and Fei, 1987).

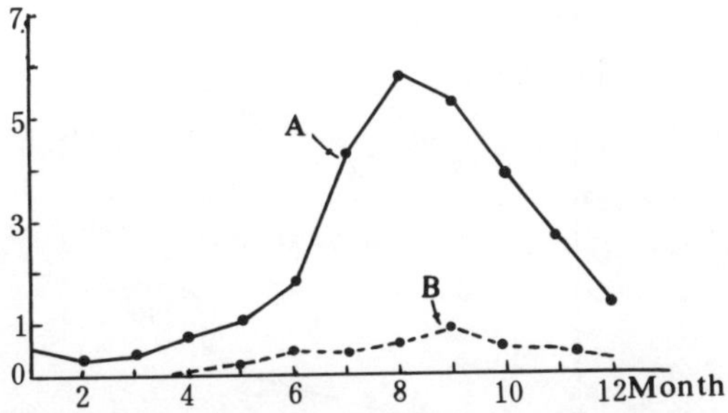

Figure 3.55. The monthly frequency of the occurrence of typhoons over the western North Pacific Ocean (A) and the South China Sea (B) for 1949–1980 (Wang and Fei, 1987).

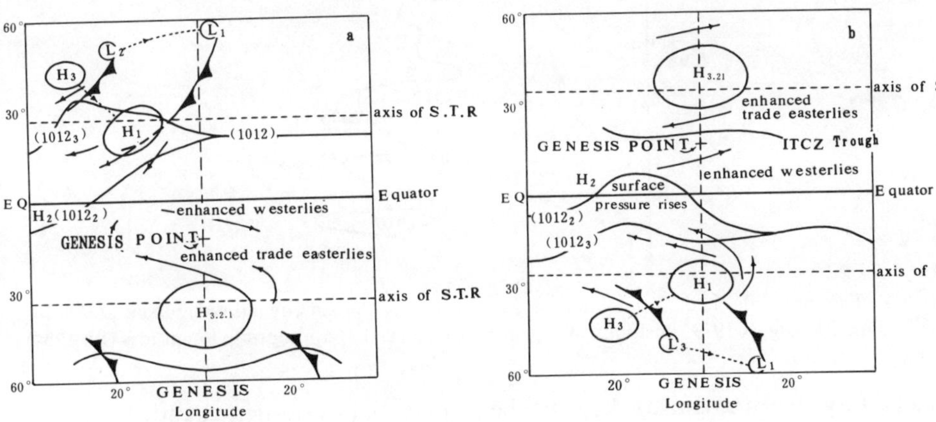

Figure 3.56. Idealized surface chart showing the positions of important synoptic scale features 3 and 1 day before cyclone genesis in the Northern Hemisphere. Subscripts on the highs (H) and lows (L) denote days before genesis (Love, 1985 a, b).

zonally-oriented ridge moves sufficiently poleward to establish strong, deep trade easterlies. The strength of the trades along the equatorward side of the subtropical ridge has long been used as an indicator of potential for cyclone genesis in the western North Pacific (Chen and Ding, 1979).

As indicated in Chapter 1, the enhancement of the ITCZ may be also caused by the eastward propagation or impulse of equatorial westerlies or the South Asian monsoonal airflows, thus creating active spells of typhoons over the western North Pacific. Li (1984) has indicated that during and prior to the formation of typhoons the monsoonal aircurrent originating in the Arabian Sea or the southern part of the Bay of Bengal rapidly extended eastward. It was observed that there was a couple of jet streaks of 15–20 m s^{-1} in this westerly airflow. When this airflow arrives in the western or central northern Pacific and met with the easterly trade wind, the typhoons may sucessively form along the convergence zone. Figure 3.57 indicates this situation.

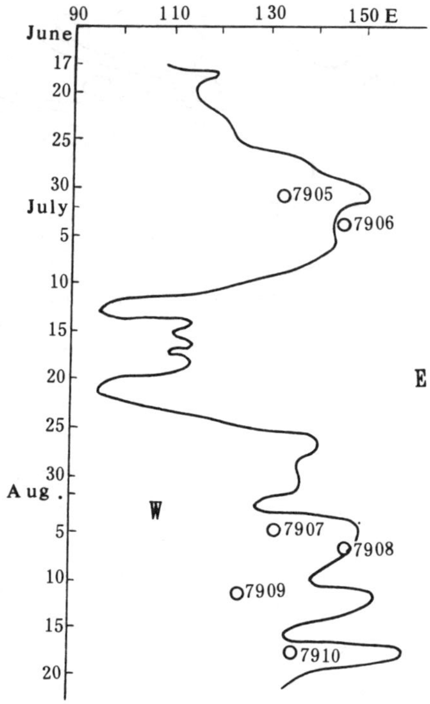

Figure 3.57. The longitude-time cross-section of the zero line of the zonal wind component at 700 hPa during 17 June-21 August 1979. Positions of typhoons are marked at appropriate longitudes (Li, 1984).

Around 3 July 1979 and during August when equatorial westerlies extended eastward down to 140–150° E, active spell of typhoon formation were observed. Figure 3.58 is the pattern of the 700 hPa zonal wind component for the first spell of active typhoon formation. The westerlies were very strong, with several wind maxima exceeding 15 m s^{-1} observed. Note that typhoons formed along the convergence zone of the westerlies and the easterly trade wind.

Ding *et al.* (1977) and Gray (1979) believe that periods of 1–3 weeks with hemispherically enhanced and suppressed tropical cyclone activity are indicative of alterations of the tropical general circulation with similar periodicity, leading to the clustering of tropical cyclones in time and space. The event is especially clearly shown in the western North Pacific. Figure 3.59 is the mean 500 hPa synoptic map on day 5 prior to the simultaneous formation of multi-typhoons composited for five cases. Quite apparently, the ITCZ or monsoon trough extended from the Indo-China Peninsula to 160° E. Farther eastward, one may observe the trade convergence zone, consisting of trade winds coming from the Southern and Northern Hemispheres, respectively. The western part of the ITCZ was connected with the Indian-Burmese monsoon trough. Therefore, the ITCZ in which at least three typhoons simultaneously spawn and develop has its easternmost position.

To the north of the ITCZ, there is a zonally oriented belt of a subtropical high, with

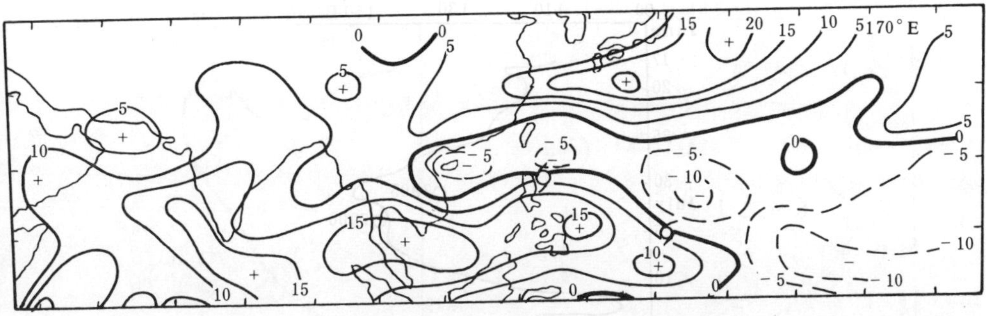

Figure 3.58. The pattern of the zonal wind component at 700 hPa for 3 July 1979. The thick solid line denotes the zero line; the thin solid lines, the westerly wind component; and the thin dashed lines the easterly wind component. Units: m s^{-1} (Li, 1984).

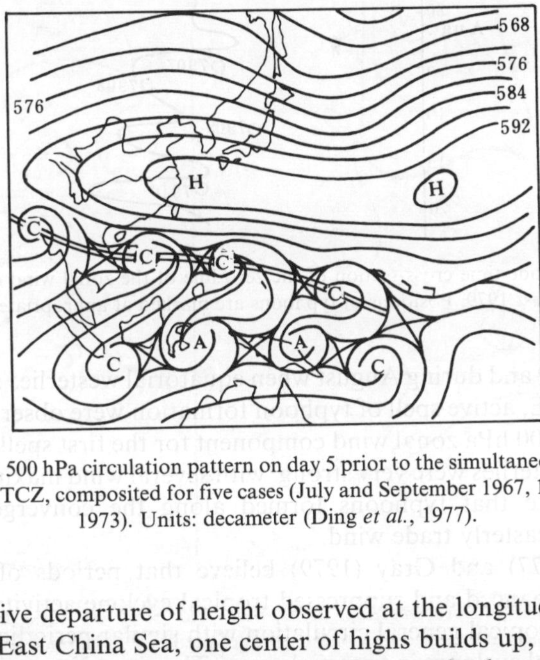

Figure 3.59. The mean 500 hPa circulation pattern on day 5 prior to the simultaneous formation of at least three typhoons in the ITCZ, composited for five cases (July and September 1967, 1968, and 1972, and July 1973). Units: decameter (Ding *et al.*, 1977).

a significant positive departure of height observed at the longitudinal range of 110–140° E. Over the East China Sea, one center of highs builds up, which is associated with the enhancement of the easterly wind. Farther north, the mid-latitude assumes the circulation pattern of a high index. The aforementioned circulation is very favorable for the enhancement and steadiness of the easterlies to the north of the ITCZ.

To the south of the ITCZ, there is an extensive area of southwesterly monsoonal airflows. The simultaneous development of multi-typhoons within the ITCZ is closely related to the "bursting" of this kind of westerly airflow, which may result from the extensive push or surge of the westerly wind in the equatorial area or originating in the Southern Hemisphere. Figure 3.60 is an example. Before 20 July 1968, the easterlies prevailed over the South China Sea and the tropical western Pacific.

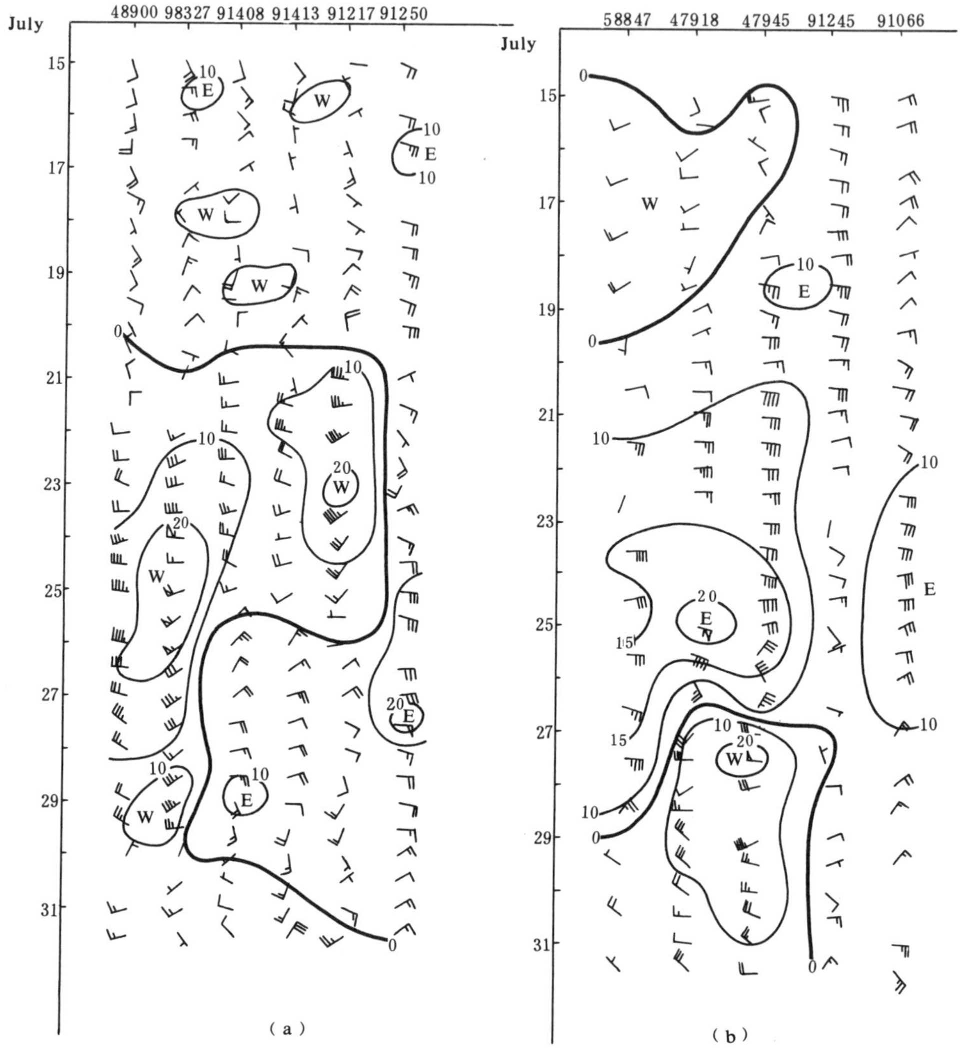

Figure 3.60. The longitude-time cross-sections of 700 hPa wind for 15–31 July 1968, (a) the condition to the south of the ITCZ, and (b) the condition of the north of the ITCZ. The thick solid lines represent zero lines of the wind and the thin solid lines isotachs. Units: m s^{-1} (Ding *et al.*, 1977).

Starting from this day onward, an abrupt change in the wind field was noticed, with westerly airflows rapidly building up from the southern part of the South China Sea to the region to the east of Guam. And the southwesterly monsoon at the longitudinal range around Guam was stronger than that of the western part, with a 700 hPa wind maximum of 20 m s^{-1} first observed there. The fact that almost in one day the westerly wind rapidly established itself over the extensive area of 6000–7000 km, may imply that this kind of westerly wind is likely to result from a drastic change or adjustment of circulation pattern in the Southern Hemisphere rather than from the eastward propagation of South Asian monsoon. The ending of the westerlies was also quite sudden. After 26 July the easterlies rapidly extended westward in the same manner as the westerlies did.

At the same time, the establishment of the easterlies was also very rapid. As the subtropical high intensified and extended westward, the easterlies rapidly propagated. Until 20 July, the easterlies had established to the north of the ITCZ, while the westerlies had established to the south. Just then, the ITCZ and its attendant vortices were produced, thus beginning the event of the simultaneous development of multi-typhoons. From the above description, it can be concluded that the unusually active bursting of the westerly monsoonal airflow over the extensive area may be a very important prerequisite for the occurrence of the spell of multi-typhoons over the tropical western Pacific, no matter how this condition is caused (either the eastward propagation of the South Asian monsoon or the cross-equatorial airflow coming from the Southern Hemisphere).

As discussed previously (See Fig. 1.57), the location and intensity of low-level monsoon troughs or the ITCZ has a significant influence on tropical cyclone frequency and the areas of development over the northern Pacific. For months with many typhoons the ITCZ is located near 20° N whereas for months with few typhoons, the ITCZ is found near 10° N. With many typhoons, it extends eastward to 160° E, whereas with few typhoons it ends at about 130° E. Over the Indo-China Peinisula and the tropical western Pacific equatorward of 20° N, stronger westerly winds were observed during months with many typhoons, whereas during months with few typhoons, stronger easterlies were encountered over the tropical and equatorial Pacific east of 150° E. Over the subtropical Pacific poleward of 20° N, stronger westerlies (easterlies) were observed with few (many) typhoons. Over the South Asian continent, the monsoon airflow over central and southern India was stronger with few than with many typhoons, whereas over northern India and the foothills of the Himalayas, the easterlies were stronger with few typhoons. Therefore, the characteristics of wind patterns during months with few typhoons are broadly similar to the conditions of active monsoons over India. Contrary to this, the characteristics of wind patterns with many typhoons are similar to break monsoon conditions over India. These findings were based on the analysis of low-level airflow fields.

For further documentation of the relationship, additional evidence was derived from rainfall data and height fields. Figure 3.61 a and b reveal almost opposite patterns of monsoon rainfall (Ding and Reiter, 1983). With few typhoons, above-normal rainfall occurs over the Indian Peninsula, especially in the west coast region, whereas with many typhoons, the west coast experiences drought conditions. Statistical significance tests for the differences between the two mean departures of rainfall over central and southern India were carried out, using the t-test. The results show that the pattern differences between summers with many and few typhoons are significant at the 5% level.

Different monsoon conditions over India and neighboring regions also show up in major changes in mid-tropospheric circulation patterns. Figure 3.62 indicates that the monsoon trough over India is slightly deeper and larger in summer with few

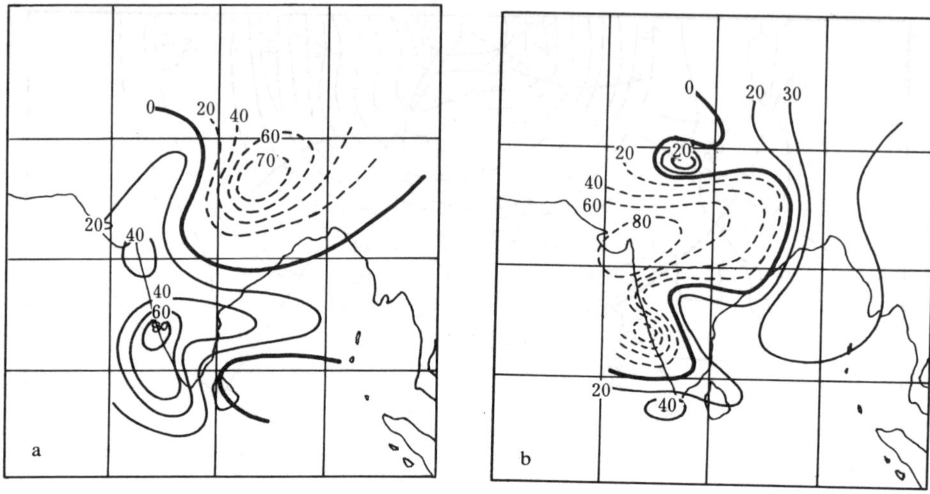

Figure 3.61. Departures from normal monsoon rainfall amounts in India (a) for summers with many typhoons and (b) for summers with few typhoons. Negative departures are denoted by dashed lines; positive departures by full lines. Units: mm (Ding and Reiter, 1983).

typhoons. In the latter case, the main body of the weaker monsoon trough tends to be displaced from the Indian Peninsula to the Bay of Bengal. This displacement is more or less similar to the characteristic circulation conditions of a break monsoon when a 500 hPa ridge, often located over the Indian Peninsula, can suppress convection in shallow westerlies. The height anomalies of the opposite sign at 500 hPa are also in good agreement with the above discussion. With many typhoons a large area of positive anomalies at 500 hPa is located over the Indian Peninsula, whereas with few typhoons, negative anomalies dominate the same region. In addition, the patterns of circulation over Tibet and its neighborhood during the periods of different typhoon activities are also strikingly different. With many typhoons, there exists a negative height anomaly, whereas during summers with few typhoons, an area of positive departure is evident. Correspondingly, in the former case, an upper trough lies over Tibet, whereas in the latter case, an enhanced high pressure region is present.

Ding and Reiter (1981) have found that the zonal Walker circulations have different positions for years with many typhoons and years with few typhoons. In 1975, when 11 typhoons were observed, the Walker circulation was centered over the equatorial central and western Pacific, with a strong subsiding branch over the equatorial eastern Pacific. In 1972, a year with many typhoons (18 typhoons), the Walker circulation was displaced towards the eastern Pacific and, at the same time, a part of the zonal monsoon circulation dominated the western Pacific. The center of the Walker circulation was shifted eastward by about 60 longitudes. In agreement with the predominance of large-scale upward motion over most of the tropical Pacific Ocean, the typhoon frequency was high during the summer of that year either as an effect or as a cause of these mean vertical motions. This Walker cell behavior agrees with the increase in the wind speeds of the low-level westerlies and upper-level

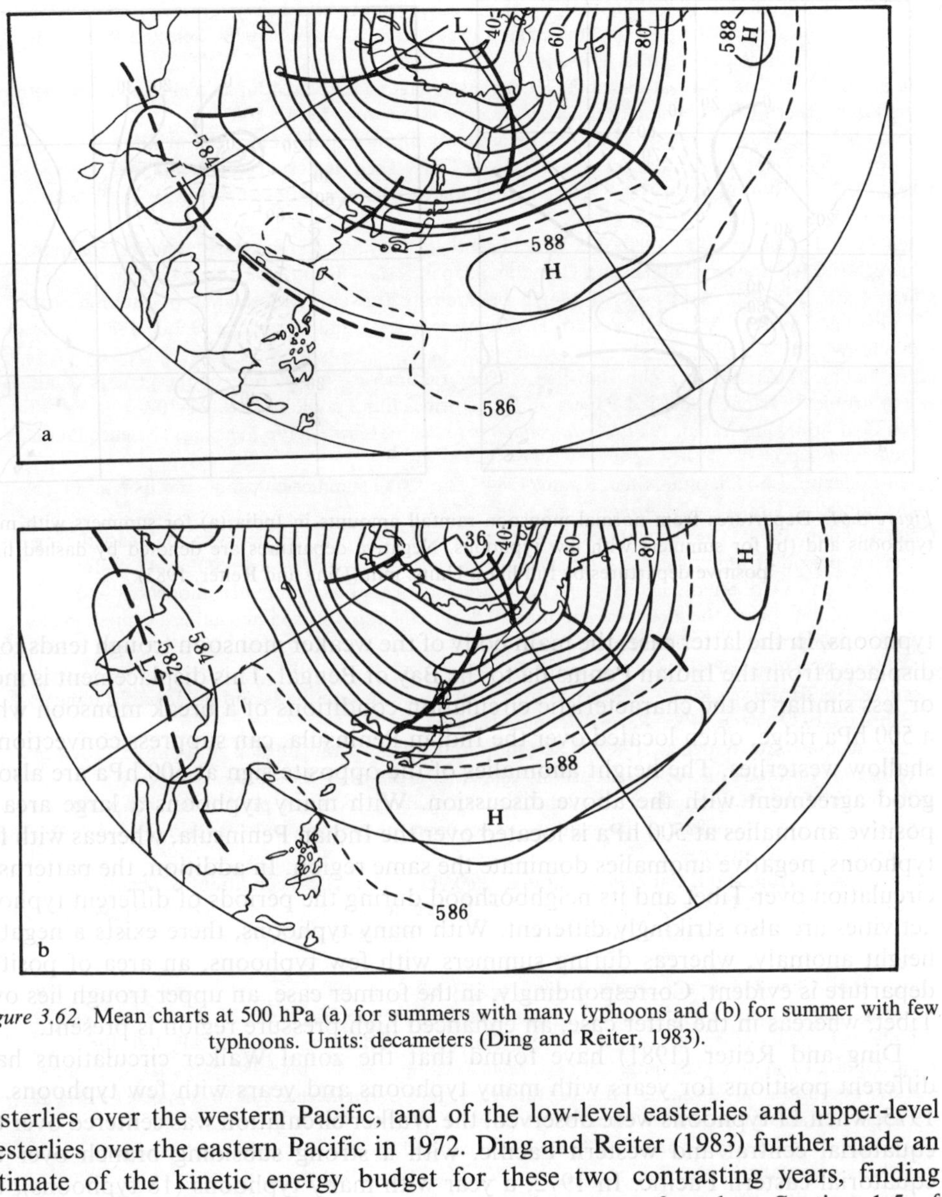

Figure 3.62. Mean charts at 500 hPa (a) for summers with many typhoons and (b) for summer with few typhoons. Units: decameters (Ding and Reiter, 1983).

easterlies over the western Pacific, and of the low-level easterlies and upper-level westerlies over the eastern Pacific in 1972. Ding and Reiter (1983) further made an estimate of the kinetic energy budget for these two contrasting years, finding significant differences in their energetics. Readers are referred to Section 1.5 of Chapter 1.

References

Browning, K. A., Hardman, M. E., Harrold, T. W., Pardoe, C. W., 1973: 'The structure of rainbands within a mid-latitude depression'. *Quart. J. Roy. Meteorol. Soc.*, **99**, 212–221.

Cai Zeyi, Zhang Mingli, 1978: 'An analysis of excessively convective heavy rain over the broader area between Inner Mongolia and the Shanxi Province on 1 August 1978'. Selected Papers on Rainstorm. Jilin People's Press, 138–146.

Cao Shuyi, 1990: 'The observations of the Huanghai Bohai and the Sea of Japan highs and the heavy rain in northern China in summer'. *Quart. J. Appl. Meteorol.*, **1**, 162–168.

Chen, G. T. J., 1983: 'Analysis of the Meiyu system and its application in the aviation weather forecast (I)'. Dept. of Atmos. Sci., Natl. Taiwan Univ. Tech. Rep.NTUATM–1983–08,73pp.

Chen, G. T. J., Wu, C. C., 1978: 'On the climatological characteristics at five cities in Taiwan'. *Atmos. Sci.*, **5**, 1–16.

Chen, G. T. J., Chang, C.-P., 1980: 'The strucure and vorticity budget of an early summer monsoon trough (" Meiyu) over southeastern China and Japan'. *Mon. Wea. Rev.*, **108**, 942–953.

Chen, G. T. J., 1989: 'Overview of Meiyu research in Taiwan'. Proceedings of International Conference on East Asia and Western Pacific Meteorology and Climate, 6–8 July 1989, Hong Kong, World Scientific, 14–37.

Chen Jiabin, Ji Zhangzhen, Zhu Baozhen, 1980: 'A numerical simulation of the genesis of an extratropical cyclone'. Selected Papers on Rainstorms. Jilin People's Press, 94–102.

Chen Lianshou, Ding Yihui, 1979: '*An Introduction to the Western Pacific Typhoons*'. Science Press, Beijing, 491 pp.

Chen Longxun, Luo Shaohua, Shen Rugui, 1984: 'The Asian summer monsoon and its relation to the rainfall in China'. *Adv. Atmos. Sci.*, **2**, 263–276.

Chen Shoujun, 1989: 'The vertical coupling of upper and lower jets in a heavy rainfall case during the late period of Meiyu period – Numerical experiment'. *Acta Meteorol. Sinica*, **47**, 8–16.

Chen Shixun, Gao Shaofeng, 1983: 'The characteristics of the summer monsoon and the relationship between the monsoon and rainfall during the rainy season in southern China'. Proc. of the Symposium on the Summer Monsoon in South and east Asia, 15–21 August 1980, Hangzhou, Yunnan People's Press, pp. 129–141.

Cram, R. S., Tatum, H. R., 1979: 'Record torrential rainstorms on the Island of Hawaii, January-February 1979'. *Mon. Wea. Rev.*, **107**, 1653–1662.

Chen Yuxiang, Zhu Baozhen, 1985: 'The synoptic features of rain season in Yunnan Province before the outbreak of tropical monsoon'. *Sci. Atmos. Sinica*, **9**, 101–106.

Ding Yihui, Feng Huijun, Hsue Chufan, Chen Kueishang, 1977: 'A preliminary study on the simultaneous development of the multiple typhoons in the intertropical convergence zone'. *Sci. Atmos. Sinica*, **1**, 89–98.

Ding Yihui, Li Hongzhou, Cai Zeyi, Li Jishun, 1980: 'On the physical conditions of occurrence of heavy rainfall and severe convective weather'. Proceedings of Eighth Conference on Weather Forecasting and Analysis, June 10–13, 1980, Denver, Colo., Amer. Meteorol. Soc., Boston (12.8), pp. 371–377.

Ding Yihui, Chen Guomian, Zhang Anqing, Gao Congqing, 1980: 'A preliminaty analysis of a heavy rainfall caused by the extratropical cyclone over the Big Bend of the Yellow River'. Collected Papers of Rainstorms and Severe Convective Weathers, No. 9 of IAP, 100–108.

Ding Yihui, Li Jishun, Sun Suqing, Cai Zeyi, Zhao Shixong, Tao Shiyan, 1980: 'An analysis of some kinds of synoptic scale systems affecting the heavy rains in northern China in summer. Collected Papers of Rainstorms and Severe Convective Weathers, **No. 9** of IAP, 1–13.

Ding Yihui, Sun Suqing, Li Jishun, Zhang Qingyun, 1980: 'A meso-scale analysis of several cases of heavy rainfalls in the Beijing area in 1977'. Collected Papers of Rainstorms and Severe Convective Weathers, No. 9 of IAP, 137–150.

Ding Yihui, Reiter, E. R., 1981: 'Some conditions influencing the variability of typhoon formation over the western Pacific Ocean'. *Arch. Meteorol. Geophys. Biocl., Ser. A*, 327–342.

Ding Yihui, Reiter, E. R., 1982: 'A relationship between planetary waves and persisitent rain and thunderstorms in China'. *Arch. Meteorol. Geophys. Biocl., Ser. B*, **31**, 221–252.

Ding Yihui, Reiter, E. R., 1983: 'The kinetic energy budget over the western Pacific during dominant trade wind and active monsoon regimes'. *Arch. Meteorol. Geophys. Biocl., Ser. A*, 201–229.

Ding Yihui, Reiter, E. R., 1983: 'Large-scale hemispheric teleconnections with the frequency of tropical cyclone formation over the Northwest Pacific and North Atlantic Oceans'. *Arch. Meteorol. Geophys. Biocl., Ser. A*, **32**, 311–338.

Elsberry, R. L., Frank, W. M., Holland, G. J., Jarrell, J. D., Southern, R. L., 1987: *A Global View of Tropical Cyclones*. WMO Publication, 185 pp.

Gao Youxi, Xu Suyin, 1962: 'Some problems of the monsoons in East Asia'. In collected papers of Institute of Geophysics and Meteorology, Academia Sinica, No. 5, Science Press, Beijing, 1–106.

Gray, W. M., 1967: 'Global view of the origin of tropical disturbances and storms'. *Mon. Wea. Rev.*, **96**, 669–700.

Gray, W. M., 1979: 'Hurricanes: Their formation, structure and likely role in the tropical circulation', in: Shaw, D. B. ed: *Meteorology over the Tropical Ocean* Roy. Meteorol. Soc., Bracknell, U.K.

Guangdong Observatory, 1976: 'A process of low-level jet during the pre-summer rainy season'. Proceedings of Conference on Tropical Weather. Science Press, Beijing.

He Shixiou, Yang Zhufang, 1981: 'The relationship between the intensity of summer southwest monsoon over the Northwest Pacific and the circulation patterns of the Southern Hemisphere'. *Sci. Atmos. Sinica*, **5**, 50–59.

Huan Anli, Gao Kun, 1982: 'A dynamic analysis of coupling effect of upper-level jet and low-level jet'. *J. Univer. Hangzhou*, **9**, 356–364.

Huang Anli, Gao Kun, Li Yufang, 1987: 'A preliminary simulation of a heavy rain process during Meiyu season'. *Acta Meteorol. Sinica*, **45**, 426–436.

Huang Ronghui, 1985: 'The numerical simulation of three-dimensional teleconnections in the summer circulation over the Northern Hemisphere'. *Adv. Atmos. Sci.*, **2**, 81–92.

Huang Shisong, 1986: *The Pre-summer Heavy Rainfall over southern China*. Guangdong Science and Technology Press, 243 pp.

Jiang Jixi, Ye Huiming, Chen Meizhen, 1990: 'Investigation of meso-scale convective cloud cluster in southern China'. *Quart. J. Appl. Meteorol.*, **1**, 232–241.

Jin Zuhui, 1982: 'A study of the meso-scale system of heavy rainfall over Valleys of the Changjiang-Huaihe Rivers'. *Sci. Atmos. Sinica*, **6**, 187–194.

Kuo, Y.-H., Chen, G. T. J., 1990: 'The Taiwan Area Mesoscale Experiment (TAMEX): An overview'. *Bull. Amer. Meteor. Soc.*, **71**, 488–503.

Lei Yushun, Wu Zhenhua, 1980: 'Basis of energy analysis of static stability'. *Acta Meteorol. Sinica*, **38**, 380–386.

Li Zengzhong, 1984: 'Suddenly eastward stretching of equatorial westerlies over the Northwest Pacific Ocean related to typhoon activities'. Proceedings of Conference on Tropical Circulation and Systems, 1982. Ocean Press, Beijing, pp. 129–139.

Li Zhenguang, Deng Lianyan, Xue Huixian, 1988: 'A composite analysis on widespread torrential rainfall during first rainy season in southern China'. *J. Tropical Meteorol.*, **4**, 97–106.

Liang Biqi, Bao Chenglan, 1980: 'The mesoscale analysis of the pre-summer rainstorms over southern China'. Selected Papers on Rainstorms. Jilin People's Press, 259 pp.

Liang Pingde, 1987: The developing process of monsoon in eastern China and the change of stratification in the atmosphere. *Weather in northern China*. Beijing University Press, 73–84.

Liang Pingde, 1988: 'The Indian summer monsoon and rainfall in northern China in summer'. *Acta Meteorol. Sinica*, **46**, 75–81.

Lin Chunyu, Lu Juzhong, Xu Guiyu, 1983: 'A tentative study on the characteristics of the summer monsoon activity over the middle and lower Yangtze River Valley'. Proc. of the Symposium on the Summer Monsoon in South and East Asia, 15–21 August, 1980, Hangzhou, China, Yunnan People's Press, 254 pp.

Love, G., 1985 a: 'Cross-equatorial influence of winter hemisphere subtropical cold surges'. *Mon. Wea. Rev.*, **113**, 1487–1489.

Love, G., 1985 b: 'Cross-equatorial interactions during tropical cyclogenesis'. *Mon. Wea. Rev.*, **113**, 1499–1509.

Lu Hancheng, Wang Jinwu, 1988: 'The feature and stability of meso-scale disturbance in the Meiyu front'. *Sci. Atmos. Sinica*, **12**, 182–190.

Ma Henian, 1980: 'Subsynoptic Omega systems and the region of occurrence of rainstorms'. Selected Papers on Rainstorms. Jilin People's Press, 171–176.

Matsumoto, S., Ninomuya, K., 1971: 'On the meso-scale and medium-scale structure of a cold front and the relevant vertical circulation'. *J. Meteorol. Soc. Japan*, **49**, 648–662.

Ninomiya, K., Akiyama, T., 1974: 'Band structure of meso-scale echo cluster associated with low-level jet stream'. *J. Meteorol. Soc. Japan*, **52**, 300–313.

Paulhus, G. L. H., 1965: 'Indian Ocean and Taiwan rainfall set new records'. *Mon. Wea. Rev.*, **93**, 331–335.

Qiao Quanming, Zhang Li, Tao Yueyin, Zhang Xiang, 1986 a: 'The compositing analysis of cloudiness and precipitation during the Meiyu rainstorms'. Proceedings of The Meso-scale Weather Experiment in East China, Vol. III, pp. 19–25.

Qiao Quanming, Zhang Li, Zhang Xiang, Deng Bin, 1986 b: 'Large-scale compositing analysis and model of the Meiyu in China. Part II'. Proceedings of The Meso-scale Weather Experiment in East China, Vol. III, pp. 11–18.

Qiu Binhuan, Ding Yihui, 1979: The circulation structure during the Meiyu period of 1973. *The Formation and Prediction of Precipitation During the Meiyu in Summer*. Science Press, Beijing, 23–32.

Rong Guanxun, Wang Yintong, Peng Hong, Xue Hueixian, 1987: 'On the major monsoon weather systems in the southern China Sea and their relationship with precipitation in southern China'. Collected Papers on Meteorological Science and Technology, No. 10 (Monsoon Meteorology). China Meteorol. Press, Beijing, 221–233.

Sato, K., 1985: 'On the abrupt change in the strucature of Baiu front over the China continent in late May of 1979'. *J. Meteorol. Soc. Japan*, **63**, 20–36.

Shen Rugui, Tao Quanzhen, Lai Yingying, Guo Shihua, 1983: 'The circulation change in lower and upper troposphere in lower latitude and the rainfall during the pretyphoon rain season'. Proc. of the Symposium on the Summer Monsoon in South and East Asia, 10–15 October 1982, Kunming, China, Yunnan People's Press.

Shi Shangwen, Cao Junmin, 1983: 'The summer monsoon in North China'. Proc. of Symposium on the Summer Monsoon in South and East Asia, 10–15 October 1982, Kunming, China, Yunnan People's Press, pp. 72–85.

Si Gongwang, 1976: 'A numerical analysis of dynamic factors leading to genesis and development of a Changjiang-Huaihe River cyclone'. *Sci. Atmos. Sinica*, **2**, 6–17.

Si Gongwang, 1988: '*The Heavy Rainfall and Severe Convective Circulation Systems*'. China Meteorol. Press, Beijing, 350 pp.

Si Gongwang, 1989: 'On the large-scale circulation of Meiyu system over East Asia'. *Acta Meteorol. Sinica*, **47**, 312–323.

Si Gongwang, Yu Zhangxiao, Li Faran, Yan Juezheng, 1982: 'A case analysis on the formation of low-level jet stream associated with "Baiu front"'. *Sci. Atmos. Sinica*, **6**, 165–170.

Sun Suqing, 1986: 'The background field of large-scale low-level jet in East Asia and the cross-equatorial airflow in the Eastern Hemisphere'. *Acta Meteorol. Sinica*, **44**, 55–62.

Tao Shiyan, *et al.*, 1957: 'A process of outbreak of cold air in East Asia during the breakdown of blocking situation'. *Acta Meteorol. Sinica*, **28**, 63–74.

Tao Shiyan, Ding Yihui, Sun Suqing, Cai Zeyi, Zhang Mingli *et al.*, 1980: *The Heavy Rainfall in China.* Science Press, Beijing, 255 pp.

Tao Shiyan, Ding Yihui, 1981: 'Observational evidence of the influence of the Qinghai-Xizang (Tibet) Plateau on the occurrence of heavy rain and severe convective storms in China'. *Bull. Amer. Meteorol. Soc.*, **61**, 23–30.

Tian Shenchun, Du Changxuan, 1983: 'Some statistical features of temperature, humidity and wind during rainstorms in Beijing'. *Sci. Atmos. Sinica*, **7**, 68–77.

Tsay, C.Y., Chen, G. T. J., 1981: 'Dynamic processes for vertical motions in a Meiyu system'. *Paper Meteorol. Res.*, **3**, 67–77.

Uccellini, L., Johnson, D. R., 1979: 'The coupling of upper and lower tropospherci jet streak and implications for the development of severe convective storm'. *Mon. Wea. Rev.*, **107**, 682–703.

Wang Jizhi, Li Maicun, 1982: 'Cross-equatorial flow from Australia and monsoon over China'. *Sci. Atmos. Sinica*, **6**, 1–10.

Wang Zhilie, Fei Liang, 1987: *Manual for Typhoon Prediction*. China Meteorol. Press, Beijing, 360 pp.

Wu Tsungyao, 1990: 'A review of the Taiwan Area Mesoscale Experiment (TAMEX) and post-field phase activities'. Proceedings of Workshop on TAMEX Scientific Results. 24–26 September, 1990, Boulder, NSF, NSC, NCAR, pp. 1–6.

Wu Zhenhua, Bai Yurong, 1982: A comparative analyses of the Ω-shape energy system prior to the occurrence of severe hailstorm and heavy rainfall. *Weather in Northern China*, No. 3, Beijing University Press, 51–58.

Xian Yuanzhen, Bao Chenglan, 1986: *The Weather for Four Seasons over the Middle and Lower Valleys of the Yangtze River*. China Meteorol. Press, Beijing, 270 pp.

Xie Yibing *et al.*, 1956: 'A study of analysis of rain-producing synoptic systems in summer half year in China'. *Acta Meteorol. Sinica*, **27**, 1–24.

Xue, Z., Neumann, C. J., 1984: 'Frequency and motion of western northern Pacific tropical cyclones'. Tech. Memo. NWS NHC 23, NOAA Washington, DC, 80 pp.

Xu Xianan, Jiao Peijin, 1984: 'The structure of cyclone over the Changjiang-Huaihe River Valley in summer'. *Sci. Atmos. Sinica*, **8**, 189–196.

Yeh Tucheng, Tao Shiyan, Li Maicun, 1958: 'The abrupt phenomena of the general atmospheric circulation in June and October'. *Acta Meteorol. Sinica*, **29**, 246–263.

Zhang Kesu, 1988: 'On mesoscale instability of a baroclinic flow. I: symmetric instability'. *Acta Meteorol. Sinica*, **46**, 258–266.

Zhao Li, Qiao Quanming, Deng Bin, Zhang Xiang, 1986: Large-scale compositing analysis and model of the Meiyu in China. Part I: Proceedings of The Meso-scale Weather Experiment in East China. Vol. III, 1–10.

Zhang Mingli, Li Weiliang, Zhang Jiacheng, Li Maicun, 1979: A case study of the Meiyu rainstorm; *The Formation and Prediction of Precipitation during the Meiyu in Summer*. Science Press, Beijing, 1–22.

Zhong Renggen, Liang Biqi, 1984: 'The low-level jet stream over southern China and its effect on heavy rainfall'. Proceedings of Symposium on Tropical circulation and System. Ocean Press, Beijing.

Zhao Sixiong, Zhou Xiaoping, Zhang Kesu, Liu Suhong, 1982: 'Numerical simulation experiment of the formation and maintenance of meso-scale low'. *Sci. Atmos. Sinica*, **6**, 109–117.

Zhao Sixiong, 1988: 'Energetics of cyclogenesis on Meiyu (Baiu) front'. *Sci. Atmos. Sinica*, Special Issue, 191–201.

CHAPTER 4

THE MEDIUM- AND LONG-RANGE FLUCTUATIONS OF MONSOONS AND THEIR ASSOCIATION WITH FLOODS AND DROUGHTS OVER CHINA

4.1. Low Frequency Oscillations of Monsoons

Low-frequency oscillations (LFO) are closely related to the slowly varying motion in the atmosphere, especially in the tropical area. Their importance has been greatly stressed since Madden and Julian (1971) originally discovered the 40 to 50 day oscillation. This so-called 40 to 50 day oscillation actually covers a wide range of frequencies and should be taken in a generic sense to collectively denote all observed periods between about 30 and 60 days. The low-frequency oscillations so far discovered may be divided into two types: intra-seasonal and inter-annual. The intra-seasonal oscillations mainly include those in the 30–50 day, the 10–20 day, and biweekly timescales. Among them, the 30–50 day oscillation is most important for monsoon activity, although the latter two are also important events in the monsoon region. The interannual oscillation include annual, the quasi-biennial (QBO) and the Southern Oscillations (mainly 3.5-yr and 5-yr oscillations, respectively). These essentially reflect the oscillatory behavior of the air-sea system.

As to the 30–50 oscillation, the major findings may be summarized in the following ways:

(1) It is an eastward propagating planetary scale oscillation (mainly the wave with wavenumber 1) in surface zonal wind and other fields, with the large amplitude confined to the equatorial region. The oscillation appears to propagate with a constant phase speed of 5 to 10 m s^{-1} over the Indian Ocean and the western Pacific, and also shows a meridional propagation (mainly northward in the Northern Hemisphere) with a phase speed of 1–3 latitude degrees per day.

(2) The 30–50 day oscillation is a global phenomenon, with its amplitude on the 30–50 day timescale not only dominant over the region of the tropics, but also at higher latitudes. Associated with this finding is the divergent motion on the planetary scale on the 30–50 day timescale which clearly shows up in the propagation of a planetary scale (mostly zonal wavenumber 1) wave from west to east in roughly 30–50 days (Krishnamurti and Gadgil, 1985; Lorenc, 1984)

(3) In the monsoon regions, the 30–50 day oscillation propagates meridionally from the equatorial latitudes to the Himalayas (Yasunari, 1981), exhibiting a near-steady meridional propagation of a train of zonally oriented troughs and ridges. The troughs were associated with rising motions and clouds, while the ridges were essentially cloud free. The phenomena of onset, active and break monsoon appeared to be related to the passage of these low frequency systems (Krishnamurti and Subrahmanyan, 1982).

(4) The ENSO event may result from the amplification and decrease in the period of the 30–50 day oscillation through the coupled interaction of air and sea. In this process, the feedback of evaporation-SST and wind stress are extremely important for the onset and evolution of the ENSO event (Lau and Chan, 1987).

(5) Theoretical and modelling studies suggest two possible mechanisms for these oscillation. The first mechanism is a Kelvin wave response forced by external sources (Chang and Lin, 1987). The second mechanism involves an interaction with the stationary oscillations of the basic state (Webster, 1983). For the former, the cumulus heating is a primary factor. For the latter, actually the generation of these oscillations is attributed to the nonlinear interaction in the atmosphere.

The 30–50 day oscillation has been specifically investigated by Chinese meteorologists over the East Asian monsoon region. Chen and Jin (1984) pointed out that the 30–50 day oscillation in the East Asian region is characterized by the large-scale break-active cycle of the monsoon as well as corresponding variations of intensity of monsoon meridional cell and the subtropical high. Chen and Ke (1983) made a detailed analysis of the feature of low-frequency oscillation feature of the break-active cycle of the East Asian monsoon by using a composite method. The variations of the 850 hPa wind pattern may be identified as four stages, each of which lasts for about 10 days:

(1) The stage of the active westerly wind over southern China: The salient feature is characterized by the westward extension into the South China Sea of the subtropical high over the western Pacific, with its main body located farther south than normal. The monsoon trough over the Bay of Bengal runs nearly meridionally, with the southwesterly airflow in advance of the trough flowing across the Indo-China Peninsula up to southern China. The monsoon in southern China has its peak activity while the southwest monsoon over India is inactive. The anticyclone dominates the South China Sea and there is no ITCZ present in this region.

(2) The stage of weakening westerly wind over southern China: The equatorial westerlies over South Asia starts to enhance significantly and extends eastward. Meanwhile, the subtropical high over the western Pacific shifts northward, with southern China gradually turning into being dominated by the easterly wind at the southwestern flank of the subtropical high. The monsoon trough begins to form over the South China Sea. Then, it shifts northward and extends eastward as the equatorial westerlies augment.

(3) The stage of the active easterly wind over southern China: This is the most active stage of the monsoon trough over the South China Sea, and also the major episode of typhoon activity. The nearly zonal equatorial westerlies prevail over the region from the Bay of Bengal to the South China Sea. At the same time, the cross-equatorial airflow at 105° E enhances the southwest monsoon over the southern part of the South China Sea. The easterly wind to the north of the monsoon trough also has its maximum intensity. A 6000–7000 km long monsoon trough extends from the Bay of Bengal to Guam.

(4) The stage of weakening easterly wind over southern China: The monsoon over the southern part of the Bay of Bengal significantly weakens. Due to the presence of the cross-equatorial airflow, the southwest monsoon over the southern part of the

South China Sea is still active, with its northern boundary shifting northward. The subtropical high is farther east than normal. At this time, the monsoon trough is mainly active over the western Pacific where typhoon formation is still observed. The Indian monsoon is not generally active, even at the break stage.

In short, the 30–50 day oscillation shows up in an alternation of northward advance of the monsoon trough as the equatorial westerlies enhance and move northward, and the southward retreat and dissipation of the monsoon trough as the equatorial westerlies weaken. Over the coastal area of southern China, one may observe a corresponding alternative change of the southwest monsoon surge and the easterly wind.

Figure 4.1 apparently shows the low frequency oscillation of the low-level zonal wind compoment at the longitudinal range of the Asian monsoon region (Yang, 1989). It can be seen that (1) from May to August, 1986, the low-level equatorial westerlies underwent four cycles of propagation of the 30–50 day oscillation, with the mode of eastward propagation dominanting, except for the time period of 2nd–4th

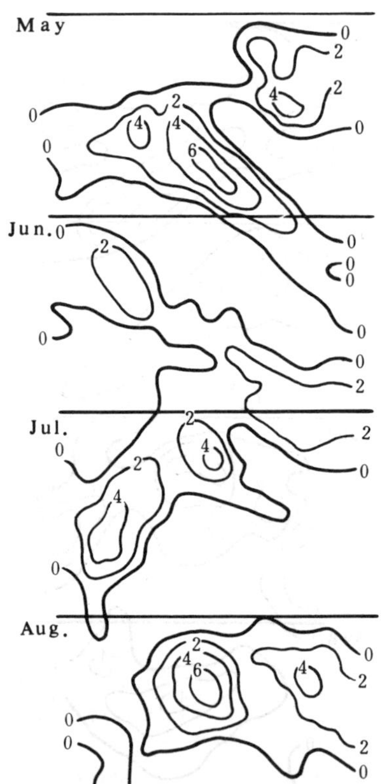

Figure 4.1. The longitude-time cross-section of 30–50 day filtered equatorial westerlies at 850 hPa for May-August, 1985. The negative values represent east wind. Units: m s^{-1} (Yang, 1989).

pentad of July that was characterized by the westward propagating low-frequency oscillation, and (2) the major oscillation centers of low-frequency activity of the equatorial westerlies were located at 105–115° E, with the filtered zonal wind component being 4–7m s^{-1}. In addition, there are other oscillation centers at 65–85° E and 155° E which are broadly consistent with the preferred channels of cross-equatorial airflows.

The 30–50 day filtered meridional wind component at 200 hPa also shows four propagating processes during this time period (Fig. 4.2). A very prominent feature is the eastward propagation of the wave phase. The oscillation centers are located at 70° E, 90–100° E and 150° E, with the maxima in the range of 3–5 m s^{-1}. The above result indicates that the near-equatorial airflows in the monsoon region assume an obvious low-frequency oscillation feature. Comparing Figures 4.1 and 4.2, one may find their correspondence.

The propagation of the low-frequency oscillation in the East Asian region has been addressed by some investigators (Zhu and He, 1985; He, 1988). In general, the 30–50

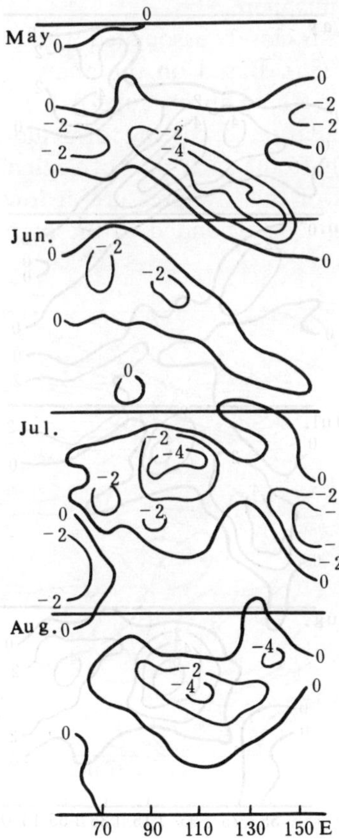

Figure 4.2. The same as Fig. 4.1, but for the 200 hPa meridional wind component (negative values: north wind). Units: m s^{-1} (Yang, 1989).

day oscillation in the East Asian region propagates eastward and northward, basically similar to the situation in the South Asian region. He (1988) has studied the meridional propagation of the quasi-40 day oscillation based on ECMWF gridded data sets for June-September 1981. He has revealed that for the East Asian area (about 120° E), there is a confluent belt for the meridional propagation of low-frequency meridional wind at lower levels, while there is a diffluent belt of the meridional propagation of low-frequency zonal wind at higher levels. This belt is found at 20–30° N. The low frequency systems on either side of this belt greatly differs in properties. Therefore, this belt may be taken as a dividing zone between the tropical and mid-latitude low-frequency oscillation. Figure 4.3 shows that north of 30° N, the low frequency oscillation moves northward during most of this period under study while the easterly jet oscillation moves southward to 20–30° S, sometime even to the mid-latitudes in the Southern Hemisphere. It seems that the low-frequency oscillations of the zonal wind at a higher level originate in the region at 20–30° N. In contrast, at 850 hPa (Fig. 4.4), in the same region one may observe a confluent trend of low frequency oscillations of meridional wind that come from 10 ° S or the mid-latitude in the Northern Hemisphere.

The 30–50 day oscillation is closely associated with precipitation in China (Zhou *et al.*, 1983; Zhou *et al.*, 1986). Based on the analysis of the power spectrum of precipitation for 15 stations in China during June-August 1979, there were two 45 day periods of low frequency oscillation, occurring form mid-June to the last part of July, and from the last part of July to the early part of September. The wet spell in China corresponded to the period of the low pressure at low-latitude and active Indian monsoon while the dry spell corresponded to the high pressure and break in the Indian monsoon.

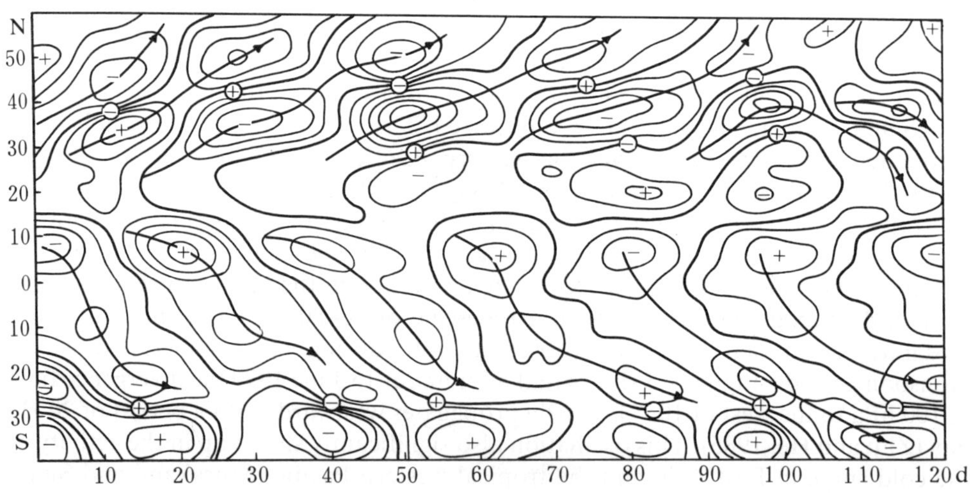

Figure 4.3. The time cross-section of the 40 day zonal wind oscillation at 200 hPa along 120° E. The "plus" and "minus" signs denote the positive and negative centers in the height field, respectively. Arrows indicate the direction of the LFO mode. Heavy sold lines indicate the zero line, with the interval of the isopleths being 2 m s^{-1} (He, 1988).

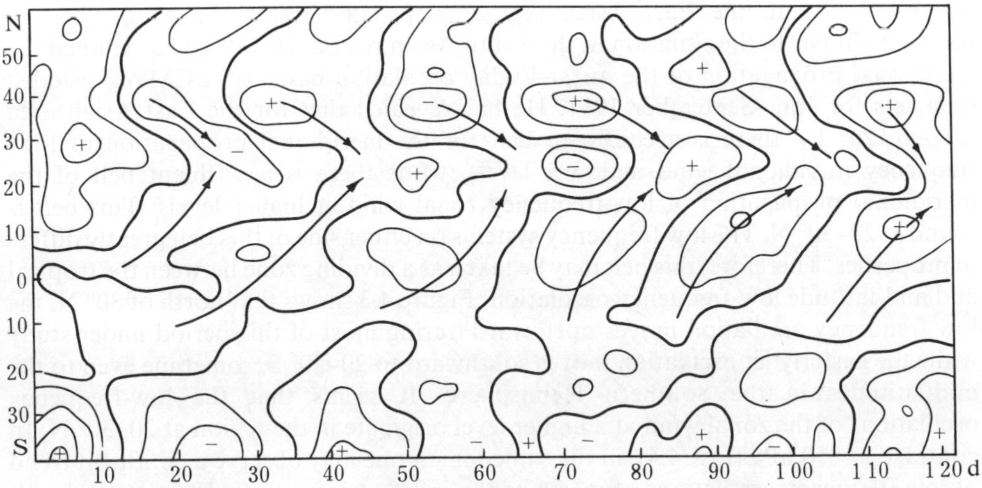

Figure 4.4. The same as Fig. 4.3, but for the 40 day meridional wind oscillation at 850 hPa. The isopleth interval is 1 m s^{-1} (He, 1988).

The wet spell in the first cycle of the low frequency oscillation in fact corresponded to the Meiyu period over the Yangtze River Valley whereas the second wet spell corresponded to the rainy season in northern China (especially in the Yellow River area). Therefore, the 30–50 day oscillation is directly related to the variations in the rainy belts in China and characters of droughts and floods over the Yangtze and Yellow River Valleys.

The low frequency oscillation shows a significant interannual variability, and its effects on the precipitation activity are rather different. For example, for 1980, a year of drought in northern China and flooding in the Yangtze River Valley, the period of low-frequency oscillation was shorter than that for 1981 when the flooding in western China and droughts in eastern China occurred. These differences clearly showed up in the activity of the ITCZ over the western Pacific (10° N). In 1980, the 20-day oscillation dominated, while in 1981 the 40 day oscillation dominated. In addition, the subtropical high over the western Pacific had different periods of oscillation, with the 40 day oscillation dominating for 1980 and the 80 day oscillation for 1981. The another important mode of low-frequency oscillation in the monsoon regions is the westward propagating systems with a timescale of 10–20 days (Krishnamurti and Ardanuy, 1980) that have been observed both during the summer and winter monsoon periods. Figure 4.5 shows that after about 15 June 1979 the divergence field at 100 hPa to the west of 105° E clearly propagates westward, with a period of 20–30 day (Zhu and He, 1985). This oscillation also significantly shows up in the activity of the cold surge in East Asia and the tropical western Pacific (Murakami and Sumi, 1981).

In contrast to the conditions in the summer monsoon regions, for the 20-day mode in the winter monsoon region, the most prominent feature is the eastward propagation of subtropical upper-level disturbances from Eurasia to the central

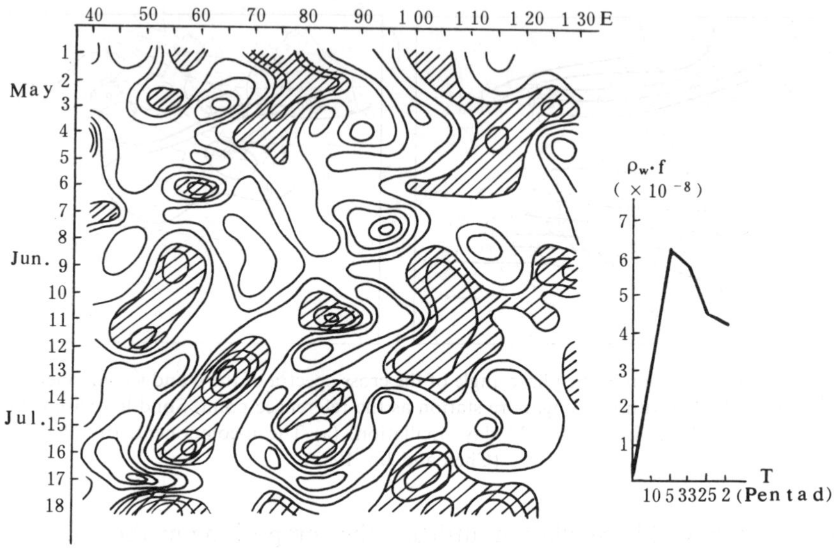

Figure 4.5. Longitude-time cross-section of the divergence field at 100 hPa along 15° N for May-July 1979. The thick solid lines represent the zero lines. The shaded area denotes the region with a divergence magnitude greater than $0.4 \times 10^{-6}\ s^{-1}$. The interval of isopleths is 0.4 Unit. The inserted diagram is the power spectrum analysis. Units: $10^{-6}\ s^{-1}$ (Zhu and He, 1985).

Pacific. On approaching the western end of the Tibetan Plateau, they intensify and reach their maximum intensity near the northern Bay of Bengal before dissipating over the South China Sea and the Philippines. As these disturbances move eastward, northerly surges are induced over Afghanistan, Japan-Philippines, and the central northern Pacific. Pan and Zhou (1985) have documented such a sequence of a high that shows a southward incursion and exits at the East Asian coast of China. Divergent circulation at 200 hPa over the tropical western Pacific intensifies and then moves eastward. This is a slowly eastward propagating low-frequency, synoptic scale disturbance, whose timescale is close to 10–20 days at 200 hPa. Ding (1990) has recently revealed the close relationship between the 20 day low frequency oscillation and outbreaks of cold air in East Asia.

The biweekly oscillation is the third kind of low-frequency oscillation in the monsoon regions. As suggested by Krishnamurti and Bhalme (1976), this reflects an inherent oscillation of the South-Asian monsoon system. They also suggested a mechanism of cloud-radiation feedback to interpret the biweekly oscillations of the various components of the Indian monsoon system. Chen and Jin (1982)revealed the existence of this kind of low-frequency oscillation in the East Asian monsoon system. They pointed out that at low-level, this originated in the Australian region and then propagated northward up to the ITCZ region, while the reverse propagation at the upper level is true, i.e., from the ITCZ region southward down to the Southern Hemisphere (Fig. 4.6).

Chen and Ke (1983) obtained the composite 700 hPa flow patterns for two extreme

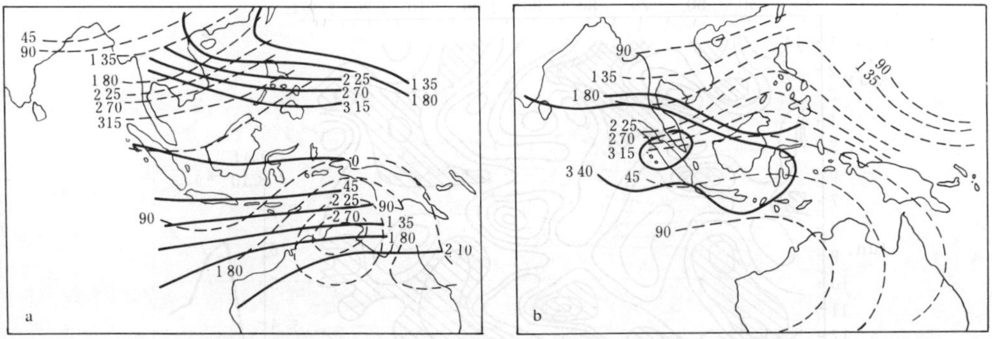

Figure 4.6. The distribution of the phase angle of the cross-spectra of the u and v components at 850 hPa in the summer of 1980 with the Singapore station as the referenece point. Solid lines are for the 40-day oscillation; dashed lines are for the 14 day oscillation. (a) The u-component. (b) The v-component (Chen and Jin, 1982).

phases of the biweekly oscillation, utilizing the compositing method for six cases of biweekly oscillation for 1979, as shown in Figure 4.7. It can be seen that for the strong southerly wind phase at Singapore the monsoon troughs over India and the South China Sea both significantly intensified. The latter shows a considerable northward advance. The subtropical high has a more northern latitude. The equatorial buffer zone in which is embedded a closed circulation feature, acquires a well-established development, thus enhancing cross-equatorial airflows at 105° E. In contrast, for the phase of weak southerly wind at Singapore station, the ITCZ over the South China Sea retreats southward while the Indian monsoon trough advances northward. The cross-equatorial airflow weakens. Similarly, Chen and Jin (1982) also obtained this type of result.

By using the dataset for the longer period (1973–1986), Mao *et al.* (1990) have recently studied the intercorrelation between the cross-equatorial airflow at 105° E and the summer monsoon over Asia, indicating the significant biweekly oscillation of the cross-equatorial airflow. At the same time and one pentad after the enhancement of the cross-equatorial airflow, the East Asian tropical monsoon tends to develop whereas the Indian summer monsoon and the East Asian subtropical monsoon tend to weaken or break. Two pentads after the enhancement of the cross-equatorial airflow, the subtropical high over the western Pacific and the subtropical summer monsoon at its western flank (near the East China Sea) tends to enhance. For the summer monsoon regions to the north of the ITCZ in East Asia, there observed the biweekly oscillation. Among them, the subtropical and South Asian highs show as the most obvious features of this mode of oscillation (Fig. 4.8) (Yu, Liu and Zhu, 1989). It can be seen that the meridional shift of position of the ridge line of the subtropical high has a dominating period of 14.2 days while the South Asian high does not show a single dominating period, even the period of 14.2 days still being very apparent. For the summer monsoon precipitation over the Yangtze River Valley one can also observe the dominating period of 14.2 days (Fig. 4.9).

In general, the oscillations of the inter-annual timescale are termed as very low-

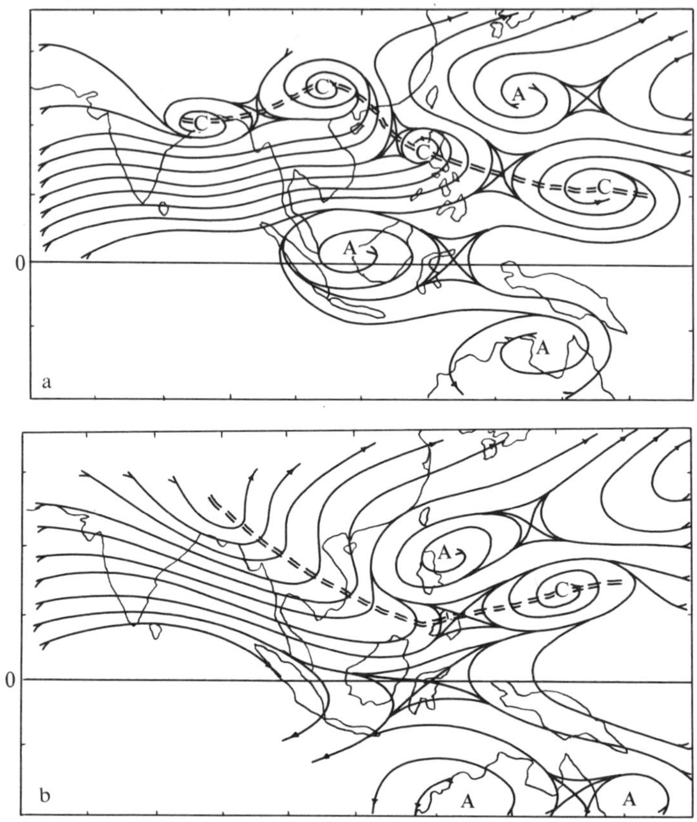

Figure 4.7. The compositing 700 hPa flow pattern for the phase of the strong southerly wind (a) and the phase of the weak southerly wind (b) of the biweekly oscillation at Singapore (Chen and Ke, 1983).

frequency oscillation, as compared to the 30–50 day low-frequency oscillation that has been discussed above. This kind of very low-frequency oscillation includes the following major modes: the quasi-biennial oscillation (QBO), 3.5-year quasi-periodic oscillation (SO), 5.5-year oscillation (FYO), and 11-year oscillation (Yan *et al.*, 1988; Lu, Ding and Gao, 1990). They, in fact, represent the characteristic oscillation modes of the sea-air system in the tropics and subtropics. Yan *et al.* (1988) point out that the QBO and SO of the zonal wind propagate eastward along the equator. Their origin may at least be traced as far as the western Indian Ocean. While propagating along the equator, they intensify at 90° E and over the western Pacific, respectively. However, the FYO of the zonal wind is somehow a standing mode. While these oscillations propagate, they may interact with those of the sea surface temperature, thus producing a good coupling relationship. When the QBO and SO of the SST move to the eastern end of the equatorial western Pacific, the QBO and SO of the zonal wind propagate to the eastern part of the western Pacific.

Also, the meridional wind was found to oscillate with the above three periods (Lu,

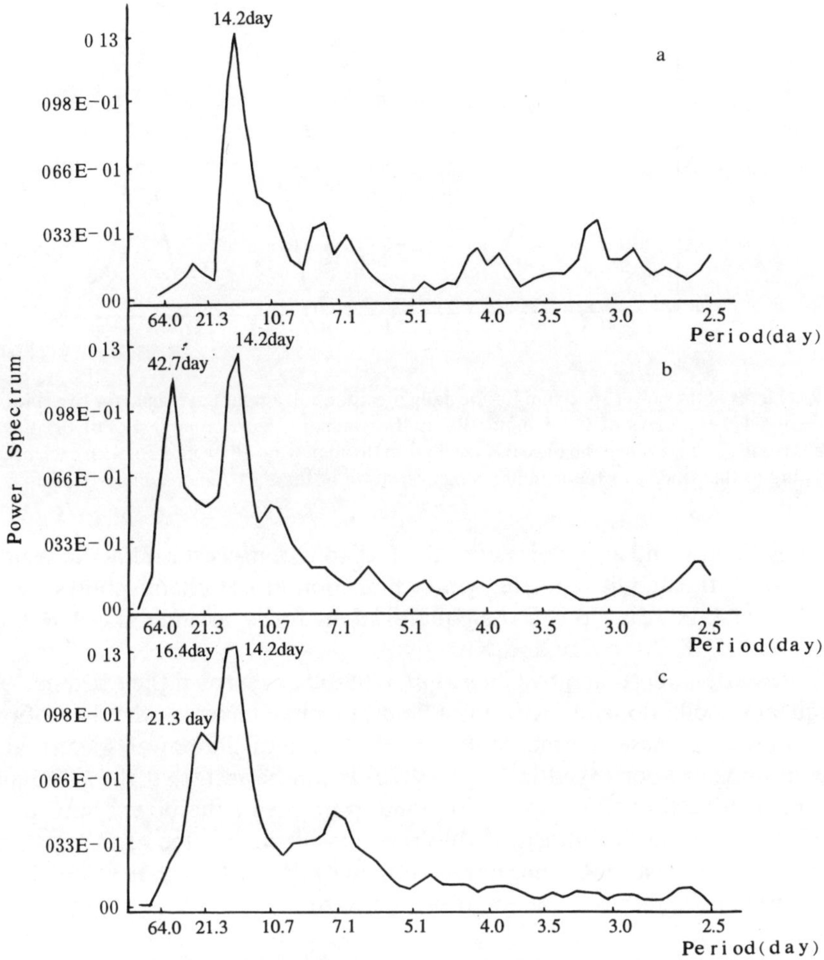

Figure 4.8. Curves of the power spectrum for the meridional shift of ridge lines of (a) the subtropical high over the western Pacific, (b) the South Asian high at 100 hPa and (c) the South Asian high at 200 hPa (Yu, Lin and Zhu, 1989).

Ding and Gao, 1990). The QBO prevails over the oceans, with the maximum amplitude over the central eastern Pacific Ocean. The SO mainly dominates the oceanic areas to the east of the dateline, while over 140° E-180 this oscillation was overshadowed by the FYO. A separate area with the FYO prevailing is the South Asian monsoon region (70–120° E). The superposition of the QBO and SO of the meridional wind may well fit the observed meridional wind fluctuation. The further addition of the FYO may improve this result. The meridional wind anomaly possesses a pronounced zonally propagating feature. It is interesting to note that it propagates eastward from the Indian Ocean in the Eastern Hemisphere, but propagates westward from the eastern coast of the eastern Pacific in the Western Hemisphere. Both meets generally at about the dateline.

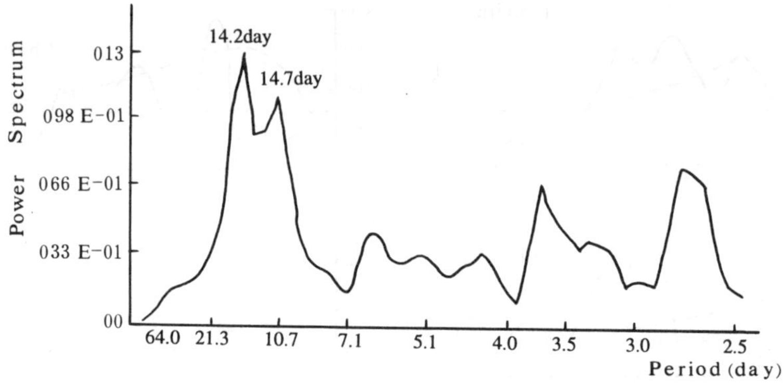

Figure 4.9. Curves of the power spectrum for the daily precipitation amount averaged for five stations over the middle and lower valleys of the Yangtze River. In summary, most components of the East Asian summer monsoon system assume the biweekly oscillation through which they interact with each other, thus leading to the process of the medium-range variation of the East Asian summer monsoon.

The meridional wind anomalies over the Indian Ocean-western Pacific seem to be associated with the activities of the winter monsoon in both hemispheres. It is very easy to understand that the cross-equatorial airflows assume the QBO mode, especially at 105° E (Mao, Xu and Xu, 1990).

A recent study undertaken by Chen *et al.* (1990) has revealed the existence of very low-frequency oscillations in the rainfall field in China, based on the data of rainfall anomalies of 48 Chinese weather stations. The curve of the power spectra at some selected stations are portrayed in Figure 4.10. It can be seen that the QBO and SO modes are relatively significant. The regional patterns of the power spectra of the QBO and SO have further indicated that the areas of high value of the QBO power spectrum are northeastern China, northern China, the eastern part of northwestern China, the middle and lower Valleys of the Yangtze River to the east of the Qinghai-Xiang Plateau and Guangdong-Hunan regions, while the regions with low values are Inner Mongolia, Xingjiang Province, Qinghai, Guangxi and Guizhou Provinces, and the Huaihe River Valley. These results show great similarity to those obtained by Huang (1988).

For the 3.5-year oscillation, 65% of the stations used in Chen *et al.*'s study have passed the significant test of 38.87 months. The 3.5-year oscillation of rainfalls is active in regions such as the western part of northernern China, the eastern part of northwestern China, the northern part of northeastern China and Guangdong Province. Therefore, both the SO and QBO are significant over northeastern China, Guangdong, and the western part of northern China.

Then, a new method used for the complex expansion of orthogonal function (CEOF) was applied to analyse the monthly rainfall data for 1951 to 1987. The results show that both the QBO and SO modes exhibit two propagating paths: the first one was characterized by the oscillation originating in northeastern China, extending southward, passing through northern China, and reaching the eastern part of Northwest China and the northern part of southwestern China; the other originating

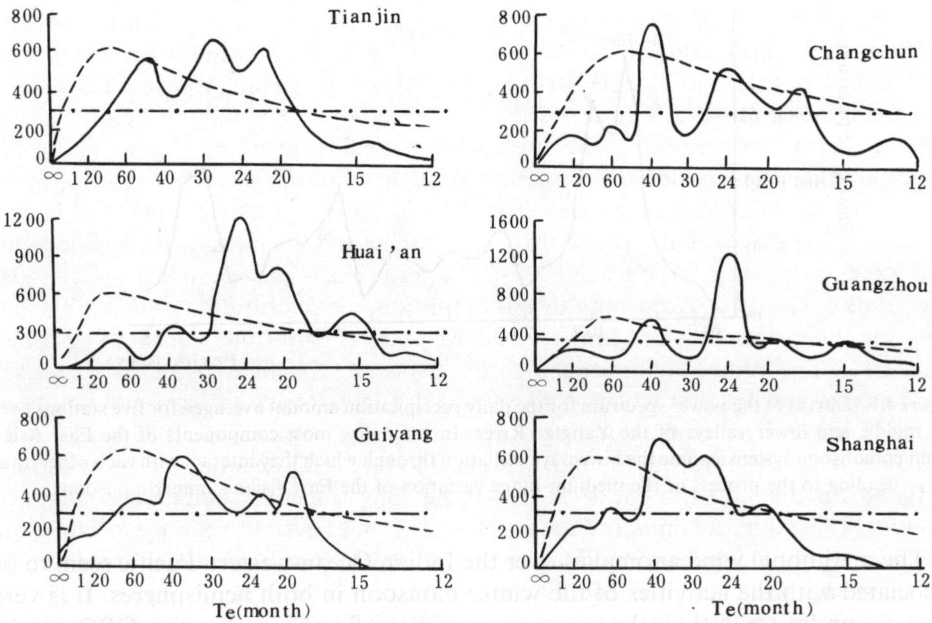

Figure 4.10. The power spectra of monthly rainfall series at Tianjin, Huai'an, Guiyang, Changchun, Guangzhou, and Shanghai. The heavy lines represent the power spectra, dashed lines and the doted and dashed lines denote,respectively, the red noise with a 0.05 significant level and the check of no-main-period (Chen *et al.*, 1990).

over Guangdong and Fujian Provinces, then moving northward and westward, respectively, to the Yellow and Huaihe Rivers Valleys and southwestern China. These two paths of propagation of the oscillation converge over northern China and the area between the Yangtze River and the Yellow River. A significant correlation relationship exists between the interannual oscillation of the rainfall in China and that of the SST in the equatorial region. Although the correlation between the rainfall over China and the SST field over the equatorial eastern Pacific is rather weak, the correlation between their oscillation components is pronounced.

4.2. Floods over China

The summer monsoon in East Asia is closely related to droughts and floods in China, due mainly to great intra-seasonal and inter-annual variability of summer monsoon activity and monsoon rainfalls. These droughts and floods, especially in the eastern part of China, are one of most severe natural disasters, occurring about once a year on an average.

From the meteorological viewpoint, droughts and floods mean an excess or deficit of the amount of rainfall relative to the normal conditions, regardless of the severity of disasters brought about by them. A number of drought and flooding indices have been designed for the climatological perspective. But, for more widespread and practical use, the departure or anomaly of rainfalls from the long-term mean is used to define droughts and floods. If the precipitation departure for summer (June-August) is in the range of – 25 to – 50%, this period or season may be defined as the drought period or season; if the precipitation departure is less than –50%, this corresponding spell is defined as severe drought. Contrary to this, precipitation departures of 25–50% and greater than 50% are defined as flooding and severe flooding, respectively. Thus, the severity of droughts and floods may be classified into five grades: grade 1: > 50%, grade 2: 50–25%, grade 3: 25–-25%, grade 4: –25–-50% and grade 5: < – 50%. According to this definition, the distribution of droughts and floods in the eastern part of China for 10 recent years (1979–1988) has been obtained (Fig. 4.11). The immediate impression after inspecting this figure is of flooding over the Yangtze and Huaihe Rivers Valleys and droughts over northern and southern China. During these years, flooding occurred in 1979, 1980, 1982, 1983, 1984, 1987 and 1988, with 1980 having the most severe flooding, only second to that of 1954.

Based on the statistics for 1951–1980, the following five regions are prone to flooding: (1) southern China (on an average, 1–2 times every 3 years), especially the coastal area of Guangdong Province and the northern part of Guangxi Province; (2) the northern part of Hunan and Jiangxi Provinces (once every 2–3 years); (3) the coastal area of eastern China (once every 2–3 years); (4) the Huaihe River Valley

Figure 4.11. The distribution of droughts (shaded area with dashed lines, grade > 3.2) and floods (shaded areas with solid lines, grade <2.8) for 1979–1988 (see the text).

(once every 2–3 years) and the Haihe River Valley (once every 3 years). The general feature of the geographical distribution of floods for the whole of China is characterized by frequent floods in the eastern part of China, especially in the coastal areas and the rare occurrence in the western part of China, especially in the Plateau regions.

China is one of the countries that suffer most from flooding disasters. For the 2125 year period from 206 B.C. to A.D. 1949, 1092 cases of relatively severe flooding occurred, once every two years on a nationwide scale.

Floods occur under special large-scale circulation conditions. As Tao and Xu (1962) point out, the persistent droughts and floods over the Valleys of the Yangtze and Haihe Rivers are closely related to the anomalous 500 hPa height fields in the mid-latitudes or tropical region, as well as the activity of the East Asian monsoon (intensity and dating of onset and retreat). No matter when droughts or floods occur, the large-scale circulation pattern must be unusually stable, i.e., persistently anomalous circulation over the extensive region is a crucial prerequisite for the occurrence of droughts or floods.

A number of investigators (Lu and Lin, 1987) have studied the relationship between monsoon circulation and doughts and floods. Figure 4.12 shows that the deficit or excess in precipitation during the Meiyu season over the middle and lower Yangtze River Velleys are consistent with the interannual variability of the south wind component in April and May, with the correlation coefficient being +0.64 (greater than 0.05 significant level). It can also be seen that the most severe flooding in the 70's and 80's occurred in 1980, and the second one in 1974, whereas the most severe drought occurred in 1978, and the second most severe in 1972. One widely used methodology is to make a contrasting study of the drought and flooding years to elucidate more clearly the respective circulation features responsible for droughts or

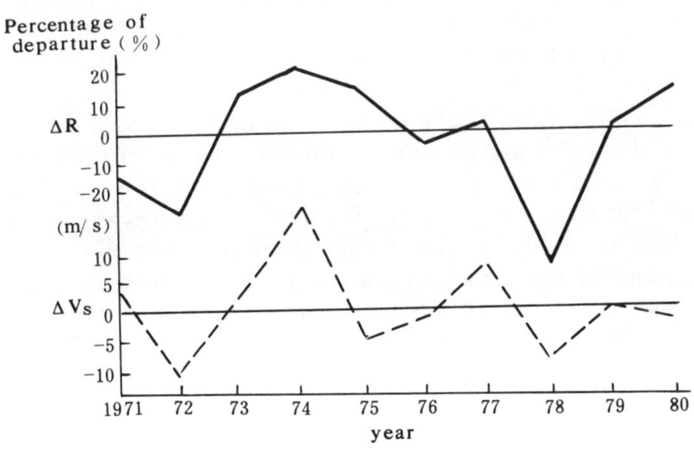

Figure 4.12. The relationship between the inter-annual variability of the precipitation anomaly (ΔR) for June and July over the middle and lower Yangtze River Valleys and the south winds (ΔV_s) in April and May (preceding months) (Lu and Lin, 1987).

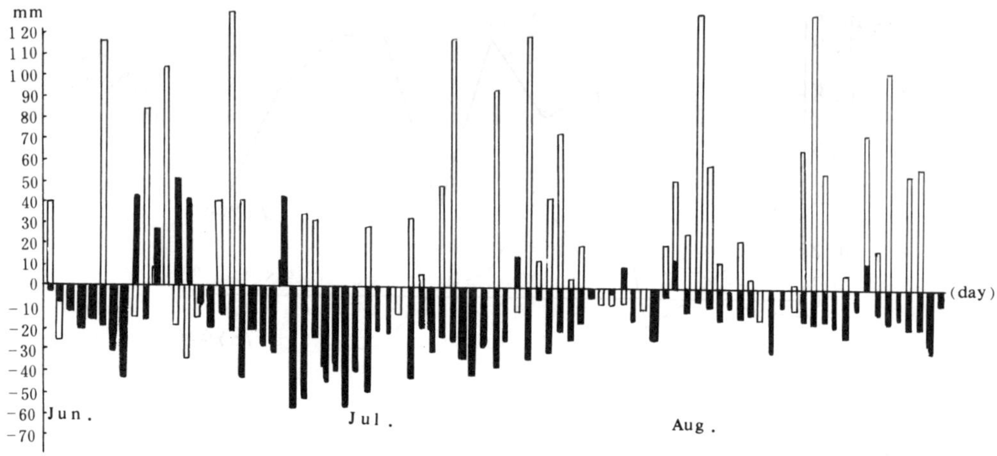

Figure 4.13. The day-to-day departure of the rainfall amount in the summers of 1980 (white rectangle) and 1978 (black rectangle) averaged for 5 stations over the middle and lower Yangtze River Velleys : Wuhan, Jiujiang, Wuhu, Nanjing and Shanghai. Units: mm (Lu and Lin, 1987).

floods. Figure 4.13 is the day to day variation of departure of the precipitation amount for 1980 and 1978 averaged for five stations over the middle and lower reaches of the Yangtze River.

In June-August of 1980, the positive departure was recorded almost every day except for one week during the last part of July. In contrast, in 1978, the daily rainfall amount was of negative departure. The condition in 1974 is similar to that of 1980. Figure 4.14 is the pentad-to-pentad variations of the meridional wind component along 10° N averaged for the longitudinal range of 10° E – 180°. It can be seen that in 1974, the flooding year, the south wind started to increase significantly from the last part of May, indicating a strong cross-equatorial airflow, while in 1978, the drought year, the south wind was very weak, less than 0.3m s^{-1}. The intensity of the south wind is associated with the transport of moisture.

Zhu, Zhao and Tao (1987) have further indicated that, as a matter of fact, the whole system of the East Asian monsoon in flooding year of 1980 shifts southward by 7–8 latitudes as compared to the condition for drought year. The upper-level divergent center for flooding year of 1980 moved farther westward than during the normal year. In this case, an apparent normal monsoon meridional cell was observed.

The circulation features at mid-latitudes for 1978 were characterized by the upper-level westerly jet located at 40–45° N, 5° latitude farther north than the position for the flooding year (1980) (Fig. 4.15), and its maximum wind speed is less than that for 1980.

In addition, the wind streaks in the upper westerly jet or the easterly jet have a more easterly position for 1980 than that for 1978.

The longwave circulation situation for 1980 at mid- and high latitudes was very stable and persistent, and the intensity of the ultra-long wave troughs and ridges was greater than for 1978 (see Fig. 4.16) (Bi, 1983). The most severe flooding over the

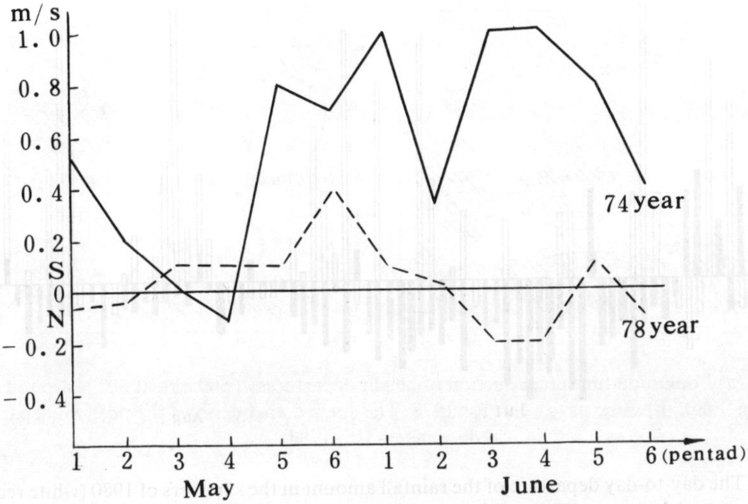

Figure 4.14. The pentad-to-pentad variations of the meridional wind component along 10° N averaged for the longitudinal range of 10° E–180.

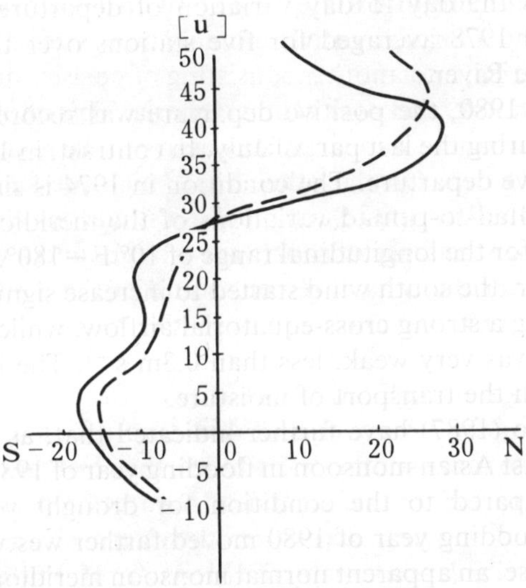

Figure 4.15. The variation of the mean zonal wind component at 200 hPa averaged for 90–150° E with the latitudes for the summers of 1980 (solid line) and 1978 (dashed line) (Zhu, Zhao and Tao, 1987).

Yangtze River Valley in June and July 1954 was caused by long and persistent heavy rainfalls. The destructive floods covered an area hundreds of miles wide along the Yangtze River. The rising water set a new high-water mark at Wuhan. This flood was caused by a Meiyu which lasted one month longer than normal (Chen, 1957). During June 1954, not only did the rain pour down almost unceasingly over the Yangtze-

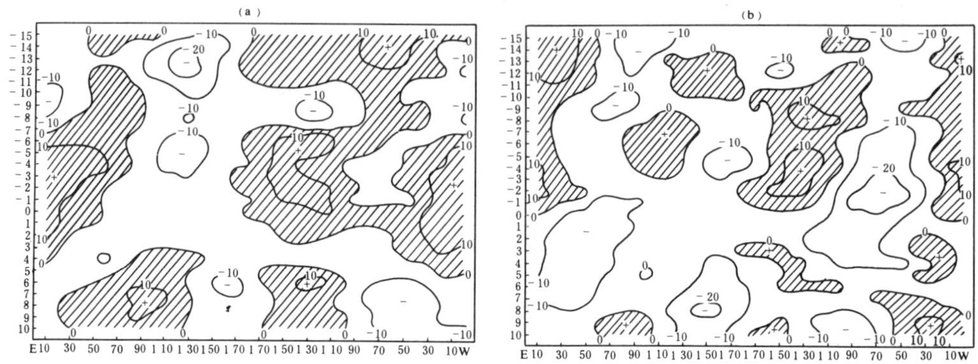

Figure 4.16. (a) Longitude-time cross-section of height departures at 500 hPa at 60° N around the onset of the Meiyu for 1980; (b) same as (a), but for 1978. The shaded area denotes the positive departure. Units: decameters (Bi, 1983).

Huaihe River regions, but also departures from the normal rainfall as high as +300 to +600% were recorded in the middle and lower Yangtze Valley. Figure 4.17 presents the 500 hPa anomalous circulation patterns in June 1954 which is typical of an enhanced Meiyu season (Ding and Reiter, 1982). In East Asia, two significant features can be noted, one being a large negative anomaly zone stretching from central Asia to western North America, and another consisting of positive anomaly areas over the Seas of Japan and Okhotsk, and over central Siberia. This anomaly pattern suggests that low-pressure disturbances were unusually active in the middle latitudes over East Asia and the Pacific Ocean. On the other hand, high-pressure ridges, or blocking highs, prevailed in higher latitudes. Such blocking situations characterize the Meiyu, together with the frequent passage of shortwave troughs and a westward extension of the subtropical high over the coastal regions south of the Yangtze River (in Figure 4.17, indicated by a positive anomaly over the western Pacific north of the Philippines).

The blocking situation in high latitudes has two effects. It maintains the stability of large-scale circulation features over East Asia through an anchoring effect, and it forces the westerlies into more southerly latitudes, thus bringing the frequent passage of shortwave troughs to the middle latitudes. A succession of such waves, coming from central Asia and moving across northwestern China, will advance into shearlines and vortices which accompany the Meiyu front. Furthermore, a southward displacement of the westerlies also enables the vortices generated over the Tibetan Plateau to be steered out of that region and to grow into major rain-producing systems over the Yangtze River.

The characteristic features of the anomalous circulation for July 1954 (Fig. 4.17) over eastern Asia were somewhat intensified in comparison to those of June: a well-developed blocking situation at high latitudes, a negative anomaly zone in middle latitudes over eastern China and Japan, and the stable subtropical high over southern China contributed to the intensive rainfall episode. The blocking highs over the Sea of Japan and over the Ural Mountains caused the main westerlies to flow about 10°

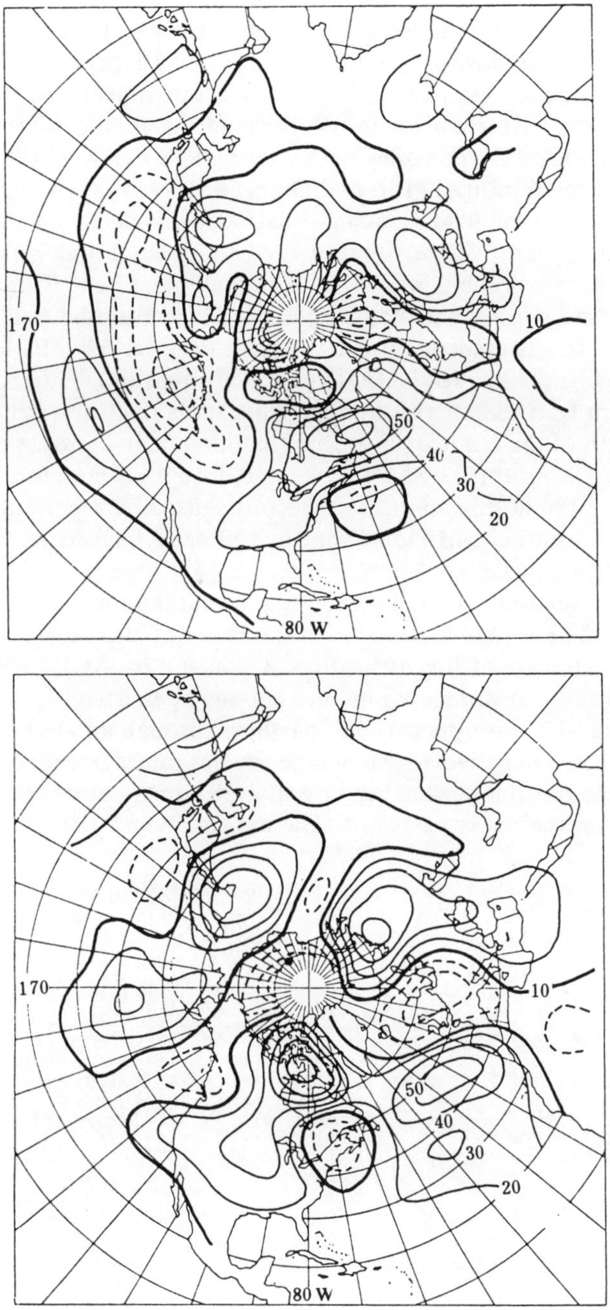

Figure 4.17. 500 hPa anomalies, truncated with wave number 4, for the period 1 to 30 June, 1954 (top panel) and for the period 1 to 31 July 1954 (bottom). Thin solid lines: positive anomalies; thin dashed lines: negative anomalies; analysis interval: 25 geopotential meters. Heavy solid lines: zero anomaly (Ding and Reiter, 1982).

latitude farther to the south than normal. As a direct consequence of this southward displacement, a sequence of wave troughs passed over the Tibetan Plateau and moved eastward as late as the second part of July. Analyses demonstrated that about eight or nine low-pressure disturbances, mostly originating in the Plateau region and passing over the Yangtze River Valley with steering troughs, were the primary cause of almost uninterrupted rainfall (Tao and Ding, 1981).

In this month, well-developed blocking situations prevailed over several regions of the Northern Hemisphere. These situations led to unusual and persistent weather types over many regions (Hawkins, 1954). For example, Great Britain and Europe experienced cold and stormy weather. The flood of the Danube was labelled as the worst in centuries. It was associated with the large negative 500 hPa height anomaly tilting from the Icelandic low to the Balkans ("Vb" situation). Late July floods were also reported from Iran, East Pakistan, and East India due to upper troughs which passed to the south of the Himalayas during that month. In contrast to these events, the United States was troubled by heat waves and drought. The central United States experienced one of the hottest months on record caused by a persistent upper-level high centered over southern Missouri that had already formed in June (Hawkins, 1954; Holland, 1954).

Longitude-time sections of daily 500 hPa height anomalies of the first four harmonic waves are presented to describe the behavior of large-scale planetary waves during the heavy rain case of July 1954 (Figs. 4.18 a and b). At 35° N over East Asia and the western Pacific, a wide low-pressure anomaly persisted for the whole month of July, suggesting frequent upper trough passages through an anomalous longwave pattern. Note that the negative anomaly area several times extended westward over the Tibetan Plateau or even farther. Apparently, the upper troughs were able to pass over the Plateau on their way eastward. The unusual heat spell and drought in the

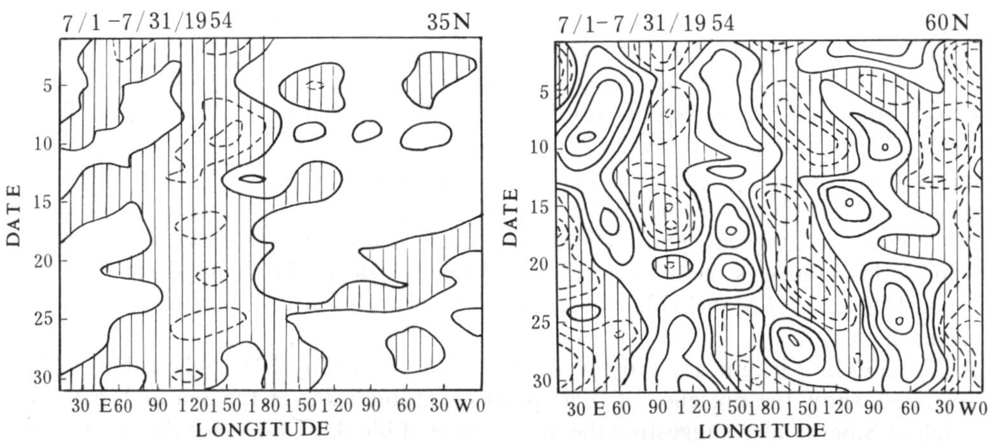

Figure 4.18. Longitude-time sections of 500 hPa height anomalies, truncated with wave number 4, for the period 1 to 31 July 1954, (a) at 35° N; (b) at 60° N. Analysis interval: 50 geopotential meters. Vertically shaded regions indicate negative anomalies. The shading interval corresponds to 10° of longitude (Ding and Reiter, 1982).

United States are revealed as a wide persistent high-pressure anomaly. The dateline divides the anomaly pattern, negative anomalies dominating the Eastern Hemisphere (0–180°), and anomalous planetary wave with wavenumber $n = 1$ dominated the subtropics in the Northern Hemisphere.

At 60° N, an anomalous planetary wave with wavenumber $n=3$ generated a very stable pattern during the first half of the month. The blocking highs over the Seas of Japan and Okhotsk (130–150° E) were unusually stable during the time period of 1–24 July. In the last week of this month, the planetary wave patterns in the Northern Hemisphere began to move slowly eastward, a sign of the end of the persistent heavy rainfall in China and of the drought in the United States.

The major flood of July 1931 in the Yangtze River Valley had a great resemblance to that of 1954, with heavy rainfall over the middle and lower parts of the valley. The number of days with heavy rain during that month exceeded 25. Based on surface data available (no upper-air data were taken at that time), this flood was also caused by five to six low-pressure disturbances which had their origin in the eastern part of the Tibetan Plateau.

Sea-level pressure anomaly charts for July 1931 showed that negative anomalies extended from the middle and lower Yangtze River Valley to Japan and a strong positive anomaly region was located over the Seas of Japan and Okhotsk, implying the presence of a stable blocking high there.

Early in July 1935 another excessively heavy rain took place in the middle Yangtze River Valley, with 1200 mm of rain falling over five days. A large-scale circulation pattern typical of the Meiyu season showed a blocking high over Lake Baikal and northern China, and an upper trough extended from the east to the delta of the Yangtze River, while an unusually stable low-pressure trough prevailed over the Bay of Bengal. This configuration apparently led to strong low and middle tropospheric southeasterly winds over the middle and lower Yangtze River Valleys and to strong southwesterly winds over much of southern China. A low-level west-east oriented shearline favored the frequent formation of vortices.

In April through June of every year, southern China has its pre-summer rainy season that is in many aspects similar to the Meiyu season in the Yangtze River Valley. This rainy season occurs before the large-scale onset of the Indian summer monsoon. Although upper troughs, accompanied by cold fronts at the surface, are responsible for the occurrence of heavy rain during this season, one can often find an involvement of tropical synoptic systems in these rain events. An excessively heavy rain of this kind was recorded in the southeast region of the Guangdong Province towards the end of May 1973, caused by low-level shearlines and stationary fronts at the surface.

The 500 hPa height anomaly patterns for 26 to 30 May 1973, when the rainfall occurred, show the presence of large positive anomalies over the Sea of Japan and much of Siberia, also suggesting the importance of blocking highs for the persistence of precipitation. A negative anomaly region dominated over China with a deep trough located at 120° E, slightly farther to the west than its seasonal mean position. Low-level cold air travelled behind this trough into southern China, where a shearline was formed against the southwest flow from the western periphery of the subtropical high.

Another significant feature was a negative height anomaly over the Bay of Bengal. On the 500 hPa mean map for the period of heavy rain, one could observe an anticyclonic cell centered over the heavy rainfall regions, with a westerly jet on its north side and an easterly jet to its south. These pressure and flow configurations favor the rapid outflow of mass and heat from the rain area to maintain the strong upward motions and thermodynamic instability. Ramage (1977) also analyzed a similar case of heavy rainfall in Hong Kong and its neighborhood which occurred on 12 June 1966, with 600 mm rainfall during 18 hours.

The period of 27 to 30 May 1973 was characterized by the establishment of a broad upper trough centered over the Bay of Bengal. During the early part of the month, this trough was located over the tropical western Pacific and then gradually retrogressed westward, finally stabilizing near 90° E as heavy rainfall broke out in southern China. As soon as the trough moved eastward, the rains in southern China ceased.

Under certain circulation conditions, low-level vortices will move northward and can generate rainstorms and severe local storms in northern China. The heaviest and most extensive rainstorms in the history of weather records of northern China occurred under such circumstances during the period 1 to 10 August 1963 with rainfall of 2005 mm during approximately one week recorded at the Zhangmao station on the eastern slope of the Taihan Mountains. The most significant feature of this rainstorm was that three low-level vortices coming from the Tibetan Plateau and its surroundings slowly moved along a southwest-northeast orientated, low-level shearline or elongated low-pressure zone. From Figure 4.19 one can clearly see the abnormal blocking high located over the Baikal region. In summer, that area usually shows a low-pressure trough. Abnormal anticyclonic conditions extended over the Seas of Japan and Okhotsk, separated from the Siberian high-pressure system by a shearline, or low-pressure zone which formed on the windward side of the Taihan Mountains (see dashed line in Fig. 4.19).

The moist and warm easterly to southeasterly flow coming from the Yellow Sea and the East China Sea, and the cold northerly to northeasterly flow guided southward by the blocking high-pressure ridge, converged towards that shearline. High pressure anomalies also dominated Tibet and the region southeast of China. The most intensive rainfall episode occurred during the passage of low-level "southwest" vortices which brought about an increase of wind speed in the easterly flow.

In a similar synoptic development, heavy rainstorms in Beijing on 1 and 2 July 1973 were caused by a northward moving "southwest" vortex under the steering influence of the very strong south-north oriented upper-level jet ahead of a deep polar trough. A south-north oriented high-pressure ridge prevailing over the circulation patterns provided the steering current for the low-level "southwest" vortex which had orginated over the Plateau of Tibet. A total rainfall amount of about 200 mm during slightly more than 24 hours was produced in Beijing, which was second to the record established on 26 June 1893 (Tao *et al.*, 1980).

During the period 14 to 19 July 1958, the middle Yellow River Valley experienced the most damaging flood of the past 50 years. Three to four days of continuous rain sent the river well over its banks with a record volume flow of 22300 $m^3 s^{-1}$ recorded at Zhengzhou. Main railroad lines were destroyed or greatly damaged and remained

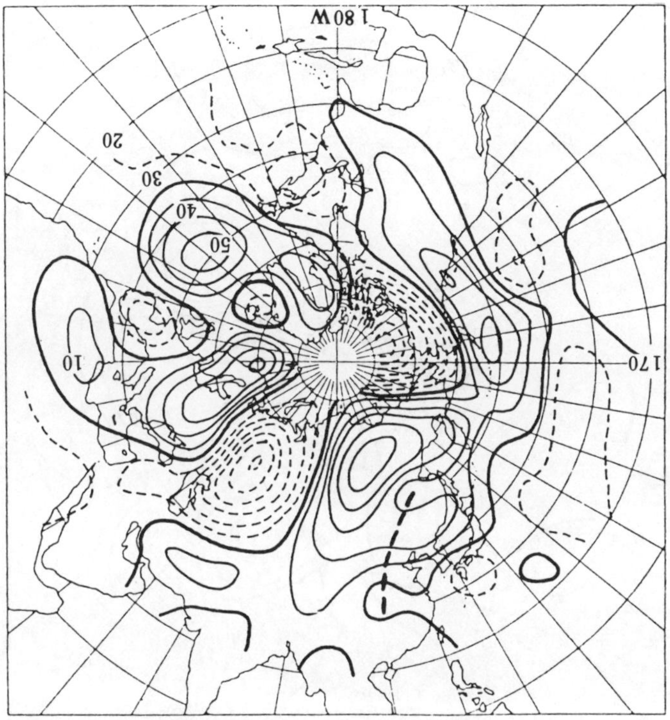

Figure 4.19. Similar to Fig. 4.17, except for the period 1 to 10 August 1963. The heavy dashed line indicates the mean position of shearline (Ding and Reiter, 1982).

cut for about two weeks. The excessively heavy rainfall was generated by a near-stationary upper trough that was blocked by a well-developed high to the east. The 500 hPa anomalous circulation pattern contained a south-north orientated, large positive anomaly zone extending from 30° N to the polar region along about 120° E (Fig. 4.20). To the south of this anomaly region, there was an extensive area of negative anomalies. Strong low-level easterly winds developed in between, which transported a great amount of moisture far westward into central China. Over the Black and Caspian Seas, another large positive anomaly can be noted, indicative of a strongly developed high-pressure ridge. Such circulation features commonly lead to heavy rainfalls in northwestern China. A negative anomaly area existed in the region between the above-mentioned two stable highs. It aided in the advection of cold air into the heavy rain region. Consequently, the excessively heavy rainfall over the middle Yellow River in 1958 occurred when the south-north oriented stationary front, or low-level shearline, separating cold air to the west from warm and moist air to the east, lingered over this region for five days.

As in the previously described case, the heavy rains of 1958 over the middle Yellow River Valley were characterized by the slow regression of a high-pressure ridge over

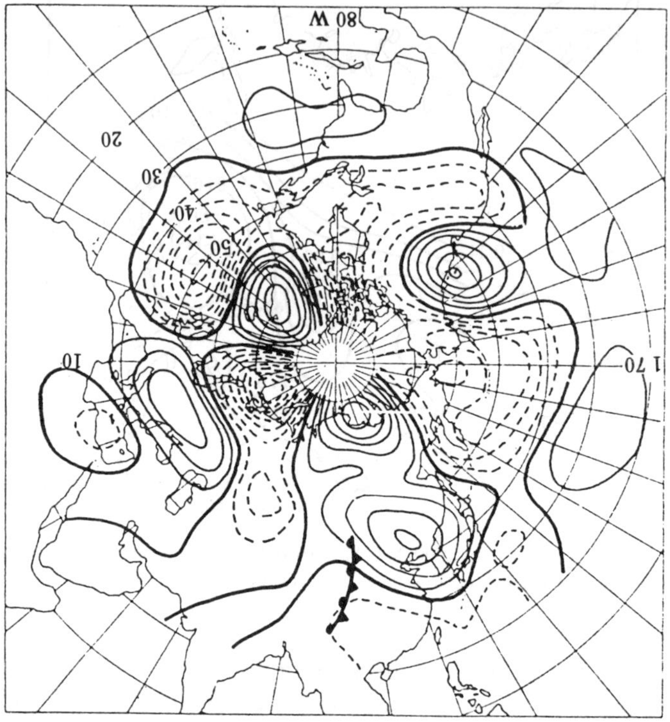

Figure 4.20. Similar to Fig. 4.17, except for the period 14 to 19 July 1958. The position of a quasi-stationary front is indicated over central China (Ding and Reiter, 1982).

the Seas of Japan and Okhotsk. The heavy rain episode commenced when the high became strongest in a nearly stationary position on 16 July. Upstream, the high over the Black and Caspian Seas was established on 11 July and persisted to the end of the heavy rainfall period. A stationary trough developed between these two high-pressure ridges. Such a trough appears to be an important circulation condition for prolonged heavy rainfall in northern China (Fig. 4.21). Note that a low-level easterly jet developed between the high over the Sea of Japan and the typhoon (Typhoon No. 10) to the south, which acted as the major channel of moisture transport to this rainstorm event.

The large-scale circulation patterns responsible for the heavy rainfall on 5–6 July 1977 basically resembled those for the 1958 rainstorms, except for a slight eastward displacement of the large anomaly centers, and for the shorter duration caused by a retrogression in the planetary wave pattern. The positive anomaly along 120° E was produced by the apparent westward displacement of the subtropical high. The high-pressure ridge over the Caspian Sea was even better developed than during the case discussed previously. Along its eastern edge, two upper troughs successively moved eastward and slowed down due to downstream blocking. These troughs greatly

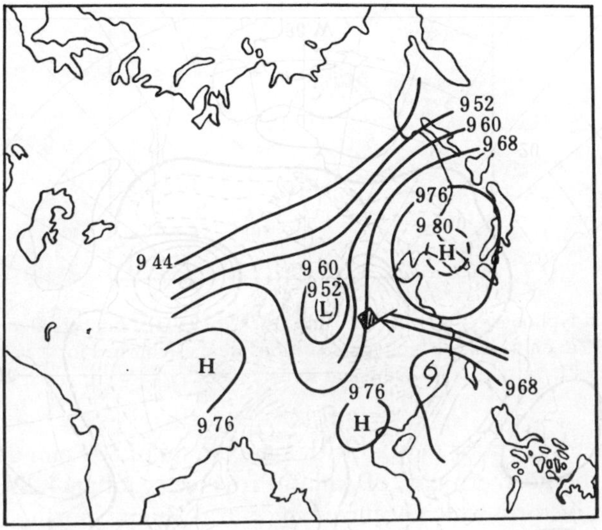

Figure 4.21. The 300 hPa weather map at 00:00 GMT, 16 July 1958. The shaded box represents the region of heavy rains. The double arrow the low-level easterly jet at 850 hPa. Units: decameter (Ding and Reiter, 1982).

enhanced the rainfall intensity, bringing about rainfall of 180 mm during 24 hr, which is a very rare occurrence in the semi-arid northwestern highlands. An immense flood, possibly the most severe in the past 100 years in northwestern China, was reported.

Occasionally, northward or northeastward-moving typhoons suddenly change their tracks, turn toward the west or west-northwest, and penetrate deeply inland if a blocking high to the north of the typhoon builds up rapidly, amplifies and retrogrades. Under such conditions, the typhoon, though usually greatly weakened, drifts in an easterly current and finally becomes nearly stationary because of the effect of the high-pressure ridge to the north of it. If other conditions are favorable, prolonged heavy rainfalls may ensue. An unprecedented case appeared early in August 1975 when typhoon Nina dumped torrential rainfall with a maximum of 1629 mm during three days over a narrow, elongated band 120 km long and 50 km wide along the windward side of the Funiu Mountain Range. This rainstorm event has been discussed in Chapter 3. Here, it should be emphasized that the occurrence and persistence of the heavy rain was closely linked to the strong low-level easterly flow established between the blocking high and typhoon Nina. The easterlies emanated from the East China Sea reached low-level jet strength. They transported a great amount of moisture into the heavy rain region (Ding *et al.*, 1978). Many analyses of heavy rainfalls in northern and central China indicate that the easterly low-level jet is of vital importance. Over northern Europe, another strong ridge was persistent. Together with the ridge over East Asia, it made the planetary circulation in the Northern Hemisphere very stable. The pattern persisted for two to three weeks. Consequently, another excessively heavy rain episode followed, caused by typhoon Ora, which resulted in weather processes surprisingly similar to those of typhoon

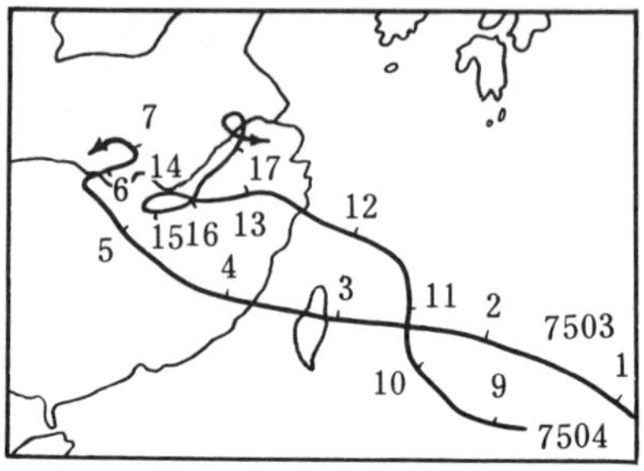

Figure 4.22. Tracks of typhoons Nina (No. 7503) and Ora (No. 7504). Numbers along the tracks indicate the dates in August 1975 on which the typhoons positions were established for 08:00 local (Beijing) time (Ding and Reiter, 1982).

Nina one week before. Major rainfall with a maximum of 809 mm during three days was observed in the northern part of Anhui Province (see Fig. 4.22).

An intensive rainstorm which occurred in central Shanxi, northern China, on 22–24 August 1966 was caused by the remnants of the weakening typhoon Tess. Again, one can observe a very large positive anomaly over a region that stretches from Lake Baikal to the Sea of Japan (Fig. 4.23). This reflects the unusual westward and

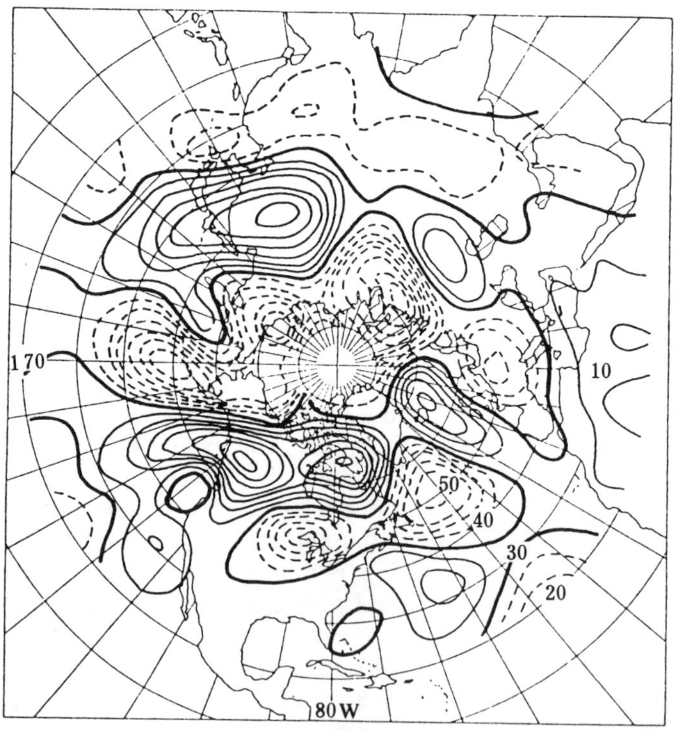

Figure 4.23. Similar to Fig. 4.17, except for the period 22 to 24 August 1966 (Ding and Reiter, 1982).

northward displacement of the subtropical high as well as the development of a strong blocking ridge along 120° E. During this period, a typhoon occurred in the region south of Japan. It acted to maintain and intensify the subtropical low-level easterly jet, thus favoring the transport of abundant moisture into the heavy rain region. After the middle of that month, a high-pressure anomaly formed over the central Pacific, and immediately started to retrograde with a phase speed of 5 to 6° longitude/day. It finally replaced the trough over East Asia. Meanwhile, a reverse change in the planetary waves took place in the subtropics and much of the tropics. In lower latitudes, a widespread large negative anomaly began to establish itself on 17 August and started to dominated the entire Eastern Hemisphere. The south-north juxtaposition of large negative and positive pressure anomalies caused the low and middle tropospheric easterlies to intensify, and to steer the weakening tropical depression far westward into the Shanxi highland. A large amount of moisture was transported into the heavy rain region by these easterlies. The above-described events indicate that excessively heavy rains caused by typhoons are the result of the interaction of circulation features in the low- and middle latitudes. Such pattern interactions may cause the typhoon vortex to become quasi-stationary.

4.3. Droughts over China

Drought is another kind of severe weather disaster that may persistently affect extensive areas of China. According to the statistics of the impact of droughts on China for the last 30 years (Table 4.1) (Bi, 1990), the Huaihe, the Yellow River, and Haihe River Valleys (including most of northern China) are the region most vulnerable to droughts, with accounting for about half of the drought-affected and drought-damaging areas for the whole of China (64.5% and 50.5%, respectively). In particular, in the 80's prolonged droughts in northern China severely affected and

Table 4.1. The statistics of the impact of droughts on China for the last 30 years (1970–1979) (Bi, 1990)

Regions	Ratio of the drought affected area to total drought-affected area for the whole of China (%)	Ratio of the drought damaging area to total drought-devastating area for the whole of China (%)
Huaihe, the Yellow and Haihe Rivers	46.5	50.5
Middle and lower Valleys of the Yangtze River	22.0	19.2
Northeastern China	11.6	11.4
Northwestern China	4.7	5.1
Southern China	25.2	4.4
Southwestern China	10.0	9.4

threated the agricultural growth, supply of water, and river transportation. Thus, research into the droughts in northern China has already become extremely urgent.

Almost every year, droughts occur which affect different areas of northern China in varying degrees of severity. Based on the statistics for 1951–1980 (Fang *et al.*, 1985), the droughts occurred over extensive regions in 1951, 1953, 1956, 1957, 1962, 1966, 1971, 1972, 1974, 1978, 1979, and 1980. The droughts in moderately affected regions occurred in 1955, 1959, 1960, 1963, 1965, 1968, 1970, 1973, 1975, 1976, and 1977. The remainder of the studied droughts were of a local scale. From the viewpoint of the severity of the drought, the following years may be classified as relatively severe: 1955, 1957, 1960, 1962, 1965, 1968, 1972, 1978, and 1980. As a matter of fact, relatively severe drought may occur about once every three years and an extensive severe droughts once every 10 years. In northern China, only in 1958 did drought not occur. Therefore, in this region the frequency of occurrence of droughts is very high. In the 80's, droughts occurred very frequently, with successive four years of 1980 to 1983. During 1980, 1983, and 1986, droughts affected the widespread regions with considerable impact.

Precipitation in northern China mainly occurs in summer season. Therefore, the amount of summer precipitation in northern China is closely related to drought and floods. In this region, the summer precipitation amount almost accounts for 65–75% of the annual precipitation total. In other regions, for instance, Shanghai and Guangzhou, it only accounts for 37.3 and 43.3%, respectively, which is much lower than the percentage for northern China (Table 4.2). The high percentage of summer precipitation is associated with the active summer monsoon that may bring a large amount of moisture into the northern China region.

Table 4.2. Percentage of the summer precipitation amount over the annual totals (%) for some selected stations

Summer Precipitation	Precipitation amount for June-August (mm)	Percentage acoounting for annual totals (%)
Beijing	482.6	74.9
Baoding	419.3	74.0
Shijiazhuang	356.3	64.9
Anyang	391.2	64.5
Jinan	448.2	65.4
Dezhou	408.1	69.2
Taiyuan	274.5	59.7
Zhengzhou	341.7	53.3
Shanghai	419.1	37.3
Guangzhou	733.0	43.3

The inter-annual variability of annual precipitation in northern China is very significant, ranging from 20 to 30%, and even greater than 30% for some areas (Table 4.3). The inter-annual variability represents the unstable or highly variable precipitation amount, thus a high potential possibility of the occurrence of drought/floods.

Table 4.3. The relative inter-annual variability of annual precipitation (%)

Location	Relative inter-annual variability (%)
Beijing	29
Baoding	32
Shijiazhuang	25
Dezhou	24
Jinan	25
Anyang	23
Kefun	20
Taiyuan	22
Shanghai	14
Guangzhou	17
Shenyang	18

In general, if the inter-annual variability of precipitation exceeds 25%, agricultural crops would be affected by drought or floods (SSTC, 1990). Figure 4.24 indicates a considerable inter-annual variability of summer precipitation departure for northern China for 40 recent years (1951–1988). Overall, the 50's was a wet period, with a relatively infrequent occurrence of droughts. However, in the 60's the severe droughts frequently occurred, especially in 1965 and 1968, which were the years of the most severe droughts during the last 40 years. In 70's, the wet period seemed to revive, except in 1972 and 1975.But in the 80's, northern China was again dominated by severe and successive droughts. Thus, droughts appears to demonstrate a quasi-20-year period (Bi, 1990).

The droughts in northern China are characterized by the following remarkable features:

(1) Prolonged duration or long persistent. Severe droughts in northern China generally occur in summer and are often persistent in spring and summer or summer

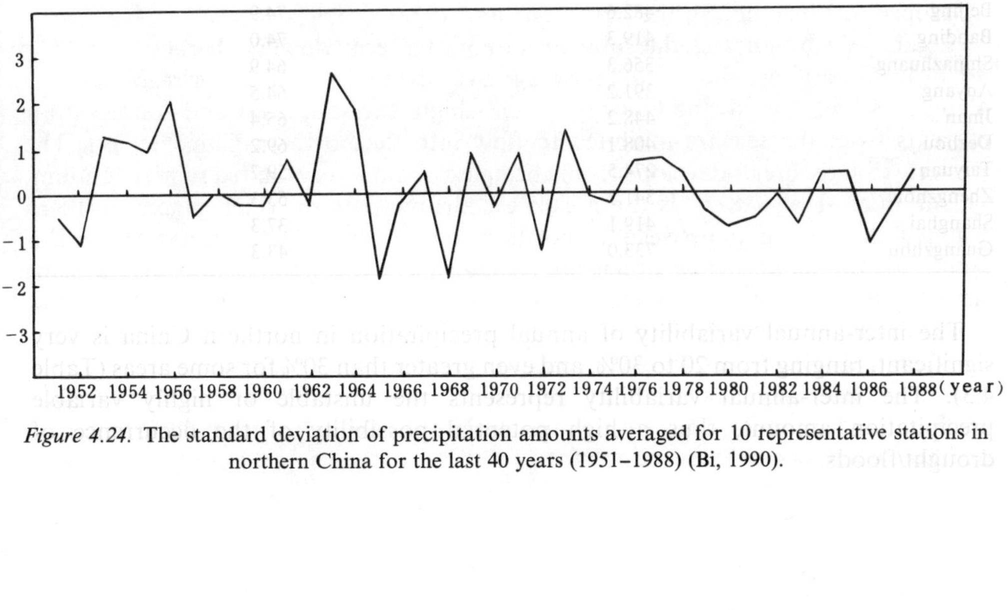

Figure 4.24. The standard deviation of precipitation amounts averaged for 10 representative stations in northern China for the last 40 years (1951–1988) (Bi, 1990).

and autumn, or even over three seasons (spring, summer and autumn). In 1957, 1965, 1968, 1972, and 1986, the droughts lasted for between 5 and 7 months.

(2) Extensive affected areas. For example, in 1965, drought most severely affected about 8 provinces. Most droughts may affect considerable parts of northern China, and sometimes even most of it.

(3) Great severity and degree of impact. In general, the precipitation amount of severe droughts is 50–70% less than normal, even 70–80% (for instance, 1968). 1980 was a year of severe drought. In Beijing only 35 mm precipitation amount was recorded during 40 days from the middle part of July to the last part of August, almost accounting for one-tenth of the normal precipitation amount for the same period. In 1983, only 27 mm of rainfall for July was recorded, 90% less than normal. This extreme minimum monthly rainfall amount was the lowest since 1870.

(4) The simultaneous occurrence of high temperatures, accompanying droughts. For instance, in 1965, the mean air temperature for May-October was 1–2°C higher than normal, with the number of days of high temperatures (> 35°C) being 10–35 days, 5–18 days more than normal. During the period of May-July 1971, the number of days of high temperatures for the eastern part of northern China was 5–10 days, 5–10 days greater than normal, with temperatures higher than 40°C being observed a number of times. Owing to the concurrent occurrence of drought and high temperatures, the evaporation is greatly augmented, so that the drought could be further enhanced or aggrevated. In the 30's, Zhu (1979) studied the cause of drought in northern China. Later, as the more and more data became available, a number of researchers have investigated this matter mainly from the perspective of large-scale circulation conditions. In summary, three types of circulation patterns are identified that may be favorable for the occurrence of drought in northern China (Fig. 4.25). The high pressure pattern (Fig. 4.25 a) is characterized by a warm high-pressure dominating the region from the Great Bend of the Yellow River to northern China. This warm high-pressure system is of the subtropical continental high which often moves from the Xinjiang Province. Under the action of the downward motion of the warm high pressure, weather conditions of clear skies and high temperatures developed, for instance, in the years 1968 and 1972.

Figure 4.25 b indicates the subtropical high pattern which is characterized by a westward extending subtropical high ridge over the middle and lower reaches of the Yangtze River. Due to the blockage of this high, the southwest and southeast air currents from the sea are inhibited to flow into the northern China region. The westerly troughs are ill-developed, only bringing about a very limited rainfall amount.

The third flow pattern is the northwest airflow pattern which places northern China in the gap region between the northern and southern frontal zones. Under this condition, the summer monsoon is prohibited from moving northward. At the same time, there is the large-scale dynamic descending motion in the northwesterly airflow. These combined effects lead to the development of drought in northern China.

Numerous studies have indicated that the position of the subtropical high over the western Pacific may exert an important effect upon drought over northern China. Statistics produced by Bi (1990) demonstrate that droughts generally occur over northern China when the position of the subtropical high is farther to west and/or to

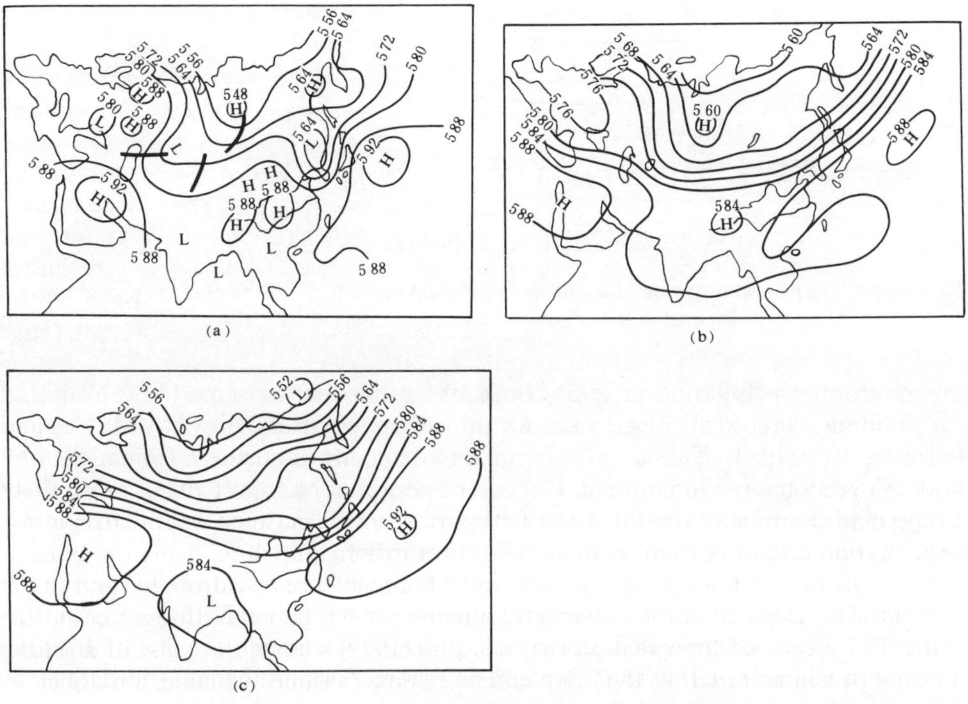

Figure 4.25. (a) The high-pressure pattern (500 hPa weather map at 00:00 GMT, 10 August 1972), (b) the subtropical high pattern (500 hPa mean chart for July 1972), and (c) the northwest airflow pattern (mean 500 hPa chart for July 1980). Units: decameter (Bi, 1990).

south than normal. However, there are some exceptions, for instance, for 1980, 1983, 1963, and 1965 when the subtropical high significantly shifted westward, thus cutting off the inflow of moist and warm summer monsoon airflows into northern China.

The stable and persistent blocking situation in East Asia is also favorable for the occurrence of drought over northern China. If it is so, the westerly jet stream splits into two branches in the western flank of the blocking high, and northern China is situated in the weak frontal zone between the northern and southern major frontal zones. The northwesterly airflow prevails over this region. Later on, a remarkable example will be discussed to demonstrate the importance of the blocking high for the development of drought over northern China.

Drought over northern China is not a regional event, but is closely linked to the anomalous general circulation in the Northern Hemisphere. Studies have indicated that the occurrence of historically significant droughts in the United States is usually consistent with that of droughts in northern China. For instance, in 1930, these two countries both underwent a prolonged 10-year severe drought. Also, in 1980 and 1983, the Great Plains of the United States and northern China both had significant droughts.

As previously suggested, droughts in northern China are closely correlated with

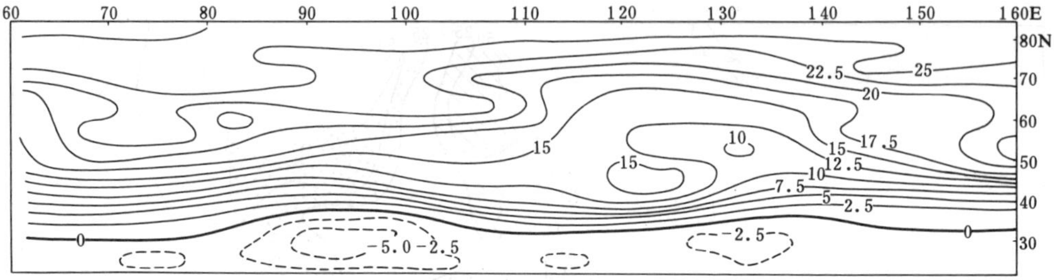

Figure 4.26. The pattern of the quasi-geostrophic potential vorticity at 300 hPa for the drought period from 1 July to 10 September 1980 in northern China. Units: $10^{-5}\ s^{-1}$ (Bi and Ding, 1992).

the monsoon precipitation in India. Both has an almost the same trend of spatial distributions. Out of 11 drought years for the Indian region as defined by Mooley and Parthasarathy (1984), 8 cases corresponded to droughts in northern China; out of 9 flooding years for the Indian region, 5 cases corresponded to wet years. Thus, there has been an almost stable positive correlative relationship between the Indian summer precipitation and the precipitation amount in northern China.

Below, a case of the drought in northern China will be presented to illustrate the role of the blocking situation in the development of the regional drought. Figure 4.26 is the distribution of the quasi-geostrophic potential vorticity (q) at 300 hPa for the drought period from 1 July to 10 September 1980 in northern China. It is defined as follows:

$$q = \zeta + f - f_0 \frac{\partial}{\partial p}\left(\frac{\delta T}{S(p)}\right) \tag{4.1}$$

where $s(p) = -T_0(p)\partial/\partial p\ [\ln\ \theta_0(p)]$ is the static stability and f is the mean Coriolis parameter in the domain under study. T_0 and θ_0 are the mean temperature and potential temperature during the studied period and in the studied domain; δT is the departure from T_0; ζ is the relative vorticity on the isobaric surface; and $f = 2\Omega \sin \phi$.

It should be pointed out that the blocking high is characterized by a region with a very low value of quasi-geostrophic potential vorticity. From Fig.4.26, it can be seen that to the north of 50° N, the blocking region at 110–140° E has a very low q, with the high value observed at either flank of the high. The air of high q coming from a long-wave trough over the Ural Mountains area flows southeastward and wedged into the region of 40–45° N to the south of the blocking high, thus creating the mean pattern of high value in the north and low q in the south. This q pattern is opposite to the climatological pattern of q. Therefore, one may use the quasi-geostrophic potential vorticity to more clearly describe the blocking system. The above q pattern demonstrates that during the blocking period a large amount of low q air is transported northward and cut off in the blocking region in the north; on the other hand, the air of high q is transported southward and acummulated in the low-pressure region to the south of the blocking high, thus creating reverse q gradient as compared to the climatological condition. Note that there are regions of low q in the tropics and subtropics as well as over the Tibetan Plateau. It can be speculated that the low q air

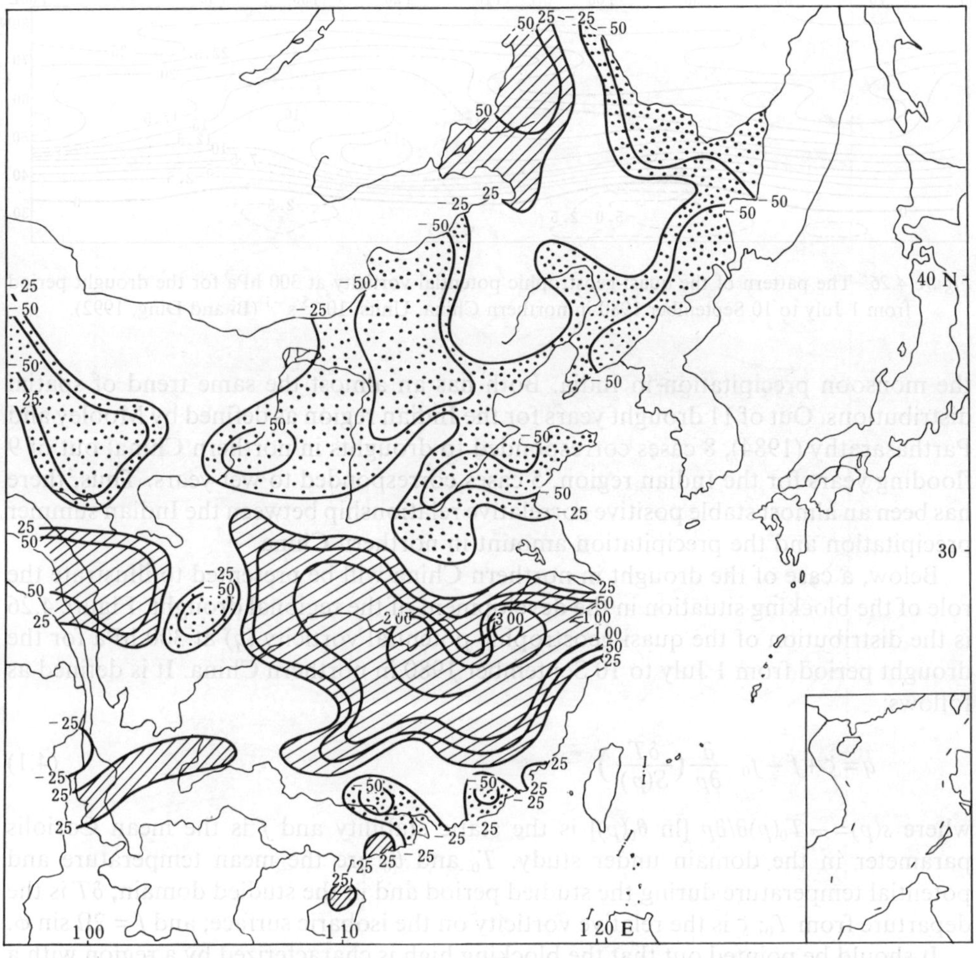

Figure 4.27. The percentage of precipitation departure (%) for August 1980 (Bi and Ding, 1991).

trapped in the blocking high in the north could come from the origin region of low q air in the tropics and subtropics.

Under the condition of this blocking situation, the westerly jet stream splits into two branches with the southern branch flowing over the Yangtze and Huaihe Rivers Valleys to produce prolonged precipitation and flooding (Fig. 4.27). Northern China is located in the weak frontal zone between the southern and northern jet streams, with frequent high-pressure activity, thus leading to severe drought lasting for 2½ months. The negative departure of the monthly total precipitation amount is about −50% (July and August). Figure 4.28 shows the long-persistence of the blocking high over northern China which is characterized by a stable zone of low q air at the longitudinal range of 110–140° E from 1 July to 10 September. When the blocking high broke down shortly after 10 September, the circulation pattern underwent an abrupt adjustment, thus ending in prolonged drought in the north.

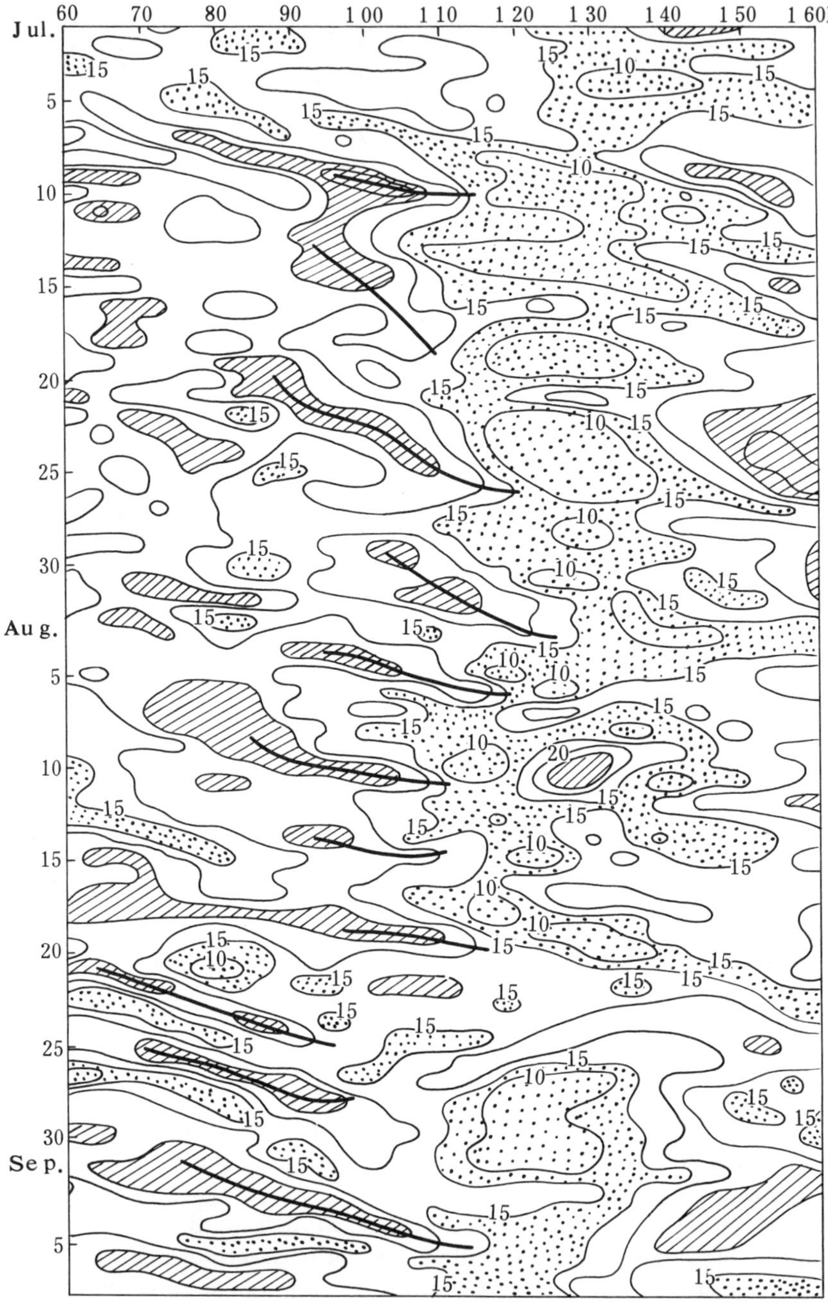

Figure 4.28. The longitude-time cross-section of 300 hPa quasi-geostrophic potential vorticity (q) along 60° N for the time period from 1 July to 10 September 1980. The shaded area denotes the high q air and the stippled area the low q air (Bi and Ding, 1992).

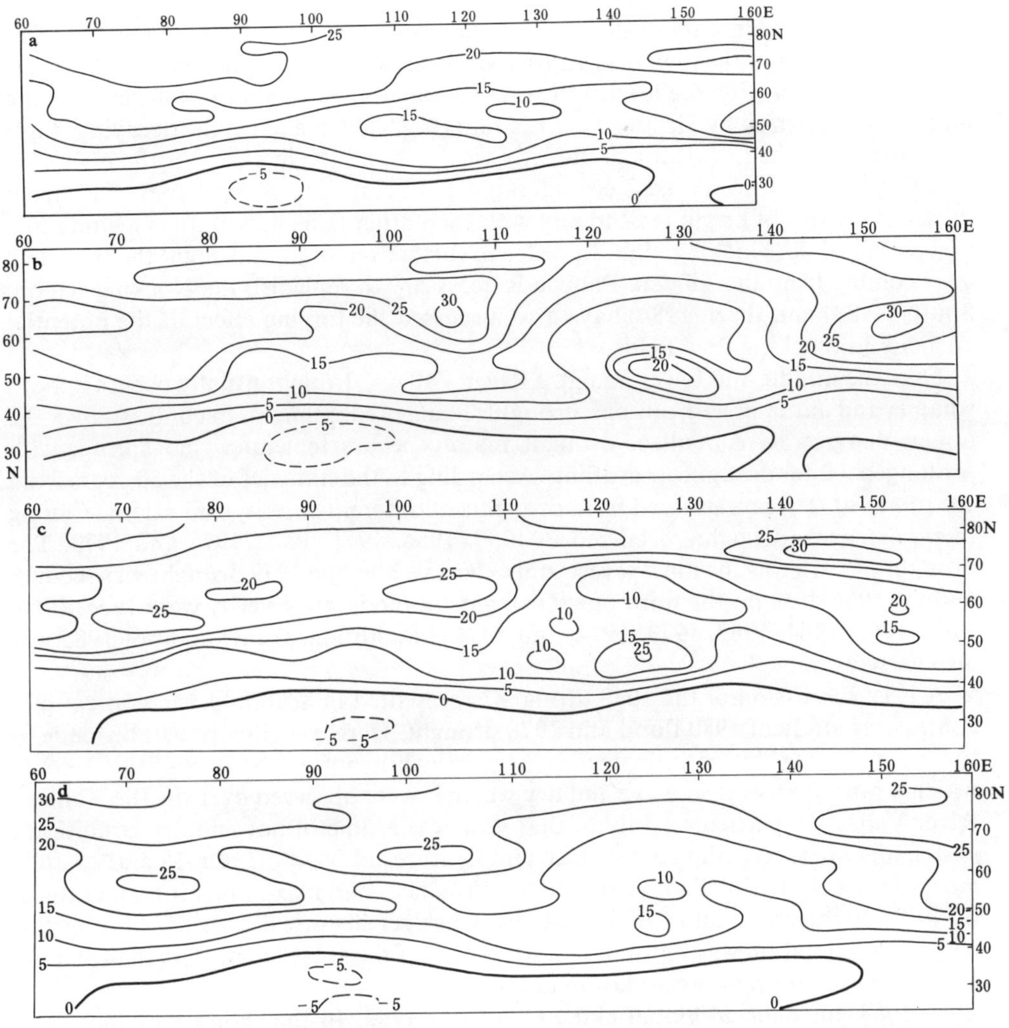

Figure 4.29. The compositing schematic of the evolution process of the quasi-geostrophic potential vorticity field for different stages of the blocking high, averaged for episodes during the time period from 1 July to 10 September 1980. (a) The initial stage, (b) the development stage, (c) the mature stage and (d) the decaying stage. Units: 10^{-5} s^{-1} (Bi and Ding, 1991).

The maintenance of the blocking high is closely related to the northward transport of low q air which mainly originated in the Tibetan Plateau region. During the time spell of blocking, about seven episodes of northward transport of low q air were observed, each of which experienced duration of about 10 days. Figure 4.29 is a schematic diagram of the evolutive process of the mean q field for different phases of the blocking high. (a) The initial stage: the low q air over the Tibetan Plateau starts to flow northward and, at the same time, the high quasi-geostrophic potential vorticity in the north enhances (Fig. 4.29a); (b) the developmental stage: the low q air extends northward into the region at 60° N and stagnates over the region of 120–130° E, and the region of the high q air moves southeastward and begins to cut off the low q air in the north (Fig. 4.29b); (c) the mature stage: the distribution of the blocking high is characterized by the low q in the south (Fig. 4.29c); and (d) the decaying stage: the intensity of the low q air significantly weakens and the intensity of the high q air to the south of it also weakens considerably. As a result, these two types of air from different regions of origin tend to mix with each other (Fig. 4.29d), thus leading to a weakening of the blocking high. In order to further maintain this high, the fresh low q air coming from the Tibetan Plateau is necessary to replenish the blocking region. Shutts (1983) and Iliari (1984) have also discussed the forcing effect of the potential vorticity in blocking flow fields.

Over the middle and lower Yangtze River Valleys, drought usually occurs in both summer and autumn, with autumn droughts being predomiment. In contrast, over the upper Yangtze River Valley, drought mainly occurs in winter and spring, with prolonged winter droughts prevailing. According to the statistics of the 30 years from 1951 to 1980, 28 years suffered from droughts of different severity and extent. Among them, extensive droughts occurred in 1959, 1966, 1971, 1972, 1978 and 1979. The drought of 1978 was the most severe and extensive one, the 1979 drought was second, and the 1966 drought the third one. Droughts of moderate severity were 1955, 1960, 1967, 1963, 1964, 1967, 1973, 1974, and 1976. In addition, the drought of 1985 was also very severe.

A brief discussion of the 1978 drought is presented in Section 4.3 in context of a contrasting study of 1980 flood and 1978 drought. Here, no attempt will be made to further elaborate this case. Instead, it is worthwhile noting the drought of 1988 in which a rare summer heat wave and dry weather were observed over the the Yangtze River Valley. In particular, July of that year was a time of hot and dry conditions, with many areas recording air surface temperatures of 39–40°C. On 18 and 19 July, most of the counties of Jiangxi and Anhui Provinces had maximum air temperatures of 40°C. Eighteen counties of Jiangxi Province set historical records. During this period, precipitation was very limited, with most of the areas having less than 50 mm, 50% below the normal precipitation amount.

In 1988, an unusual global climate was observed. In the summer of that year, record-breaking droughts also occurred in the United States. But the cause of this global climate change is not yet clear. There is insufficient evidence to support the claim that this unusal climate is associated with the human activity (CO_2-induced greenhouse effect).

Drought is also disastrous in southern China. During 40 recent years (1951–1988), at least three severe droughts occurred (1955, 1963 and 1977). Persistent spring drought occurs most frequently and is the most damaging. For instance, for the 30 years of 1951–1980, eight years experienced extensive spring droughts, with the 1963 drought being the most severe. The precipitation amount was only 30–50 mm higher than that over the central and southern parts of Guangdong Province. However, the evaporation amount at the same period was as high as 600–900 mm, thus producing the most severe drought of the last 60 years.

4.4. Interannual Variability of the Monsoon and Monsoon Precipitation in East Asia and the ENSO Event

The summer monsoonal rainfalls in East Asia show a significant interannual variability, thus causing a climate of droughts and floods in China, which have a great impact on the national economy. Therefore, understanding the interannual variability in monsoon rainfalls and the predictability mechanism of this variability has long received special attention of Chinese meteorologists. In the 1930s, the noted meteorologist Tu Changwang (1937) related the weather in China to the global atmosphere oscillation and discussed the problem of long-range forecasting of summer droughts and floods. In the 1960s, Zhang *et al.* (1963) examined monthly precipitation grades by using precipitation data from 86 surface stations for 1901–1960, on the basis of which, they studied the problem of the large-scale precipitation anomaly in 60 recent years of this century.

As previously indicated, Meiyu is the major rainy season in early summer over East Asia. It not only exerts an important effect on the drought and flooding conditions over the Yangtze and Huaihe River Valleys, but also, as an integrating system of summer monsoon circulation and important heat source, it exerts a significant effect on large-scale changes. The interannual variability of Meiyu is significant, with some years being "bad Meiyu" and other years "good Meiyu". The timing of the onset of Meiyu is variable, with some years being described as "early onset" and other years "late onset".

Tao and Chen (1987) analysed the beginning and ending dates of the Meiyu period over the Yangtze River Valley and the percentage departure of rainfall amounts for June-July 1954–1983. The earliest date of onset was 1 June (in 1954) while the latest date of onset was 7 July (in 1982). The average duration of the Meiyu period is 23 days, with the very short period of Meiyu for some years (6 days for 1964, 10 days for 1959, and no Meiyu for 1958 and 1978) and a very long period for others (63 days for 1954, 45 days for 1956, 43 days for 1980, and 39 days for 1974).

Xu (1965) analysed the long-term variability of Meiyu over the middle and lower Yangtze River Valleys utilizing the data of 1885–1963, and obtained a considerable inter-annual variability, with periods of 2.3 and 11–13 years, which are likely to be related to solar activities.

In recent years, as the study of dynamic climate has progressed internationally, investigators have made attempts at seeking physical factors influencing long-term changes and the characteristics of the underlying surfaces have been noticed. These include the gradual changes in sea surface temperature, and of sea ice and snow cover in extensive regions. From this viewpoint, Tao *et al.* (1980) have recently discussed inter-annual variability in Meiyu. Figure 4.30 shows the long-term variability of the standard deviation of precipitation departure for Shanghai for 1874–1986. To a certain extent, this curve may represent the Meiyu precipitation over the middle and lower Yangtze River Valleys. The black rectangles in this figure indicate El Nino years. From it, it may be seen that, of 24 El Niño years, 15 cases had positive departures of precipitation and 9 cases negative departures, but with smaller magitudes of departure. It seems to us that the rainfall amount of Meiyu in this region

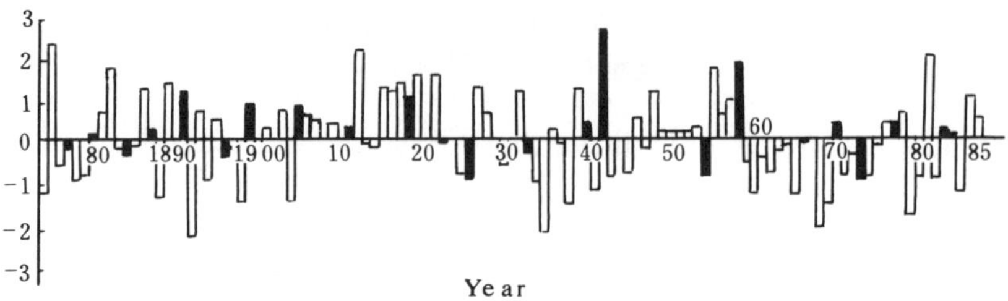

Figure 4.30. The standard deviation of precipitation departure for June-August in recent 100 years (Tao *et al.*, 1980).

above-normal for El Niño years. This result is opposite to that obtained by Rasmussen and Carpenter (1983) in a study of inter-annual variability of monsoon rains using Indian rainfall data for 1875–1979. It may be seen from this study that the inter-annual variability in monsoons in China and India is likely to have an opposite tendency of change.

In fact, the relationship between Meiyu and the ENSO event greatly depends upon the timing of onset of El Niño. Fu and Teng (1988), Zhao *et al.* (1989), and Lu and Lin (1990) have recently studied this problem and obtained similar results. From Table 4.4 (Zhao, Zhang and Ding, 1989), it is found that when the warming of the El Niño event occurred in fall and winter, the rainfall amount during the Meiyu period in the middle and lower Yangtze River Valleys in either the El Niño onset year or the following year is usually greater than normal. The probability of a pluvial year is 88%. Especially, in the following year of the El Niño onset, the Meiyu rainfall is fairly obvious and the probability of a pluvial period is 100%; when the warming onset of the El Niño event occurrs in spring and summer, the rainfall amount during Meiyu period is usually less than normal for either the El Niño onset year or its following year. The probability of a rainfall deficient year is 80%. So the rainfall tendency of the

Table 4.4. The relationship between the El Niño event and the Meiyu tendency in the Yangtze Valley (Zhao, Zhang and Ding, 1989)

Season of warming onset	Year of event	Meiyu rainfall anomaly*	
Warming onset in autumn or winter	1953–1954	1953 (0^+),	1954 (+)
	1969–1970	1969 (+)	1970 (+)
	1982–1983	1982 (0^+)	1983 (+)
	1986–1987	1986 (0)	1987 (+)
Warming onset in spring or summer	1957–1958	1957 (0^+),	1958 (+)
	1963–1964	1963 (−)	1964 (0^+)
	1965–1966	1965 (−)	1966 (0^-)
	1972–1973	1972 (−)	1973 (−)
	1976–1977	1976 (0^-)	1977 (−)

* Numbers in parenthesis represent above-normal (+), slightly greater than normal (0^+), slightly less than normal (0^-), and below-normal (−) rainfall anomalies, respectively.

Table 4.5. Drought/flood condition frequency of occurrence/total studied cases for years following the ENSO event over the middle and lower Yangtze River Valleys (1951–1984)

ENSO intensity	Drought year	Flooding year	Normal year
Strong	6/7	1/7	0/7
Moderate	4/8	2/8	2/8
Weak	4/6	1/6	1/6
Average (%)	69%	19%	14%

Meiyu period in the middle and lower Yangtze River Valleys are related to the warming onset timing of the El Niño year.

Fu and Teng (1988) related the interaunual variability of the rainfall amount for the time period of the summer monsoon over the middle and lower Yangtze River Valleys for 1951–1983 to the ENSO onset. It is evident that, except for 1969, almost all the ENSO onset years corresponded to the below-normal rainfalls in this region (Table 4.5). But the summer of the following year after the onset of the ENSO event corresponded to a high probability of flooding conditions (6/9).

This result is in general agreement with the result obtained by Zhao *et al.*(1989). Lu and Lin (1990) made a more detailed classification of the timings of the onset of the ENSO event and related them to the rainfall anomaly in the Meiyu period over the middle and lower Yangtze River Valleys (Table 4.6). When the warming onset occurred in the preceding winter and spring, the precipitation anomaly was below normal in the same year. When the warming onset occurred in the early summer, less Meiyu precipitation was recorded, but with more precipitation in June of the following year and less precipitation in July of that year. When the warming onset occurred in autumn, the following June and July generally received an above-normal amount of precipitation. Thus, these relationships are in good agreement with those obtained by Zhao *et al.*(1989).

Table 4.6. The relationship between ENSO events and the Meiyu precipitation anomaly over the middle and lower Yangtze River Valleys (Lu and Lin, 1990)

Type	Year of event	Precipitation anomaly in the same year		Precipitation anomaly in next year	
		June	July	June	July
Type I Warming onset in preceding winter and spring	53	−4	−52	339	434
	57	−115	10	−193	−65
Type II Warming onset in early summer	63	−197	−63	150	−124
	65	−90	−33	−19	−54
	72	−96	−126	48	35
	76	18	−76	25	−14
Type III Warming onset in autumn	68			41	350
	82			139	118
	86			−117	186

The anomalies of the sea-surface temperature over the northern Indian Ocean and the South China Sea have been revealed to be related to the drought/flood over the Yangtze River Valley during the Meiyu period (Luo and Jin, 1987). When the SST is relatively warm in these regions, the middle and lower Yangtze River Valleys have an above-normal precipitation amount. Zhang and Zhao (1990) found an inverse relationship between the latitude of the summer rainbelt and SST over the equatorial eastern Pacific (Fig. 4.31). When the SST in the equatorial eastern Pacific is high, the July rainbelt moves southward to eastern China, whereas when the SST is low, it moves northward. In the 80's, while the SST over the equatorial eastern Pacific was, on an average, getting warmer, the location of the rain belt in China was found to be farther south than normal.

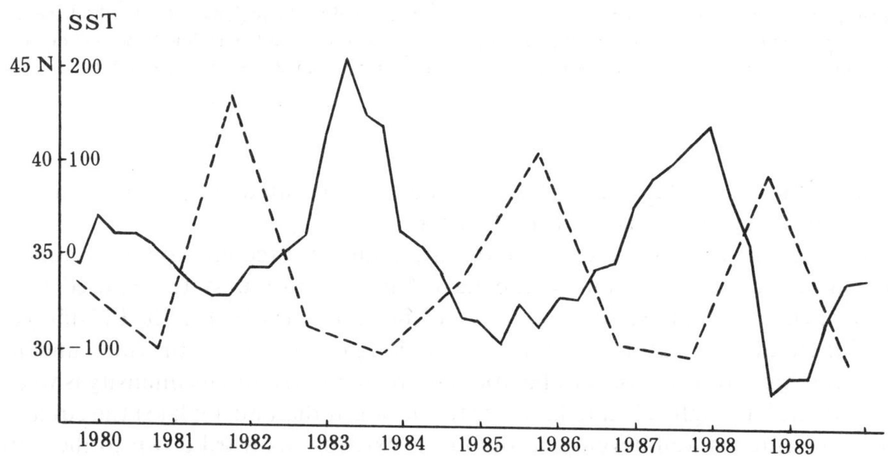

Figure 4.31. The seasonal mean SST (solid line) for the equatorial eastern Pacific and the location of the rainbelt in July in eastern China (dashed line) (Zhang and Zhao, 1990).

Zhang and Zhao (1990) further studied the relationship between the summer rainbelts in China and El Niño events (Table 4.7). It can be seen that during the onset year of the El Niño, rain pattern type II was often observed (5/7), while during the onset year of anti-El Niño event, rain pattern types I and III were often observed. For the year following an El Niño event, type II often occurred, whereas for the year following an anti-El Niño, the types II and III often occurred. It seems that the El Niño and anti-El Niño events can exert an effect on the summer rainbelts in China. But, this result is rather tentative. Further study is needed to document this relationship. Recently, Liang (1990) used the dataset for 1951–1986 to study the relationship between El Niño and summer rainfall in northern China. In this study, it is shown that when the SST over the equatorial eastern Pacific region is warmer (the onset year of El Niño), the rainfall during June-September over the vast area in northern China is less than normal; if the SST over the equatorial eastern Pacific region becomes colder in the following year (the following year of El Niño), the

Table 4.7. The relationship between the types of rainbelt in summer (June-August) in China, and El Niño and anti-El Niño events (Zhang and Zhao, 1990)

El Niño event	Onset year	Following year	Anti-El Niño event	Onset year	Following year
1957–58	II*	I	1955–1956	III	II
1963–64	II	I	1964–1965	I	II
1965–66	II	I	1967–1968	I	III
1969–70	III	III	1970–1971	III	II
1972–73	II	I	1973–1974	I	III
1976–77	I	I	1975–1976	II	I
1982–83	II	III			

*Rain pattern type I: the major rainfall region or belt is located in the regions to the north of the Yellow River; rain pattern type II: the Major rainbelt is located in the area between the Yellow River and the Yangtze Rivers; and rain pattern type III: the major rainbelt is located in the Yangtze River Valley and the region to the south of it.

summer rainfall is greater than normal over an extensive area of northern China. This result is more or less consistent with Table 4.7.

The effect of the El Niño event on the summer precipitation in China is a manifestation of the variation of the East Asian monsoon system caused by this oceanic phenomenon. Based on the study by Mao and Xu (1985), in the El Niño years, the major circulation features assume the following variations: (1) the ridge-line of the subtropical high over the western Pacific is more northerly and its intensity is weaker; (2) the Australian high is located farther to the south than normal; (3) the equatorial buffer zone ridge is well developed, the ITCZ extends eastward down to the central Pacific, and the trade wind coming from the Southern Hemisphere crosses the equator at 105° E, then deflects toward the east at 10° N and flows into the western Pacific; and (4) the Indian summer monsoon is at break stage, the monsoon coming from the Bay of Bengal cannot easily to penetrate northward into the eastern part of China. Therefore, in an El Niño year, it is vulnerable to droughts in eastern China (including the middle and lower Valleys of the Yangtze River). This situation may be used to interpret the result indicated in Table 4.4.

Figure 4.32 (a) shows the inter-annual variation of the intensity of the subtropical high at 500 hPa in summer. It can be seen that in most onset years of the El Niño event, the subtropical highs are weaker than normal (6/8) (Fu and Teng, 1988). Further, in most onset years of the El Niño event, the position of the subtropical high is farther to the east (6/8) while in the summer of the following year, it generally shifts farther westward than normal (7/9) (Fig. 4.32 (b)). The mean circulation charts shown in Figure 4.33 clearly demonstrate these differences. These may be used to explain the observational evidence of the effect of the ENSO event on the climate anomaly in summer in the eastern China previously indicated. In the onset years of ENSO events, the southerly airflow at the western flank of the subtropical high is inactive due to a weak subtropical high that has a more easterly position. This is unfavorable to the

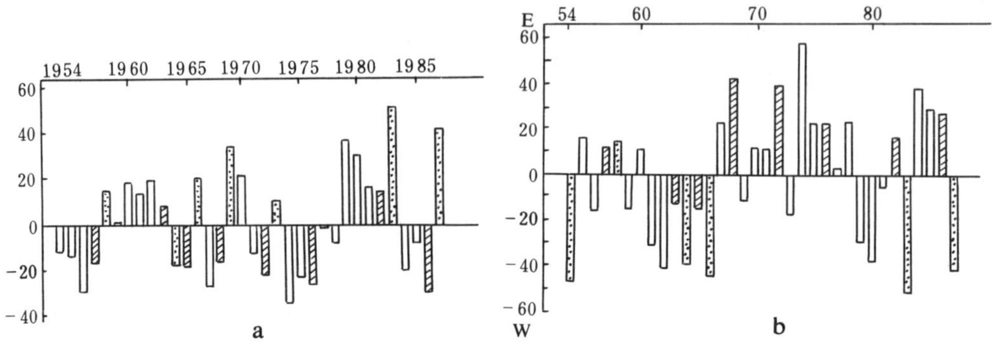

Figure 4.32. Inter-annual variability of the intensity (represented by the areal index) (a) and the westward extending extent (b) of the subtropical high in summer (1954–1987). The rectangles with shaded area denote the summers for onset year of ENSO and the rectangles with stipled area the summers for following year of ENSO event (Fu and Tang, 1988).

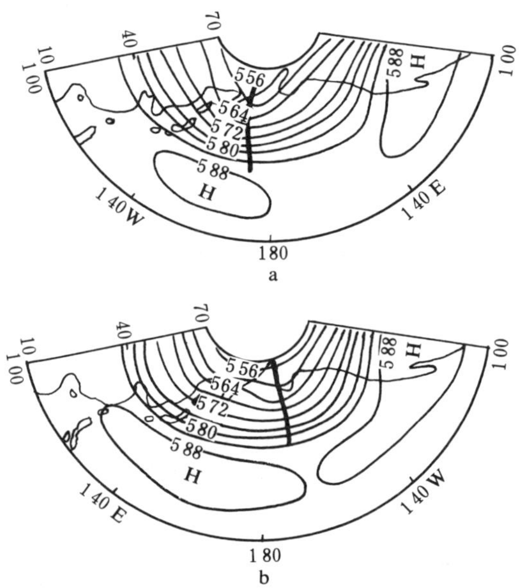

Figure 4.33. The 500 hPa mean circulation charts for the onset year (a) and the following year (b) of an ENSO event compiled for 8 cases of ENSO events. Units: decameter (Fu and Tang, 1988).

northward transport of moisture, and meeting with the cold air coming from the mid-latitudes, and thus leading to below-normal precipitation conditions. However, in the following summer, the situation is opposite, being favorable for the development of precipitation processes. The rainbelt often swings around the region of the Yangtze and Huaihe River Valleys.

For the 1986/1987 ENSO event, the monthly mean intensity index and the areal index of the subtropical high followed the above-described process. They were much greater than normal for the following year (Table 4.8). The position of the westernmost point of the ridge line of the subtropical high markedly extended westward. It is worth noting that the change in the intensity of the subtropical high from weak to strong conditions and the change of the equatorial eastern Pacific SST anomaly from negative to positive values are almost synchronous. It is not found that a change in the subtropical high lags behind SST by several months. This result is not in agreement with the results given by Chen (1982) and Fu (1979).

This simultaneous change may be understood in relation to the propagating mode of warming. In several past El Niño events, the propagating mode of warming was from east to west. The abnormal SST warming firstly occurred along the western coast of South America and then extended gradually northward and westward. When the warming peak arrived in the equatorial central Pacific about six months later, the main body of the subtropical high over the northern Pacific located to the north of the equator turned from weak into strong conditions, affected by oceanic anomalous heating. Therefore, the variation of the subtropical high has an obvious lagging response. However, for the 1986/1987 El Niño event, the propagating mode of warming was from west to east; the abnormal warming first appeared in the equatorial central Pacific. The anomalous heating on the oceanic surface rapidly affected the main body of the subtropical high, so that the variation of that high coincided with that of the SST. The same process happened during the 1982/1983 El Niño event, with a variation between the subtropical high and the SST in the equatorial Pacific being almost simultaneous.

Table 4.8. The variation of the areal and intensity indexes of the subtropical high over the northern Pacific from August 1986 to August 1987 (Zhang and Zhao, 1990)

Month	Aug.	Sep.	Oct.	Nov.	Dec.	Jan.	Feb.	Mar.	Apr.	May	Jun.	Jul.	Aug.
Areal index	0	14	27	18	19	19	16	19	23	21	30	29	31
Normal condition	19	18	15	12	9	6	6	7	9	12	19	18	19
Intensity index	0	25	62	36	25	31	24	32	43	32	58	73	68
Normal condition	31	35	26	21	14	8	9	9	12	18	37	31	31

Numerous investigators have revealed the close relationship between the cool summer in northeastern Asia and the ENSO event (Wang and Zhu, 1986). The former may usually lead to extreme cold injury in northeastern China, the Far East, Korea, and northern Japan in summer (June-August). During the 90 years from 1899 to 1988, 22 cases of cool summer occurred, of which 9 cases were severe (Fig. 4.34). The frequency of the occurrence of a cool summer is once in 4 years. It is very interesting to note that 13 of the 22 cases of cool summer occurred in the onset year of an El Niño event, 6 cases in the following year, and 3 cases in the preceding year. Only for the cool summer of 1908, there was no direct link to the El Niño event detected. In 1983 and 1986, cool summers occurred in northeastern China. Therefore, the link between the cool summer in northeastern Asia and the ENSO has been documented with high reliability. However, it is not yet clear how the ENSO event exerts an effect on the cool summer in northeastern Asia.

As shown in Chapter 2, the Indonesian low and associated winter monsoon convection and the South Pacific convergence zone (SPCZ) migrate eastward from their normal position over Indonesia to the vicinity of the dateline in the El Niño years (Lau and Chan 1983). The resulting reduction in convection over Indonesia and northern Australia and an enhancement in convection over the equatorial central Pacific, create the appearance of a dipole-like anomaly in rainfall over the two regions. Accompanying this eastward shift is a reduction of surface northeasterlies over the western Pacific and the South China Sea and a weakened East Asian Hadley circulation with less intense (or below-normal intensity) cold surges (Lau and Chang, 1987).

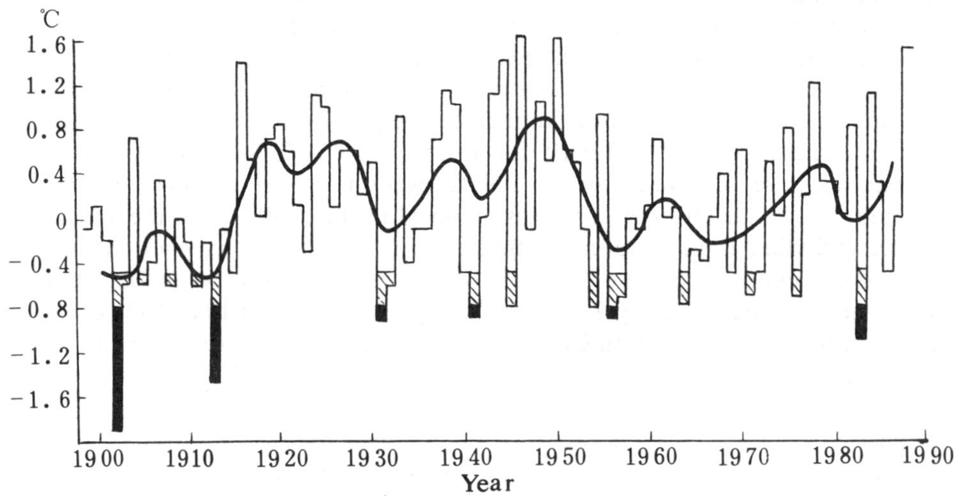

Figure 4.34. The departure curve of air temperature in northeastern Asia in summer (June-August) from 1899 to 1988. The shaded and black areas denote low temperatures and significantly low temperatures, respectively (Wang and Zhu, 1986).

Recently, Li (1988) suggested that the frequent activity of cold waves in East Asia (strong winter monsoon) may cause a prolonged weakening of the trade wind over the equatorial central and western Pacific and the enhancement of convection. The former would lead to a change in height of the sea level and the anomaly of the Kelvin wave, thus playing an important role in triggering an El Niño event. The latter may lead to the enhancement of a 30–50 day oscillation and then, through the interaction of the coupled air-sea system, exert an important effect on the onset of the El Niño event. But, a systematic investigation will be needed to establish the statistical significance of this relationship.

The effect of the ENSO event on the climatic anomaly in China appears to be also shown in the climatic conditions over southern China in winter and spring. But, the results obtained in this aspect are far from conclusive. For instance, the 1982/83 and 1986/87 ENSO events had a strikingly different effect on the weather-climatic conditions in southern China in the winter (Zhao, Zhang and Ding, 1989). In the winter of 1982–1983, the precipitation in a widespred area of southern China was one to three times greater than normal, especially in the regions of the central Guangdong, eastern Guangxi, and the western Sichuan Provinces. Owing to continuous heavy rainfalls, it turned out that rare winter flooding occurred in the region of southern China. At the same time, the temperature over an extensive area of southern China was continuously lower than normal in winter, with the seasonal mean temperature being 10°C lower than normal. In the winter of 1986–1987, however, the precipitation in a large area of southern China was lower than normal, and the regions of southern and southwestern China had severe winter drought. At the same time, the seasonal mean temperature was unusually high, causing the warmest winter since 1949. Especially in southern and western China, the seasonal mean temperature was 3 to 4°C higher than normal. This warm winter persisted for a long duration, was extensive in area, and significant in intensity. It is seen that the effects of the two El Niño events on the climatic anomaly of the winter season in southern China are fairly striking: flooding and cold season from December 1982 to February 1983 and a dry and warm winter from December 1986 to February 1987. This result implies that the regional response to forcing of an ENSO event is rather complex, with other important factors or physical processes involved in the evolution of the anomalous climate, including large-scale circulation conditions. Therefore, caution must be exercised in seeking the effect of an ENSO event on the subtropical and mid-latitude weather and climate. Recently, based on a study of the effect of 10 El Niño events on the air temperature of southern China in winter and spring between 1950 and 1987, in the onset year of an El Niño event, the temperature tended to be above-normal (warm winter) (Zhao *et al.*, 1989).

An ENSO event has a remarkable impact on the activity of tropical cyclones over the tropical western Pacific. Table 4.9 presents the statistics of tropical cyclones in 1949–1987 in relation to the ENSO years, indicating that a below-normal frequency of occurrence of tropical cyclones was observed in the ENSO years, over both the tropical western Pacific and the South China Sea. The same conclusion was reached by Gray (1984) in his statistics of Atlantic hurricanes. For La Niña years, an above-normal frequency of tropical cyclones was observed.

The anomalous behavior of the general circulation in East Asia and associated drought/floods are not only related to ENSO events, but also to the convective activity over the South China Sea and the region around the Philippines. Based on the studies of Nitta (1986) and Huang and Li (1988), these heating forces, caused by heat sources in this region, may excite a stationary wave train, thus producing a teleconnection pattern. The immediate downstream effect of the propagation of this wave train is exerted upon the behavior of the subtropical high over the western Pacific, and especially on its position (Fig. 4.35). Then, the summer rainfall will be influenced by the anomalous behavior of the subtropical high. Huang and Sun (1990), by use of the data for 12 summers from 1978 to 1989, studied the relationship between the subtropical high and the anomaly of summer precipitation in eastern China, and

Table 4.9. The number of occurrences of tropical cyclones over the western Pacific an their relationship to ENSO events (Zhao *et al.*, 1989)

Region	El Niño or La Niña	Typhoon		TS* + STS + TY	
		Total	Mean	Total	Mean
Western Pacific	Normal condition	668	17.6	1093	28.0
	El Nina	210	17.5	309	25.8
	La Niño	301	18.8	445	29.6
South China Sea	Normal condition	37	0.9	135	3.5
	El Nina	13	1.0	43	3.6
	La Niño	19	1.2	65	4.1

* TS: tropical storm; STS: severe tropical storm; TY: typhoons

the convective activity in the warm pool over the tropical western Pacific. Moreover, they also analysed the relationship between the conditions of anomalous summer precipitation in the eastern China and the temperature in surface and subsurface layers of the warm pool at depths of between 50 and 300 m. Results have shown that when the sea-water temperature in the surface and subsurface layers of the warm pool is above-normal, warm water is accumulated in the tropical Pacific, and a cold tongue extends westward from the coast of Peru along the equatorial eastern Pacific. This generally happens in post-ENSO years, such as in the midsummers of 1977, 1978, 1981, 1984, 1985, 1988, etc. The convective activities are intensified from the Indo-China Peninsula to the area around the Philippines. Through the teleconnection pattern described above, the subtropical high over the western Pacific often shifts northward, and the summer rainfall is below-normal in such regions as the Yangtze and Huaihe River Valleys and south of Japan, thus possibly leading to droughts in these regions. In contrast, when the temperature in the surface and subsurface layers of the warm pool is below-normal, the warm water extends eastward from the warm pool along the equatorial western Pacific, generally in the developing stages of ENSO events, such as during the midsummers of 1980, 1982, 1983 and 1987. The convective activity is weak and the subtropical high abnormally shifts southward. The summer rainfalls in the Yangtze and Huaihe River Valleys, and south of Japan are above normal, thus floods may be caused.

The land surface process is also an important factor affecting the anomaly of climate in East Asia through land-air interaction and the anomalous heating field.

Snow cover and its melting may influence the surface heat balance and moisture budget. Therefore, the radiation of areas of snow-cover may affect the general atmospheric circulation. Tao *et al.* (1988) examined the relationship between the rainfall amount for June-August of 1967–1977 over the Yangtze River Valley and the area of snow-cover equatorward of 52° N over the Eurasian continent in the preceding winter, and concluded that they had an opposite relationship of correlation. Hahn and Schukla (1976) also indicated the opposite correlation between the area of snow-cover over the Eurasian continent in winter and the Indian monsoon rainfalls in the following year. Ye (1987) studied the relationship between the inter-

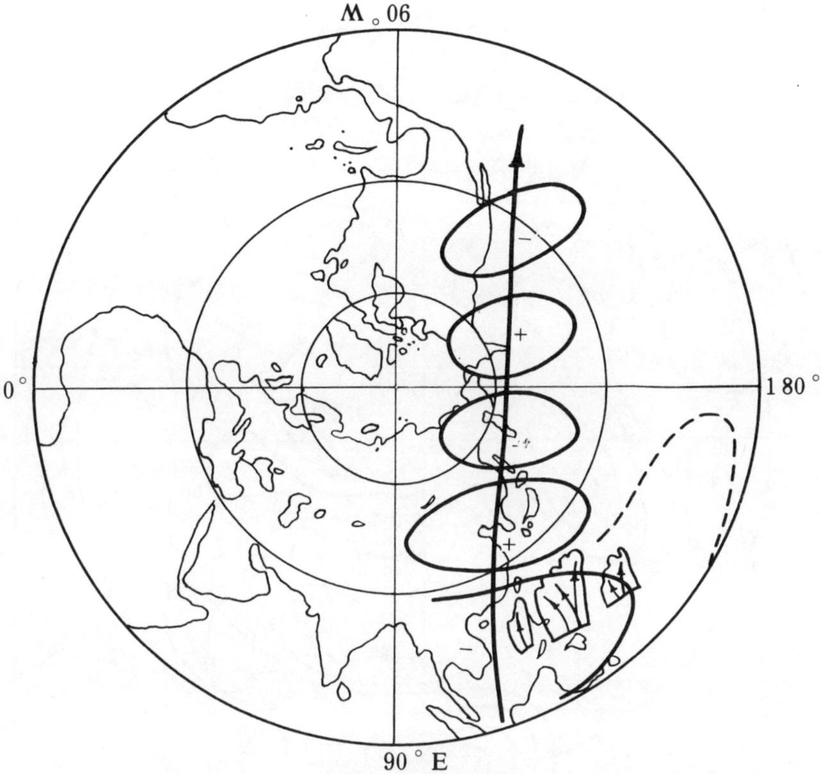

Figure 4.35. A schematic map of the pattern of anomalous departures of the general circulation in East Asia and the northern Pacific after the enhancement of convective activity around the Philippines (Nitta, 1986).

annual variability of summer monsoon rainfall in Hunan Province and the winter Eurasian snow-cover. His result is in good agreement with that obtained by Hahn and Shukla (1976), i.e., the Eurasian snow-cover exhibits a similar interaction with monsoon rainfall over both India and central China, including Hunan Province. These results appear to suggest that the inter-annual variability of monsoons over China and India probably has the same tendency. In addition, Chen and Yan (1981) also obtained a positive correlation between the snow-cover over the central Tibetan Plateau in spring and winter, and precipitation over the region to the south of the Yangtze River for May and June. The physical factors relating to the production of the interannual variability of Meiyu remain to be studied.

The thermal condition on the land surface, in particular the Tibetan Plateau, may exert a significant effect on the general circulation on a large-scale basis. If the heat source over the Plateau becomes enhanced, the Tibetan high will, in turn, intensify. Then the trough to the north of it will deepen, thus causing a chain of developing events downstream (Fig. 4.36) (Gambo and Kudo, 1983). This teleconnection pattern may be used to predict drought/flood in northern China and northern Japan, based

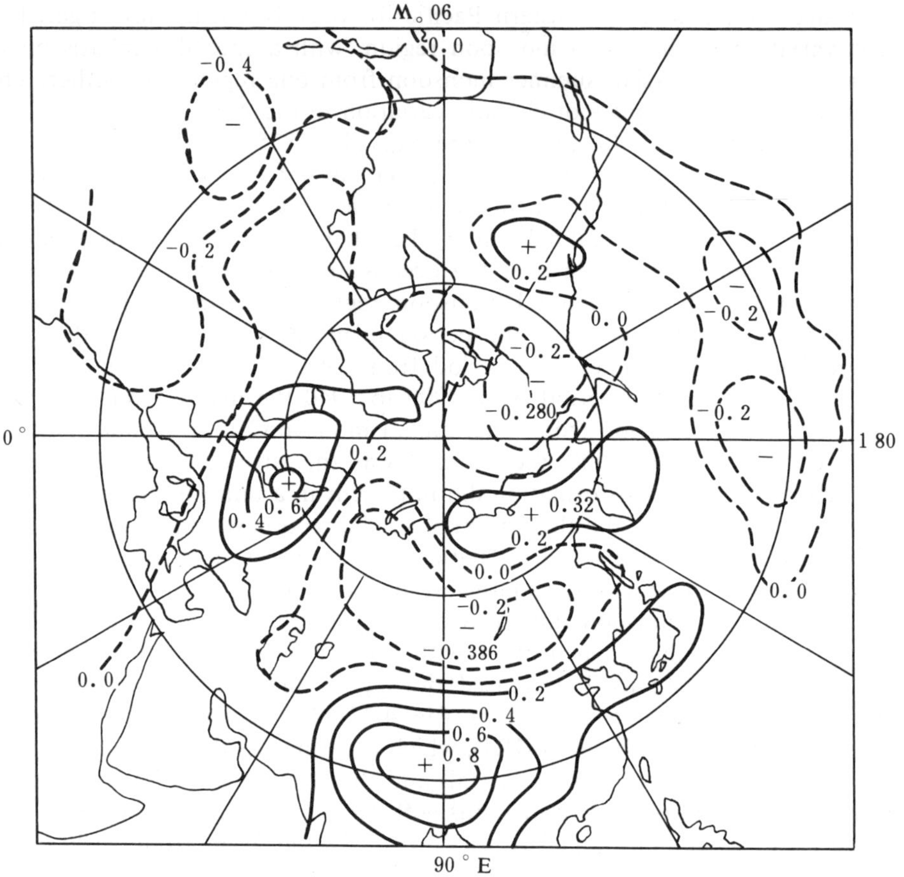

Figure 4.36. A correlative map of the departure of the height field at 500 hPa for the Northern summer, with 90° E, 30° N as the reference base point. The solid lines denote the positive correlation and the dashed lines the negative correlation (Huang, 1985).

on an anomalous change in the heating field or snow-cover field over the Tibetan Plateau. Huang (1985) simulated the effect of the change in heat sources over the Tibetan Plateau on the general circulation in the Northern Hemisphere, obtaining a result similar to the observed pattern.

4.5. Fluctuations of the Asian Monsoon and Drought/Flood in China in Historical Times

Only very few works on the long-term variation of Asian or East Asian monsoons have appeared so far, due mainly to the sparcity or unavailability of datasets for longer time periods, especially wind field data. Among them, two works have been devoted to this matter. Fu and Fletcher (1988), by use of surface wind data over the

North Indian Ocean and the western Pacific for over 100 years, investigated the climatic variation in the Asian monsoon region, with a special emphasis on the transformation of the Asian summer monsoon from one regime to another. They revealed that the large signals of climatic variation in the Asian summer monsoon region occurred in 1900, 1940, and 1960 (perhaps also 1875). The signal of 1900 was the strongest one and has been detected on the data of both the surface wind over the oceans and rainfalls over the continents, when a sudden change also occurred in the global climatic system, as pointed out by Fletcher *et al.* (1982). The long-term variability of summer monsoon over the oceans are mainly characterized by the alteration between the "meridional monsoon regime" and the "zonal monsoon regime". During the last century, the period from 1875 to 1900 was the major "meridional monsoon regime", the period from 1900 to 1940 the major "zonal monsoon regime" and the period from 1900 to 1940 the major "zonal monsoon regime". The transformation from "meridional monsoon" to a "zonal monsoon" regime occurred in about 1900, therefore, it was the strongest signal of the last century. However, this transformation did not occur in a uniform manner over the Asian monsoon region, but the alternation of wind direction was much more obvious in the ITCZ region over the western Pacific. The southwest monsoon in the Indian Ocean oscillates in phase with the southwest monsoon in the South China Sea.

The variability of monsoon rainfall over India and eastern China also exhibits some large signals which are synchronous with those of the wind field over the oceans: the monsoon rainfall increased (decreased) during the "meridional monsoon period" (zonal monsoon period) over the oceans (Fig. 4.37). The accumulative curve of the Indian Monsoon Rainfall Index (IMRI), which is defined as the difference between the number of regions with excess rainfall and the number of regions with deficit rainfall, shows several obvious transitions with large signals at around 1878, 1898, 1940, and 1964. A wet monsoon lasted for 20 years for the period from the 80's of the last century to around the turning point of this century (1878–1898), and then the climate entered into a relatively dry monsoon period for about 42 years (1899–1940), with some short periods of wet monsoon in 1907–1910, 1915–1917, and 1930–1933. Afterwards, a wet monsoon occurred again for 25 years (1940–1964). By 1974 there had been another dry monsoon period which seemed to last for 20 years. The variations of the Indian monsoon rainfall appeared to be synchronous with the variations of the summer monsoon over the Bay of Bengal. There were three signals in the accumulative curve of the anomalies of the U-component of the surface wind over the Bay of Bengal, occurring in about 1870, 1898 and 1944, which divide the whole curve into three periods: 1870–1898 with a decrease in the U-component of monsoons, corresponding to the "meridional monsoon period"; 1898–1944, with increase in the U-component of monsoons corresponding to the "zonal monsoon period", and 1944–1960, with the normal condition of monsoon. As compared with the curve of IMRI, it appeared that the monsoon rainfall in India increased during the "meridional monsoon period" (1878–1898), while the monsoon rainfall decreased during the "zonal monsoon period" (1898–1940).

Figure 4.38 indicates a comparison between the variation of monsoon rainfall in East Asia and the variation of the summer monsoon in the South China Sea. The

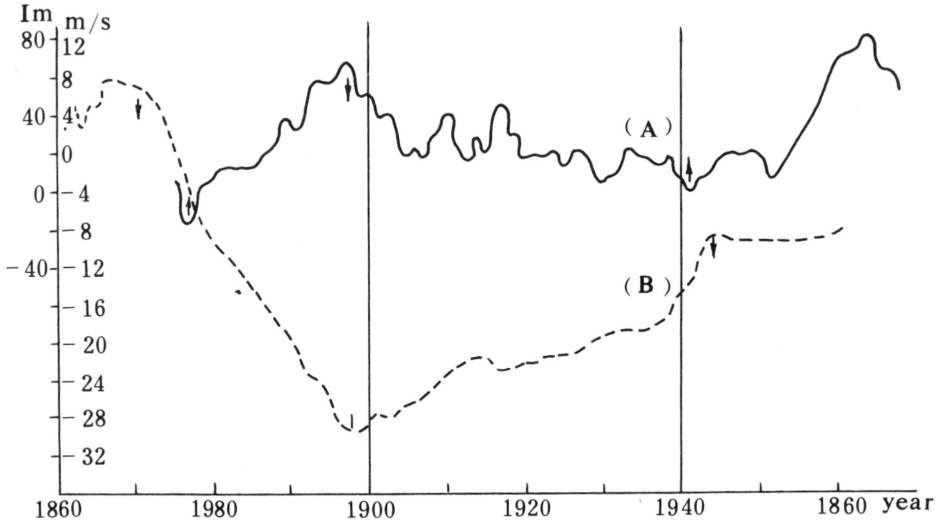

Figure 4.37. The accumulative curves of the Indian monsoon rainfall index (I_m) (A) and the zonal component (U) of monsoons over Bay of Bengal (June to August) (B) (Fu and Fletcher, 1988).

accumulative curve of the length of Meiyu (days of pluviosity for each year) in the Yangtze and Huaihe River Valleys from 1875 to 1975 can be divided into 4 regimes: 1875–1887, wet monsoon; 1887–1905, dry monsoon, 1906–1958, rapid increase in monsoon rainfall (for about a 53 year period); and after 1958, dry monsoon. Two significant signals in 1905 and 1958 are considerably different from those of IMRI in about 1898 and 1940. This evidence seems to imply a certain degree of time-lag between the variation in East Asia and that in South Asia.

However, the variation of Meiyu in eastern China is closely linked to that of the summer monsoon in the South China Sea. The accumulative curve of the U-component of the surface wind in the South China Sea in the northern summer in Figure 4.38 shows four signals occurring in 1875, 1882, 1905, and 1958, respectively. The variation of this curve is almost out of phase with that of the Meiyu, i.e., when the U-component of the summer monsoon in the South China Sea increased or became more zonal, the Meiyu in eastern China decreased and vice versa. It is interesting that the apparent decrease in Meiyu since 1958, a fact which has been well known in China, coincided with a sudden increase in the U-component of the summer monsoon in the South China Sea. It has a direct effect on the Meiyu in the Yangtze River Valley, being a major moisture supplier for that area. Based on previous discussions, the variation of the summer monsoon in this region is closely linked to the condition of the cross-equatorial airflow, therefore, it is very interesting to seek the association between the cross-equatorial aircurrent and the variability of the Meiyu from the perspective of a long-time series of data. Figure 4.39 demonstrates this point. Before 1900, there was an enhanced cross-equatorial flow in the Indian Ocean which was related to the meridional monsoon in the region, while it was below normal in most parts of the South China Sea and the western Pacific which was

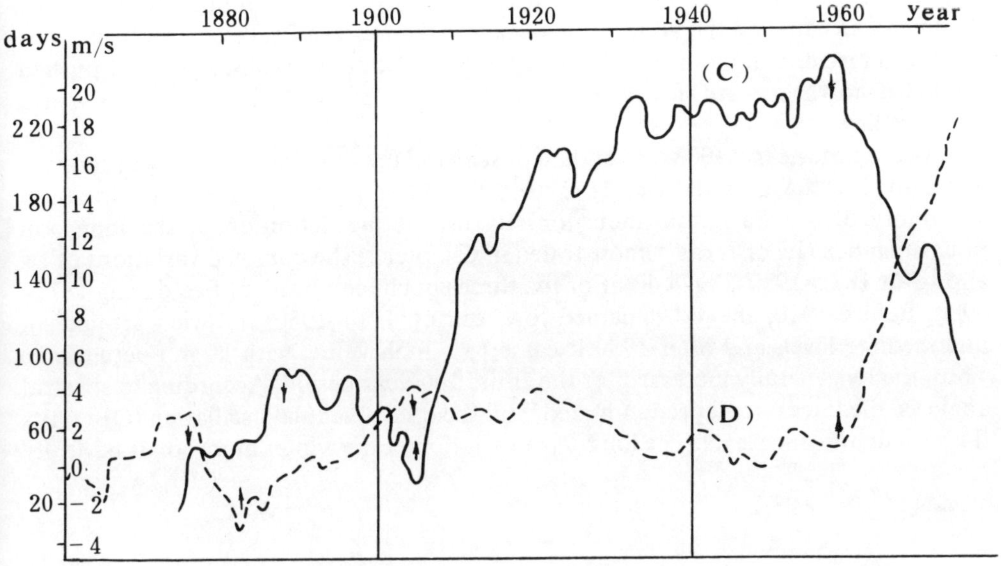

Figure 4.38. The accumulative curves of the Meiyu index in eastern China (C) and the U-component of the monsoon over the South China Sea (June to August) (D) (Fu and Fletcher, 1988).

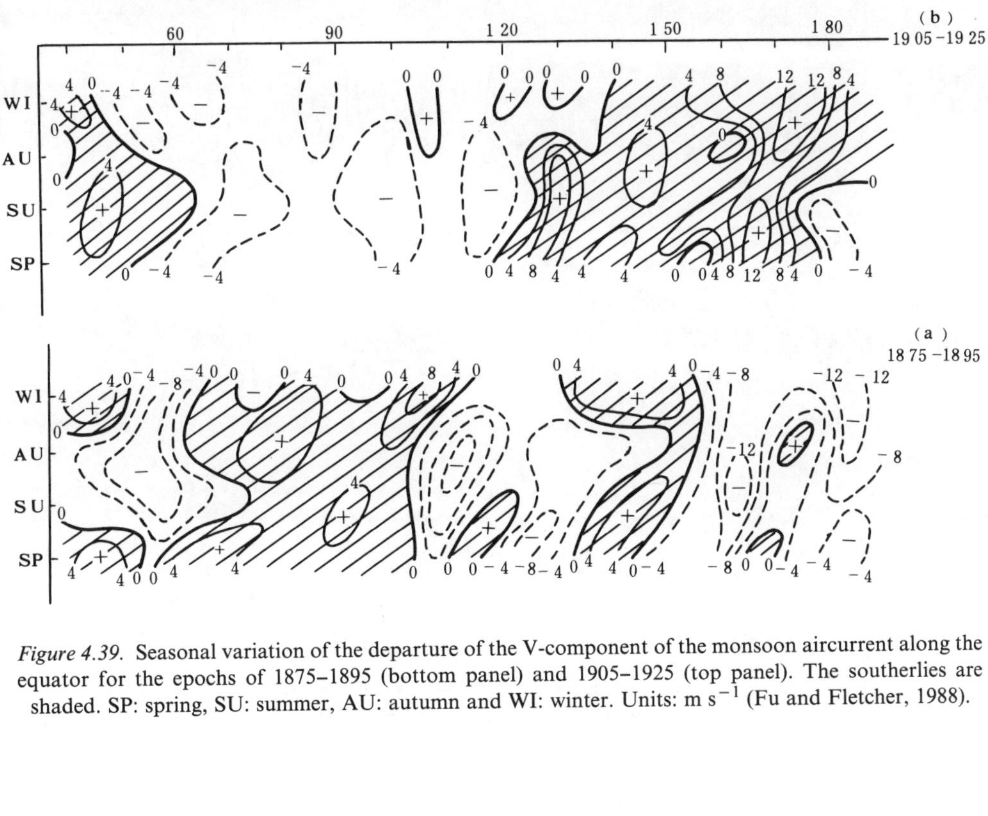

Figure 4.39. Seasonal variation of the departure of the V-component of the monsoon aircurrent along the equator for the epochs of 1875–1895 (bottom panel) and 1905–1925 (top panel). The southerlies are shaded. SP: spring, SU: summer, AU: autumn and WI: winter. Units: m s^{-1} (Fu and Fletcher, 1988).

characterized by the zonal monsoon. After 1900, however, the opposite was true for the Indian Ocean, and the South China Sea and the western Pacific, respectively. The enhanced meridional monsoon aircurrent over the South China Sea possibly pushed the ITCZ northeastward, favoring the enhancement of the Meiyu over the Yangtze River Valley.

Based on the nearly 100 years of data of sea level pressure (50° S–70° N) in January and July for the period from 1871 to 1980, Guo (1987) studied the long-term variability of the East Asia monsoon. By use of the definition of the monsoon intensity index (Iw or Is) as demonstrated in Chapter 1, she obtained variations of Iw and Is for 1871–1980 (Fig. 4.40). For Iw, three epochs can be identified during 1871–1980. Before 1910, the Iw remained low, during 1910–1950, it varied around an intermediate level, and after 1950 it came to a high value, with large fluctuations. Thus, the Iw generally increased for the entire 100-year period. According to spectral analysis, the linear trend predominated, and the quasi-biennial oscillation (QBO) fits the 95% significance level. The long-term variation of the winter monsoon is basically

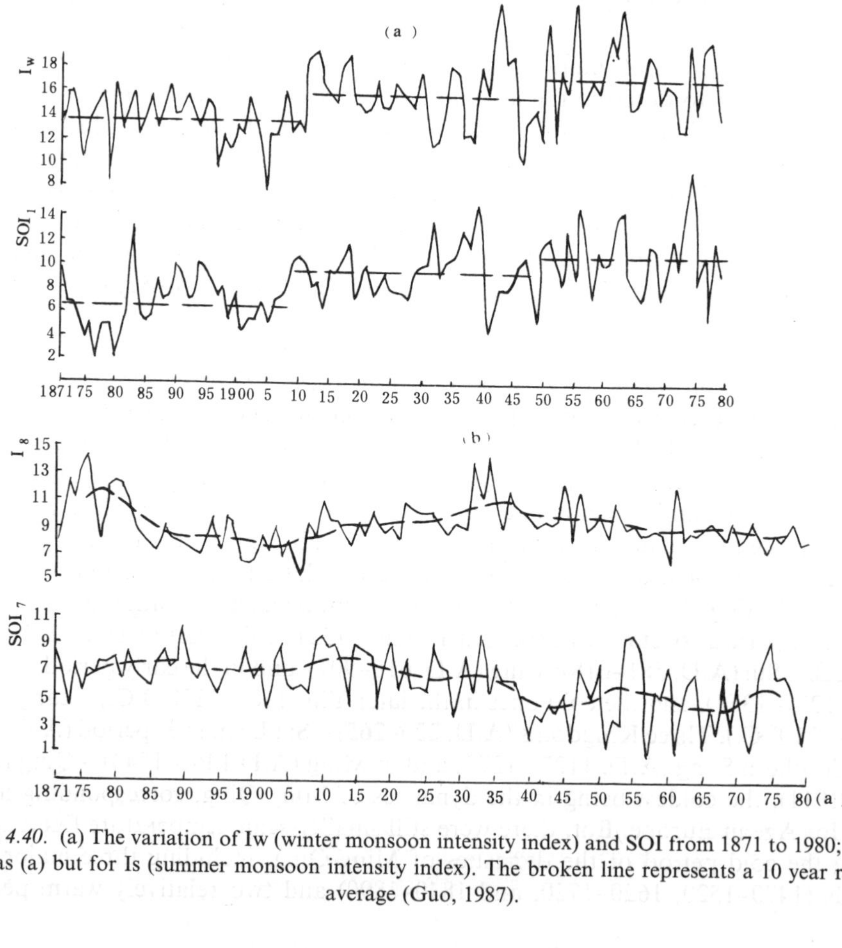

Figure 4.40. (a) The variation of Iw (winter monsoon intensity index) and SOI from 1871 to 1980; (b) the same as (a) but for Is (summer monsoon intensity index). The broken line represents a 10 year running average (Guo, 1987).

similar to that of the Southern Oscillation Index (SOI), with their phase differences being negligible.

The variation of Is shows that the summer monsoon started with high Is, and then decreased down to a minimum around the turning point of the 20th century. After about 1905, the Is increased up to a maximum at 1940, and decreased again after 1940. This kind of variation of the East Asian summer monsoon is in many ways coincident with that of the summer monsoon over the South China Sea as indicated above. The large signals were also identified as occurring in 1900 and 1940. The intensity of the SOI decreased gradually for the whole 110 year period, but one has not found any definite association between Is and SOI from the viewpoint of long-term variation.

Longer time series are essential for documenting the natural variability on historical timescales (dating back about 5000 years). Because a relatively widespread instrumented network, as described above, has been available for only about the last century, it is necessary to resort to various proxy climate data, e.g., ice cores, tree rings, diaries, ancient books and other data sources. Chinese historical records represents one of best datasets available for the study of climatic trends over the past 5000 years, and are by far the most extensive of any historical datasets. The scientific research done by Chinese climatologists using these data sources has increasingly attracted the attention of international academic communities. Here, a brief review will be presented to describe the advances in research on Chinese major historical results related to climatic change, especially drought and flood in China that were strongly affected by the variability of monsoon activity in East Asia. Main materials that will be given below are based on Zhang (1988), Zhang and Crowley (1989), and Ye, Fu, Chao and Yoshino (1987). The methods of processing historical climatic records and the reconstructon of climates in historical times will not be discussed here. The reader is referred to Zhang (1988), and Zhang and Crowley (1989) for this information.

Figure 4.41 is the long-term variability of air temperature in China over the past 5000 years based on "phenological" observations (Zhu, 1973). This index is based on the timing of recurrent weather-dependent phenomena, such as dates of flowering of shrubs or the arrival of migrant birds, or the distribution of climatically sensitive organisms. Zhu concluded that 3000 years ago in the Yellow River Valley, the annual mean temperature and winter temperature were 2–3 and 3–5°C higher than at present. There were significant northward displacements of subtropical fauna and flora during this time. Since then, a series of temperature oscillations has occurred, with each oscillation lasting 400–800 years and having an annual mean variation of about 1°C. The warm periods occurred in the Qin Dynasty (221–207 B.C.) to Han (206 B.C.–A.D. 220), Sui (A.D. 581–618); Tang (A.D. 618–907), and in the early period of Yuan (A.D. 1271–1368), and the cold ones in the later Yin (1600–1100 B.C.) – early Zhou (1100–771 B.C.), Three Kingdoms (A.D. 220–265) – Six Dynasties period (A.D. 265–580), Southern Song (A.D. 1127– 1279) and in Ming (A.D.1368–1644) – Qing (A.D. 1644–1911), the coldest being in the dynasties of Ming-Qing, corresponding to the Little Ice Age in Europe. But, there were still smaller-scale temperature fluctuations during the cold period of the dynasties of Ming-Qing, including three major cold periods (1470–1520, 1620–1720, and 1840–1890) and two relatively warm periods

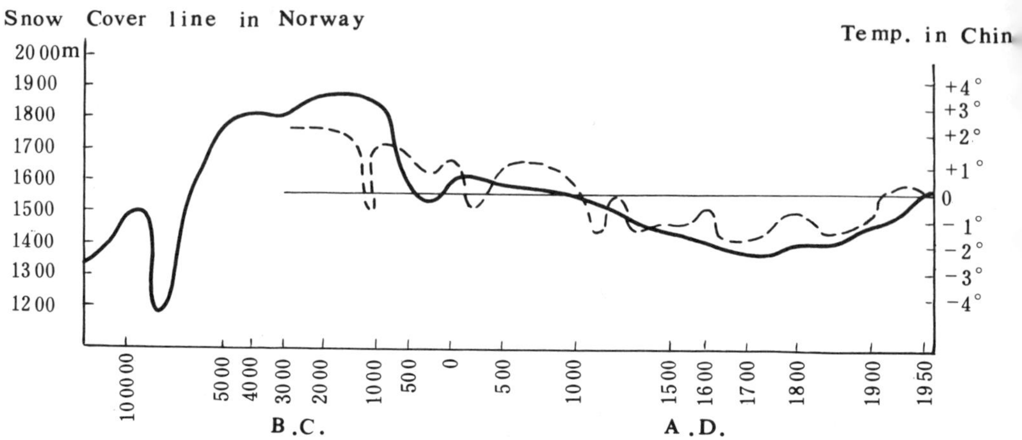

Figure 4.41. The phenological temperature (dashed line) in China over the past 5000 years. This index is based on the timing of recurrent weather-dependent phonomena, such as dates of flowering of shrubs or arrivals of migrant birds, or distributions of climatically sensitive organisms. The curve of the height of the snow line in Norway for the past 10,000 years is also given (solid line) (Zhu, 1973).

(1550–1600 and 1770–1830) in between, with the temperature amplitude being 0.5~1.0°C higher. The coldest interval occurred in the mid-17th century.

From the end of the 19th century, the temperature began to rise. Over the past 100 years (Fig. 4.42), the general trend of variability of the mean air temperature, obtained from averaging the data of 137 meteorological stations, is in agreement with that of the Northern Hemisphere, even in the global region. Specifically, a significant temperature increase began in the early 20th century and reached a peak in the 1940s. And then, it started dropping, reaching its trough in 1950s. There was a transient rise of temperature in the early 1960s, but soon it dropped again, more significantly in summer than in winter. From the mid-1970s the temperature started to rise significantly again, but in the 1980s the increase was far smaller in China than for the Northern Hemisphere, with its peak slightly lower than that of the early 1960s. Bradley *et al.* also indicated a similar result (Fig. 4.43).

From the Chinese historical records shown above, one may conclude that there are significant century and decadal scale temperature trends, with temperature fluctuations of the orders of 1.0–1.5°C. The records are characterized by peak warm events in the 12th and 20th centuries, with a series of cold and less-warm events in between.

As in the case of temperature, precipitation in China during the last 5000 years has fluctuated with the alternating occurrence of dry and wet periods. Analyses of the excavations and records in the historical writings show that most of the time before Shang, 4000–1388 B.C., the precipitation over the Zhongyuan area (the Henan Province and its surroundings) was plentiful, while most of time from the 12th century B.C., the early Zhou Dynasty to 269 B.C., the last stage of the Warring States Periods, there was very little precipitation. In the period of 2200 years since 268 B.C., the Qin

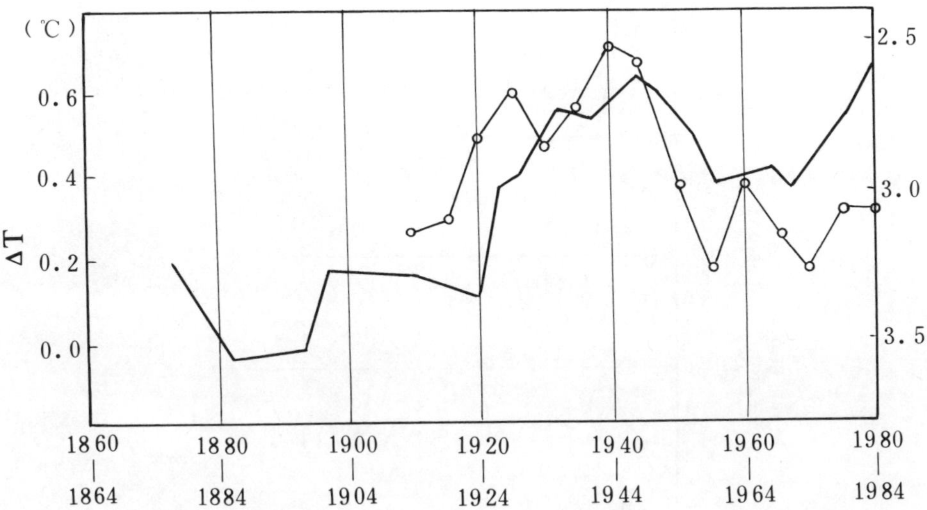

Figure 4.42. The long-term variability of air temperature for the past 100 years in China (fine sold line) and the Northern Hemisphere (thick solid line). The mean air temperature is divided into 5 grades (GT): 1 -warm, 2 -slightly warm, 3 -normal, 4 -slightly cold, and 5 - cold (SSTC, 1990).

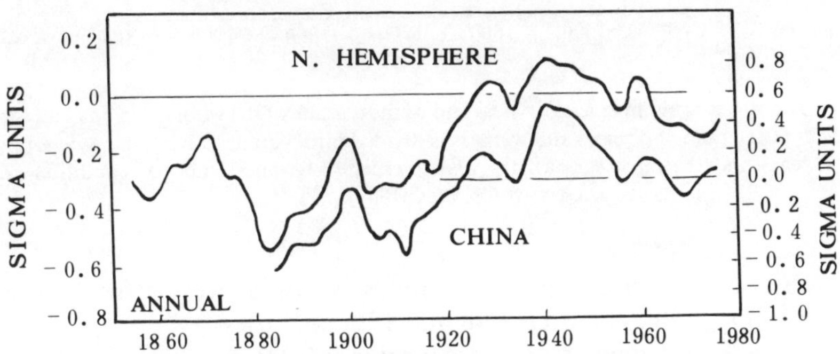

Figure 4.43. Annual smoothed temperature anomalies for China and the northern Hemisphere land areas compared. The left scale is for Northern Hemisphere and the right scale is for China. Zero reference point has been arbitrarily displaced to permit comparison (Zhang and Crowley, 1989).

Dynasty, 9 cases of severe drought and 8 cases of long wet intervals have been recorded, with the wettest occurring in the period of A.D. 634–834 during the Tang Dynasty and the driest in the period of Eastern Jin-early Tang Dynasty (A.D. 301–629). Another relatively drier period has started since the 1930s. A more detailed description of the dry and wet conditions for the last 500 years is shown in Figure 4.44. As the figure indicates, the wet and dry periods shifted almost 400 years. Each period consists of several shorter wet and dry episodes. The dry episodes last for an average of 57 years, while the wet episodes last 33 years. So, besides the long cycle of about 400 years, cycles also exist with periods of about 90 years. From Figure 4.44, it can be also seen that the climate in the 20th century is not in an extreme condition, but an

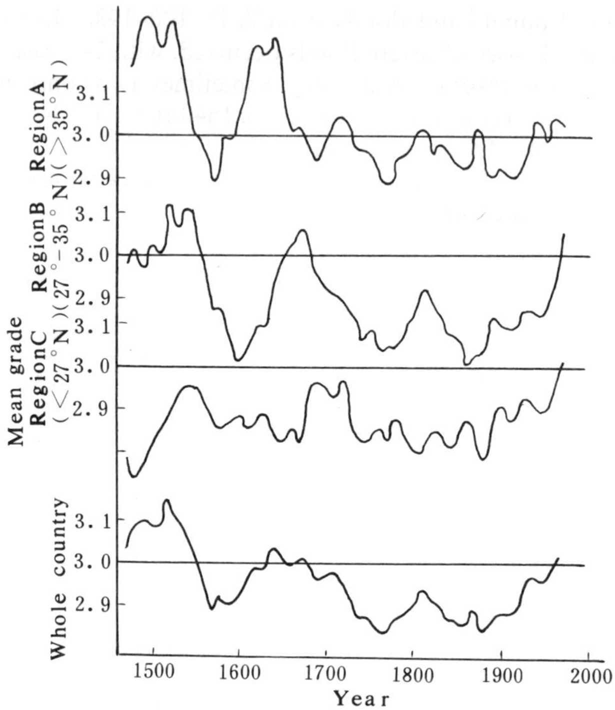

Figure 4.44. The 50-year running mean dryness and wetness grade of the whole of China and three regions for 1470–1799. The grade of dryness and wetness is divided into 5 grades with 1 and 5 denoting the most serious wet years and the most serious dry years, respectively, and 3 normal condition (Zhang and Crowley, 1989).

intermediate one. It is clear that this century is not so wet as the 18th and 19th centuries, but not so dry as in the 16th and 17th centuries.

In the history of China, significant events of drought and floods frequently occurred. As indicated previously, the extensive droughts and floods are associated with the activities of the summer monsoon. When the summer monsoon sets in earlier than normal and lately advances northward, the major rainbelt associated with the summer monsoon is stationary in southern China, thus leading to floods. To the contrary, when the summer monsoon is active and rapidly advances northward, it often brings about abundant rainfalls in northern China while in southern China the severe drought dominates. Therefore, the occurrence of droughts and floods is, to a greater degree, a manifestation of the activity of the summer monsoon in China.

Based on the statistics of the frequency of occurrence of severe droughts and floods over the Yangtze and Yellow River Valleys for 2188 years (200 B.C.–A.D. 1988), for the latter, there were 196 cases of severe drought and 147 cases of severe floods, averaging some 8.95 and 6.72 times per 100 years, respectively. Roughly speaking, severe drought occurred once every 11 years and severe flooding occurred once every 14 years. Thus, flooding occurred less frequently than drought. Table 4.10 lists the nationwide severe droughts and floods from A.D. 100–1988. In total, 40 cases of severe drought and 39 cases of severe floods occurred, with 2–3 cases of droughts or floods recorded on an average every century. Sometimes, two consecutive drought or flooding years occurred, for instance, 1074, 1075; 1640 1641; or 1848, 1849. In general, there were three periods of frequent occurrence of floods i.e., the 3rd–4th, 8th–10th, and 17th–19th centuries. The periods of frequent occurrence of droughts were the 3rd–4th, 11th, 13th, 15th, 18th, and 20th centuries.

Table 4.10. The nationwide severe droughts an floods from A.D. 100 to 1988 (SSTC, 1990)

Century	Drought year				Flooding year			
2		134			102	107		
3	236	255	266		254	292	298	
4	301	309	330		320	371	395	
5	464	473			440	499		
6	537	573			585			
7	617	668			636	682		
8	790				726	786	792	
9	862	868			828	830	856	858
10	992				931	977	983	
11	1033	1074	1075		1056			
12	1197				1188	1192		
13	1208	1215	1218	1253	1227			
14	1306	1329			1319			
15	1433	1467	1484		1409			
16	1528				1569			
17	1640	1641			1663			
18	1721	1728	1785		1709	1725	1761	
19	1835				1848	1849	1889	
20	1900	1928	1972		1954			

The distributions of droughts and floods in China have some characteristic modes which are related to the activity of the summer monsoon (Wang and Zhao, 1981). An eigenvector anlysis of rainfall data from 25 stations indicates that the dryness/wetness index of each year in the recent 500 years can be put into one of 6 types, which are three eigenvectors of opposite sign: I: flood in most parts of China, II: drought in most of parts of China, III: drought in the Yangtze Valley; flood in other parts, IV: floods in the Yangtze Valley; drought in other parts, V: floods in southern China; drought in northern China, VI: drought in southern China; floods in northern China.

Figure 4.45 gives an example of the eigenvectors of drought and flood for type III. The relative frequency of these modes have varied with time (Ronberg and Wang, 1986). A comparison of spatial precipitation patterns for the past century with patterns of the 17th century illustrated regional variations in drought frequency, with some indication of a more meridional orientation of trends in the 17th century. Especially for the driest interval (1601–1650), reduced precipitation was more common in the north and west of China, and conditions were unchanged or slightly wetter in the southeast.

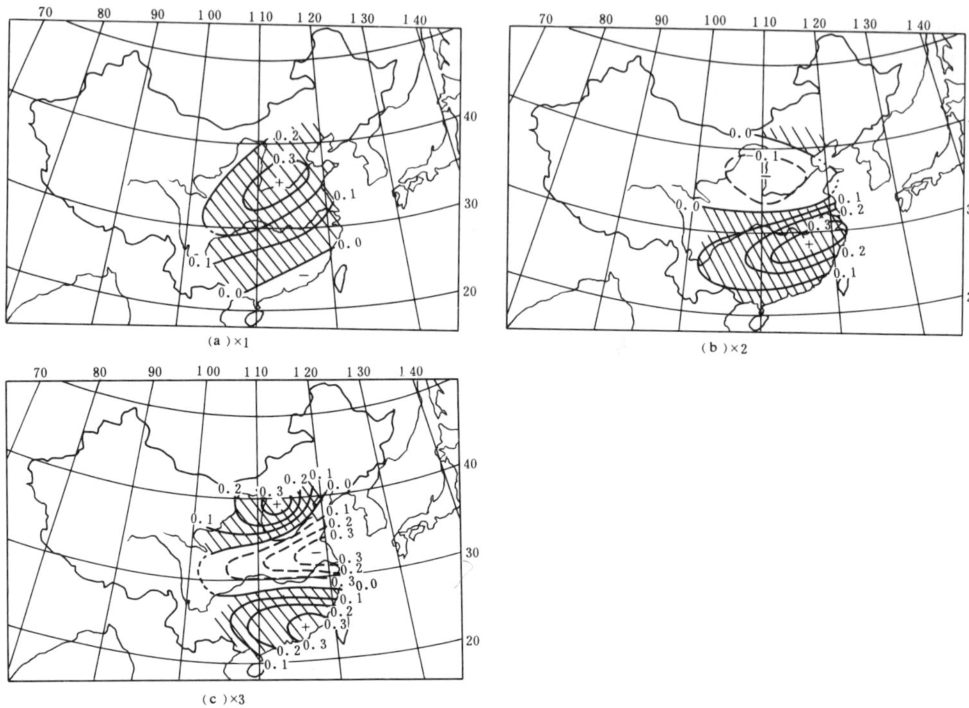

Figure 4.45. The first three eigenvectors (a, b and c) of drought and flood using historical data from A.D. 1470 to A.D. 1977 (Wang and Zhao, 1981).

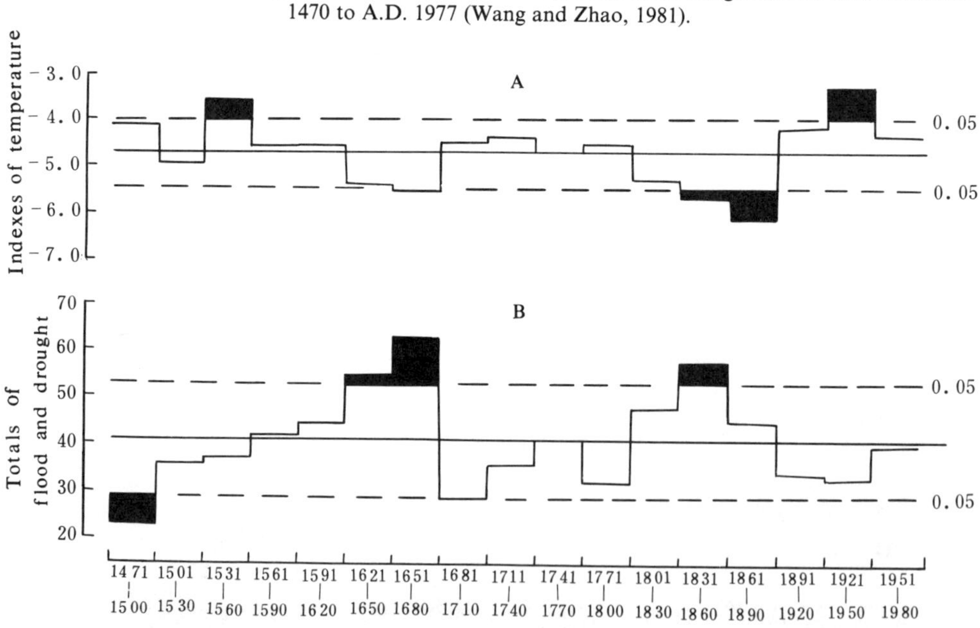

Figure 4.46. Comparison of 30 year average of winter temperature record from the middle and lower Yangtze River Valleys with record of drought and flood from the same region (Zheng and Feng, 1986).

Although there was an increased frequency of drought during the cool periods 1470–1979, there was also a considerable frequency of flooding (Fig. 4.46; Zheng and Feng, 1986), with the net result being a greater variability in parts of China during the cool periods. The catastrophic floods in the Yellow River during last 500 years (1482, 1662, and 1843) occurred in the cold periods; A.D. 1761 is an exception. The great floods of the Yangtze River in A.D. 1153, 1368, and 1870 were also observed in or near the cold perios. Analysis of rainfall patterns during the 17th century indicates that drying conditions were most prevalent in the north and west of China. This pattern could result from a suppressed development of the later stages of the summer monsoon, when monsoon rains normally shift to northern China. This decrease may be related to expanded Eurasian show-cover (Zhang and Crowley, 1989).

As pointed out by Zhang and Crowley (1989), past Chinese droughts may have been due to a modified development of the different phases of the summer monsoon. For example, in both the 17th and 20th centuries, the first phase of the monsoon (rainfall in southern China) appears to be normal. However, the second (Meiyu) and third (northern China) phases were weaker than present during the 17th century. But, it is not yet clear why this monsoon pattern could occur.

Finally, a comparison was made of droughts and floods in China with the precipitation of different parts in the world was made. Figure 4.47 (Zhang, 1988)

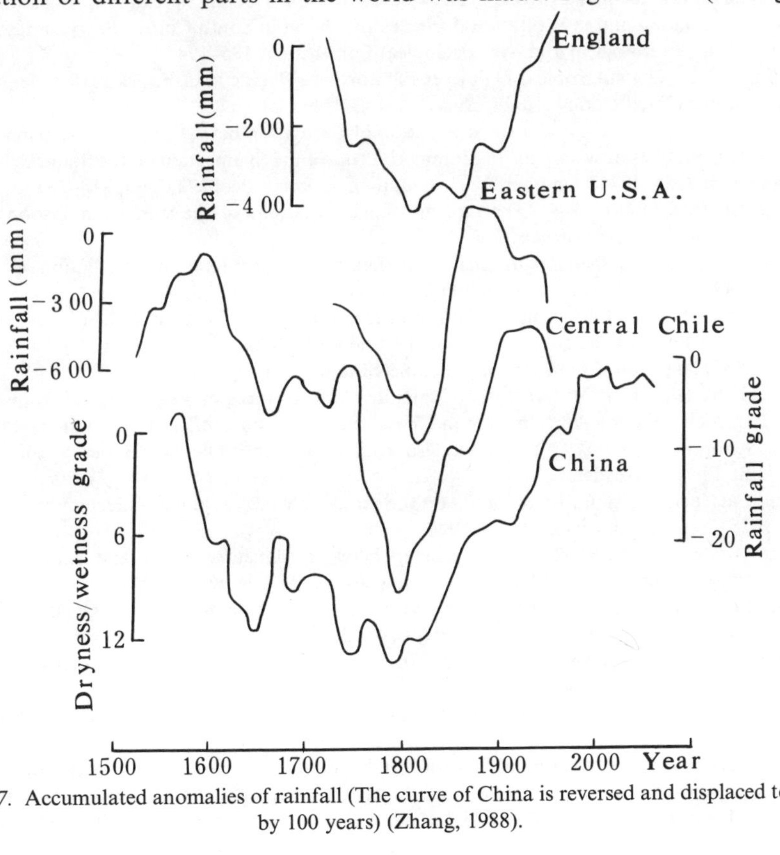

Figure 4.47. Accumulated anomalies of rainfall (The curve of China is reversed and displaced to the right by 100 years) (Zhang, 1988).

shows that the trend of precipitation in eastern U.S. and central Chile as well as in Britain almost coincides with that in China. The general trend of precipitation in all these regions showed a decrease before the 1800's, but an increase afterwards. In China, the turning point from dry to wet stages occurred in the 1700's, about a century earlier than that in Europe and America.

References

Bi Muying, 1983: 'Behaviour of the ultra-long wave in the Northern Hemisphere around the onset and end of Meiyu in the lower and middle reaches of the Yangtze River'. Collected Papers of the Meteorological Science and Technology, No. 4, Climate and Drought/Floods, 120–129.

Bi Muying, 1990: 'Features and causes of droughts in northern China in recent 40 years'. In: Ye Duzheng and Huang Ronghui, eds: *Advances in the Disastrous Climate Research Series*. China Meteorol. Press, Beijing, 23–32.

Bi Muying, Ding Yihui, 1991: 'A dynamic study of the drought in northern China in 1980'. *J. Appl. Meteorol.*, **3**, 145–156.

Chang, C.-P., Lim, H., 1988: 'Kelvin wave-CISK: A possible mechanism for the 30–50 day oscillations'. *J. Atmos. Sci.*, **45**, 1709–1720.

Chen Hanyao, 1957: 'The characteristics of circulation during the period of the flood event in the Yangtze and Huaihe Rivers in 1954'. *Acta Meteorol. Sinica*, **38**, 142–149.

Chen Lieting, Yan Zhixin, 1981: 'Effects of snow cover over the Tibetan Plateau in spring and summer on the general circulation and precipitation during rainy season in South China'. Proceedings of Medium and Long-Term Hydrological and Meteorological Forecasting, 185–194.

Chen Lieting, 1982: 'The subtropical high over the northern Pacific and the sea surface temperature of eastern equatorial Pacific'. *Sci. Atmos. Sinica*, **6**, 148–156.

Chen Longxun, Jin Zuhui, 1983: 'On the interaction of circulation between the two hemispheres in East Asia monsoon circulation system during summer'. Proc. of the Symposium on the Summer Monsoon in South and East Asia, 10–15 October, 1982, Kunming, Yunnan People's Press, 218–231.

Chen Longxun, Jin Zhuhui, 1984: 'The medium-range variations of the summer monsoon circulation system over East Asia'. *Adv. Atmos. Sci.*, **2**, 224–233.

Chen Longxun, Chen Duo, Shen Rugui, Zhang Qingfen, 1990: 'The interannual oscillation of rainfall over China and its relation to the interannual oscillation of air-sea system'. *Acta Meteorol. Sinica*, **4**, 589–612.

Chen Shixun, Ke Shizhao, 1983: 'The oscillation of the tropical circulation and the convergence zone in monsoon area'. Proc. of the Symposium on the Summer Monsoon in South and East Asia, 115–21 August, 1980, Hangzhou, China. Yunnan People's Press, 1–14.

Ding Yihui, Feng Huijun, 1974: 'Analysis of satellite cloud pictures of some types of tropical synoptic systems influencing the low-latitude area in China during the summer'. Selected Papers on Studies of Tropical Synoptical Systems Over the western Pacific Ocean. Publication Series of the Institute of Atmospheric Physics, Academia Sinica, No. 2.

Ding Yihui, Cai Zeyi, Li Jishun, 1978: 'Study on the excessively heavy rainfall in Henan province early in August 1975'. *Sci. Atmos. Sinica*, **2**, 276–289.

Ding Yihui, Reiter, E. R., 1982: 'A relationship between planetary waves and persistent rain and thunderstorms in China'. *Arch. Meteorol. Geophys. Biokl., Ser. B*, **32**, 221–252.

Ding Yihui, 1990: 'Build-up, air mass transformation and propagation of Siberian high and its relations to cold surge in East Asia'. *Meteorol. Atmos. Phys.*, **44**, 281–292.

Feng Peizhi, Li Cuijing, Li Xiaoquan, 1985: *An Analysis of Major Meteorological Disasters (1951-1980)*. China Meteorol. Press, Beijing.

Fletcher, J. O., Radok, U., Slutz, R. J., 1982: 'Climatic signals of the Antarctic Ocean'. *J. Geophys. Res.*, **87**, 4269–4276.

Fu Zhongbin *et al.*, 1979: 'The long-range variation of the sea surface temperature over the northern Pacific and its impact on subtropical high'. Collected Papers on Geography, **11**, 146–157.

Fu Zhongbin, Fletcher, J., 1988: 'Large signal of climatic variation over the ocean in the Asian monsoon

Region'. *Adv. Atmos. Sci.*, **5**, 389–404.

Fu Zhongbin, Teng Xinlin, 1988: 'Climate anomalies in China associated with El Niño/Southern Oscillation'. *Sci. Atmos. Sinica*, Special Issue, 133–141.

Gambo, K., Kudo, K., 1983: 'Teleconnection in the zonally asymmetric height field during the Northern Hemisphere summer'. *J. Meteorol. Soc. Japan*, **61**, 829–837.

Gray, W. M., 1984: 'Atlantic seasonal hurricane frequency, Part I: El Niño and 30 hPa QBO influences'. *Mon. Wea. Rev.*, **112**, 1649–1668.

Hahn, D. G., Schukla, J., 1976: 'An apparent relationship between Eurasian snow cover and Indian monsoon rainfall'. *J. Atmos. Sci.*, **33**, 2461–2463.

Hawkins, H. F., Jr., 1954: 'The weather and circulation of July 1954– One of the hottest months on record in the central United States'. *Mon. Wea. Rev.*, **82**, 209–217.

Holland, J. Z., 1954: 'The weather and circulation of June 1954– Illustrating the birth and growth of a continental anticyclone'. *Mon. Wea. Rev.*, **82**, 163–171.

Huang Jiayou, 1988: 'The manifestation of quasi-biennial oscillation in the monthly rainfall over China'. *Sci. Atmos. Sinica*, **12**, 267–273.

Huang Ronghui, 1985: 'The numerical simulation of the three dimensional teleconnections in the summer circulation over the Northern Hemisphere'. *Adv. Atmos. Sci.*, **2**, 81–92.

Huang Ronghui, Li Weijing, 1988: 'Influence of heat sourceanomaly over the western tropical Pacific on the subtropical high over East Asia and its physical mechanism'. *Sci. Atmos. Sinica*, Special Issue, 107–116.

Huang Ronghui, Sun Fengying, 1990: 'The impacts of the western Pacific warm pool on the summer climate anomaly in East Asia'. Training Workshop on Diagnosis and Prediction of Monthly and Seasonal Atmospheric Variations, Nanjing, China, 15–19 October, 1990. 71–74.

Iliari, L., 1984: 'A diagnostic study of the potential vorticity in a warm blocking anticyclone'. *J. Atmos. Sci.*, **41**, 3318–3526.

Krishnamurti, T. N., Bhalme, H. N., 1976: 'Oscillation of a monsoon system. Part I: Observational aspects'. *J. Atmos. Sci.*, **33**, 1937–1954.

Krishnamurti, T. N., Ardanuy, P., 1980: 'The 10–20 day westward propagating mode and breaks in the monsoon'. *Tellus*, **32**, 15–26.

Krishnamurti, T. N., Gadgil, S., 1985: 'On the structure of the 30 to 50 day mode over the globle during FGGE'. *Tellus*, **37A**, 336–360.

Lau, K.-M., Chang, C.-P., 1987: 'Planetary scale aspects of the winter monsoon and atmospheric teleconnections'. In: C. P. Chang and T. N. Krishnamurti, eds: Chapter 6, *Review of Monsoon Meteorology*, Oxford University Press, New York.

Lau, K.-M., Chan, P. H., 1983: 'Short-term climate variability and atmospheric teleconnection inferred from satellite observed outgoing longwave radiation. I: Simultaneous relationships'. *J. Atmos. Sci.*, **40**, 2735–2750.

Lau, K.-M., Chan, P. H., 1987: 'Interannual and intraseasonal variations of tropical convection: a possible link between the 40–50 day oscillation and ENSO'. *J. Atmos. Sci.*, **44**, 506–521.

Li Chongyin, 1988: 'Frequent activity of strong East Asian major trough and occurrence of the El Niño'. *Sci. Sinica (B)*, **31**, 667–674.

Liang Pingde, 1990: 'Relations between the El Niño, Indian summer monsoon and the summer rainfall in northern China'. *Marine Forecasts*, **7**, 17–23.

Lorenc, A., 1984: 'The evolution of planetary scale 200 hPa divergence during the FGGE year'. *Quart. J. Roy. Meteorol. Soc.*, **110**, 427–442.

Lu Jingxi, Ding Yihui, Gao Shiying, 1990: 'Interannual low frequency oscillations of meridional winds over the equatorial Indian and Pacific Oceans'. *Acta Meteorol. Sinica*, **4**, 586–597.

Lu Juzhong, Lin Chunyu, 1987: 'The relationship between the circulation change over lower latitudes of the Eastern Hemisphere and drought /flood in the Meiyu period'. Collected Papers on Meteorological Science and Technology, No. 10, 69–79, China Meteorol. Press, Beijing.

Lu Juzhong, Lin Chunyu, 1990: 'The influence of the El Niño on the monsoon circulation and precipitation in the Meiyu period over Changjiang River'. Extended abstract of WMO Traning Workshop on Diagnosis and Prediction on Monthly and Seasonal Atmospheric Variations, Nanjing, China, 15–19

October, 1990, 121–124.
Madden, R., Julian, P. R., 1971: Detection of a 40–59 day oscillation in the zonal wind in the tropical Pacific'. *J. Atmos. Sci.*, **28**, 702–708.
Mao Tiansong, Xu Naiyou, 1985: 'A preliminary study on the El Niño phenomenon and the influence of it on the summer monsoon circulation over East Asia'. *J. Tropical Meteorol.*, **1**, 19–26.
Mao Tiansong, Xu Qichun, Xu Naiyou, 1990: 'The intercorrelation between the cross-equatorial airflow near 105° E and the summer monsoon over Asia'. *Marine Forecasts*, **7**, 1–7.
Mooley, D. A., Parthasarathy, B., 1984: 'Fluctuations in all India summer monsoon rainfall during 1871–1978'. *Climate Change*, **6**, 287–301.
Nitta, Ts., 1986: 'Long-term variations of cloud amount in the western Pacific region'. *J. Meteorol. Soc. Japan*, **64**, 373–390.
Pan Hualu, Zhou, F. X., 1985: 'The 10–20 day tropical midlatitude interaction during winter monsoon season'. *J. Meteorol. Soc. Japan*, **63**, 829–843.
Ramage, C. S., 1977: *Monsoon Meteorology*. Academic Press, New York, 296 pp.
Rasmussen, E. M., Carpenter, T., 1983: 'The relationship between eastern equatorial Pacific sea surface temperatures and rainfall over India and Sri Lanka'. *Mon. Wea. Rev.*, **111**, 517–528.
Shutts, G. J., 1983: 'The propagation of eddies in different jet streams eddy vorticity forcing of blocking flow fields'. *Quart. J. Roy. Meteorol. Soc.*, **109**, 737–761.
State Science and Technology Commission, The Peoples Republic of China, 1990: Climate, No. 5 Blue of Science and Technology of China. China Meteorol. Press, Beijing, 367 pp.
Tao Shiyan, Ding Yihui, Sun Suqing, Cai Zeyi, Zhang Mingli, Fang Zhongyi *et al.*, 1980: *The Heavy Rainfall in China*, Science Press, Beijing, 225 pp.
Tao Shiyan, Ding Yihui, 1981: 'Observational evidence of the influence of the Qinghai-Xizang (Tibet) plateau on the occurrence of heavy rain and severe convective storms on China'. *Bull. Amer. Met. Soc.*, **61**, 23–30.
Tao Shiyan, Chen Longxun, 1987: 'A review of recent research on the East Asian summer monsoon in China'. In: C. P. Chang and T. N. Krishnamurti, eds: Chapter 8, *Review of Monsoon Meteorology*, Oxford University Press, New York.
Tao Shiyan, Zhu Wenmei, Zhao Wei, 1988: 'On the interannual variability of Meiyu'. *Sci. Atmos. Sinica*, Special Issue, 13–21.
Tu Changwang, 1937: 'Weather in China and the atmospheric oscillation over the world and its application to summer droughts in China'. *J. Meteorol.*, **13**, No. 11.
Wang Shaowu, Zhu Hong, 1986: 'The El Niño and cool summer in East Asia'. *Kexue Tongbao*, **31**, 474–478.
Wang, S.-W., Zhao, Z. C., 1981: 'Droughts and floods in China, 1470– 1979. *Climate and History*'. In: T. M. L. Wigley, M. J. Ingrasham and G. Farmer, Eds. Cambridge University Press, p. 271–288.
Webster, P. J., 1983: 'Mechanisms of monsoon low frequency variability: Surface hydrological effects'. *J. Atmos. Sci.*, **40**, 2110–2124.
Xu Qun, 1965: 'Meiyu over the Valley of the Changjing River during recent 80 years'. *Acta Meteorol. Sinica*, **23**, 507–518.
Yan Jinhua, Chen Longxun, Wang Gu, 1988: 'The propagation characteristics of interannual low frequency oscillation in the tropical air sea system'. *Adv. Atmos. Sci.*, **5**, 405–420.
Yang Jingzun, 1989: 'Low-frequency oscillation of summer nonsoon in East Asia and its relation with the transport of angular momentum at 500 hPa'. *J. Tropical Meteorol.*, **5**, 295–304.
Yasunari, T., 1981: 'Structure of an Indian summer monsoon system with around 40 day'. *J. Meteorol. Soc. Japan*, **59**, 336–554.
Ye Duzheng, Fu Chongbin, Chao Jiping, Yoshino, M., 1987: *The Climate of China and Global Climate*. China Ocean Press, Springer-Verlag. 441 pp.
Ye Yuyuan, 1987: 'A preliminary study of the interannual variations in Hunan monsoon rainfall associated with the El Niño and winter Eurasian snow cover'. *J. Tropical Meteorol.*, **3**, 163–167.
Yu Shihua, Liu Zhonghui, Zhu Wuxiu, 1989: 'A self-adjusting mechanism of the quasi biweekly oscillation of East Asia summer monsoon circulation system'. *J. Ins. Meteorol. Air Force*, **9**, 1–8.
Zhang Jiacheng, Crowley, T. J., 1989: 'Historical climate records in China and reconstruction of past

Climate'. *J. Climate*, **2**, 833–849.

Zhang Jiacheng, 1988: *The Reconstruction of climate in China for Historical Times*. Science Press, Beijing. 174 pp.

Zhang Xiangong *et al.*, 1963: 'A preliminary study of years of unusual precipitation over extensive regions in the 60s of this century'. *Acta Meteorol. Sinica*, **33**, 64–77.

Zhang Xiangong, Zhao Hanguang, 1990: 'The El Niño event in 1986–1987 and its impact on summer rainbelt and temperature in China'. *Meteorol. Monthly*, **14**.

Zhao Hanguang, Zhang Xiangong, Ding Yihui, 1989: 'The El Niño and the anomalous climate in China'. *Acta Meteorol. Sinica*, **3**, 471–481.

Zhao Yongping, Chen Yonglu, Zhang Bicheng, Yang Liansu, 1989: 'Temperature of southern China in winter and spring'. *J. Tropical Meteorol.*, **5**, 358–363.

Zheng, S., Feng, L., 1986: 'Historical evidence of climatic instability above normal in cool periods in China'. *Sci. Sin. (B)*, **29**, 441–448.

Zhou Jinjia, Yang Dasheng, Huang Jiayou, 1986: 'The periodic oscillations of circulation systems in tropics and subtropics and power spectrum of precipitation in China in summer'. *J. Tropical Meteorol.*, **2**, 195–203.

Zhou Jinya, Zhang Tang, Wang Jinyi, 1983: 'A preliminary study of medium-range oscillation of tropical monsoon circulation in summer of 1979'. *Acta Oceanographica Sinica*, **5**, 28–34.

Zhu Kezhen, 1973: 'A preliminary study on the climatic fluctuations during the last 5000 years in China'. *Sci. Sin.*, **16**, 226–256.

Zhu Kezhen, 1979: *Droughts in northern China and their causes and consequences. Selected Works of Professor Zhu Kezhen*. Science Press, Beijing.

Zhu Qiangen, He Jinhai, 1985: 'On features of the upper circulation in the establishment of Asian monsoon in 1979 and its medium-range oscillation'. *J. Tropical Meteorol.*, **1**, 9–18.

Zhu Wenmei, Zhao Wei, Tao Shiyang, 1987: 'Some characteristics of the East Asia monsoon circulation during anomalus droughts and floods in 1987 and 1980'. Collected Papers of Meteorological Science and Technology, No. 11 (Summer Monsoon and Drought/Floods). China Meteorol. Press, Beijing, pp. 44–53.

CHAPTER 5

EFFECTS OF THE TIBETAN PLATEAU (QINGHAI-XIZANG) ON THE MONSOON

5.1. Introduction

The Tibetan Plateau is one of the most complex geographical features in the world, with an average elevation of about 4 km. Thus, it not only exerts a significant mechanical barrier effect, but also serves as an elevated heat source that generates a temperature contrast with the surrounding free atmosphere. These two important effects of the Tibetan Plateau to a great extent determine the monsoon circulation over South and East Asia, with the mechanical barrier effect mainly affecting the Asiatic winter monsoon circulation and the thermal effect mainly controlling the summer monsoon over Asia.

In 50's of this century, a network of surface and upper-air stations was established with most of the stations situated above an altitude of 3000 m. One may use the data obtained from this network to study the large-scale aspects of the circulation systems over the Tibetan Plateau. These achievements are mainly reflected in the poineering book on the Meteorology of the Tibetan Plateau (Yang *et al.*, 1960). During the period from 1961 to 1970, data coverage and quality continued to improve with the establishment of the Chinese National Project on the Meteorology of the Tibetan Plateau Region. The main results obtained from this project were well summarized in the book on the Meteorology over the Tibetan Plateau (Yeh, Gao *et al.*, 1979). In the FGGE year, the observational network was further upgraded with new and some temporary stations set up over the central and eastern Tibetan Plateau. In particular, Chinese scientists carried out the Qinghai-Xizang Plateau Meteorology Experiment (QXPMEX) during summer of the 1979 (May to August), the same period as the MONEX. The main objectives of QXPMEX were to study the thermal and dynamic effects at the surface of the Tibetan Plateau on the general circulation in East Asia, the radiation and heat budgets at the surface of the Plateau, and the weather systems under various synoptic conditions.

The experimental area includes nearly all the whole Plateau, i.e., the region from 80 to 105° E in the latitude band of 28-40° N (Fig. 5.1). During the QXPMEX period, the density and frequency of observations in the experimental area were enhanced. There were 223 surface stations, 83 upper-air stations and 37 pilot balloon sites with 4 upper-air stations located over the previously data-void region of the western and central Tibetan Plateau. In addition, 6 radiation stations were set up and distributed in the some typical regions with different vegetations.

Utilizing the QXPMEX data with a number of additional observations included,

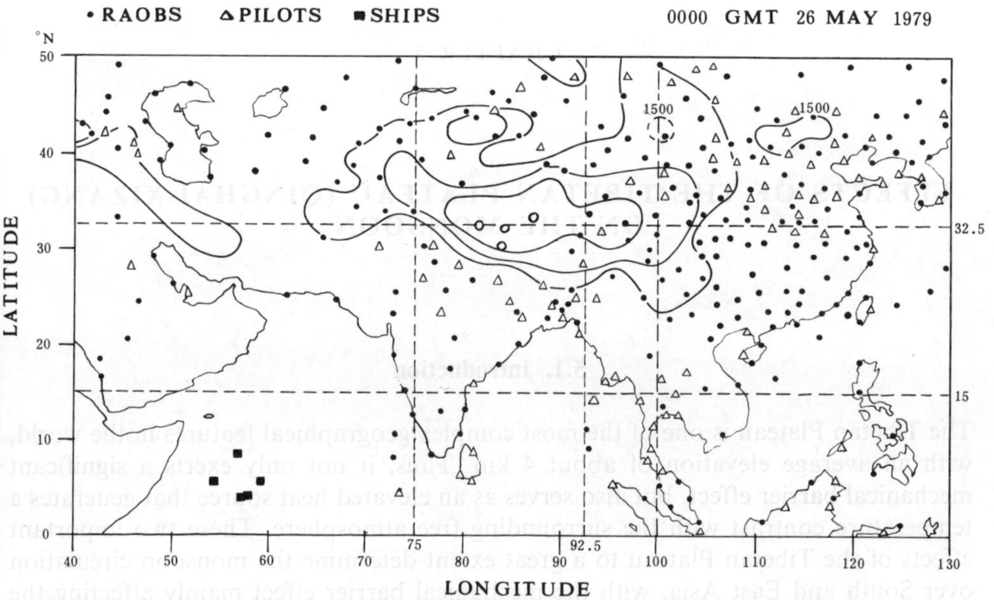

Figure 5.1. The 2.5° × 2.5° grid, smoothed topography (full contours in meters) and FGGE Level II-b data distribution at 00:00 GMT, 26 May 1979. Additional QXPMEX radiosonde data are shown by open circles (He, McGinnis, Song and Yanai, 1987).

investigators at the first time are able to analyse and study the heat budget and circulation features over the Tibetan Plateau with the greater reliability and details than ever, this significantly improving our understanding of the orographic influence of the Tibetan Plateau upon monsoon circulations.

Following the QXPMEX, Chinese meteorologists once again carried out another field experiment over the Tibetan Plateau (Qinghai-Xizang Land Surface Heat Source Experiment) in the period from August 1982 through July 1983, with a special emphasis on the geographic distributions of the heat sources and sinks over the Tibetan Plateau and their seasonal variations. For this purpose, 4 Heat Source Observing Stations (HSOS) were specially set up over the Plateau (Fig. 5.2), at each of which the components of the surface radiation balance were directly measured. Based upon these data, one may readily calculate the surface radiation balance over the Plateau, that is part of the heat sources. In this chapter, we shall concentrate on the thermal and dynamic effects of the Tibetan Plateau on the winter and summer monsoon circulation on a large-scale basis.

5.2. Heat Sources of the Tibetan Plateau and Neighboring Areas

Efforts to investigate diabatic heat sources over the Tibetan Plateau have been made by several researchers by employing FGGE/MONEX and QXPMEX data on both

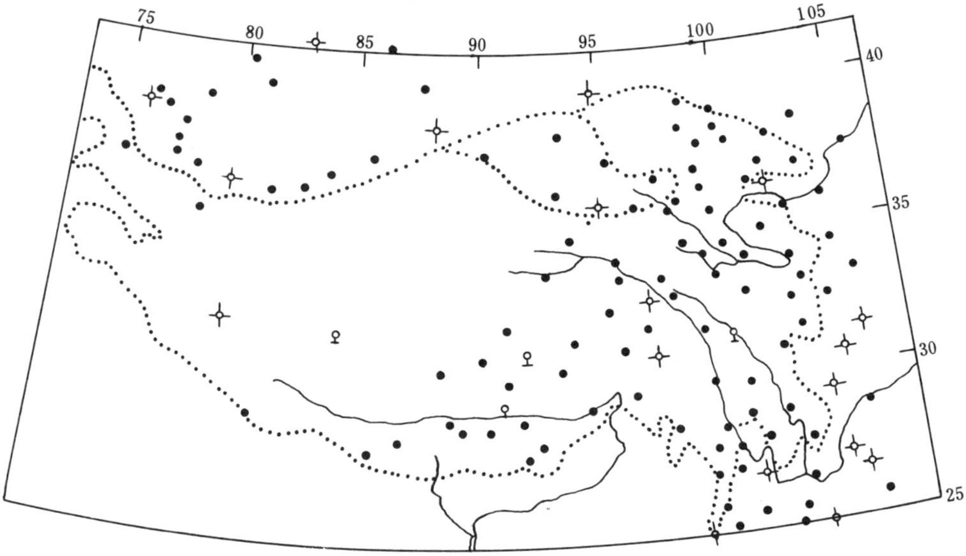

Figure 5.2. The distribution of four heat source observing stations and pyranometer stations, and weather stations over the Qinghai-Xizang Plateau and its neighborhood. ♀ – Heat Source Observation Station (HSOS); • – pyranometer station; -ф- – weather station.

global and regional scales. Two fundemental problems have been dealt with: (1) is the Tibetan Plateau heat sources or a heat sink and (2) is it sensible or latent heat to make a dominating contribution to the heat sources and sinks? Evidence that the Tibetan Plateau is a heat source in summer has been previously presented, but this evidence was not conclusive for the winter situation. According to recent studies (Yeh, Gao *et al.*, 1979), the Tibetan Plateau, on an average, is deemed to be the heat source relative to the atmosphere above it for both winter and summer. Namely, during the whole year, the heat is transported from the surface of the Plateau upward to the atmosphere in different forms. Here, the heat source is defined as the upward heat transport from the underlying surface to the atmosphere. Three types of heat may be transported to the atmosphere from the surface: the effective radiation, latent heat, and turbulent sensible heat. During the whole year, the turbulent sensible heat transfer from the Earth's surface is the major component of the Plateau heat source; the effective radiation is next to the sensible heat; the evaporation is at a minimum. In July and August, the surface evaporation reaches its peak, but is still much less than the turbulent sensible heat. In the remaining months, the latent heat carried upward by the evaporation may be negligible. During the winter, the surface effective radiation becomes the maximum component and the turbulent sensible heat transfer is of secondary importance. The long-term monthly mean values of the above three components are given in Table 5.1.

For the annual mean, the upward heat transfer from the surface of the Tibetan Plateau to the atmosphere is in the order of nearly 340 cal cm^2 day^{-1}, which accounts for 65% of the global maximum surface heat source near the eastern coast of North

Table 5.1. Long-term (1961–70) monthly mean heat transfer from the surface of the Tibetan Plateau to the atmosphere. Units: Cal cm^2 day^{-1} (Yeh, Gao *et al.*, 1979)

	Jan.	Feb.	Mar.	Apr.	May	Jun.	Jul.	Aug.	Set.	Oct.	Nov.	Dec.
Sensible heat	43	89	162	255	300	291	240	198	164	130	66	27
Effective radiation	167	167	167	167	167	162	162	162	162	162	162	167
Surface evaporation	2	4	21	21	29	78	86	75	29	4	4	2
Total heat transfer from the surface to the atmosphere	212	260	350	443	495	481	418	385	355	296	232	196

America ($\sim$575 cal cm^2 day^{-1}). This fact demonstrates the importance of the Tibetan Plateau for the general circulation from the viewpoint of energetics. Table 5.1 also indicates that the total energy supplied by the Tibetan Plateau has its maximum in late spring and early summer, with a peak of about 500 units occurring in May. A minimum heat transfer of nearly 200 units is observed in December. From winter to spring, the total upward energy transfer increases rapidly, due mainly to the contribution of the increase in turbulent sensible heat transfer.

The whole heat amount transported upward from the surface of the Tibetan Plateau cannot be used for heating the atmosphere above the Tibetan Plateau. For instance, most evaporation would not condense locally and is transported outside the Plateau. The heat to be actually used to warm the local atmosphere of the Tibetan Plateau has 5 components: The effective radiation coming from the surface(LR_1), the surface turbulent sensible heat transfer (SH), the solar radiation (SR), the condensation heat released through local precipitation (LP), and the outgoing longwave radiation at the top of the atmosphere (LR_2). The sum of these 5 components of diabatic heating (cooling) may be written as follows:

$$E = SH + LR_1 + LP + SR - LR_2, \qquad (5.1)$$

where E is defined as the atmospheric heat source. If $E > 0$, it is a heat source; if $E < 0$, it is a cold source. The computational results are shown in Table 5.2. For the annual mean, the tropospheric air column per squared centimeter can receive heat of 40–50 cal day^{-1}. Therefore, the atmosphere over the Tibetan Plateau is a heat source on an annual basis, although significant seasonal variations can be found. From March through September, the atmosphere over the Plateau has a net heat gain, showing the nature of the heat source. A part of the heat thus acquired is used to heat the atmosphere over the Plateau and another part of it is transported outside the Plateau. In late fall and winter, the atmosphere of the Tibetan Plateau is a heat sink, with the maximum observed in December and January. One may note that neither July nor August the are months when the atmosphere above the Plateau has a maximum net heat gain.

Finally, the monthly heat transfer from the earth-atmosphere system of the Tibetan Plateau to the surrounding atmosphere (*F*) may be easily estimated:

$$F = \mathrm{SH} + \mathrm{Le} + \mathrm{LR}_1 + \mathrm{SR} + \mathrm{LP} - \mathrm{LR}_2 - \mathrm{C}_p \mathrm{M} \Delta \mathrm{T} \tag{5.2}$$

Here Le is the surface evaporation latent heat, M is the total air mass above the Tibetan Plateau in the layer of 600–100 hPa, and ΔT is the month-to-month variations of temperature. The calculated results are also presented in Table 5.2. On average for the whole year, the earth-atmosphere system of the Tibetan Plateau transports 80 cal $\mathrm{cm}^2\ \mathrm{day}^{-1}$ outside to the surrounding atmosphere. Therefore, this atmosphere system is a heat source on annual basis. In winter (December through February), it is a heat sink, with the maximum observed in December and January when the heat of 140–150 cal $\mathrm{cm}^2\ \mathrm{day}^{-1}$ is transported into the earth-atmosphere system from the outside. In spring and summer, a huge heat source exists over Tibet, with the maximum observed in June when heat of 210 cal $\mathrm{cm}^2\ \mathrm{day}^{-1}$ is transported outside. Thus, one may conclude that the thermal effect of the Tibetan Plateau should be very significant.

Table 5.2. The atmospheric heat source *E* and energy budget of the earth-atmosphere system (F) of the Tibetan Plateau. Units: cal $\mathrm{cm}^2\ \mathrm{day}^{-1}$ (Yeh, Gao *et al.*, 1979)

	Jan.	Feb.	Mar.	Apr.	May	Jun.	Jul.	Aug.	Sept.	Oct.	Nov.	Dec.	Annual average
E	−48	−87	51	124	193	224	208	152	91	−20	−112	−159	43
F	−147	−88	44	119	181	207	203	154	101	−9	−98	−152	43

For a long time there have been different opinions about the nature of the heating of the Tibetan Plateau or about the question as to which component of diabatic heating is the dominating one. Previously, Flohn (1968) suggested that the Tibetan Plateau may be considered as an elevated sensible heat source. He also estimated the input of sensible heat into the atmosphere from a semi-arid surface in western Tibet based on the surface heat budget, and obtained the heat flux of about 120 w m^{-2}. Flohn, however, then changed his viewpoint with special emphasis on the release of condensational heat by convection, especially over the southeastern region of the Tibetan Plateau. This is consistent with Riehl's (1967) speculation that latent heating, rather than sensible heating, is the primary factor in atmospheric heating sources over the eastern Tibetan Plateau. Thus, these works emphasized importance of latent heating. Based on stations data for 10 years (1961–1970), Yeh, Gao *et al.* (1979) computed various components of long-tèrm mean heat balance over the Tibetan Plateau and their seasonal variations, and pointed out that there is a marked difference in the nature of the diabatic heating between the eastern part and western part of the Tibetan Plateau. In Figure 5.3 the mean values of the sensible heat transfer from the surface (SH), the latent heat released by condensation (LP) and the total diabatic heating (E), are shown. One immediately notes an extremely large sensible heat flux over the arid western Tibetan Plateau with a maximum of 450 cal

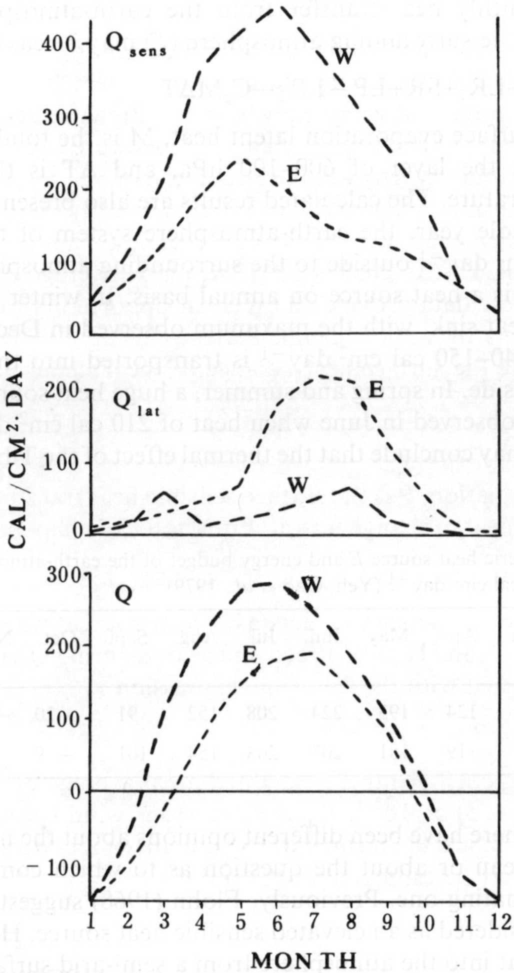

Figure 5.3. The mean sensible heat flux at the surface (Q_{sens}), latent heat of precipitation (Q_{lat}), and net atmospheric heat source (Q) over the western (W) and eastern (E) Tibetan Plateau averaged for 10 years (1961–1970) (Yeh, Gao *et al.*, 1979).

cm^{-2}.day^{-1} ($\sim$219w m^{-2}). This is about twice the magnitudes of Flohn's estimate of 250 cal cm^{-2}.day^{-1}.

Because of the great height of the Plateau, this sensible heat supply is used directly to warm up the middle and upper troposphere. As compared with SH over the western Tibet, SH over eastern Tibet is much less pronounced; yet it exceeds LP until June, which indicates that the sensible heating predominates over entire Tibet until the onset of summer monsoon or the rainy season occurs.

Then, during the height of summer (July and August), LP becomes slightly larger than SH. Because of the predominance of SH, the net heating E is quite large over the western Tibetan Plateau, with the result that this region contributes most to the net heat balance over the entire Plateau. The contribution from the eastern Tibetan Plateau is much less significant. Although computed heating components obtained by Yeh, Gao *et al.* provide useful information on the heat budget over the Tibetan Plateau, some of the estimated values may be of questionable accuracy. For example, the observational error in the ground surface temperature can be substantially large and may result in significant errors in the estimation of SH, Le, and radiative heating (or cooling). Furthermore, SH and Le are quite sensitive to the eddy viscosity coefficient (C_d), which may vary from one region to another over the Tibetan Plateau. During FGGE years and QXPMEX, data coverage over the Plateau was much improved. These data provide an excellent opportunity to reestimate the heat and moisture budgets over the Tibetan Plateau. Weng (1986) used the following equation to calculate the surface heat balance of the Tibetan Plateau from May to August of 1979:

$$B=P+LE+H, \tag{5.3}$$

where B is the net radiation, P is the surface sensible heat transfer, LE the latent heat transfer, and H the heat exchange in soil. Equation (5.3) may be rewritten as

$$B-H=P+LE. \tag{5.4}$$

When $(B - H) > 0$, the surface acts as a heat source; when $(B - H) < 0$, the surface is a heat sink. B and H were directly measured during QXPMEX. The sensible heat may be determined with the following expression:

$$P=\mathrm{p}C_pC_DV_{10}(T_8-T_a) \tag{5.5}$$

where C_D is the drag coefficient, the weighted areal average of which for the entire Plateau is about $(4 - 5) \times 10^{-3}$, V_{10} is the average wind speed at 10 m above the ground, and T_g and T_a are the mean surface and air temperatures, respectively.

Latent heat LE may be estimated as the residual,

$$LE=(B-H)/1+\beta, \tag{5.6}$$

where β is the Bowen ratio. H may be calculated with a certain empirical relation.

In May, which is the dry season, the whole Plateau acts as a strong heat source with a maximum exceeding 105 w m^{-2} found at the surface of central and southern Tibet, where an abundance of solar radiation exists due to a lack of rainfall and cloudiness (Fig. 5.4).

In June, the general pattern of the heat source is similar to that of May, but with maximum increasing up to 120 w m^{-2}. In July when the rainy season sets in the surface heat source further intensifies all over Tibet, except over the eastern Plateau (Fig. 5.5). Note the distinct increase in the Yarlung Zanbo River in the southern Tibet and the Sichuan Basin. It is worthwhile pointing out that this increase is due to an increase in the latent heat transfer that may be caused by nocturnal rainfalls and the moist ground surface there.

Figure 5.6 shows that a maximum latent heat component was observed over the Yarlung Zanbo River which is coincident with the maximum of the surface heat source (Fig. 5.4). In contrast to the seasonal variation of the latent heat, the sensible

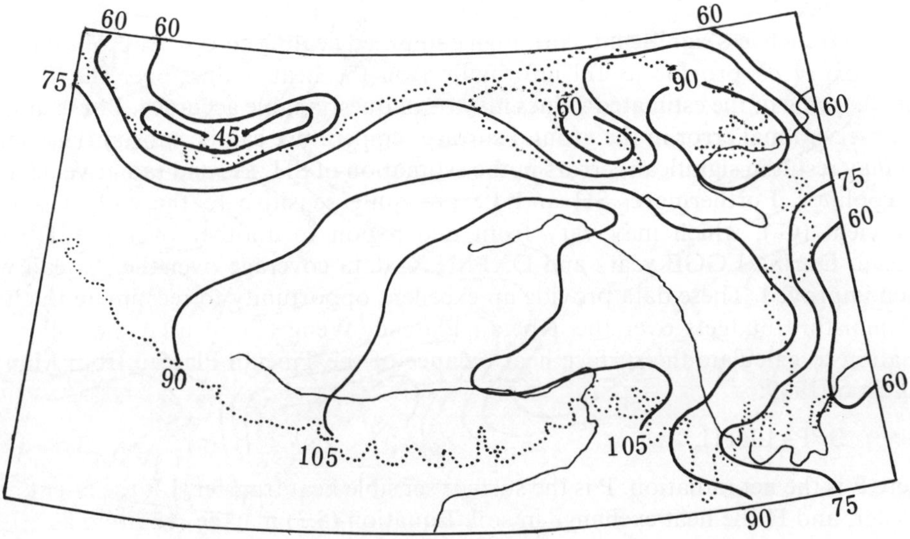

Figure 5.4. Distribution of the surface heat source in May 1979. Units: w m^{-2} (Weng, 1986).

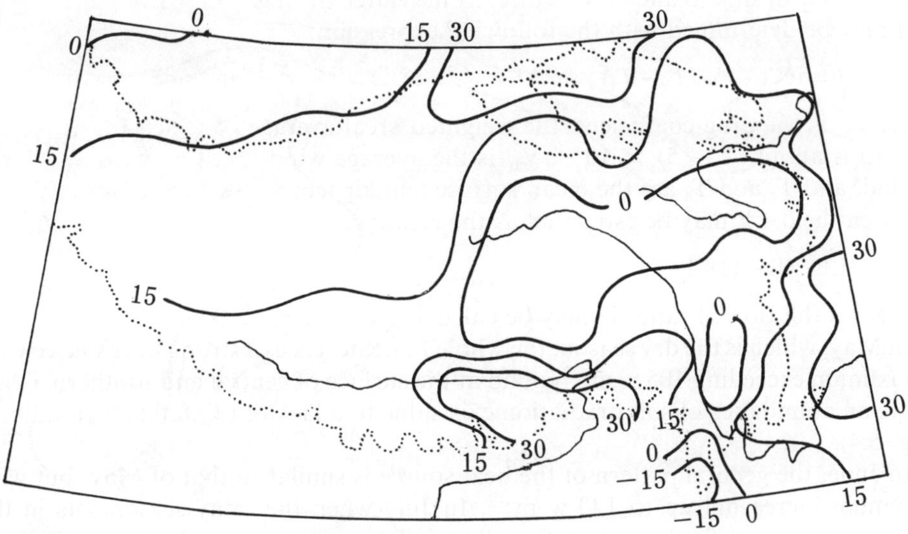

Figure 5.5. Distribution of the difference in the surface heat source between July and May 1979. Units: w m^{-2} (Weng, 1986).

heat greatly decreased from the dry season (May) to the wet season (July), especially over eastern Tibet. Figure 5.7 clearly indicates this variation, with the maximum drop occurring in the Yarlung Zanbo River. This trend of seasonal variation is in broad agreement with that of the previous result for the long-term mean obtained by Yeh,

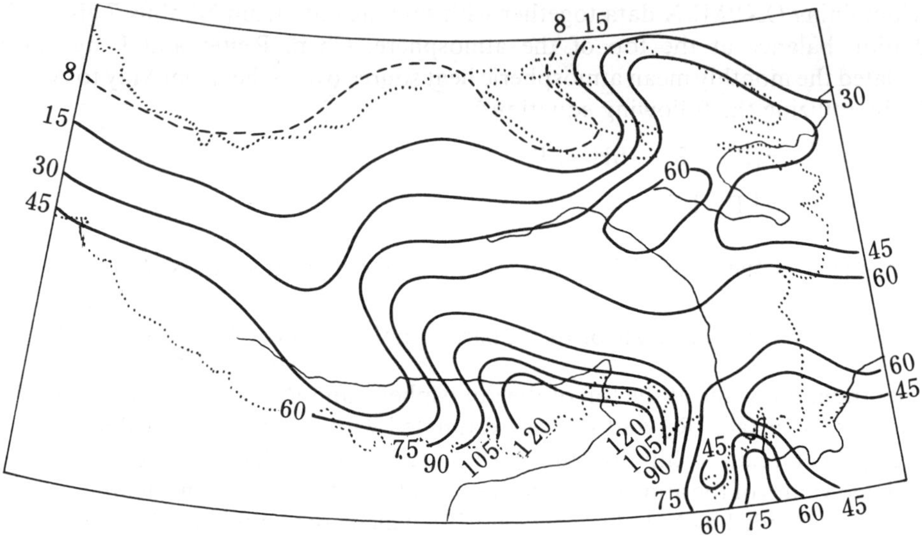

Figure 5.6. Distribution of the surface latent heat transfer for July 1979. Units: w m^{-2} (Weng, 1986).

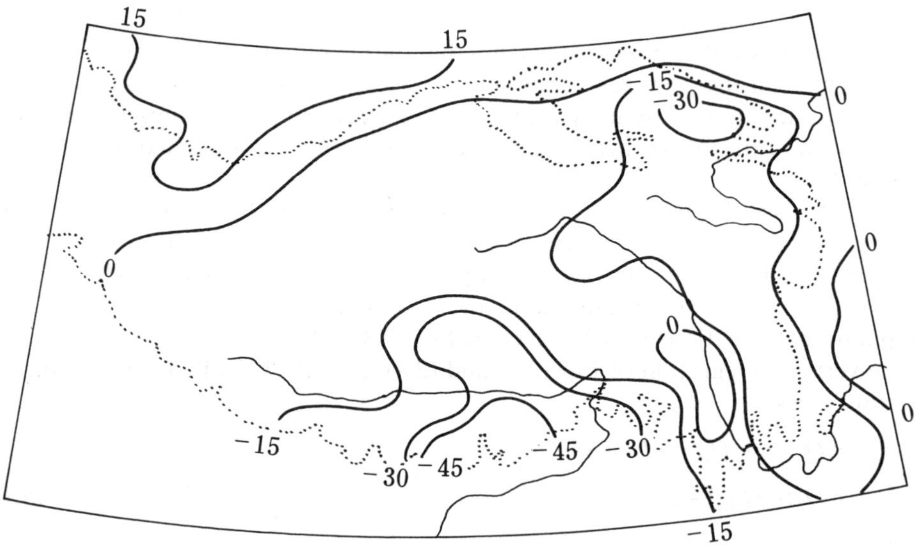

Figure 5.7. Distribution of the differences in the surface sensible heat between July and May 1979. Units: w m^{-2} (Weng, 1986).

Gao *et al.* (See Table 5.1), but with different magnitudes. For example, the calculated values of latent heat are considerably larger than Yeh, Gao *et al.*. In addition, surface sensible heat transfer is found to be the primary process for heating the atmosphere during the dry period, especially over the western Tibetan Plateau.

Combining QXPMEX data together with satellite data from Nimbus-7 giving the radiation balance at the top of the atmosphere, Chen, Reiter and Feng (1985) calculated the monthly mean atmospheric heat source over Tibet from May to August 1979 by utilizing the following equations:

$$<Q_l>=(R_\infty-R_0)+SH+LP, \tag{5.7}$$

$$<Q_2>=LP-LE, \tag{5.8}$$

where $<Q_l>$ is the vertical integrals of the apparent heat sources $\frac{1}{g}\int_{P_t}^{P_s} Q_1\,dp$, $<Q_2>$ is the vertical integrals of the apparent moisture sink $\frac{1}{g}\int_{P_t}^{P_s} Q_2\,dp$, R_∞ is the net radiation value at the top of the atmosphere, and R_0 is the net radiation value at the earth's surface. R_∞ and R_0 were estimated with the Nimbus-7 ERB data and the empirical relationship for the surface radiation budget, respectively. Their calculated values of the atmospheric heat source turned out to be considerably smaller than those provided by Yeh, Gao *et al.* (1979), which are about half of the previously estimated values (64 w m^{-2} for the average over the whole Tibet; 82 and 46 w m^{-2} for eastern Tibet and western Tibet, respectively). The main reason for the large discrepancy may be the different assumption of the drag coefficient, which was estimated to be 4×10^{-3} during QXPMEX. This is nearly half of the value used by Yeh, Gao *et al.* Another possible reason is that the net radiation observed by Nimbus-7 is smaller than the previous value, but the net radiation at the ground surface is greater than the previously used value. But, Yeh, Gao *et al.* also obtained a similar result that the atmospheric heat source over Tibet is mainly modulated by the release of latent heat.

Tables 5.3 and 5.4 compare the total heat and moisture budget over the western and eastern Tibetan Plateau obtained by various investigators. Over western Tibet, Luo and Yanai(1984), using a different approach, obtained a smaller SH value (109 w m^{-2}) than the June mean value of Yeh, Gao *et al.* (219 w m^{-2}). Their computational equations may be given as Eqs. 2.3 and 2.4.

The vertical integration of Eqs.(2.3) and (2.4) from P_t=125 hPa to P_s (the surface pressure) gives

$$<Q_1>=<Q_R>+LP+S \tag{5.9}$$

$$<Q_2>=L(P-E) \tag{5.10}$$

where

$$<>=\frac{1}{g}\int_{P_t}^{P_s} (Q_2)\,dp \tag{5.11}$$

Note that Eq. (5.7) is identical with Eq. (5.10). The latent heating there is very insignificant, so that the net heating $[Q_l]$ ([ ] represents the areal averaging) value (142 w m^{-2}) derived by Yeh, Gao *et al.* is one third larger than that obtained by Luo and Yanai (101 w m^{-2}). For the moisture balance, the evaporation is the maximum component, with the value over western Tibet is very close to the average value for the whole of Tibet for June. Over the eastern Tibetan Plateau, the SH derived by the three groups are nearly identical to each other. The latent heating obviously becomes large, but still smaller than SH. The net heating $[Q_l]$ is in the range of 94–120 w m^{-2} with the contributions of LP and SH both being rather significant. For the $[Q_2]$, the contribution from LP exceeds that made by LE, thus creating positive $[Q_2]$. From the above descriptions, one may clearly conclude that the sensible heat transfer from the surface is the dominating factor in the heat balance over the whole Plateau, in particular over the western part of Tibet in June 1979; the latent heating is very important for the heat budget over the eastern Tibetan Plateau, especially during and after the summer rainy season.

Table 5.3. A Comparison of heat and moisture balances over western Tibetan obtained by various investigators. Units: w m^{-2}

Investigators	$<[Q_l]>$	$<[Q_R]>$	[LP]	[SH]	$<[Q_2]>$	[LP]	[LE]
Luo and Yanai. Dry season prior to rainy season	101	−77	9	(169)*	−22	9	(31)
Yeh and Gao (June)	142	−94**	17	219	−22	17	39**

* Estimated as the residual of heat balance equation
** Average value for the whole Plateau.

Table 5.4. The same as Table 5.3, but for the eastern Tibetan Plateau. Units: w m^{-2}

Investigators	$<[Q_l]>$	$<[Q_R]>$	[LP]	[SH]	$<[Q_2]>$	[LP]	[LE]
Luo and Yanai (39-day mean)	113	−62	71	(104)	44	71	(27)
Yeh and Gao (June)	(94)	−94**	86	(102)	(47)	86	39**
Nitta (1983) (100-day mean)	120	−75	90	(105)	(25)	90	(65)

Figure 5.8 shows the horizontal distributions of the vertically integrated heat source $<Q_l>$ and moisture sink $<Q_2>$ averaged for the 40-day period from 26 May to 4 July 1979 (Luo and Yanai, 1984). The most pronounced heat source of ~300 w m^{-2} is not located over the Tibetan Plateau as previously thought, but over the Assam-Bengal region. Consistent with the moisture sink of the same order of magnitude, this maximum heat source is undoubtedly related to the heavy rains in this region. This result is in basic agreement with that obtained by Chen and Li (1985) who estimated the atmospheric heat source by use of the direct approach. Their maximum heating rate of 8 °C day^{-1} is located over the northeastern part of the Bay of Bengal, whereas the long-term mean heating rate (1961–1974) over the Tibetan Plateau is only of about 1 °C day^{-1} (Fig. 5.9), which cannot be regarded as a major heat source in the summer monsoon region. Yao and Luo (1982) also estimated the monthly mean heating fields of the atmosphere over the Tibetan Plateau and its surrounding areas, along with their annual variation, utilizing the climatological data of 1961 to 1970 for

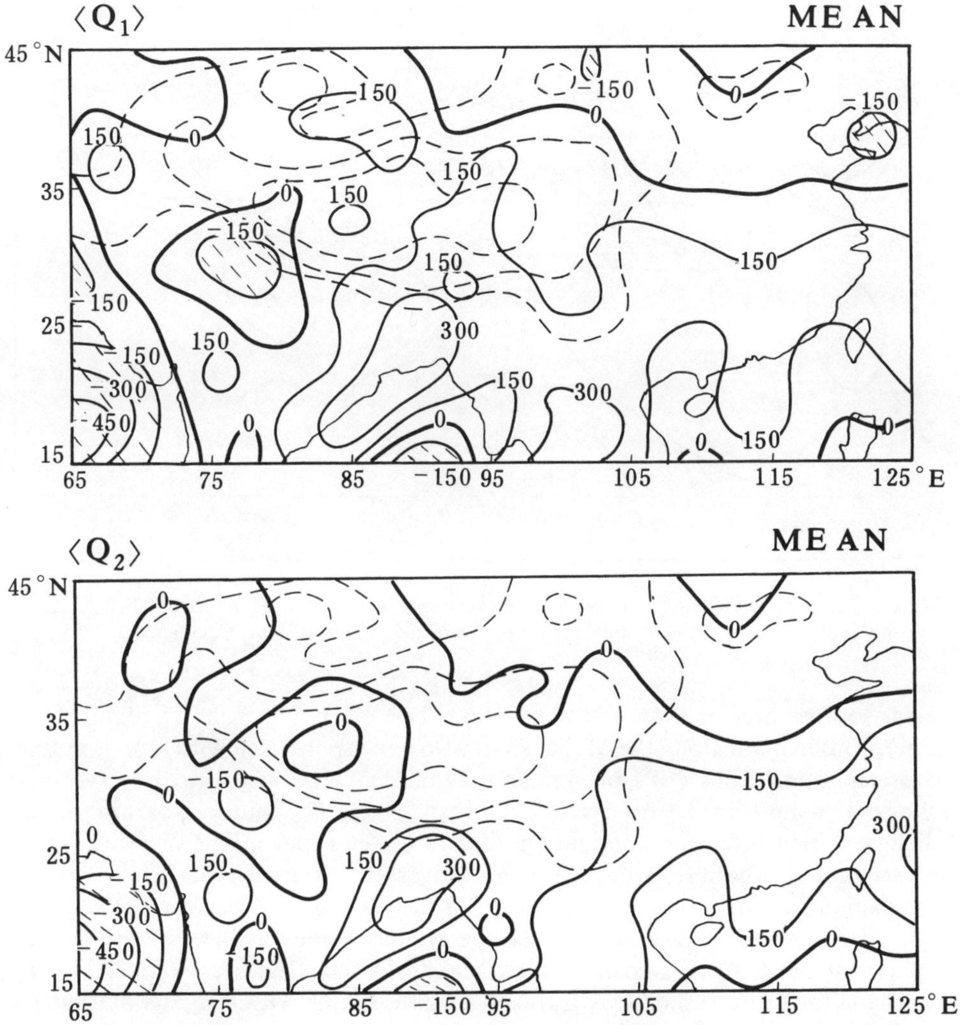

Figure 5.8. 40-day mean values (units of w m^{-2}) (a) for the vertically integrated apparent heat source $<Q_1>$ and (b) for the vertically integrated apparent moisture sink $<Q_2>$ (Luo and Yanai, 1984).

more than 180 meteorological stations. Figure 5.10 is the long-term mean horizontal distribution of the atmospheric heat source and sink over East Asia for winter and summer, respectively. A major heat source is located over the region of the 60 – 100° E longitudinal range to the south of 45° N, with the maximum over the northern part of the Bay of Bengal, Assam, and northern India. The average heating rate in this region is about 8 °C day^{-1}, which is nearly the same as the value obtained by Chen and Li. The heating rate over the Tibetan Plateau is only of 1.0–1.5 °C day^{-1}. In winter, the heat sink over the Tibetan Plateau seems to be weaker than those to the south and north of the Plateau.

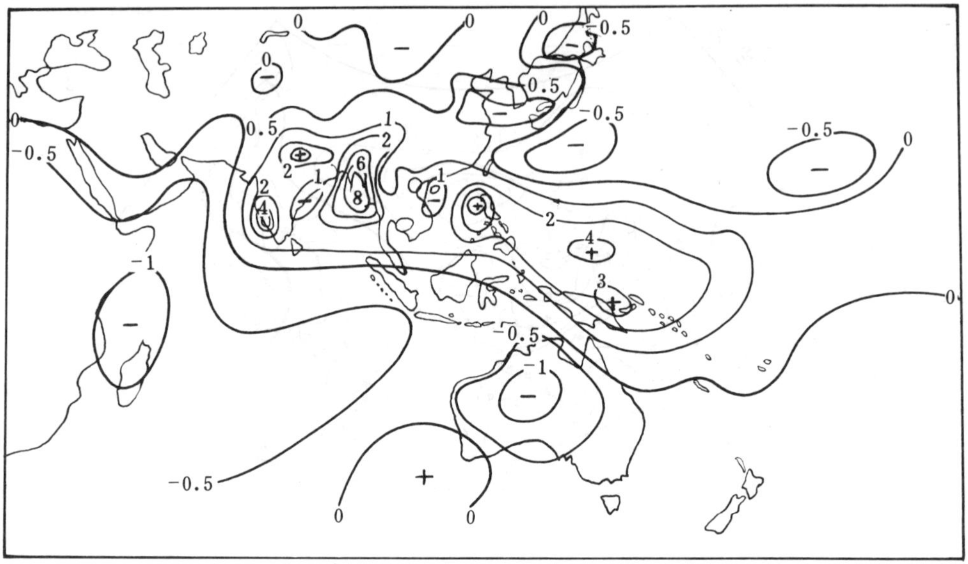

Figure 5.9. The horizontal distribution of the mean atmospheric heat sources for July. Units: °C day^{-1}. Positive values denote the heat source and negative values, the heat sinks (Chen and Li, 1985).

Heat sources greater than 150 w m^{-2} in a belt extending from Thailand to the southern China plain along the Meiyu front also correspond well with moisture sinks and precipitation rates (LP) of similar magnitudes. One may also observe large cooling with a moisture source over the Arabian Sea and the southern part of the Bay of Bengal which indicates dominant radiative cooling and intensive evaporation. Heat sources of about 100 to 150 w m^{-2} over the eastern Tibetan Plateau are accompanied by moisture sinks of about 100 to 150 w m^{-2}. Heat sources of about 100 to 150w m^{-2} are also found over the western Tibetan Plateau, the western periphery of the Plateau, and the Takla Makan Desert. The heat sources in these regions are not accompanied by significant moisture sinks. This suggests that intense sensible heating from the dry ground surface exceeds the net radiative cooling.

The meridional profiles of $<Q_1>$ and $<Q_2>$ averaged over an 80-day period from 16 April to 4 July 1979 at 75° E and 100° E further demonstrate the distinct difference in differential heating between western and eastern Tibet (Fig. 5.11). At 75° E, maximum heating with smaller $<Q_2>$ values occurs on the western Tibetan Plateau region (23–32° N). The condensation heating is evident only along the western coast of India (south of 23° N). On the other hand, the heating distribution along 100° E is dominated by strong condensation heating on the south side of the eastern Plateau and cooling on the north side. Maximum heating (270 w m^{-2}) is located over Thailand (17.5° N).

The areal mean vertical profiles of ω, Q_1, and Q_2, i.e., $[\omega]$, $[Q_1]$, and $[Q_2]$ show more distinctly the regional differences of heating processes between eastern and western Tibet, the Yangtze River Valley, and the Assam-Bengal-North India regions

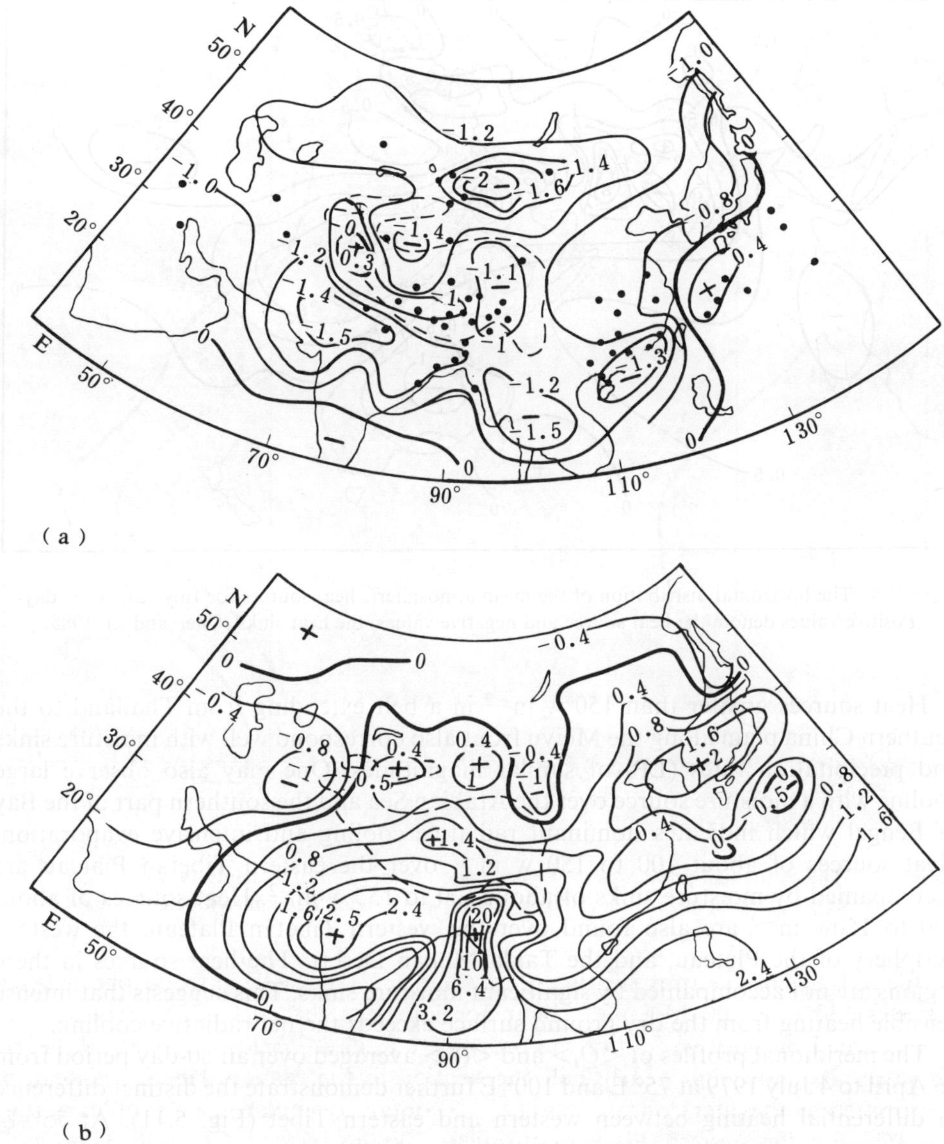

Figure 5.10. Long-term mean distributions of atmospheric heat sources and sinks (1961–1970) (a) for winter and (b) for summer Units: °C day^{-1} (Yao and Luo, 1982).

(Fig. 5.12). Over the western Tibetan Plateau, the mean vertical motion is weakly upward, and the heating (1 to 3 K day^{-1}) in the whole layer is accompanied by very small values of $[Q_2]$. The heating is more intense in the layer above the 500 hPa level than in the layer near the ground surface. Over the eastern Tibetan Plateau, the mean vertical motion is distinctly upward with a peak value at 400 hPa. The mean vertical

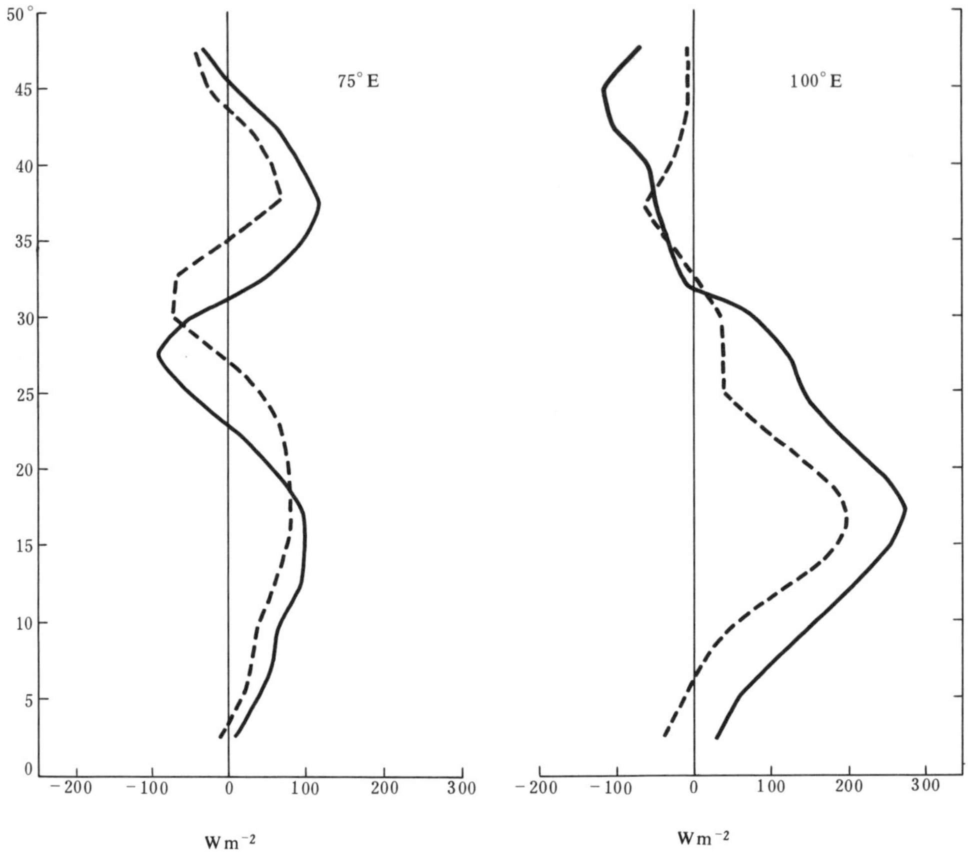

Figure 5.11. The meridional profiles of the 80-day period from 16 April to 4 July 1979, for mean $\langle Q_1 \rangle$ (solid) and $\langle Q_2 \rangle$ (dashed) at 75° E and 100° E. Units: w m^{-2} (He, McGinnis, Song and Yanai, 1987).

profile of $[Q_1]$ shows strong heating (2 to 3 K day^{-1}) in the deep tropospheric layer between 150 and 500 hPa and slight cooling near the surface. The heating is accompanied by positive $[Q_2]$ of the order of 1 K day^{-1} below the 800 hPa level. These results show that nearly half the heating in the eastern Tibetan Plateau is contributed by the latent release. Nitta (1983) obtained similar mean profiles of $[Q_1]$ and $[Q_2]$ for the eastern Plateau during a 100-day period from 30 May to 7 September, 1970.

The mean vertical profiles of $[Q_1]$ and $[Q_2]$ for the Yangtze River Valley are remarkably similar to each other, showing that the heating in this region is mostly the result of frontal precipitation. On the other hand, the mean $[Q_1]$ and $[Q_2]$ profiles for the Assam-Bengal-North India region show large differences in both their magnitudes and the levels of maximum values. The upward shift of the $[Q_1]$ profile from the $[Q_2]$ profile is a phenomenom typical for a cumulus convective atmosphere.

Figure 5.13 shows the 40-day mean vertical distributions of the heating rate Q_1/C_p

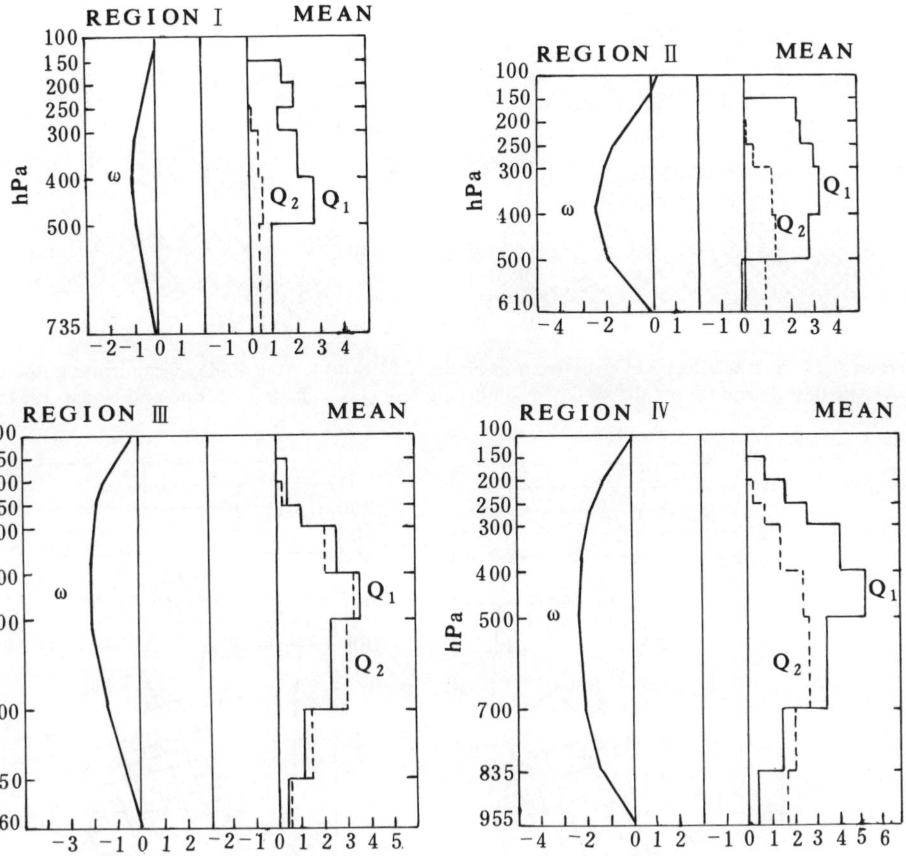

Figure 5.12. The 40-day mean vertical distributions of the areal mean vertical *P*-velocity (hPa h^{-1}), heating rate $[Q_1]/C_p$ (K day^{-1}), and drying rate $[Q_2]/C_p$ (K day^{-1}) for the western (I), eastern Tibet (II), the Yangtze River Valley (III), and the Assam-Bengal-North India regions (IV) (Luo and Yanai, 1984).

and the drying rate Q_2/C_p in the longitudinal plane along 32.5° N. Over the Tibetan Plateau between 82.5 and 97.5° E, a deep layer of heating occupies the whole troposphere except near the ground surface. The heating over the Plateau appears to consist of two distinctly different regimes. The heat source over the eastern Plateau with peak values of ~5 K day^{-1} in the 300–400 hPa layer is accompanied by a moisture sink with peak values of ~4 K day^{-1} in the 400–500 hPa layer. On the other hand, the heat source over the western Plateau is associated with a weak moisture source in the lower layer.

Figure 5.14 shows the 40-day mean meridional vertical distributions of the heating rate Q_1/C_p and the drying rate Q_2/C_p along 92.5° E. Over the eastern Tibetan Plateau, there is a strong heating rate with a maximum of 5.4 K day^{-1} found at 350 hPa. Correspondingly, the separation of the height of the peaks of Q_1 and Q_2 indicates the presence of active cumulus convection. To the south of the Tibetan Plateau, one may

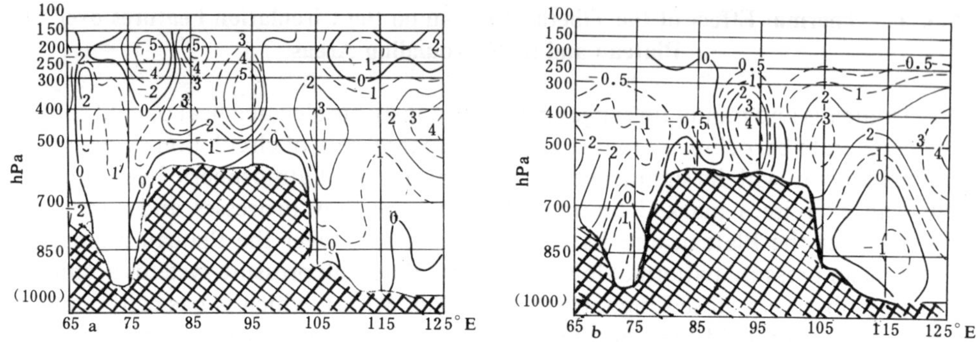

Figure 5.13. East-west vertical cross-sections along 32.5° N (a) for the 40-day mean heating rate Q_1/C_p (K day^{-1}) and (b) for the 40-day mean drying rate Q_2/C_p (K day^{-1}) (Luo and Yanai, 1984).

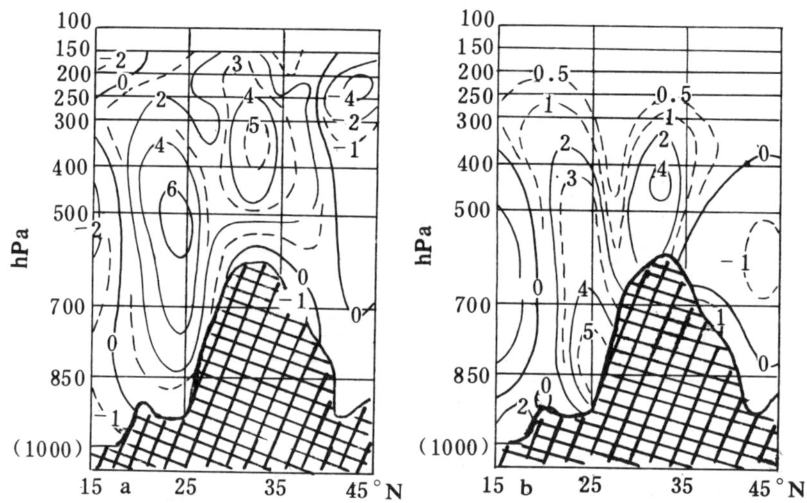

Figure 5.14. South-north vertical cross-sections along 92.5° E (a) for the 40-day mean heating rate Q_1/C_p (K day^{-1}) and (b) for the 40-day mean drying rate Q_2/C_p (K day^{-1}) (Luo and Yanai, 1984).

observe a very deep heating layer that is related to the monsoon rainfalls at the southern foothills of the Himalayas and the Assam-Bengal region. The height of the maximum heating rate is found at 500 hPa, which is 300 hPa higher than the height of the maximum moisture sink (5.7 K day^{-1}, 800 hPa). This indicates the severity of the convective activity of monsoon rainfalls in this region. The mean heating rate in the layer of 200 – 500 hPa over the eastern Tibetan Plateau is ~3 K day^{-1}, which is very similar to the magnitude of the heating rate in the Assam-Bengal region.

The foregoing results show that the sensible heat flux from the ground surface is the dominant factor in the heat budget of the Tibetan Plateau, especially for the western Plateau during the early summer of 1979. The contribution of condensation heating is also significant in the heat budget of the eastern Plateau. These finding confirm the previous results of Flohn (1968) and Yeh, Gao *et al.* (1979).

5.3. The Thermal Effect of the Tibetan Plateau on the Circulation Fcatures over the Plateau and its Surrounding Areas

Large plateaus and mountain ranges generally constitute elevated heat or cold sources which generate baroclinicity that results in a great variety of circulation systems. The Tibetan Plateau with its significant seasonal variations of atmospheric heat source or sink, can exercise a marked thermal effect on the flow processes and circulation features over the Plateau and its surrounding areas. Over the Plateau region, the winter mean pressure pattern at low level shows that high pressure systems dominate. One may observe a closed high at 600 hPa (Fig. 5.15a), which is basically separate from the Siberian cold high on the long-term mean chart. Therefore, they are two independent circulation systems. This fact reflects the prominent thermal impact of the Tibetan Plateau on the regional circulation system. In summer, there is a closed

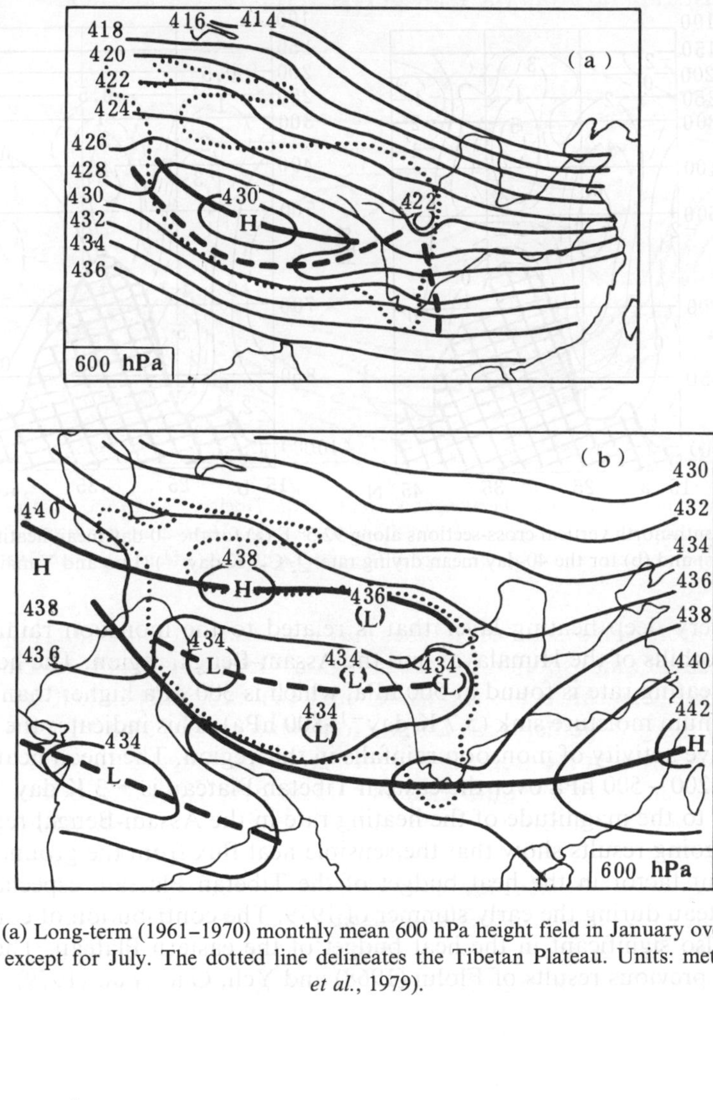

Figure 5.15. (a) Long-term (1961–1970) monthly mean 600 hPa height field in January over Asia; (b) the same as (a), except for July. The dotted line delineates the Tibetan Plateau. Units: meters (Yeh, Gao *et al.*, 1979).

heat low over Tibet at low level (Fig. 5.15b). This heat low is also independent of the heat low or monsoon trough over India on the long-term mean 600 hPa chart. There are semi-permanent meso-scale features in the low pressure area whose impact is felt in the development of convective precipitating systems. The apparent seasonal variation of the height patterns at low level described above (the dominance of low-pressure conditions in summer and of high-pressure conditions in winter, respectively) can cause a monsoonal flow around the Plateau that may be considered as a self-maintaining circulation system. The layer characterized by a monsoonal wind reversal (> 120 degrees between winter and summer) is rather distinct over the Tibetan Plateau, with the thickness of the monsoon layer greater than 4 km over the main body of the Plateau (Fig. 5.16). The high elevation of the Tibetan Plateau obviously generates a great degree of baroclinicity with its surroundings, causing a vigorous circulation system to establish itself. The active layer of monsoonal flow may extend downstream far from the Plateau region (up to central China).

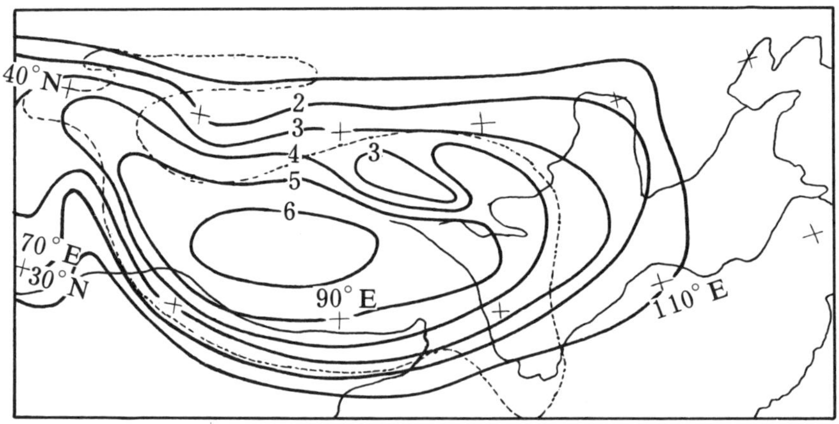

Figure 5.16. The height of the top of the monsoon layer over the Tibetan Plateau and its surroundings. Full lines indicate the depth of the monsoon layer above the terrain. Heavy line denotes the extent of the monsoon region caused by the Tibetan Plateau. Units: km (After Gao *et al.*, 1981).

Thus, the monsoon system over Tibet may also exert an important impact on the winter and summer monsoon in the Asian region. Occasionally, the two independent cold highs on the mean chart for winter and the two independent heat lows for summer may merge together, thus leading to the blowing of a northeast monsoon in winter and a southwest monsoon in summer up to the Tibetan Plateau. Under these conditions, one may observe outbreaks of cold air over Tibet in winter and vigorous transport of abundant moisture in summer. In winter, there is a belt of low pressure surrounding the surface cold high over the Plateau, whereas there is a belt of high pressure surrounding the heat low over Tibet in summer.

There are four types of weather systems at low and middle levels over the Tibetan Plateau in summer, i.e., the low-level vortex, wind shearline, moving anticyclone, and cold air outbreak. The vortex at 500 hPa is one of the sub-synoptic scale rain-bearing weather systems on the Plateau in summer. The vortex occurred mainly in the central and western Plateau during QXPMEX with a total number of 54 cases. The vortex generally moved eastward along the shearline and dissipated in the eastern Plateau and at its eastern periphery. Among them, only a few can move eastward out from the Plateau. Figure 5.17 shows the distribution of vortices during QXPMEX. One may immediately see that most vortices originated in the central Plateau. Table 5.5 is the statistics of the frequency of occurrence and eastward movement of vortices over the Plateau for May through August of 1975–1982. It is seen from Table 5.5 that there were 54 vortices during QXPMEX, out of which, only 7 moved out from the Plateau, with 5 dissipating at 102–110° E and 2 vortices moving to the east of 110° E. This is broadly similar to the mean conditions for 1975–1982, where there were 37 vortices per year, out of only 9 vortices could move out from the Plateau. Seven vortices further dissipated at 102 –110° E and 2 moved to the east of 110° E.

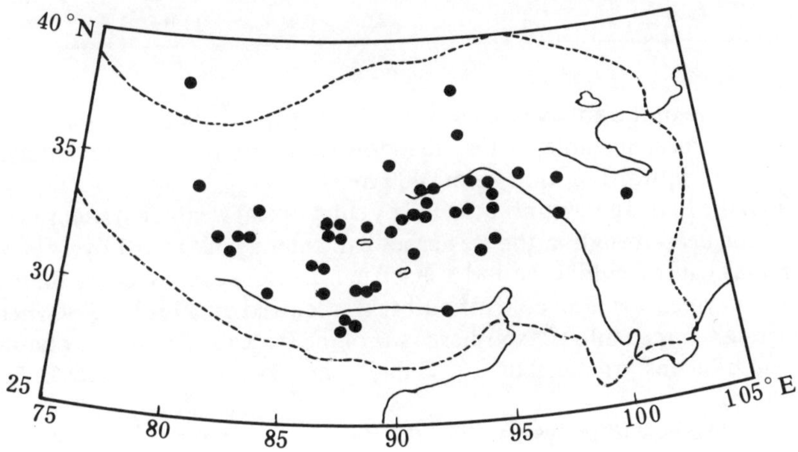

Figure 5.17. The distribution of Plateau vortices at 500 hPa for May through August of 1979 (Zhang *et al.*, 1989).

The following six synoptic factors generally contribute to the creation of vortices: large static stability near the surface, large barotropic instability, small vertical wind shear, large relative vorticity, large land-air temperature difference, and high relative humidity.

The shearline is also one of the sub-synoptic scale, rain-bearing systems over the Plateau in summer. There is no distinct discontinuity in the temperature and pressure gradient field across the wind shearline which is quite shallow, only about 2 km deep. This is mainly caused by the dynamic effect of the Plateau, although the thermal effect also plays a strengthening role based on a simple numerical simulation of the shearline over the Plateau.

Sometimes the subtropical high moves eastward from the Iranian Plateau or westward from the western Pacific over the Tibetan Plateau. This is a sort of dynamic warm high. But, owing to the intensive surface heating effect over the Plateau, the

Table 5.5. Frequency of occurrence and eastward movement of the Plateau vortices during summer (May through August) of 1975–1982 (Zhang *et al.*, 1989)

Year	Total number of occurences	Number of vortices dissipating at 102–110° E				Number of vortices dissipating to the East of 110° E			
		May	Jun.	Jul.	Aug.	May	Jun.	Jul.	Aug.
1975	42	0	3	3	0	1	0	0	0
1976	32	1	2	2	0	2	0	0	0
1977	36	1	1	1	0	2	1	0	0
1978	25	2	0	3	0	2	2	0	0
1979	54	2	2	1	0	1	1	0	0
1980	49	1	4	6	0	0	3	0	0
1981	28	4	6	3	0	0	0	0	0
1982	27	1	0	2	0	0	1	0	0
Average	37	1.5	2.3	2.6	0.4	1.0	1.0	0	0

thermal structure of the anticyclone below 2 km changes rapidly and the temperature lapse rate becomes unstable. Thus, the downward motion in the anticyclone is changed into an upward motion below 400 hPa.

Although the Plateau is mainly affected by subtropical weather systems in summer, sometimes the deep trough in the westerlies can extend southward over its southern part. This may cause a cold wave and a sharp temperature drop over the Plateau, even in summer. A typical event of cold air outbreak occurred on 30 July 1979 when a deep westerly trough dipped into the southern part of the Plateau. The 24 hour temperature drop at 500 hPa was greater than 16 °C day^{-1} and at the surface was 15 °C day^{-1} (Fig. 5.18)

The upper level weather system over the Plateau in summer is predominantly the Tibetan high that is part of the huge South Asian high. The Tibetan high is of a thermal nature, with its peak intensity, horizontal extent, and high stability at 100 hPa. It has a very great effect on the circulation conditions in the Northern Hemisphere. In general, when the Tibetan high at 200 or 100 hPa superposes over the high at the lower and middle level (500 hPa), the Plateau would have a dry season or spell, whereas there would often be a rainy season or spell when the upper-level Tibetan high lies over a low-level low. The Tibetan high is the major center of action in the Northern summer and its formation and maintenance is to a great extent related to the effect of heat sources over Tibet and its surrounding areas (Fig. 5.19)

In winter, this high is located over Southeast Asia-Indonesia-New Guinea where the greatest heat source is also observed. This situation may exist until May. In June, the center rapidly moves over the northern head of the Bay of Bengal. In July and August the high jumps farther northward over the Tibetan Plateau and its surrounding areas and then demonstrates a quasi-periodic oscillation about these positions, with a biweekly oscillation often being observed. The activity of the Tibetan high is markedly related to other synoptic systems, especially those in the westerlies. Its activity may be divided into eastern and western patterns (Fig.5.20).

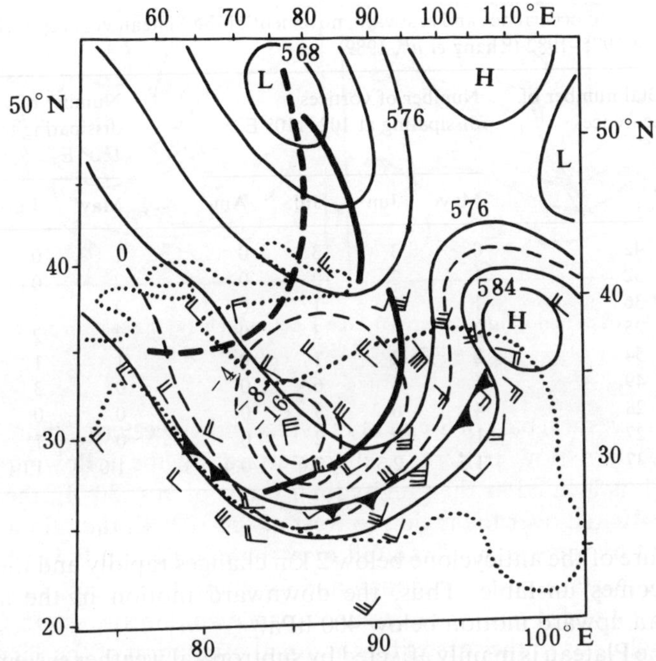

Figure 5.18. 500 hPa synoptic chart at 00:00 GMT, 30 July 1979. The heavy lines indicate the trough lines and the heavy dashed lines indicate those on the previous day. The thin solid lines are contour lines and the thin dashed lines are isopleths of 24 hour temperature variation (Zhang *et al.*, 1989).

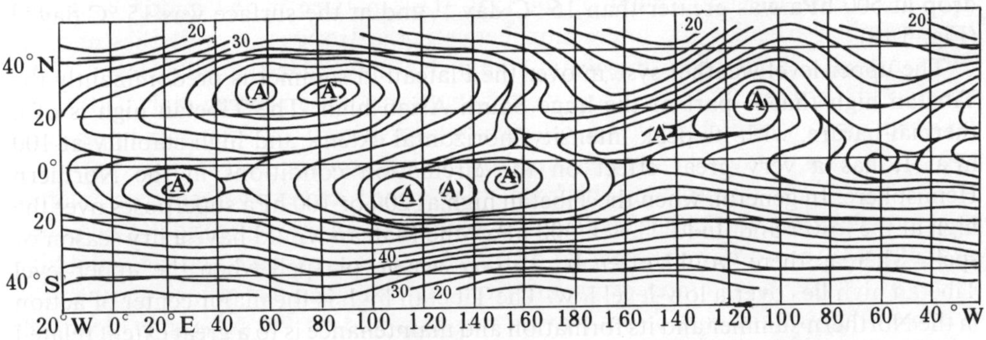

Figure 5.19. The long-term mean (1968–1980) 200 hPa streamline field for summer (June through August). Full lines denote streamlines. A: anticyclonic circulation; C: cyclonic circulation (Ding, Zhao and He, 1984).

The eastern flow pattern is characterized by the following features: the westerly trough is located at the longitudinal range of 0–60° E; the rainfall distribution is characterized by a deficit of rainfall over the middle and lower the Yangtze River Valleys, and the eastern part of Sichuan and Guizhou Provinces, and the excess of

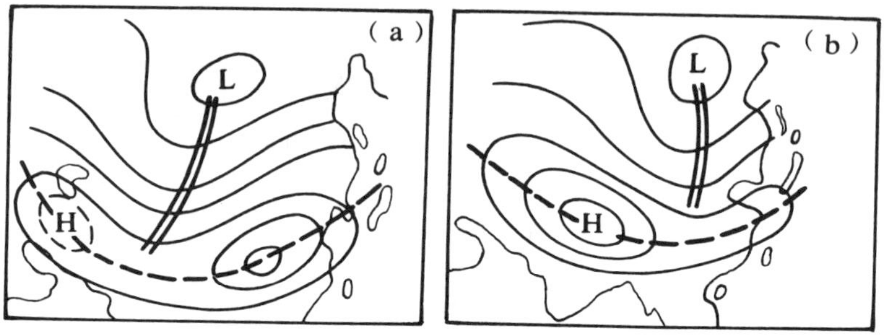

Figure 5.20. Two kinds of major flow patterns for the Tibetan high: (a) the eastern pattern; (b) the western pattern (Yeh, Gao *et al.*, 1979).

rainfall over the western part of Sichuan Province, northwestern China, and northern China. The western flow pattern is characterized by the following features: the westerly trough is located at the longitudinal range of 90–130° E; the center of the major high is situated over the region to west of the 100° E; the rainfall distribution is characterized by the excess of rainfall over the middle and lower Yangtze River Valleys and the eastern part of Sichuan and Guizhou Provinces, and the deficit of rainfall over the western part of Guizhou Province and northwestern China.

It is not yet clear as to the reason for the zonal oscillation of the Tibetan high. On the one hand, this oscillation reflects the thermal effect of the high, mainly related to the distribution of latent heat and sensible heat over Tibet and its eastern periphery. Previously, some investigators stressed the role of the release of latent heat over the Tibetan Plateau. On the other hand, the zonal oscillation of the high is believed to be associated with the adjustment of the long wave pattern in the middle and high latitudes, in particular the interaction between the troughs in westerlies and the Tibetan high. Based on a diagnostic study, one may infer the following transfer mechanism between the eastern and western patterns of the Tibetan high at 100 hPa. Upward motion and upper-level divergence (downward motion and upper-level convergence) is produced and maintained by the atmospheric heat source (heat sink). At the top of the upward and divergent (downward and convergent) air column, negative (positive) vorticity is created, thus leading to an intensification (weakening) of the high. According to the actual distribution of heat source and sink, the high center at 100 hPa would move westward from the east when the surface sensible heat flux over the western part of the Plateau increases and, at the same time, decreases over the eastern part of the Plateau. In this case, the eastern pattern is changed into western pattern. On the contrary, when the opposite heating condition is true, especially when strong condensation heating exists over eastern Tibet, the center of the high at 100 hPa accordingly moves eastward from the west, thus causing a transformation from the western pattern into the eastern pattern.

Starting from September, the Tibetan high begins to retreat southward and moves back to Southeast Asia in October and November. According to recent studies (Murakami and Ding, 1982; Yanai, 1987), the seasonal advance and retreat of the Tibetan high has a more complicated behavior. From both the climatological mean and case study (Ding *et al.*, 1988), it has been found that, while the Tibetan high advances northward, there is a center of high moving over Iran-Afghanistan and the western part of Tibet from the Arabian Peninsula and the latter has a greater intensity and a farther northern position than the eastern one. On the long-term 200 hPa flow charts in summer, there exist actually two huge highs over Asia, with the high center at 85–95° E being the stronger. This result, on the one hand, indicates the fact that the Tibetan high should be viewed as one cell out of the upper tropospheric belt of anticyclonic circulation. This cell is regionally intensified due to the effect of the Plateau. On the other hand, the thermal effect of the Tibetan Plateau on the formation of the Tibetan high should not be overemphasized. At present, it is not clear what causes the formation of the high center over the western part of the Iranian Plateau. This is an important problem to be further studied.

The meridional circulation over the Tibetan Plateau and its surrounding areas is closely associated with the characteristics of the distribution of the heat source and sink over the Plateau and continental region. Figure 5.21 is the cross-sections of the mean meridional circulation along 90° E, showing the conditions across the Tibetan Plateau (Yeh, Gao, *et al.*, 1979). It may be seen that the Hadley cell is very vigorous in winter, extending northward all the way to about 30° N. This indicates that the atmosphere over the Tibetan Plateau acts as a cold source in winter. Further, this heat sink is found in the middle troposphere, thus greatly enhancing the intensity of the Hadley circulation in the monsoon region. As a result, one may observe a strong Hadley meridional circulation at the longitudinal range of the Tibetan Plateau. Eastward from this region, the Hadley circulation gradually weakens and is not detectable in the central Pacific. Therefore, the mean Hadley circulation in the Northern winter is likely to be mainly a phenomenon over the continent.

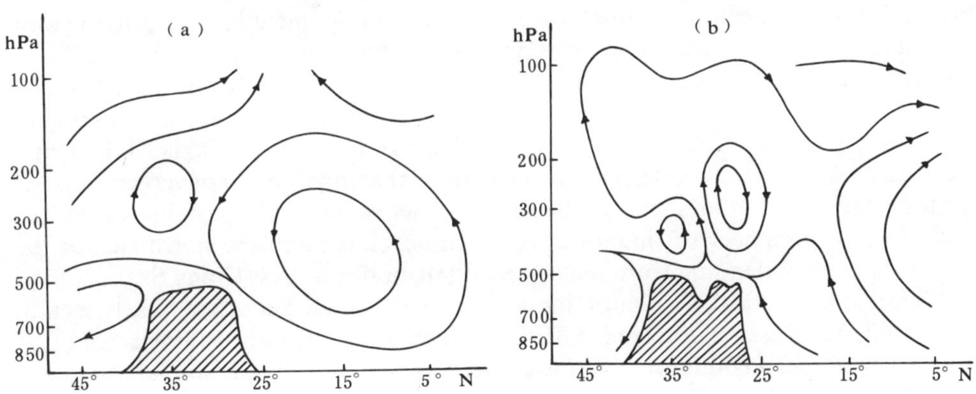

Figure 5.21. Cross-sections of the mean meridional circulation along 90° E: (a) for January; (b) for July (Yeh, Gao *et al.*, 1979).

In summer, the Tibetan Plateau is a huge heat source. The upward motion predominates in the atmosphere over the Plateau (Fig. 5.22). Mean upward motion associated with the low-level southwesterlies covers a large domain to the east of 65°

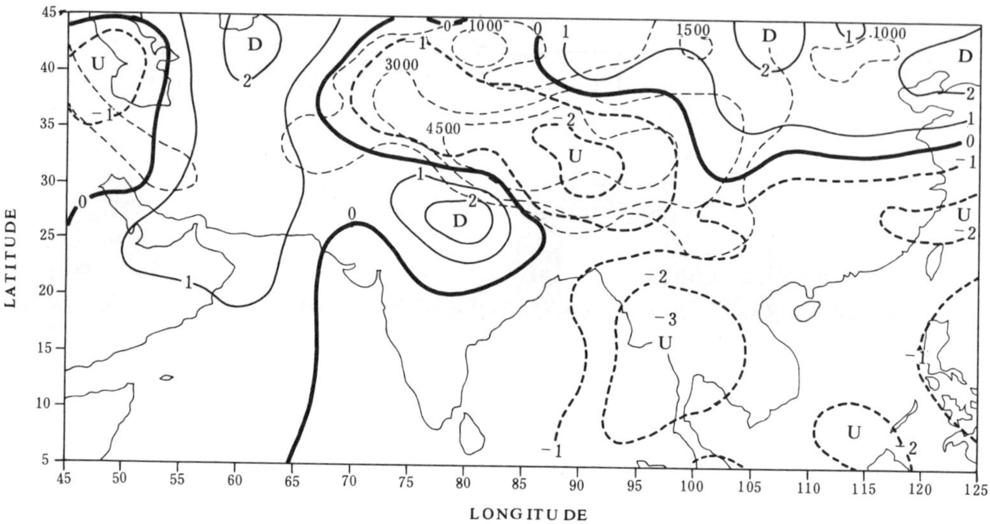

Figure 5.22. The 80-day mean vertical *P*-velocity (hPa/h) at 400 hPa for the period from mid-April to early July 1979 (He, McGinnis, Song and Yanai, 1987).

E and south of 35° N, except North India, but including the eastern half of the Arabian Sea, the Bay of Bengal, Southeast India, the southern China plain, and the South and East China Seas. The maximum upward velocities (-2 to -3 hPa h^{-1}) are located over the Burma-Thailand region and along a belt extending from the southern China plain to the East China Sea.

The ascending motion with its maximum (-2 hPa h^{-1}) over the central Tibetan Plateau, and the descending motions over its surroundings including areas to the west of the Plateau, North India, and the northeastern China Plain, are a manifestation of the vertical circulation induced by the thermal and mechanical effects of the Plateau. It is noted that the areas of descent correspond remarkably well to the desert areas of Turkestan, Iran, and Saudi Arabia, and the Great Indian (Thar), Takla Makan, and Gobi deserts. The areas of descent correspond to the upper-level convergence to the west of the speed minimum in the westerly jet stream, the confluence at the northwestern periphery of the South Asian anticyclone, and the north side of the westerly jet axis extending from the eastern Plateau to the East China Sea.

The summer vertical circulation induced by the Tibetan Plateau is clearly seen in the meridional cross-section (Fig. 5.21 (b)), which is significantly different from that for the winter mean condition. On the whole, a huge monsoon meridional cell exists, with the upward flow of its northern leg extending up to 40–50° N. The descending motion is predominantly observed in the Southern Hemisphere. Also, it may be seen that there are two smaller meridional cells on the northern and southern slopes of the Plateau, respectively. The northern slope only reaches 300 hPa in the vertical extent while the southern one extends upward up to over 200 hPa. Note that these two meridional cells only show up at the longitudinal range of the Tibetan Plateau, indicating that they result from the heating effect of the Plateau. The above condition

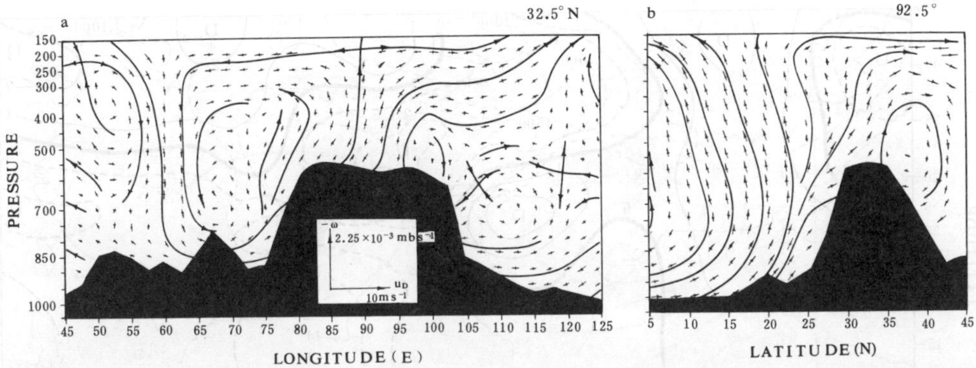

Figure 5.23. (a) The east-west vertical cross section showing the 80-day mean vectors (u_D,ω) along 32.5° N (from 16 April to 4 July 1979), and (b) the north-south vertical cross-section showing the 80-day mean vectors (V_D, ω) along 92.5° E. V_D (u_D, V_D) is the divergent part of wind vectors (He, McGinnis, Song and Yanai, 1987).

is also true for individual years. For instance, a similar meridional circulation was observed for the period before and after the onset of the summer monsoon of 1979 (from May to the early part of July). The east-west and north-south vertical cross-sections of the 80-day mean meridional circulation (Figs. 5.23 a, b) show that the intense ascending motion over the Plateau is accompanied by subsidence over its vicinity, i.e., Iran and Afghanistan (55–65° E) and the eastern periphery of the Plateau and the China Plain. The distribution of the vertical circulation in this plane is similar to that of the daily mean vertical motions along 35° N for July simulated by Kuo and Qian (1981).

The meridional cross section along 92.5° E (Fig. 5.23b) shows that the intense ascending motion over the Plateau is accompanied by subsidence over the Tarim Basin (40–50° N). There is also a narrow zone of weak ascent between the stronger ascending motion over the Plateau and the vigorous ascent over Bangladesh (~20° N) which is associated with the southwesterly monsoon. These are similar to the distribution of the mean July vertical circulation along 90° E presented by Yeh, Gao *et al.* (1979) (Fig. 5.21b) and simulated by Kuo and Qian (1981).

As indicated above, the maximum sensible heating occurs during the seasonal transition period from spring to summer (April to June). Affected by the heat source, the atmosphere over the Plateau would warm up rapidly, so that the temperature gradient on both sides of the Plateau can be changed. Correspondingly, the flow pattern around the Plateau would also be changed. Figure 5.24 presents the long-term mean monthly variation of the 300–500 hPa thickness field along 30° N. It may be seen that the atmosphere over the Plateau in the mid-summer has abnomally high temperatures, with the warmest region in July and August found in the region of the longitudinal range of 50–110° E, including the Tibetan and Iranian Plateaus. During the transition season from spring to summer, the warming in this region also occurs earlier than other zones of the same latitude. In March, the increase in thickness is very evident and attains its maximum in May and June. This drastic change in temperature during the transition season should cause a corresponding change in the wind field.

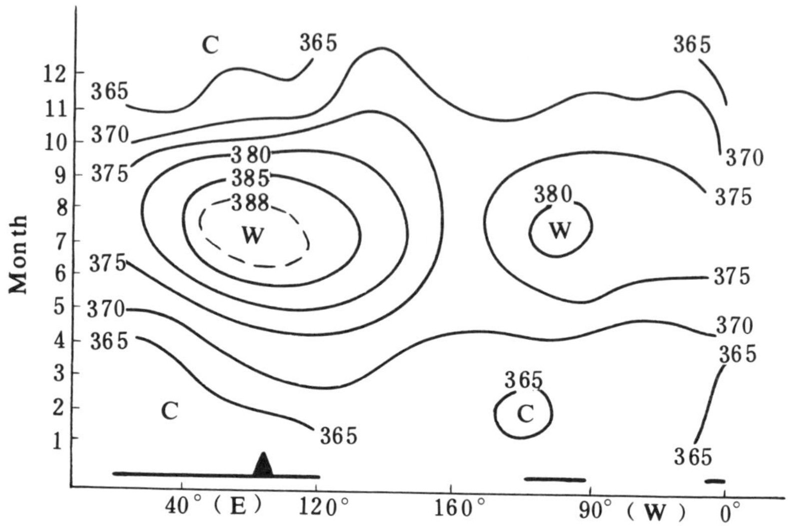

Figure 5.24. Long-term mean monthly variation in thickness between 500 and 300 hPa along 30° N. Units: decameters (Yeh, Gao *et al.*, 1979).

Figure 5.25 further illustrates the changes in the tropospheric temperature based on the data of 1979. A sudden temperature increase over the eastern Plateau and the central China Plain (85–115° E) occurred during the period from 6 May to 15 May (first transition). From 11 May to 4 June, there was a remarkable temperature contrast between the eastern and western longitudes bordering at 85° E. The 200–500 hPa layer mean temperature over Irag, Iran, Afghanistan, and the western Tibetan Plateau (40–85° E) increases rapidly during the period from 5 June to 29 June (second transition) and afterwards with the weakening of the cold temperature trough over the western Plateau (Fig. 5.25 b, c). After the second transition, the upper tropospheric temperature over the eastern Plateau and the China Plain also increases.

The drastic changes of the mean tropospheric temperature along 32.5° N are contrasted with the much weaker temperature changes along 15° N shown in Figure 5.25 b. The warm temperature over Southeast Asia (east of 95° E) does not show appreciable time change. Figure 5.25 c shows the longitude-time section of the mean meridional gradient of the 200–500 hPa layer mean temperature between 5 and 25° N. The reversal of the meridional temperature gradient first occurs over the longitudes east of 85° E and then over the longitudes west of 85° E. The two stages of the onset of the reversal of the temperature gradient coincide with the two stages of the onset of the low-level southwesterlies and organized rains over the Bay of Bengal and the Arabian Sea. The dominant role played by the temperature increases over the land areas including the Plateau in this reversal is evident from Figure 5.25 a, b.

Yin (1949) earlier pointed out that the westerly wind at the upper level to the south of the Tibetan Plateau assumes the feature of sudden change. In the early part of June, the upper-level westerly wind drastically decreases. This decrease is related to the

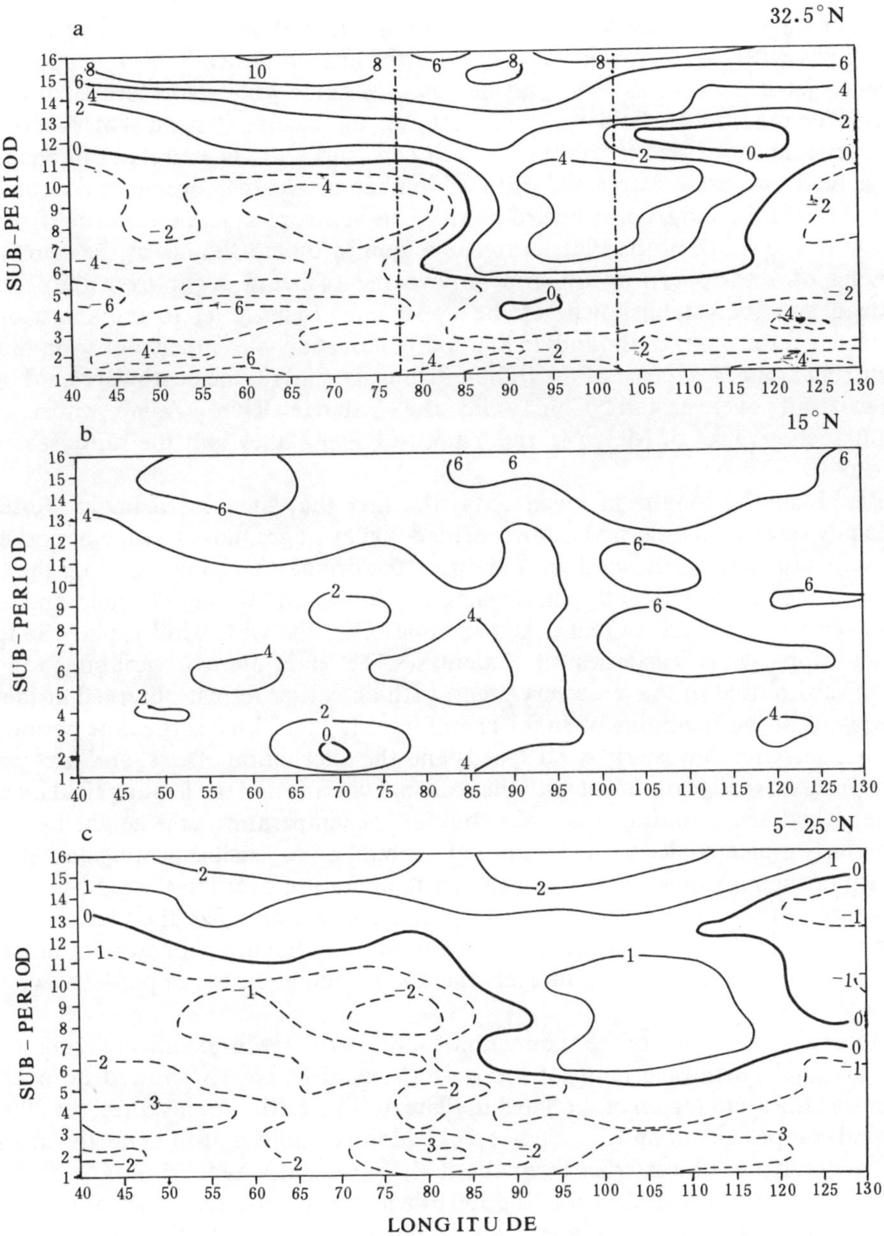

Figure 5.25. Longitude-time sections for (a) the 5-day mean deviation of the 200–500 hPa layer-mean temperature (K) from the time and domain average at 32.5° N, (b) as in (a), but at 15° N and (c) the 5-day mean meridional gradient of 200–500 hPa layer mean temperature (10^{-6} K km^{-1}) between 5 and 25° N. The vertical dashed-dotted lines in (a) indicate the boundaries of the Tibetan Plateau (>3000 m) (He, McGinnis, Song and Yanai, 1987).

onset of the Indian southwesterly monsoon. Later, Yeh *et al.* (1958) found that the above-described sudden decrease and northward withdrawal of the upper-level westerly jet-stream in East Asia may be observed in the whole Northern Hemisphere, but this change in wind field occurs earliest over the Tibetan Plateau. This fact reflects the particularity of the Plateau region. Yeh *et al.* further indicated that this change in wind field is closely associated with a number of weather events and processes, including (1) the sudden northward jump of the subtropical jet in westerlies from the region to the south of the Plateau to the region to the north and, at the same time, moving of the upper-tropospheric anticyclone (Tibetan high) over the Tibetan Plateau and the establishment of the upper-level easterly jet to the south of the Plateau; (2) the onset of the Indian southwest monsoon with much monsoon rainfall occurring along west coast of the Indian Peninsula; and (3) the northward shift of the polar front over eastern China and the Yangtze River Valley, causing the simultaneous onset of Meiyu in the Yangtze River Valley and the rainy season in Japan.

Based on the long-term mean data, the fact that the jet stream in westerlies suddenly retreats, has been also documented. On average, June is a time period when the subtropical westerly wind jet undergoes the dramatic change. During the third pentad, the jet significantly diminishes to the south of the Plateau, and soon disappears (northward retreat). At the same time, the west wind rapidly shifts to steady upper-level easterlies at Calcutta. The inter-annual variability in the northward retreat of the jet is very great, with an earlier retreat observed at the end of May or at the beginning of June. Therefore, May and June is the time period of a sudden decrease in westerly wind speeds and the alternation of west and east winds. This process is fully consistent with the seasonal variation of the heating field over the Plateau and the resulting seasonal evolution of temperature and height fields. As previously indicated, May is the month of the maximum sensible heating and also the month of the maximum increase in the air temperature over the Plateau. Under the influence of this very heating, the atmospheric temperature over the Plateau sharply increases, leading to a weakening of the normal meridional temperature gradient or in the reverse direction, and, further, the above change in the upper-level westerly wind jet.

According to a study of the summer monsoon of 1979 (Murakami and Ding, 1982), this seasonal northward jump of the westerly wind jet clearly showed up over the extensive Eurasian region of the Tibetan Plateau (Fig. 5.26). For instance, the 300 hPa U wind component along 75° E (Fig. 5.26, middle) exhibits a distinct northward shift of strong (20 m s^{-1}) westerlies from about 30° N during May to 35–40° N after mid-June. A similar northward shift of the 300 hPa jet stream also took place far upstream (55° E) from the Tibetan Plateau on about 3 June (Fig. 5.26, left). Yin (1948) postulated that the onset of the Indian monsoon occurs almost simultaneously with a sudden shift in the location of the upper-tropospheric subtropical jet stream from the southern to northern periphery of the Tibetan Plateau. However, this is not evident in their study as the latitude-time section along 95° E (eastern Tibetan Plateau) does not exhibit a similar northward shift of the 300 hPa U jet stream (Fig. 5.26, right). This indicates the possible existence of a significant inter-annual

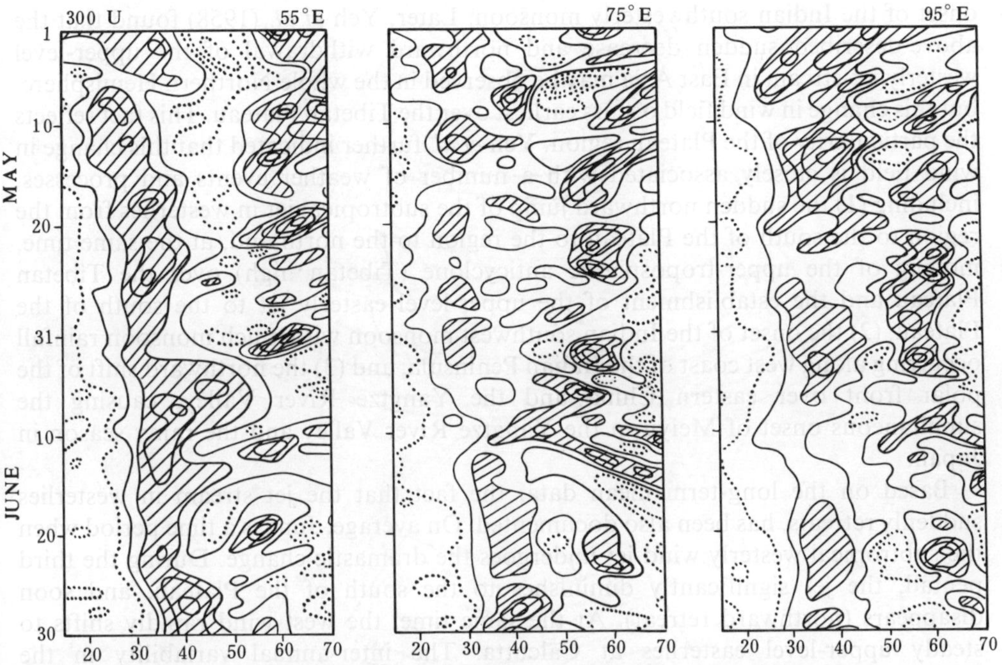

Figure 5.26. Latitude-time sections of 300 hPa U, averaged over a region extending 10 degree of longitude east and west of 55° E (left), 75° E (middle) and 95° E (right),respectively, during the early summer (1 May to 30 June) of 1979. Interval are 10 m s^{-1} with full and dashed lines representing positive, and zero and negative wind, respectively. Hatching denotes regions of greater than 20 m s^{-1} westerlies. Dashed hatching denotes regions of greater than 10 m s^{-1} easterlies (Murakami and Ding, 1982).

variability of the seasonal change, which is a problem to be further studied. Other problems associated with the seasonal change in monsoon activity will be dealt with in Sections 5.5 and 5.6.

In order to study the role played by the heating field over the Plateau and its surrounding areas in the build-up of mean circulation in winter and summer, a number of numerical simulations on this problem have been done by using 5-level p-σ incorporated coordinates (Qian *et al.* 1988). Major achievements may be summarized as follows:

(1) Figure 5.27 a is the 100 hPa weather map for July simulated only with the inclusion of the large-scale topographic effect in the model. There is no Tibetan high detectable on it. However, in Figure 5.27 b which represents the simulation with the combined effect of thermal and dynamic effects taken into consideration, one may clearly see a closed huge Tibetan high at 100 hPa. This result implies the dominating effect of the thermal factor of the Plateau.

(2) As demonstraed in (1), the contribution of the summer heating field to the horizontal flow pattern seems to more significant in the upper troposphere than in the middle and lower troposphere. If the heating effect is not included in summer, the huge anticyclone in the upper troposphere cannot be simulated. This may result from

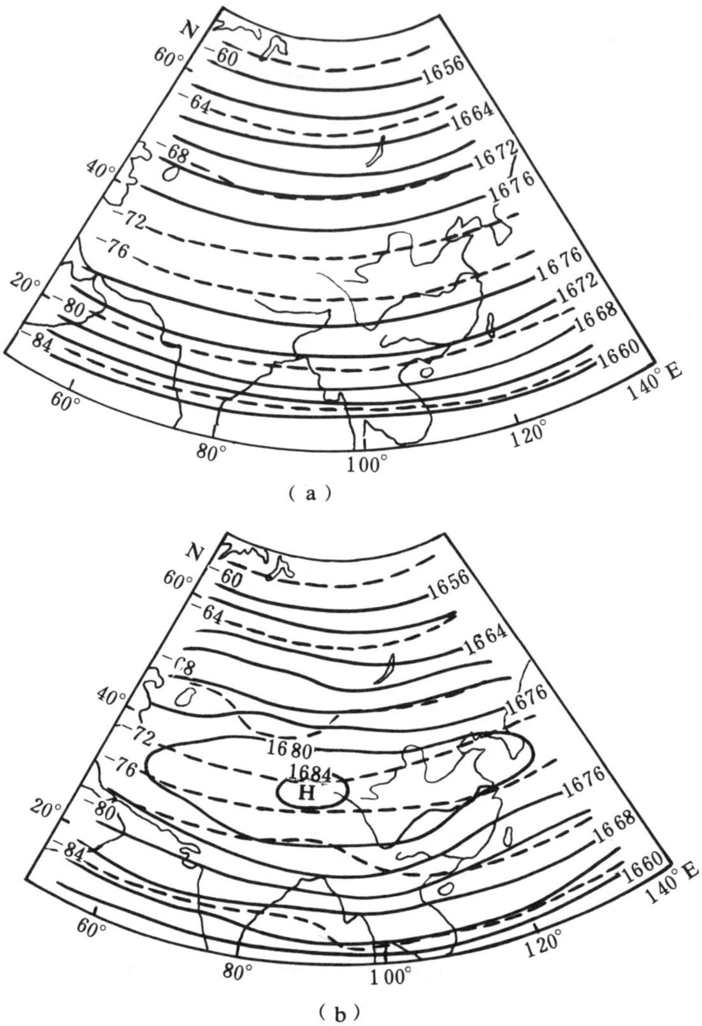

Figure 5.27. (a) The 100 hPa weather map for July simulated only taking into consideration the large-scale topographic effect in the model with the integration of 120 h; (b) the same as Figure 5.27 (a), except for the case of the inclusion of combined dynamic and thermal effects. Units: decameter (Qian *et al.*, 1988).

the fact that the vertical flux of the latent and sensible heat may reach a very high level. However, the case for winter is opposite, with the effect of the heating field (heat sink) being more significant in the middle and lower troposphere, i.e., the heat sink over the Plateau causes the deepening of a major trough off the coast of East Asia and a weakening of the topographic ridge over the Plateau.

(3) Figures 5.28a, b are sea-level weather maps simulated by including the dynamic effect and the combined dynamic and thermal effects, respectively. The effect of heating in winter is to strengthen the surface cold high, especially in the region to the

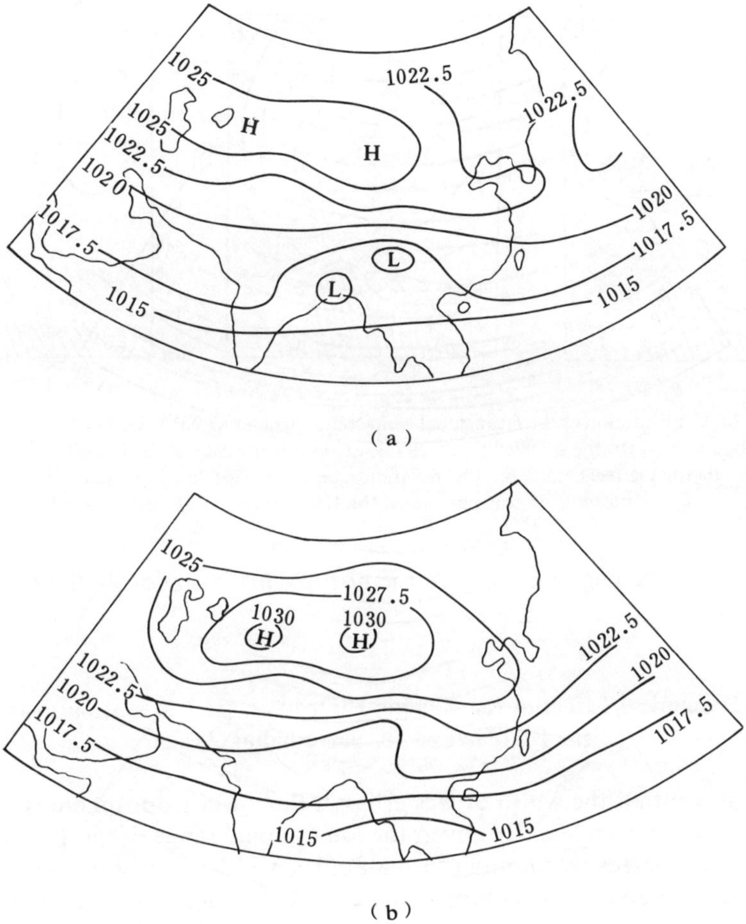

Figure 5.28. (a) January sea-level weather map simulated by including topographic effect in the model. Integration is performed for 120 hr; (b) the same as Figure 5.28 (a), except for the case of combined dynamic and thermal effects (Qian *et al.*, 1988).

east of the Plateau. This apparently results from outbreaks of cold high along the southeastward paths after the cold high intensifies.

(4) Figures 5.29a, b show the simulated cross-sections of the January mean zonal wind component along 90° E for the case with only the large-scale topography included and the topographic and thermal effects included together, respectively. Figure 5.29 (c) is the difference between Figure 5.29 (a) and Figure 5.29 (b), representing the thermal effect of heat sink over the Plateau. Results have shown that although the heating field is not so important as described above, it may influence the intensity and position of the jet stream in westerlies. It can be seen that in the later case the southern branch of the jet is stronger, which is basically consistent with observations. From Figure 5.29c, one may observe the positive value to the south of

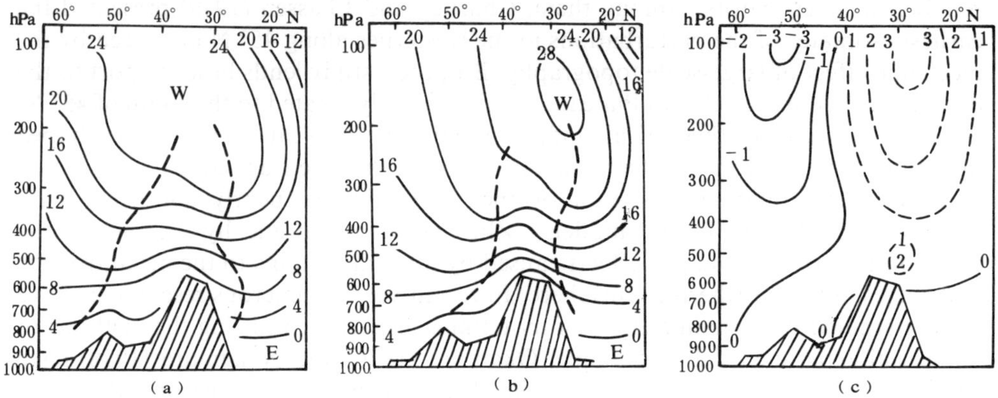

Figure 5.29. (a) Cross-section of the mean zonal wind component along 90° E for January considering only large-scale topography, (b) the same as Fig. 5.29 (a), except for the case of the inclusion of the combined dynamic and thermal effects, and (c) the difference between Figs. 5.29 (a) and (b). Units: m s^{-1}. Intergration was performed for 120 hr (Qian *et al.*, 1988).

the Plateau and the negative value to the north. This is obviously the result of heat sink over Tibet.

5.4. The Dynamical Effect of the Tibetan Plateau on the Circulation Features Over the Plateau and Its Surrounding Areas

It is well known that the width of westerlies in the lower troposphere is only 30–40 latitude degrees or even less. However, the longitudinal range of the Tibetan Plateau has 15 latitude degrees, accounting for half of the width of the westerlies. The zonal extent of the Tibetan Plateau is about 3000 km, accounting for one tenth of the entire latitude circle; vertically, the average height of the Tibetan Plateau is about 4 km, nearly one half to one-third of the tropospheric layer. It is very natural to believe that such a huge obstacle is able to exert a significant dynamic effect on the westerlies.

The large-scale dynamic effect of the Tibetan Plateau is another important problem associated with the topographic effect of the Plateau. It is related to the formation of the jet stream in westerlies in East Asia, the establishment of the major trough in East Asia, and so on. The dynamic effect of the large-scale topography includes the following two aspects: (1) the mechanic barrier effect that can force the airflows to cross over or go around the Tibetan Plateau. This forced flow, either going over or around, is not only associated with the shape of the large-scale topography, but also associated with the strength of the airflows that cross over the mountains. Therefore, although the mountain ranges are fixed and unchanged as geographical features, the dynamic effect of the mountains on airflows assume seasonal variations. (2) The inhomogeneous distribution of friction caused by the large-scale topography. In this section, only (1) will be discussed.

The dynamic effect of the plateau and mountain ranges have been studied by a number of investigators. Among them, Charney and Eliassen (1949) presented the earliest estimates of the perturbations in the westerlies along 45° N produced by the mechanic effect of large-scale topography. But, their study lends little support to the dynamic effect of the Tibetan Plateau, as this is mainly located to the south of 45° N. Bolin (1950) studied a theoretical model of the flow pattern caused by a circle-shaped mountain in the uniform westerly winds and found that the ridge over the mountain and the first trough downstream of the mountain are nearly consistent with observations. Using linear models, Zhu (1957) theoretically discussed the contributions of the dynamic effects caused by huge mountains to the mean troughs and ridges in the Northern Hemisphere. In recent years, numerous researchers have further studied the dynamic effect of mountains, especially utilizing the methods of numerical experiments. For example, Okamura (1976) and Nakamura (1978) have studied the dynamic effects of various ideal terrains in different seasons. Recently, Chinese meteorologists have studied the effects of the Tibetan Plateau on various barotropic zonal flows with annulus experiments (Yeh, Gao *et al.*, 1979).

When the airflows impinge on mountains, they may go over or go around them. These two dynamic effects are different. Based on the studies of numerical experiments, the purely dynamic effect of the Tibetan Plateau in summer shows up mainly in a splitting of the airflows while going around the Plateau. Many circulation features are associated with the effect of airflows going around the Plateau, for instance, the mean flow pattern of two troughs and one ridge in between to the north of 35° N, the intensity and location of major trough in East Asia, low-pressure systems, and the belt of high pressure to the north (Fig. 5.30a). The effect of sloping and crossing of airflows over the mountain is of secondary importance, but it may make a greater contribution to the breaking up of the subtropical high. In winter (Fig. 5.30b), in the latitude belt to the south of 40° N where highs exist, the flow patterns are considerably distorted due to the influence of the topography. Over the main body of the Plateau, the flow pattern assumes a significant anticyclonic curvature (ridge), thus creating trough zones. The trough at the eastern periphery of the Plateau is of a travelling character, and this movement is related to the new development of a low trough in the northwesterlies to the northeast of the Plateau. This kind of process is somewhat similar to the developmental process of the major trough in East Asia. The trough in the southern branch of the westerlies is numerically well simulated and its genesis and amplification mainly come from the contribution of the effect of flow around the Plateau, while the anticyclonic circulation feature in the western part of the Plateau seems to result from the nonlinear interaction between the air flows going over and around the Plateau. Thus, either sloping effect or flow around the Plateau cannot appropriately reflect the influence of the topography on the circulation features.

The dynamic forcing can also influence the vertical structure and wavelengths of the disturbance in various basic zonal currents (Chen, Ji and Shen, 1985). Chen, Ji and Shen found that, in a baroclinic atmosphere, the dynamic disturbances in different basic westerly currents caused by the Tibetan Plateau bear a westward tilting structure associated with a well-developed vertical circulation. The strength weakens rapidly with height. The wavelength at mid-latitudes is about 70–90° longitudes, which is longer than that of the corresponding quasi-stationary free wave.

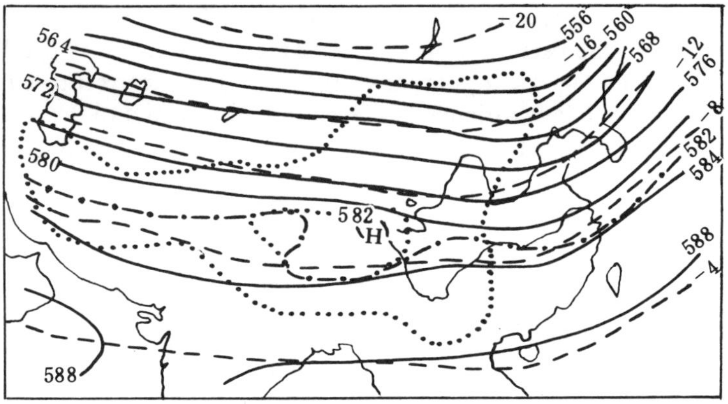

(a)

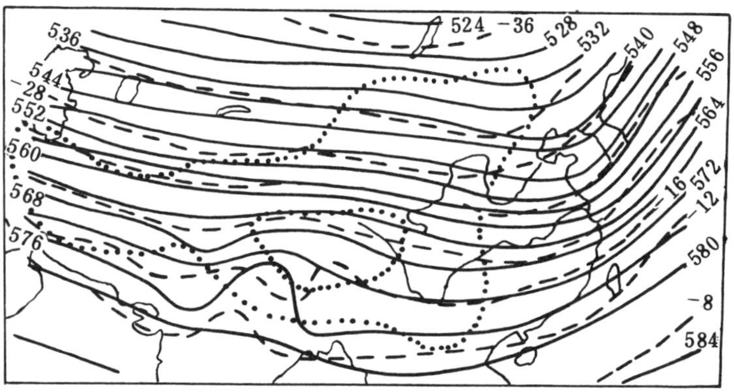

(b)

Figure 5.30. (a) The 500 hPa weather chart in summer simulated only with the inclusion of the dynamic effects of the Plateau; (b) the same as Figure 5.30 (a), but for winter. Integration was performed for three days. Solid lines indicate geopotential height. Units: decameters. Dashed lines denote isotherms. Units: °C (Qian *et al.*, 1988).

The disturbed energy propagates mainly downstream, but its amplitude decays rapidly. So, its influence is largely restricted to the west of 140° E.

Finally, a comparison of the difference in the dynamic effect of large-scale topography between winter and summer may be made as follows:

(1) The dynamic effect of the topography of the Plateau for winter and summer is different. In summer, the dynamic effect plays an important role in the maintenance of the belt of high pressure to the north of the Plateau. The belt of low pressure over the Plateau seems to be associated with the thermal effect. In winter, however, the high pressure ridge and thermal ridge over Lahsa are associated with the dynamic effect. In comparison with the climatological mean condition, the simulations for winter are more realistic than those for summer, possibly indicating that the dynamic

effect of the Plateau in winter may be more important, whereas the influence of the Plateau in summer may possibly show up in the thermal effect; (2) owing to the fact that the difference in numerical simulations between winter and summer results from different zonal mean fields at initial state, one may infer that some sort of interaction exists between the purely dynamic effect of the Plateau and the circulation conditions; (3) in summer, the dynamic effect of the Plateau is largely reflected in the flowing of airstreams around it while in winter the flows both over and around the Plateau are rather important. When a weak basic current impinges upon the Plateau, the flow over it would dominate, whereas the flow around it is very important when the aircurrents are strong; (4) the dynamic effect of the Plateau for both winter and summer has the most significant influence in the lower and middle troposphere. But this dynamic influence thus produced may propagate upward to a very high level.

In annulus experiments, the dynamic effect of large-scale topography of the Plateau has been well simulated. For the case of the westerly wind airflow field, it has been found that the Tibetan Plateau has an obvious splitting effect on the westerlies in the lower layer (equivalent to 850 hPa), with the splitting point located at the western flank of the Plateau and the confluent point in the northeast. This effect decays rapidly with height and it cannot be detected in the middle layer. In addition, a weak trough is formed upstream and downstream of the Plateau, respectively, that is similar to the situation shown in Figure 5.30b. The airflow speed in this layer begins to decrease, even sometimes forming a blocking situation. These results are in many ways similar to those obtained with numerical simulations.

5.5. Comparison of the Topographic Influences between the Tibetan Plateau and the Rocky Mountains

The Rocky Mountains, as the major mountain range in North America, can exert an important effect on weather and climate over North America and its contiguous areas. The problem of the Rocky Mountains' lee cyclogenesis has been of particular interest in past studies. Due mainly to the differences in terrain features, the topographic influences of the Rocky Mountains and the Tibetan Plateau on circulation patterns and synoptic disturbances are strikingly different. Five significant geographical differences should be mentioned:

(1) The smoothed topographic heights over the Rockies extend slightly above 850 hPa, while a substantial part of the Tibetan Plateau projects upward to almost the 500 hPa level.

(2) The north-south extension of the Rockies (10–70° N) is much larger than that of the Tibetan Plateau (30–45° N).

(3) Conversely, the east-west extension of the Rockies is less than 20 degrees, while the high (>1.5 km) Tibetan Plateau covers approximately 50 degrees of longitude.

(4) The U.S. east coast is nearly 4000 km downstream from the highest elevations of the Rocky Mountains, which is contrasted with a distance of 2000 km or less from the Tibetan Plateau to the East China Sea.

(5) The warm, moist Pacific Ocean lies upstream (west) of the Rockies, while the cold, relatively dry European land mass is found to the west of the Tibetan Plateau.

These large geographical differences may result in substantial differences between the mean circulations and synoptic processes over and around the North American and Asiatic continents. It is necessary to briefly describe the influence of the Rocky Mountains on the circulation patterns and synoptic systems in North America before a comparison can be made. The following discussion will be mainly devoted to the winter conditions.

Figure 5.31a, b are the winter mean fields at 700 and 200 hPa for 1979. One may note that the prevailing westerlies undergo substantial deceleration when approaching the windward edge of the Rocky Mountains. Perhaps, this speed convergence reflects the orographically forced ascending of these westerlies, namely, a substantial part of the westerly component appears to flow over the mountain barrier, although also exhibiting a tendency to flow around the American Rockies.

This is in sharp contrast to what is observed near the Tibetan Plateau where the low-level westerlies distinctly split into northern and southern branches and flow around, rather than up and over these barriers. Presumably, these differing patterns of airflow may be due to the smaller horizontal extent and lower topographic profile of the Rocky Mountains as compared to the Tibetan massif. Due to upstream frictional dissipation within the planetary boundary layer over Europe, low-level winds near the western edge of the Tibetan Plateau have speeds of less than 5 m s^{-1} and result in relatively weak upward motions over that region.

Significant differences also appear to exist in the frictional effects of these elevated

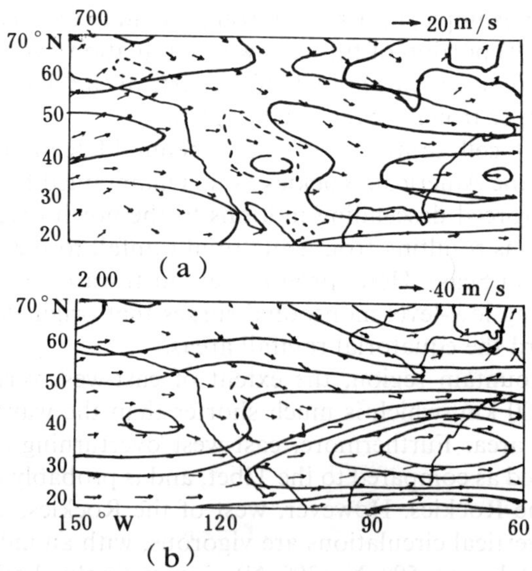

Figure 5.31. Winter mean vectors (arrows) at 700 hPa (tops; units: 20m s^{-1}) and 200 hPa (bottom; units: 40 m s^{-1}). Isotachs (solid lines) are shown at 5 m s^{-1} intervals. Dashed lines are 1500 m smoothed topographic height contours (Lyons and Murakami, 1981).

regions. The Rockies exhibit a limited zone of weak winds and weak, frictionally induced vorticity pairs at 700 hPa. In contrast, frictional effects over the Tibetan Plateau are much more pronounced, with weak winds up to the 500 hPa level, and pairs of well-organized, frictionally induced, vorticity cells both upstream and downstream at 700 hPa.

Interestingly, jet streams at 200–300 hPa with pronounced eastward accelerations occur downstream and to the south of both the Rockies and the Tibetan Plateau along 25° N. Absolute potential vorticity conserving processes, which are related to orographically forced upward and downward motions, do not appear to be a major factor in producing these characteristic features of winter mean flows near 25° N. Rather, the eastward acceleration of the jet streams seem to be primarily accomplished by upper-level southerly wind (upper-level branch of a local Hadley cell), which are enhanced by heat contrasts between cold continents (Asia and North America) and the warm equatorial regions (Malaysia and Indonesia, and the equatorial eastern Pacific and Brazil).

North-south and east-west vertical overturnings also differ over these two regions. Over Asia, orographically forced ascending (descending) motions are observed over the Kazakh (northern China) region, with a corresponding indirect (direct) north-south vertical circulation to the west (east) of the Tibetan Plateau. Associated with these orographically forced circulation is a pronounced east-west vertical overturning along about 45° N, with ascending (descending) motions near the western (eastern) edge of the Tibetan Plateau. This east-west vertical circulation is comparable in size to the wavelength of transient disturbances in the area. Thus, these winter mean vertical circulation features strongly influence transient eddy activity around the Tibetan Plateau. For example, upper-level troughs tend to dissipate as they approach the updraft (divergent) portion of this east-west circulation over the Kazakh region, while the downdraft (convergent) branch of the vertical circulation contributes to the intensification of upper troughs over northern China.

A vigorous direct cell southeast of the Tibetan Plateau resembles the direct circulation south of the American Rockies as it is maintained by radiational cooling and orographically forced descending motions to the north (over northern China), and ascending motions resulting from equatorial rainfall to the south (the southern China and Indonesian Seas). Here, pronounced northward sensible heat fluxes are associated with strong low-level monsoonal surges that originate from the Siberian high and flow toward the equatorial rainfall areas.

In the Rocky Mountain region, the extent of east-west vertical overturning is limited to about 1000 km, which is much shorter than the wavelength of transient disturbances in the area. Furthermore, east-west overturning over the Rockies is much less pronounced as compared to the Tibet, and is probably due to the relatively low elevations of the Rockies. However, west of the Rockies, along the U.S. west coast, north-south vertical circulations are vigorous, with an indirect cell having an updraft (downdraft) leg at 50° N (30° N). Interestingly, both the updraft and downdraft zones are not directly due to orographic forcing, as the updraft branch near 50° N is the result of frictional convergence along the coast and sensible heat flux convergence due to disturbances. South of 30° N, a direct cell is defined by ascending motions due to equatorial rainfall, and dynamically controlled descending air over the subtropical Pacific high pressure area.

Orographically forced sinking motions near the southeast border of the American Rockies (35° N, 105° W) contribute to the enhancement of a direct circulation south of the Rockies. However, since this direct cell weakens rapidly east of about 90° W, north-south vertical circulations near the U.S. east coast are not well organized. Northward sensible heat transport associated with transient eddies is much weaker over the U.S. east coast than over the Asiatic east coast. Thus, cold surges over the east coast of the U.S. dissipate south of 30° N (downward leg). Conversely, significant northward sensible heat fluxes from the northeastern Pacific cross over the Rockies into the central northern U.S. and Canada. This partially offsets the continental radiational cooling effect, which results in much weaker continental polar anticyclogenesis as compared to the Siberian high.

The orographical influence of the Rockies and the Tibetan Plateau on the winter mean temperature fields is quite different. Along any given latitude, 700 hPa winter mean temperature fields over the central American Rockies are lower than in surrounding areas. This is in sharp contrast to the Tibetan Plateau region, where the mean winter temperature over these mountains is higher than the surrounding air temperature (Yeh, Gao *et al.*, 1979; Murakami, 1981). Yeh, Gao *et al.* (1979) observed that the Tibetan Plateau is relatively snow free, with a measured sensible heat flux at the surface that is upward throughout the winter. Conversely, the relatively cold American Rockies experience significant amounts of snow during the winter months. South of the Rockies, another large gradient coincides with the 700 hPa jet axis over northern Mexico.

A major difference between the Rockies and the Tibetan Plateau can be seen in the frictional effects within the planetary boundary layers above these elevated land masses. Over the American Rockies, 700 hPa wind changes associated with the passage of major, eastward moving trough and ridge systems are much less pronounced than those over the orography free regions up- and downstream of mountains. These weak wind variations, as stated earlier, are probably a result of the 700 hPa level being located within the planetary boundary layer above the American Rockies. In contrast, frictional effects over the Tibetan Plateau are much more dominant, with extremely weak wind changes up to the 500 hPa level. Furthermore, mid-latitude systems do not appear to extend equatorward across the Tibetan Plateau into the subtropical troposphere below 500 hPa due to the frictional (or mechanical blocking) effects of these high mountains. This is demonstrated by the rapid weakening and dissipation of eastward propagating cyclonic (anticyclonic) cells over northern India in the lower troposphere. Conversely, low-level (even at 850 hPa) synoptic disturbances exhibit systematic eastward movement along the southern periphery of the American Rockies.

Interestingly, lee cyclo- and anticyclogenesis occurs near the northeastern borders of the American Rockies and the Tibetan Plateau. Here high (low) pressure cells develop in regions of anticyclonic (cyclonic) wind shear along the northeastern periphery of these high mountains, and are most clearly defined at 850 hPa near the American Rockies and at 700 hPa in the lee of the Tibetan Plateau. These small-scale lee anticyclonic (cyclonic) cells generally propagate southward along the eastern periphery of the mountains in both regions. However, in the American Rockies region, lee anticyclonic (cyclonic) cells merge with major anticyclonic (cyclonic) systems that have propagated eastward along the southern periphery of these mountains. The merged systems then move east-northeastward to the U.S. east coast. Conversely, southward propagating, orographically induced, anticyclonic (cyclonic) cells dissipate near the southeastern border of the Tibetan Plateau. This is due to the fact that the "meridional blocking" effect of the Tibetan Plateau inhibits the eastward movement of major low-level disturbances along the southern periphery of these mountains. Orographic enhancement of the low-level divergence (convergence) fields associated with synoptic disturbances also differs over these two regions. For example, when an eastward moving anticyclonic (cyclonic) system approaches the California coast, low-level northerlies (southerlies) flow down (up) the southern slopes of the American Rockies and enhance the local, low-level divergence (convergence) field. As a result, the anticyclonic (cyclonic) cell tends to propagate along the southern (not northern) border of these mountains and is most intense in the vicinity of Arizona and New Mexico. This disturbance then weakens sharply in response to the decreased, low-level divergence (convergence) field in the lee of the American Rockies, and becomes weakest over Texas. Similarly, when a ridge (trough) system approaches the western Tibetan Plateau, pronounced low-level, divergent outflow (inflow) occurs over the northern Pakistan-India region, with northeasterlies (southwesterlies) flowing down (up) the extremely steep topographic gradient of these highly elevated mountains. However, changes in the low-level divergence (convergence) fields, and associated divergent parts of the wind, are less substantial around the southeastern and eastern borders of the Tibetan Plateau. This may be related to the "mechanical blocking" effect in these elevated regions, and also to planetary boundary-layer friction. As described by Murakami (1981), the planetary boundary-layer effect in this region extends approximately 1000 km downstream from the Tibetan Plateau, with extremely weak winds below the 700 hPa level.

5.6. The Effect of the Tibetan Plateau on the Summer Monsoon over East Asia

As indicated in Section 5.2, the Tibetan Plateau receives a large amount of solar radiation in summer, thus causing a significant contrast of heating rate between the Plateau region and the surrounding free atmosphere. This kind of horizontal heating difference in the mid-troposphere is even greater than heating contrast between continent and ocean in the plain region. Therefore, one is naturally led to the conclusion that it is far from complete to illustrate the formation of the Asian monsoon according to Halley's classical theory. Halley's theory, only taking into consideration the continent-ocean heat contrast, interprets the Asian monsoon circulation as a huge sea-breeze system produced by the continent-ocean heat contrast. In order to elucidate the formation and development, it is necessary to consider the thermal effect of the Plateau by itself. But in recent years, the apparent effect of the Tibetan Plateau upon the summer monsoon circulation was questioned with an analysis of the 500 hPa temperatures field over Eurasia and it was found that an east-west elongated band of high temperature extended along latitudes 25–30° N from Egypt, through northern India, into southern China. This evidence implies that the Tibetan Plateau is not directly responsible for high temperatures over Eurasia through its thermal effect. Thus, one has to re-emphasize the importance of the classical theory in explaining the maintenance of the upper tropospheric monsoon anticyclonic system, namely, the Tibetan high is a part of the upper-tropospheric anticyclonic system which extends along 25–30° N over the entire Eurasian continent. Nevertheless, these high mountains undoubtedly contribute to the enhancement of local circulation around the entire highland complex. This point will be examined in the following.

Figure 5.32 shows the 80–day (for 1979) mean streamlines and isotachs at the 200, 500, and 850 hPa levels. The mean 850 hPa field shows southwesterlies over a large area extending from the Arabian Sea to the Bay of Bengal and to Indochina and the South China Sea. There is a cyclonic inflow toward the Tibetan Plateau, which is pronounced, especially along the southeastern periphery of the Plateau and the northern slope facing the Tarim Basin. This inflow is a manifestation of the Plateau monsoon (Yeh, Gao *et al.*, 1979). The inflow exhibits a remarkable diurnal variation. The mean wind fields at this level and 700 hPa, suggest the presence of a mean inflow center on the eastern Tibetan Plateau. The inflow is related to the heat low on the Plateau. On the eastern side of the Plateau, the confluent zone between the north- and southwesterlies is the dominant feature of the mean 850 hPa flow.

At 500 hPa, the mean winds are generally westerly except for the regions influenced by the subtropical ridges over Saudi Arabia and the western Pacific. The wind speed of the 500 hPa flow above the Tibetan Plateau is noticeably small, and the westerly jet stream splits into northern and southern branches between approximately 75 and 100° E. The reduction of wind speed above the Plateau extends well up to the tropopause.

A spectacular feature of the Asian summer monsoon circulation is the development of the South Asian anticyclone in the upper troposphere over the low-level southwesterlies. The 80-day mean 200 hPa wind field shows that the mean center of the South Asian anticyclone is located over the border between Burma and Thailand. Figure 5.33 shows the two components of the mean wind for July 1979 as a function of height and latitude along 90° E. These cross sections are representative of the mean winds when the summer monsoon is at its peak. The low-level southerlies are seen to be strongest near the southern margin of the Himalayas and extend upward along the southern slope. Flow against a barrier cause forced ascending motions. Similarly, there are forced ascending motions to the north slope. Where these two meridional flows converge, corresponds to the location of the upper-tropospheric Tibetan high. To the south of the Himalayas, above 200 hPa, a northerly return flow prevails. Its maximum is 4 m s^{-1} near 150 hPa at 17° N. This northerly flow represents the upper-level branch of the monsoon meridional cell along 90° E. Note that it is associated with strong zonal easterlies with a maximum of about 25 m s^{-1} near 100 hPa at 17° N. These strong easterlies are defined as the tropical easterly jet. Below this easterly jet are the low-level westerlies, which exceed 10 m s^{-1} at 850

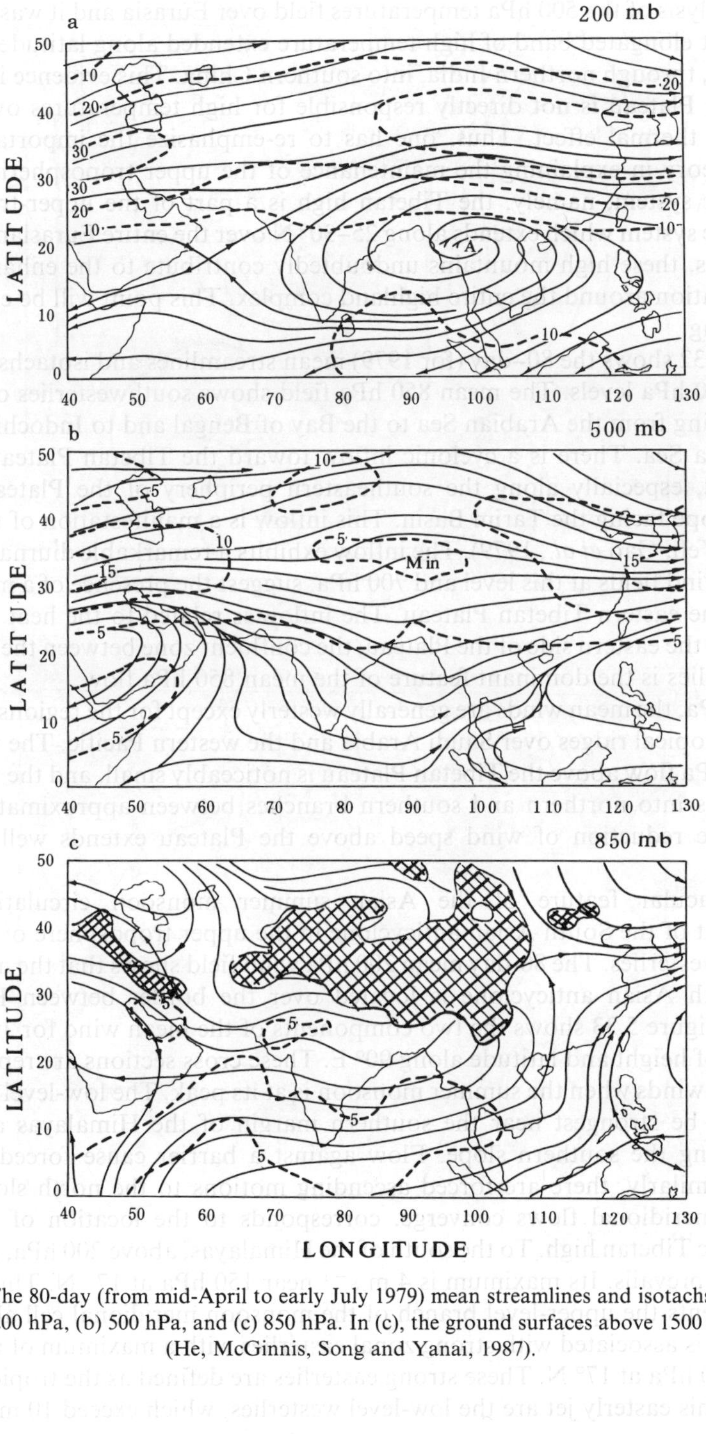

Figure 5.32. The 80-day (from mid-April to early July 1979) mean streamlines and isotachs (dashed line, m s^{-1}) at (a) 200 hPa, (b) 500 hPa, and (c) 850 hPa. In (c), the ground surfaces above 1500 m are masked (He, McGinnis, Song and Yanai, 1987).

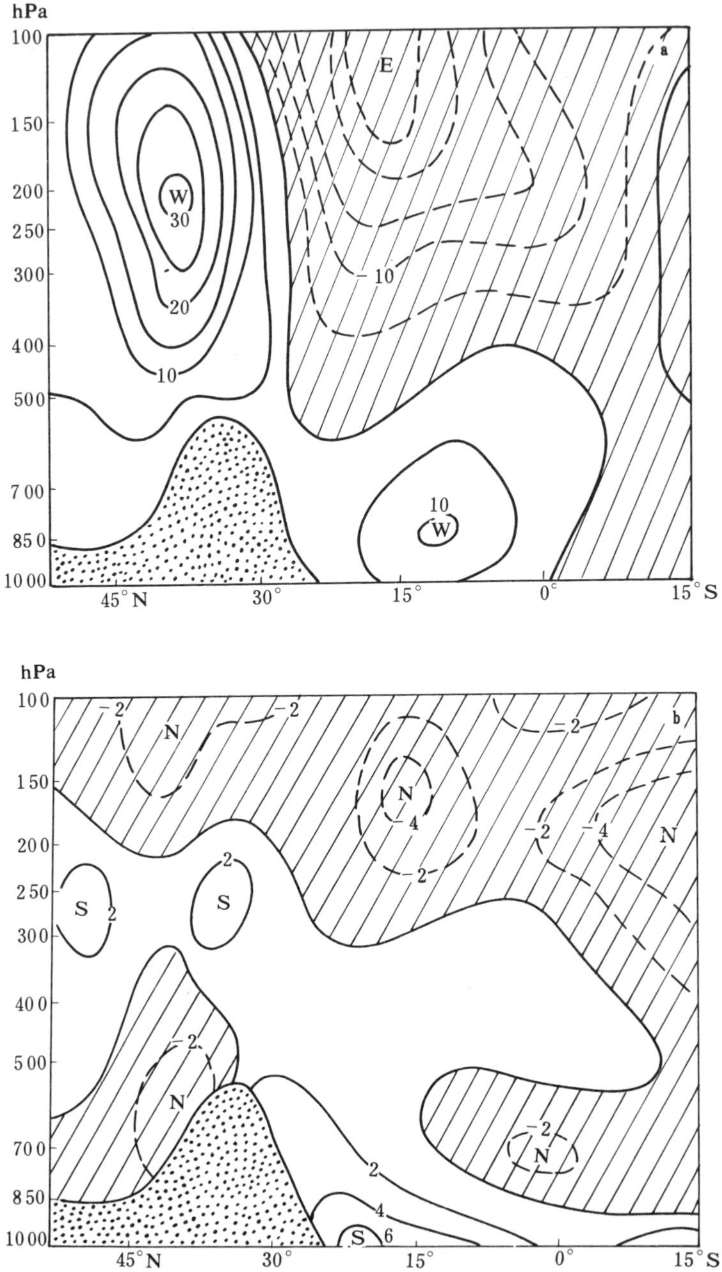

Figure 5.33. Mean zonal (east-west) wind (10m s^{-1} interval, top) and meridional (north-south) wind (2m s^{-1} interval; bottom) along 90° E for July 1979.

hPa near 13° N. These opposite currents indicate that the tropospheric temperature decreases southward from near the Tibetan Plateau to the equatorial region.

Next, it is interesting to investigate how these winds are established during the onset of the summer monsoon. In Assam and the southeastern corner of Tibet, summer rains start much earlier than in central India. The frequency and amount of rainfall during April and May in Assam is remarkable, generally exceeding 200–500 mm for these two months alone. The combined effect of the direct heating of the elevated surface and the release of latent heat in the ascending air, causes upper-tropospheric warming. This upper-tropospheric warming and its associated thermal wind balance result in a weakening of the upper westerlies which circulate around the southern periphery of the Tibetan Plateau during winter and spring. By early June, continual upper-tropospheric warming causes the westerly jet to disappear and to be replaced by the easterly jet. In general, the establishment of this upper-level easterly jet occurs nearly simultaneously with the onset of monsoon rains over central India. As indicated previously, the shift from the westerly to easterly jet takes place in an abrupt manner. Yin (1949) suggests that the thermal effects of the Tibetan Plateau are responsible for these sudden changes in the zonal wind fields. But, other authors have postulated different views, finding that the northward retreat of the westerly jet is not only observed around the Tibetan Plateau, but also nearly simultaneously over the Middle East and at locations some distance away from the Tibetan Plateau regions. Therefore, Yeh *et al.* (1958) postulated that as the elevation of the sun increases northward from winter to summer in the Northern Hemisphere, the temperature contrast from pole to equator increases until it reaches a threshold value, at which time a type of instability in the atmosphere appears, causing an abrupt change in the upper-air circulation which is not directly related to mountain effects.

Figure 5.34 shows the differences in the 300 hPa mean temperature fields (DT) between the pre- and post-onset phases of the summer monsoon of 1979. It is immediately evident that there are well-organized east-west oriented bands of positive temperature DT at approximately 30–40° N across Eurasia with two distinct centers: near Japan and the Middle East and the western part of the Tibet. Hence, these temperature changes between the pre- and post-monsoon onset periods represent a very large-scale phenomenon. The largest temperature increase occurs over the western Tibetan Plateau where the northward retreat of the westerly jet is most prominent. In contrast, over the eastern Plateau where atmospheric warming occurs much earlier than the monsoon onset over central India, the temperature increase is small and the associated northward jump of the westerly jet is not clearly defined. Murakami and Ding's study (1982) indicates the importance of the thermal effect of the Eurasian continent as a whole. However, the western Tibetan Plateau is instrumental in causing a local enhancement of the temperature increase during the monsoon onset phase.

From the above discussions, it has turned out that our observational knowledge is insufficient to adequately describe the exact role of the Tibetan Plateau in maintaining the summer monsoon circulation. No unified results and viewpoints have emerged. This may be partly due to the different data sources used in these studies; then again this fact also indicates that the study of the topographic effect of the

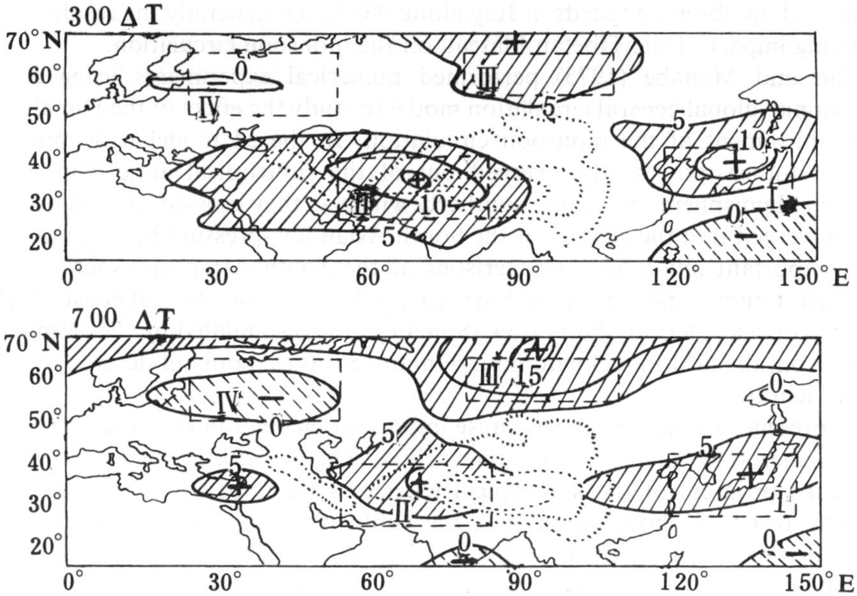

Figure 5.34. Mean temperature difference between the pre-onset (15 – 30 May) and post-onset (20 – 30 June) phases of the 1979 summer monsoon as a function of latitude and longitude at 300 hPa with 5 °C intervals. Topographical contour (1.5 km interval) are shown by the dotted lines. Areas of ΔT greater than 5 °C are hatched and ΔT less than 0 °C are indicated by dashed-hatching (Murakami and Ding, 1982).

Tibetan Plateau is a difficult problem to solve. To overcome this difficulty, one may utilize other methods. Among them, an effective method of identifying the role of mountains in the monsoon circulation is to perform numerical simulation experiments. In addition, the rotating annulus experiment is also an important method for studying this problem.

Murakami *et al.* (1970) used an eight-layer, two-dimensional model to study the summer monsoon, with the primary purpose of studying the influence of the Himalayas on the zonal wind components (the low-level westerlies and upper-level easterly jet) of monsoon circulation. The model extends from the equator to the North pole, at 80° E, with the Asiatic continent lying poleward of 15° N and the Indian Ocean lying equatorward of 15° N. The model includes the effects of radiation and condensational (latent) heating. The land-surface temperature is determined from the calculated heat balance, while the sea surface temperature is kept constant at 27 °C. The model was started with a completely calm and dry atmosphere using the observed July mean temperature distribution. The experiment was performed for a period of 80 days with 10-minute time steps. Without mountains, the low-level westerly wind maximum was less than 5 m s^{-1} and the upper-level easterly wind maximum was less than 10 m s^{-1}. When the mountains were included, the low-level westerly jet exceeded 10 m s^{-1} and the upper-level easterly jet became stronger than 35 m s^{-1}. These model-computed westerly and easterly jets are slightly stronger than the corresponding observed speeds in July along 90° E, but generally the results confirm the strong impact of the Tibetan Plateau on the monsoon circulation.

Hahn and Manabe (1975) performed numerical experiments using a three-dimensional, global general circulation model to study the effect of the Himalayas on the establishment of the monsoon circulation. In the M model with mountains, surface southeasterly winds diverged out of the subtropical anticyclones of the Southern Hemisphere, becoming southerly winds as they crossed the equator. They then converged into the vicinity of the South Asian low-pressure belt around 25° N. Other important low-level characteristics of the South Asian monsoon were also simulated. For example, along the East African highland off the east coast of Africa, a strong concentrated southerly flow (Somali jet) was simulated by the M model. In comparison, the NM model with no mountains produced only ill-defined wind field in that vicinity.

The presence of the Himalayas cause important changes in flow patterns at levels much higher than the mountain themselves. The M model simulates the rapid northward progress of the subtropical jet across the Tibetan Plateau during the monsoon onset. Without mountains, the subtropical jet does not abruptly jump northward to its summertime position, but rather it slowly moves northward throughout May and June, stabilizing its position in July at a latitude approximately 10 farther south than in the M model. Maximum temperatures found over the Tibetan Plateau in the M model agree with observations, while in the NM model, temperatures at 500 hPa over Tibet are approximately 10–12 °C lower than in the M model. This clearly indicates that the mountains, at least in the model, act as a mid-tropospheric heat source and contribute to the formation of the warm-core low pressure area near the surface of the Tibetan Plateau as well as the warm-core high pressure system aloft. The M model and NM model have significant differences in heating fields. In the M model, the latent heating tends to be most important over Tibet, while in the NM model, sensible heating tends to dominate. In short, the above two numerical experiments both emphasize the crucial role played by the Tibetan Plateau in the development of the monsoon circulation. If there were no mountain barriers, the monsoon flows in South Asia and East Asia would not be as it is now.

Qian *et al.* (1988) used a five-level limited-area model to simulate the role played by the Tibetan Plateau in the monsoon development in South Asia and East Asia. This numerical model includes the daily variation of solar insolation, large-scale condensation and cumulus covection heating, continent-ocean contrast, and topography. Figure 5.35 is the comparison of sea-level pressure fields between the simulation with mountains and that without.

Figure 5.35 (a) is very similar to the mean sea-level pressure field for July. But, after the mountains were removed from the numerical model, the thus simulated sea-level pressure field had a much weaker intensity (the maximum pressure difference being 10 hPa), although the overall picture is basically similar. The simulated near-surface wind fields for the experiments with and without mountains are basically similar to each other. For example, the location and intensity as well as the eastward extent of the low-level Somali jet are nearly the same for these two cases, indicating that the mountain barrier of East Africa is not the necessary condition to produce this jet

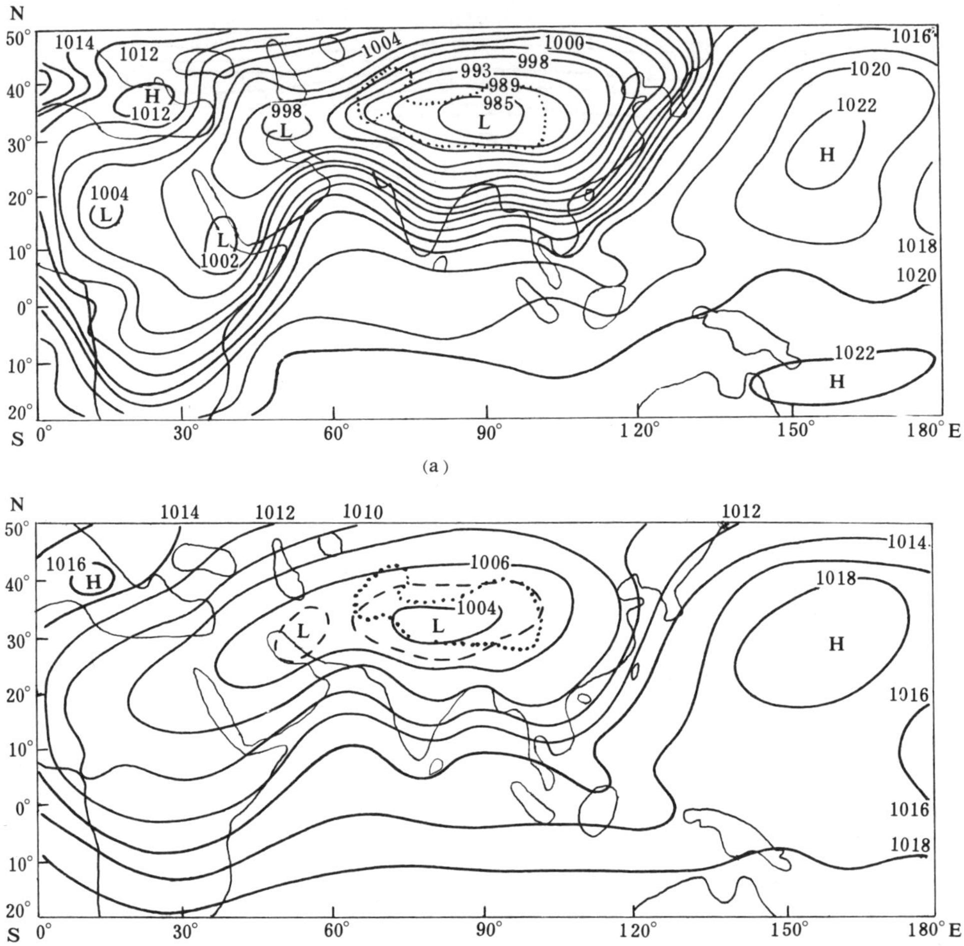

Figure 5.35. The sea-level pressure fields simulated (a) for the case with mountains included in the model; (b) for the case without mountains. The integration was performed for 8 days. Units: hPa (Qian *et al.*, 1988).

(Krishnamurti *et al.*, 1976). Therefore, one might draw the following conclusion: the sea-level pressure and near-surface wind fields are mainly determined by the diabatic heating field associated with the continent-ocean contrast, and the topography only excercises a secondary role in the development of these fields. However, a comparison of the mean precipitation rate made for these two experiments demonstrates a remarkable difference, especially in East Asia. If the Tibetan Plateau were nonexistent, the precipitation over the Bay of Bengal would disappear and cause a wetter region along the East Asian coast. This suggests the importance of the topography in the precipitation distribution of monsoon region. The effect of topography on the upper-level circulation features is also evident from Figure 5.36. It

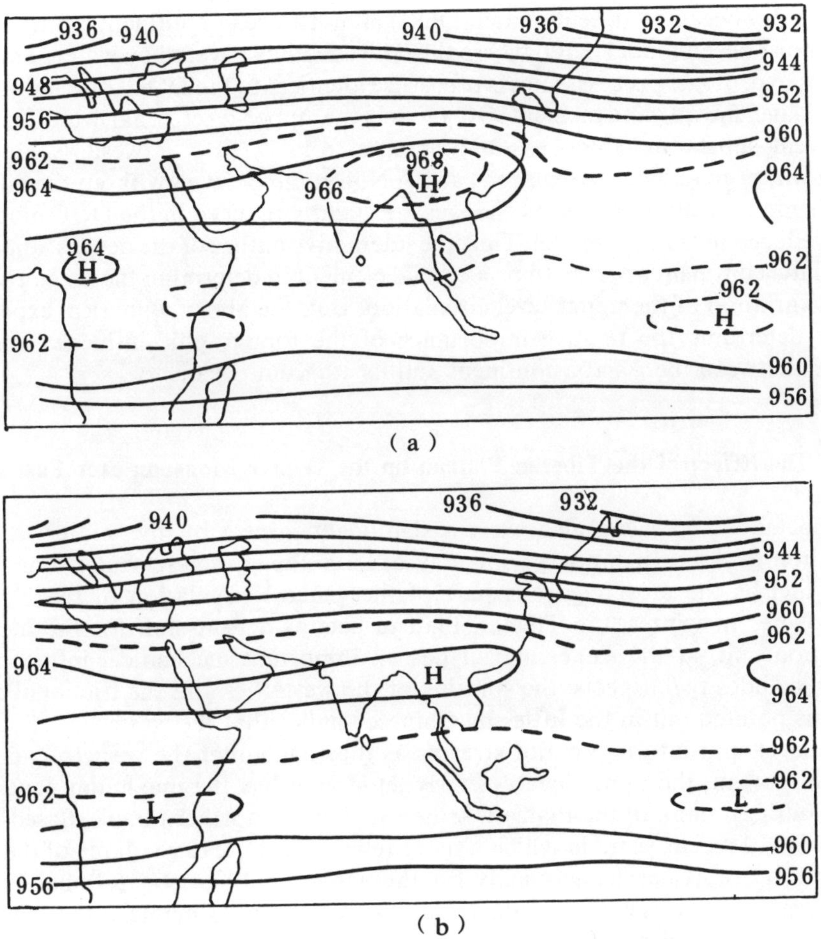

Figure 5.36. The same as Figure 5.35, but for 300 hPa geopotential fields. Units: decameter (Qian *et al.*, 1988).

can be seen that the simulation with mountains is more realistic than that without. For example, in Figure 5.36 (a), the Tibetan high over the Tibetan Plateau was well simulated, although its intensity is slightly weaker than observed and its location is 5° latitudes farther to the south. But, for the case without mountains, none of the upper-tropospheric high cells was simulated. This results indicates the important influence of topography on the basic circulation pattern at upper-level.

All of the above simulations have emphasized the remarkable importance of the Tibetan Plateau in the development of monsoon circulation. Using a different approach, Webster and Chou (1982) performed an experiment which included the Asiatic continent and adjacent oceans but not the Tibetan Plateau, thereby focusing on the heat contrast between the Asiatic continent and adjacent cooler oceans rather than on the topography. Experiments were conducted with three variations of the model: a dry ocean-continent model (DOC), a moist ocean-continent model (MOC), and a moist ocean model (MO) where the entire earth is covered by water. The MOC includes hydrologic processes but is otherwise identical to the DOC. The simulational results have shown that in the MOC model, the upper-level zonal wind (250 hPa) underwent substantial inter-seasonal changes from 37 m s^{-1} (westerly) at 30° N during winter to –25 m s^{-1} (easterly) at 10° N in summer, even without the effects of the Tibetan Plateau. The annual changes are slightly reduced in the DOC model and much reduced in the MO model. Thus the interactive nature of the oceans adjacent to the Asiatic continent appears to be a crucial element in determining the magnitude and spatial variation of the upper-level circulation. But, the above numerical experiment cannot determine the relative importance of the topographic influence and heat contrast between the Asiatic continent and its adjacent oceans.

5.7. The Effect of the Tibetan Plateau on the Winter Monsoon over East Asia

The Tibetan Plateau can also exert a significant impact on the winter monsoon circulation in Asia. In January, one may observe the strongest winter monsoon in Asia which is the most vigorous part of the general circulation in the Northern Hemisphere. In winter, the Tibetan Plateau acts as a heat sink for the monsoon circulation and, at the same time, it has an important mechanical influence that mainly includes two aspects: the splitting of the westerlies and the frictional effect.

It was pointed out in the fifties by Chinese studies that the westerlies at low and middle level split into two major streams as they encounter the western end of the Tibetan Plateau; the upper-level westerly jet stream has the maximum wind speed at the southern flank of the Plateau, rather than at the northern flank. Based on the studies for 10 recent years in which a more reliable and greater volume of data over the Plateau are available, especially for the winter of 1978-1979 (Global Weather Experiment), previous major results remain correct and reasonable, although a large amount of new evidence has been provided (Murakami, 1981).

Figure 5.37 indicates the 1978–1979 winter mean wind and vorticity fields at 700 hPa. It may be seen that the 700 hPa mean wind splits into southern and northern branches of major streams starting from the western end of the western Plateau. The northern branch is directed northeastward approximately along 2 km height contours, then turns eastward into the saddle area between the Tibetan Plateau and the Altain Mountains. The southern branch, which is slightly weaker than its northern counterpart, flows eastward nearly parallel to the southern periphery of the Tibetan Plateau. To the east of the Plateau, these branches tend to converge into a single major stream near 120° E. Of particular interest is the zone of extremely weak winds which extends approximately 1000 km downstream from the eastern slope. Associated with this wind minimum is a vorticity pair, with anticyclonic vorticity to the north and cyclonic vorticity to the south (See Fig. 5.37, bottom). A similar feature, although not as well defined, is observed near the western end of the Tibetan Plateau. These features are called Kàrmàn vortex streets. They mainly develop in the stagnation region downstream of a sharp-edged obstacle. Staffs Members of

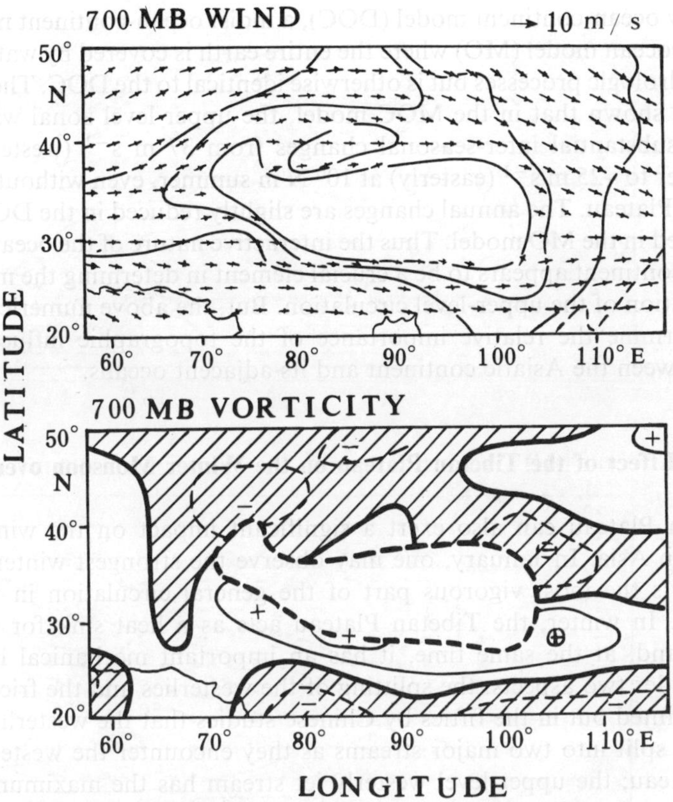

Figure 5.37. Wind and vorticity. Top: winter mean wind vectors ($10 m s^{-1}$, unit vector). Full lines are smoothed topographic contours with 1 km interval. Bottom: winter mean vorticity ($1 \times 10^{-5} s^{-1}$ interval). Negative (anticyclonic) vorticity is shaded. The heavy dashed line is the 3 km contour of the Plateau. Fields are at 700 hPa, averaged from December 1978–February 1979 (Murakami, 1981).

Academia Sinica (1956) have already indicated these phenomena based on early studies.

At 500 hPa, the splitting of the main westerly flow is still seen at both 25° N and 45° N between 70° E and 120° E (Fig. 5.38, top). An interesting feature is that the winter mean winds at 500 hPa are weakest (less than $10 m s^{-1}$) directly over the Plateau. These extremely weak winds may be an indication of the frictional effects of the Tibetan massif. Murakami (1981) estimated the average height of the planetary boundary layer to be about 1.5 km above the Plateau. Thus, over the Plateau, the top of the planetary boundary layer may reach up to about 6 km above sea level. It is natural to conceive that the some kind of weather system generated in the planetary boundary layer, mainly through the frictional effect, once brought out of the Plateau region, would have a considerable influence on the weather over the region downstream, especially when they obtain a rapid growth due to some dynamical or thermal mechanisms.

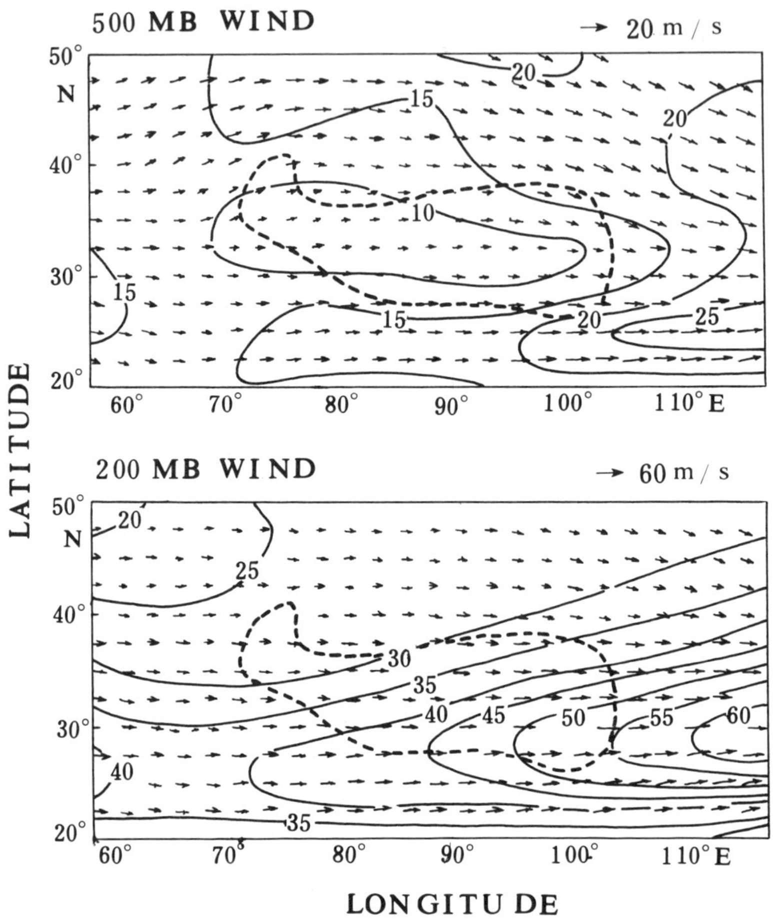

Figure 5.38. Winter (December 1978–February 1979) mean wind vectors with 5m s^{-1} isotachs for 500 hPa (top) and 200 hPa (bottom). The heavy dashed line denotes the smoothed 3 km topographic contour (Murakami, 1981).

This high planetary boundary layer acts as extending the horizontal and vertical extents of the Asian high mountain ranges. Winter mean winds at 300 hPa (not shown) are also weak over the Plateau, indicating that the mechanical effect of the Tibetan Plateau may extend that high. The 300 hPa wind field suggests that the Taylor column phenomenon, the imaginary fluid columns which extend upward from the obstacles, may occur above the Tibetan Plateau. Nakamura and Murakami (1983) simulated flow fields over and around a prescribed obstacle using a six-level primitive equation model, with flows exhibiting features that were similar to the Taylor column phenomenon; namely, lower-tropospheric winds tended to flow around the obstacle, and upper-tropospheric flows were characterized by weak winds above the obstacle. In fact, even at 200 hPa (Fig. 5.38, bottom), winds are strongest not above, but along the southern periphery of the Himalayas. This characteristic feature of the winter mean flow cannot be explained by the conservation theorem of absolute potential vorticity. According to this theory, if air encountering an obstacle rises, acquiring an anticyclonic curvature as the depth of the air column decreases on its windward side, the wind speed should increase markedly above and to the north of the barrier. This is not the case in the observed winter mean upper-tropospheric flows for 200 hPa.

Another important effect of the Tibetan Plateau on the winter monsoon circulation shows up in the distinct eastward acceleration of the upper-level jet stream south of the Himalayas at the 200 hPa level. Murakami (1981) pointed out that this substantial eastward acceleration is maintained primarily by southerly ageostrophic winds. This southerly flow may correspond to an upper branch of the thermal direct, local Hadley circulation which is associated with the heat contrast between the cooler continent and high mountains to the north and the warmer ocean and equatorial rainfall to the south. Chinese scientists (Yeh, Gao *et al.*, 1979) demonstrated the importance of a heat sink over the Tibetan Plateau in inducing a strong jet stream to its south. It is reasonable to expect that the available potential energy represented by these temperature contrasts is converted more or less in situ to fuel the acceleration of jet stream.

The existence of a direct solenoidal circulation to accomplish this conversion has been known for many years. Staff members of the Academia Sinica (1957–1958) deduced mean vertical motions indicating a large-scale direct circulation. Murakami and Unninayar (1977), using operational analyses from the National Meteorological Center (NMC), outlined a large-scale direct circulation over East Asia for the winter of 1970–1971. They showed the updraft part of the circulation over the Malaysia-Indonesia area and the sinking motion centered over northern China. They pointed to the wet and cloudy weather over the Malaysia-Indonesia area as a major heat source to maintain the supply of available potential energy. Blackman *et al.* (1977) concluded from a study of momentum flux that in the time-averaged flow there is a direct meridional circulation in the region of the jet stream acceleration. Cressman (1980) also showed a strong single-celled direct circulation centered on the East Asian-West Pacific upper front and jet stream system. Later, his estimates of kinetic energy budgets further provided a quantitative description of the sources and sinks of kinetic energy for the jet stream of that region. In the location of jet stream acceleration corresponding to a large-scale direct solenoidal circulation, the mean generation of kinetic energy was 95×10^{10} kw or 34 wm^{-2}. Kinetic energy conversion at these rates greatly exceeds that in a typical vigorous extratropical cyclone, which could be in the range of $10–20 \times 10^{10}$ kw. Orographic forcing over the eastern edge of the Himalayan Plateau is suggested as a process partly responsible for the geographic reliability of the generation region in China.

5.8. Plateau-Produced Low-Level Vortices

The low-level vortices produced locally over the Plateau and its sloping periphery are unique weather systems of a subsynoptic scale in East Asia. They are among the most important rain-bearing systems in China during spring and summer. The low-

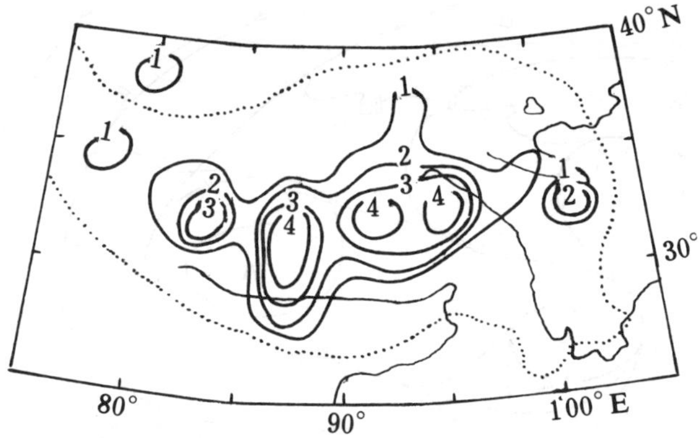

Figure 5.39. The distribution of the occurrence of the initial Plateau vortices for May through August of 1979.

level vortices have their geographical origin in three areas: the southeastern part of the Plateau (called southwest (SW) vortices); the main body of the Plateau (called Plateau vortices) and the northern part of the Plateau (termed northwest (NW) vortices because of their location in the northwestern part of China). These vortices can produce excessive rainfall after they leave the Plateau if they are supplied with ample moisture and are supported by a favorable large-scale environment. The genesis and development of these vortices, especially SW vortices, are closely related to activity of monsoonal airflows. Therefore, they may be considered, to a certain extent, as the result of interaction between the Plateau and monsoon.

Figure 5.39 is the frequency distribution of the initial detection of the Plateau vortices for May through August of 1979, showing two regions of high frequency: the region to the north of Nagqu, which was already indicated to be an origin region for the Plateau vortices in the 70's; and the western part of the Plateau between Shengza and Gaize. The frequency of occurrence of the Plateau vortices in the western part of the Plateau is greater than that in the eastern part, with 32 vortices for the former and 19 vortices for the latter. After the initial occurrence of the Plateau vortices, some only persist locally for one day, and then weaken and dissipate; some may persist for a time span longer than 36 hours. These long-lived vortices can generally develop and move even farther eastward out of the main body of the Plateau.

Figure 5.40 is an example of the development of the Plateau vortex. It may be initially detected around Zhongba where the upward sensible heat from the surface is especially significant (Zhang *et al.*, 1989). The unstable stratification causes the rapid growth of convective activity, thus leading to the formation of a Plateau vortex. Its well-organized circulation at 500 hPa is very evident, as shown in Figure 5.40 (b).

The structure of Plateau vortices in summer has been recently studied by use of a compositing method (Lu *et al.*, 1986) similar to that of Reed (1977). At the initial stage of the development of Plateau vortices, an apparent cyclonic circulation or a convergent center is observed at or below the 500 hPa level in the vicinity of the low center (Fig. 5.41 a). Above 400 hPa, an anticyclonic circulation is found. At the

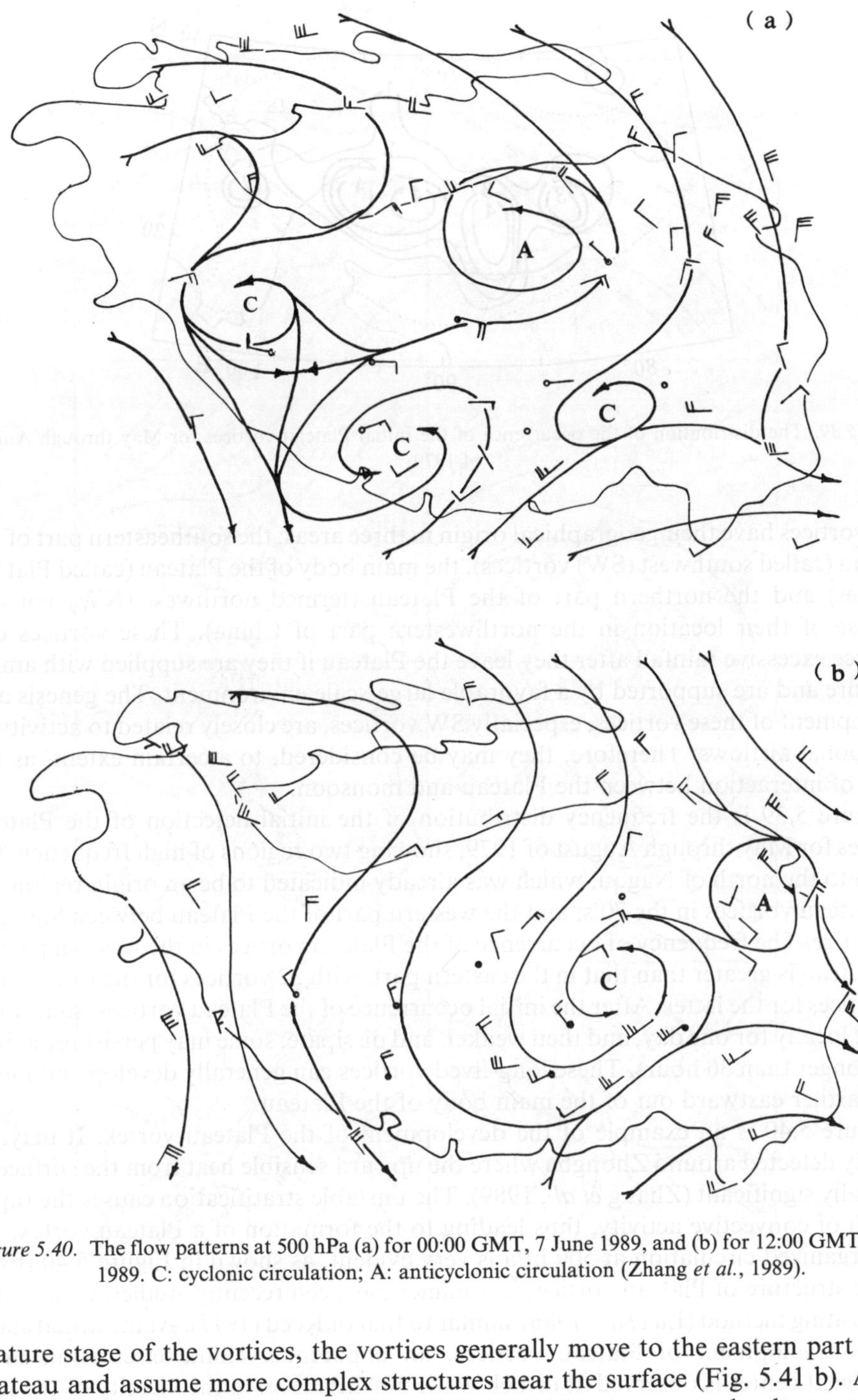

Figure 5.40. The flow patterns at 500 hPa (a) for 00:00 GMT, 7 June 1989, and (b) for 12:00 GMT, 8 June 1989. C: cyclonic circulation; A: anticyclonic circulation (Zhang *et al.*, 1989).

mature stage of the vortices, the vortices generally move to the eastern part of the Plateau and assume more complex structures near the surface (Fig. 5.41 b). At 500 hPa, the cyclonic circulation is evident and at 400 hPa one may only observe a weak

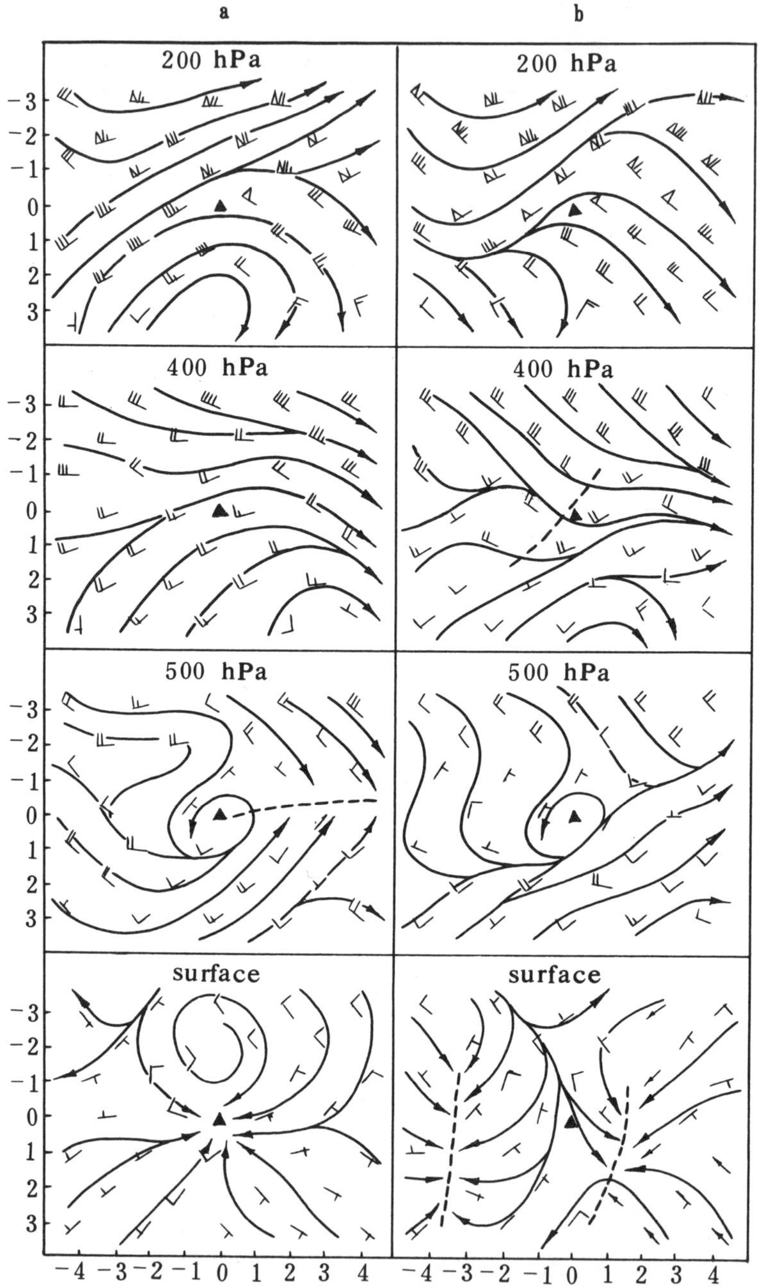

Figure 5.41. Compositing flow fields at different levels at the initial stage of the Plateau vortices (a), and at the mature stage (b). Black triangle is the center of vortex. The grid length represents 2° latitudes (Lu *et al.*, 1986).

cyclonic disturbance. In the upper troposphere, anticyclonic airflows predominate.

The above results show that the Plateau voritices at the mature stage have a deep upward extension. The vorticity fields indicate that cyclonic vorticity predominates below 400 hPa (Fig. 5.42), with a maximum of $2.2 \times 10^{-5}\ s^{-1}$ at 500 hPa. Anticyclonic vorticity prevails above 500 hPa, with a maximum around 200 hPa. The divergence field of the Plateau vortices is characterized by convergence below 400 hPa with a maximum of $-1.6 \times 10^{-5}\ s^{-1}$ and divergence above 300 hPa (Fig. 5.43). Upward motion is found throughout the entire tropospheric column from the surface up to about 100 hPa with the maximum speed occurring between 400 and 500 hPa (Fig. 5.44). The ascending motion is stronger to the south of the vortex than to the north; this is consistent with the fact that more precipitation occurs in the south of the vortex.

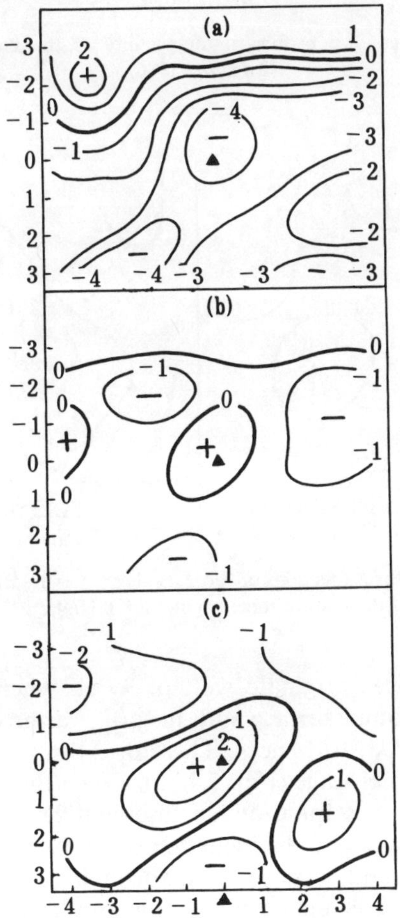

Figure 5.42. Compositing vorticity field around the Plateau vortices at the different levels. (a) 200 hPa, (b) 400 hPa, and (c) 500 hPa. Units: $10^{-5}\ s^{-1}$ (Lu *et al.*, 1986).

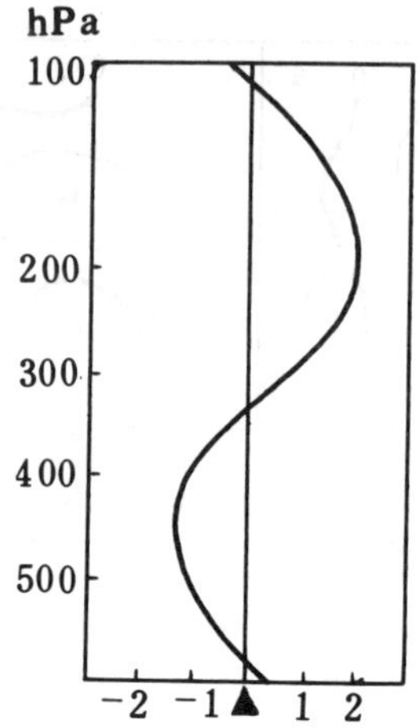

Figure 5.43. Compositing divergence profile in the vicinity of the vortex center. Units: 10^{-5} s^{-1} (Lu *et al.*, 1986).

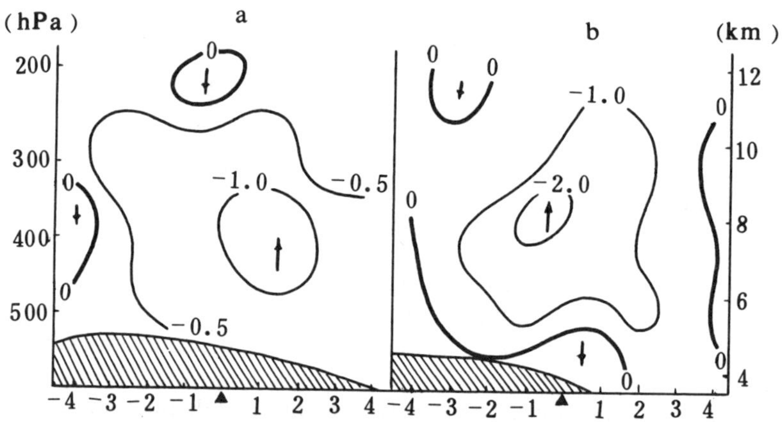

Figure 5.44. Compositing zonal cross-sections of vertical velocity (ω=dp/dt) around the Plateau vortex for the initial stage (left panel) and the mature stage (right panel). Units: 10^{-3}hPa s^{-1} (Lu et *et al.*, 1986).

Most of Plateau vortices originate west of 95° E where intense incident solar radiation heats up the ground surface and, in turn, warms up the atmosphere, thus producing instability and disturbances. Therefore, the Plateau vortices are of the warm-core type at the initial stage (Fig. 5.45 a). When the low vortices move to the eastern part of the Plateau, precipitation increases and the air at low-level is cooled by the evaporation of raindrops and the incursion of cold air, while the upper part gets warmer on account of latent heat released through condensation. As a result, a thermal structure of a warmer upper part and a cold lower part of vortices may be observed (Fig. 5.45b). The precipitation generated by the vortices is mainly concentrated over the inner area 400 km across, with the heavier rain coinciding with

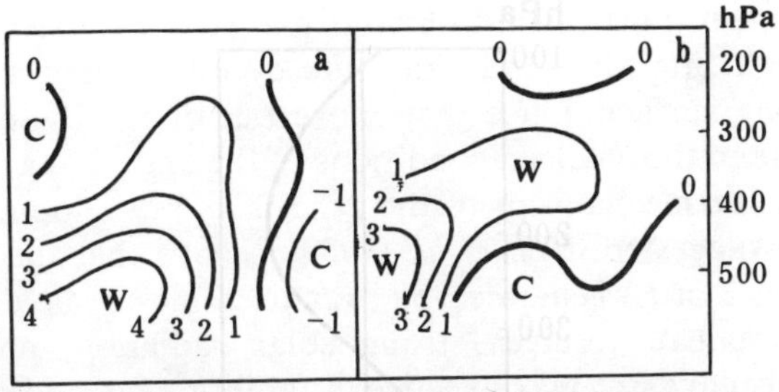

Figure 5.45. Cross-sections of temperature departure from the normal of the Plateau vortices (a) at the initial stage, and (b) at the mature stage. Units: °C (Lu *et al.*, 1986).

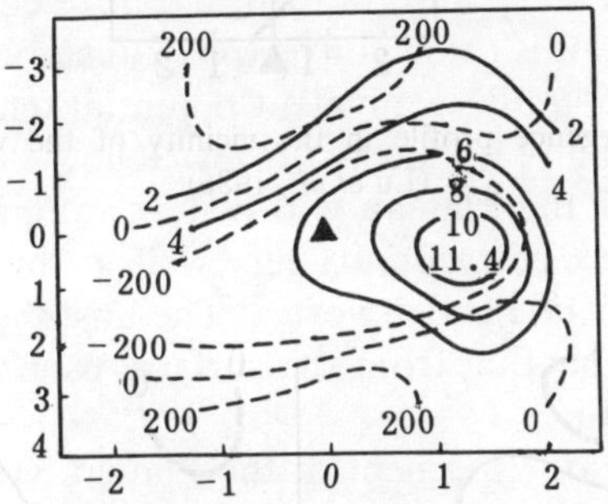

Figure 5.46. Average rainfall amount (Units: mm, solid line) around the vortex and moisture flux divergence (10^{-1} d^{-1}) at the 500 hpa (dashed lines) (Lu *et al.*, 1986).

the strong moisture convergence zone (Fig. 5.46). The daily average rainfall amounts to 6-11 mm, with the daily rainfall amount in the east of the vortices being twice as much as that in the west due to the lower elevation and, hence, thicker moist layer over the eastern part.

From the above discussions, one may feel the marked distinction of the structure between the Plateau vortices and monsoon depression. The former has an horizontal scale of 400–800 km across, with an eastward displacement speed of 2–3 longitude degrees per day and a life span of 1–3 days. The Indian monsoon depressions have horizontal scales of a few thousand kilometers and a westward movement speed of 5.7 longitude degrees per day, lasting for 3–7 days. This disturbance has a vertical extent of about 10 km while the Plateau vortices are only 2–3 km in thickness. The maximum cyclonic vorticity associated with the Indian monsoon depression may amount to 12 $\times 10^{-4}$ s^{-1} and may extend upward to 300–200 hPa, whereas for the Plateau vortices the maximum cyclonic vorticity is only of the order of 2×10^{-5} s^{-1}. However, the upper anticyclonic vorticity, with a maximum of -4×10^{-5} s^{-1}, is much stronger than the monsoon depression. The Indian monsoon depressions have a thermal structure of a cold core in the lower troposphere, and warm core in the upper troposphere while the Plateau vortex has a warm core throughout the troposphere during the developing stage.

In short, the Plateau vortices are obviously a weather system of a small scale, shallow depth, weak intensity, and short life cyclone. They are generated by a particular underlying surface. Once they move out of the Plateau, they generally dissipate rapidly due to the changed conditions of the underlying surface. According to recent studies, some mechanisms of the genesis and development of the Plateau vortices have been suggested. Yeh *et al.* (1989) believe that the Plateau-produced mesoscale vortices are often generated by organized mesoscale convection. During the summer, the Plateau receives strong solar radiation, and the overlaying atmosphere often displays strong convective instability, even under the influence of a subsiding high pressure system. With strong solar radiation, every mountain peak acts as an isolated heat island which triggers convection if moisture is available (Reiter and Tang, 1984). If the convective activity persists long enough, a mesoscale cyclonic circulation can develop and, in turn, organize cumulus convection. In this case, the air over the Plateau is often convectively unstable ($\partial\theta_e/\partial P > 0$) at low level. Figure 5.47 indicates the 500 hPa equivalent potential temperature (θ_e) distribution in relations to the formation of a Plateau vortex (Kuo, Cheng and Anthes, 1986). A maximum of θ_e exists over the Plateau with values ranging from 77 to 80 °C. The center of the Plateau vortex corresponds very well to the maxima of θ_e, suggesting the warm core structure of the Plateau vortex. The unstable air stratification may be caused by strong sensible heating from the surface, especially over the central and western parts of the Plateau.

The exact causes of the development of the Plateau vortices are not yet known. Some investigators suggest the barotropic instability to be one of the factors affecting the development of the vortices. The computation of barotropic instability for the summer of 1979 indicated that $\beta - \partial^2 u/\partial y^2 = 0$ is satisfied somewhere between 32.5° N and 34.5° N, thus the horizontal shear of the mean flow was favorable for the development of the vortices. In contrast, the vertical wind shear was very weak as the vortices developed. This implies that the baroclinic instability of the flow is not important for the development of vortices. However, recent studies (Chen and Dell'Osso, 1984; Kuo, Cheng and Anthes, 1986) suggest that both baroclinic instability and latent heat release associated with organized convection over the Plateau might play important roles in the generation of the Plateau vortices. A vortex often develops within an existing shearline that separates the warm, moist monsoon flow to the south from the midlatitude colder, drier air to the north. In fact, this Plateau-produced vortex is embedded within a sharp, short-wave trough. This vortex is also associated with strong convective activity throughout its life cycle. Chen and Dell'Osso (1984) also showed that the intensity of this vortex was significantly weakened when latent heat release was removed from the model.

Now, we shall further discuss the SW vortex. In general,the SW vortex is defined as a 700 hPa closed cyclonic circulation over southwestern China, mainly over the western part of the Sichuan Basin. It is a low-level circulation system, often only visible on 850 and 700 hPa analyses. On the surface weather map, one may often observe a negative pressure tendency during 24 hours over the low-vortex region. In this sense, the SW vortex is also called the SW low vortex. Most of the time, the vortex forms and dissipates without intense development and the severe weather associated

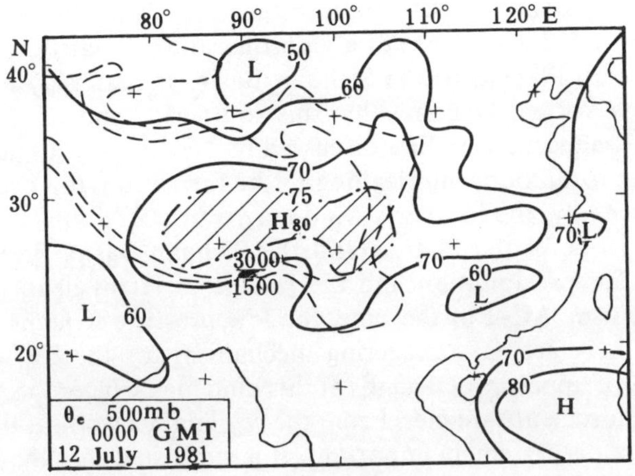

Figure 5.47. The 500 hPa equivalent potential temperature (θ_e, °C) at 00:00 GMT, 12 July 1981 (Kuo, Cheng and Anthes, 1986).

with it, because there is usually a lack of upper-level support and moisture. But if the synoptic conditions are favorable, the SW vortex may cause severe weather, especially heavy rainfalls. The SW vortex may provide strong orographic lifting to trigger convection and, consequently, a large amount of rainfalls on the steep topography surrounding the Sichuan Basin. Many cases may be examplified, for example, the heavy rainfalls in the Sichuan Basin on 11-14 July of 1981 which have been extensively studied by numerous Chinese and western meteorologists (Zhou and Hu, 1984; Chen and Dell'Osso, 1984; Hovermale, 1984; Anthes and Haagenson, 1984; Kuo, Cheng and Anthes, 1986; Wang and Orlanski, 1987). This heavy rainfall took a large toll in human life and caused much damage to property. If the SW vortex moves eastward after its formation, it can cause heavy rainfall over the middle and lower Yangtze River Valleys. Based on the statistics of 115 SW vortices for April-July from 1970 to 1974, that moved over Hunan Province, 45% of them (51 cases) brought about heavy rainfalls. Similar statistics are true for Jiangxi Province. If the SW vortices take a northward track of movement, they can bring heavy rainfall to northern China. For example, one half of the total number of heavy rainfalls affecting the Beijing region were caused by SW vortices (Tao *et al.*, 1980).

The SW vortex is often observed during the spring and summer as the southwest monsoon flows around the Plateau from the Bay of Bengal and the South China Sea, carrying warm moist air. The statistics of the seasonal distribution of the SW vortex from different sources (Lu, 1986) indicate that the maximum and minimum of frequency of occurrence of the SW vortex is observed during March-June and August, respectively. In peak month, May, 11.6 SW vortices occurred on an average. Next, the secondary peak was observed in April in which 11 SW vortices formed. In July and August, less than 8 SW vortices were generally generated. Thus, the seasonal distribution of the SW vortex has a wide spectrum which is closely related to the seasonal change in the general circulation in East Asia as well as the regional conditions. From the synoptic viewpoint, the genesis of the SW vortex needs to meet two requirements: (1) the existence of a vigorous southerly airflow from the eastern slope of the Tibetan Plateau to the Sichuan Basin. It may play a dual role in the genesis of the SW vortex. Dynamically, this southerly wind produces "differential frictional effects", a mechanism first discussed by Newton (1956) in connection with Colorado cyclone formation, thus leading to the formation of a cyclonic circulation at low level. Thermally, the southerly wind may transport abundant warm moist air into the eastern slope of the Plateau and the Sichuan Basin, providing the major moisture source for precipitation and the release of latent heat; (2) the necessary triggering mechanism. Most of the time, the low pressure troughs passing over the Tibetan Plateau may act as a triggering mechanism for the SW vortex. Statistics indicate that 73% of upper-level troughs of this kind may trigger the genesis of the SW vortices if the requirements are met. From the regional viewpoint, the topography of the Tibetan Plateau is extremely important. It is suggested that the formation of the SW vortex is a consequence of the blocking effects of the mesoscale mountain range, located at the southeastern corner of the Tibetan Plateau, on the southwesterly monsoonal flow. This is supported by the fact that the SW vortex develops mainly within the southwesterly monsoon current, that it is strongest at the lower level during both its formation and mature stages, and that low level cyclonic vorticity is present throughout the period that the southwesterly monsoon flows around the Plateau. In addition, reseachers have argued about the importance of cold air coming from the north and northeast.

The development of the SW vortex is expected to depend greatly on the effect of latent heat release, due to the fact that this vortex is usually accompaied by a large amount of rainfall and convective activity. In order to document better the effects of strong latent heat release associated with convection, Kuo, Cheng and Anthes (1986) calculated mesoscale heat and moisture budgets associated with a SW vortex, which resulted in a flood catastrophe in the Sichuan Basin, 11–15 July 1981. Figure 5.48 shows the vertical profiles of heating (Q_1) and drying (Q_2) during three time periods, averaged horizontally over the area of heavy rainfall (area of approximately 880 × 880 km^2). The radiative cooling rate averaged over the Sichuan Basin is less than 2 °C day^{-1}. Therefore, radiation is not important for the heat balance of this midlatitude heavy rainfall event. The intensity of the apparent heat source (Q_1) increases with time (Fig. 5.48a), reaching a maximum at 12:00 GMT, 13 July. The upward movement of the level of maximum heating (~20 °C day^{-1}) may be noted, implying that the stabilizing effects of the latent heat release becomes stronger with time. The vertical profiles of the convective moisture sink (Q_2, Fig. 5.48 b) are consistent with the heating profiles. Similarly, the maximum moisture sink for this heavy rainfall event gradually moves upward. One can infer the type of precipitation from the separation of the Q_1 and Q_2 profiles. Figure 5.48 shows that the separation between the level of maximum heat source and that of maximum moisture sink decreases with time. This suggests more convective precipitation at an early stage and more stratiform-type precipitation at a later stage. Although it is difficult to quantitatively assess the impact of latent heat release on the development of the SW vortex, the large amount of latent heat release at the middle levels early in the event implies a strong energy source for

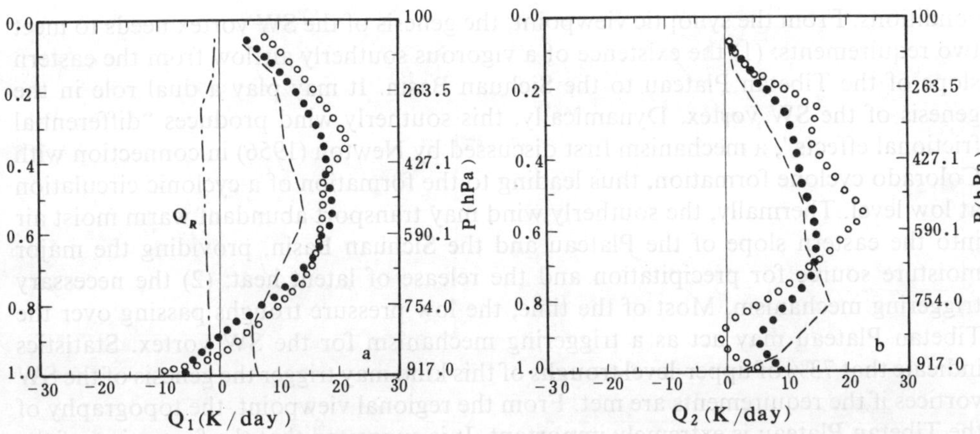

Figure 5.48. The vertical profile of (a) Q_1 and (b) Q_2 at 12:00 GMT, 11 July (dashed); 00:00 GMT, 13 July (dotted) and 12:00 GMT, 13 July 1981 (open circle) averaged over the heavy rainfall region. The radiative cooling rate (Q_R: dot-dashed) is also shown in (a). The Unit is °C day^{-1} (Kuo, Cheng and Anthes, 1986).

the SW vortex. With weak stability at the middle levels, latent heat release can induce strong, upward vertical motion, which in turn enhances low-level convergence, low-level spinup and convective cloud development, establishing a positive feedback between the circulation of the SW vortex and the cumulus convection. Along this line Wang and Orlanski (1987) used the CISK theory to interpret the development of this SW vortex. They theoretically obtained a wavelength of optimal development of a low vortex which is about 400 km. This scale is approximately equal to the observed one.

A perturbation energy equation for disturbances in a moist baroclinic atmosphere is used to study this flash flood event, indicating that the available potential energy generation is reduced by 80% when the moisture effects are excluded. Using a limited area model, Chen and Dell'Osso (1984) simulated this flood event with (wet case) and without (day case) latent heat release. They found that the vertical motion associated with the SW vortex in the wet run is 6 times larger than that in the dry run. Both the low-level cyclonic circulation and the upper-level anticyclonic circulation were considerably stronger in the wet case than in the dry case. All the above studies appear to support the hypothesis that the latent heat release contributes significantly to the development of the SW vortex.

Finally, we shall discuss the structures of the SW vortex. Figure 5.49 is the 700 hPa analysis for 00:00 GMT, 12 July 1981– 12:00GMT, 13 July 1981 showing that the SW vortex is a low-pressure system, with strong cyclonic circulation. The low-pressure feature may extend upward to 400 hPa, above which there exists a relatively high pressure region (Fig. 5.50). Correspondingly, the positive vorticity is generally observed in the layer below 400 hPa and the negative vorticity above 400 hPa. But, for some SW vortices, the positive vorticity may extend up to an even higher level. Figure 5.51 is an east-west cross-section along approximately 30° N at 12:00 GMT, 13 July 1981, indicating the maximum cyclonic vorticity associated with the SW vortex: one at 400 hPa, and the other at 850 hPa. There is very significant vertical tilt associated

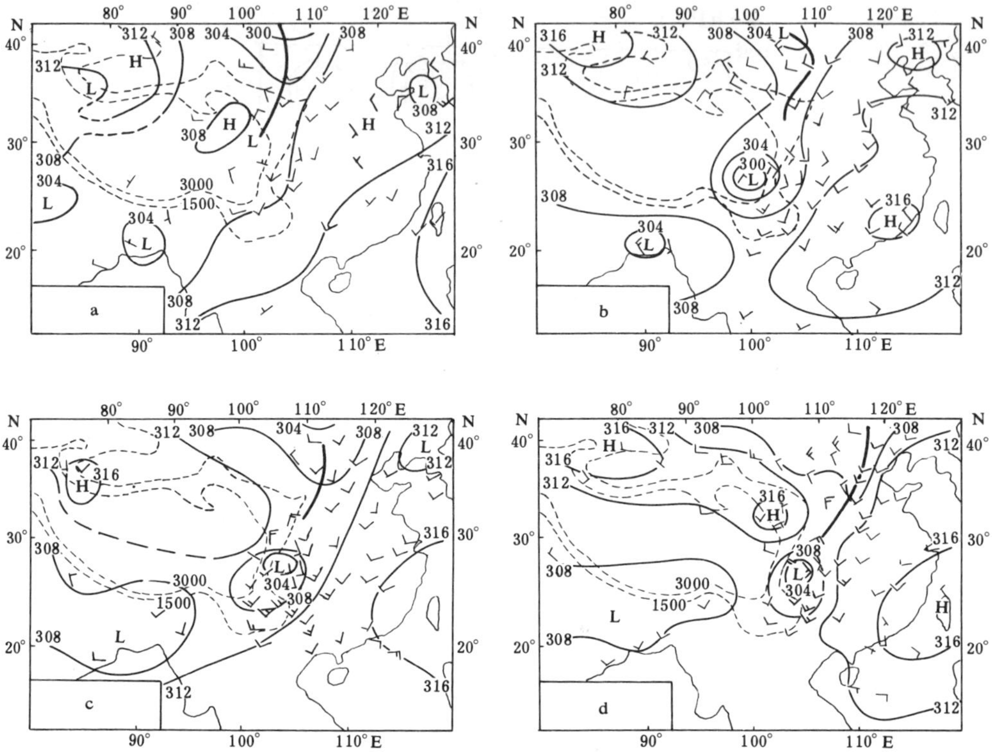

Figure 5.49. The 700 hPa analysis at 00:00 GMT, 12 July 1981 (a), 12:00 GMT, 12 July 1981 (b), 00:00 GMT, 13 July 1981 (c) and 12:00 GMT, 13 July 1981 (d). One full barb=10 m s^{-1}. Solid contours are geopotential height (dam) (Kuo, Cheng and Anthes, 1986).

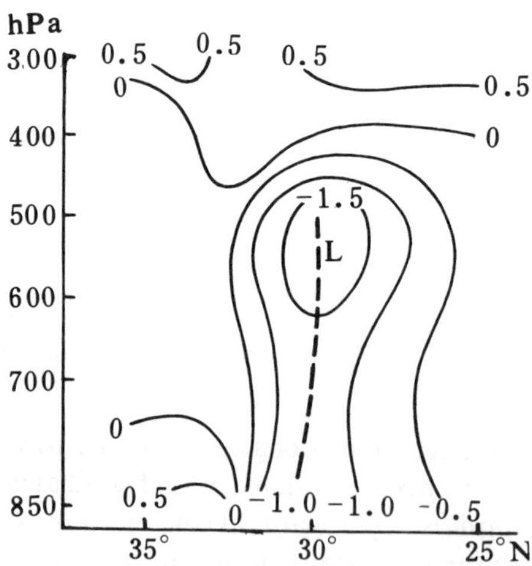

Figure 5.50. North-south cross section along 110° E of the departure of geopotential height from the area-averaged value for 00:00 GMT, 5 May 1974 (Zhang et al., 1989).

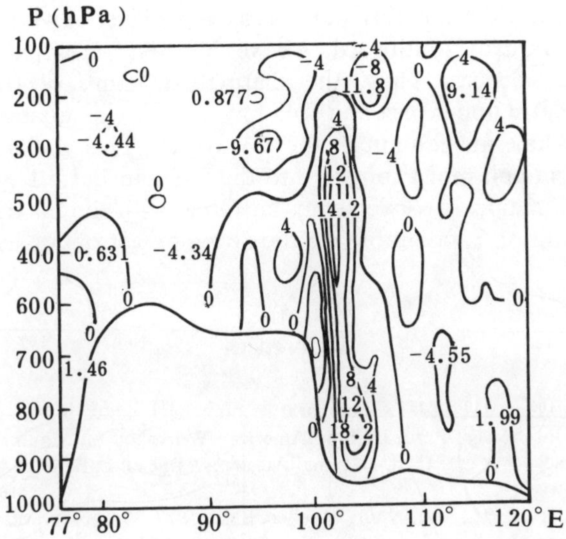

Figure 5.51. East-west cross section along 30° N of the relative vorticity at 12:00 GMT, 13 July 1981. The unit for relative vorticity is 10^{-5} s^{-1} (Kuo, Cheng and Anthes, 1986).

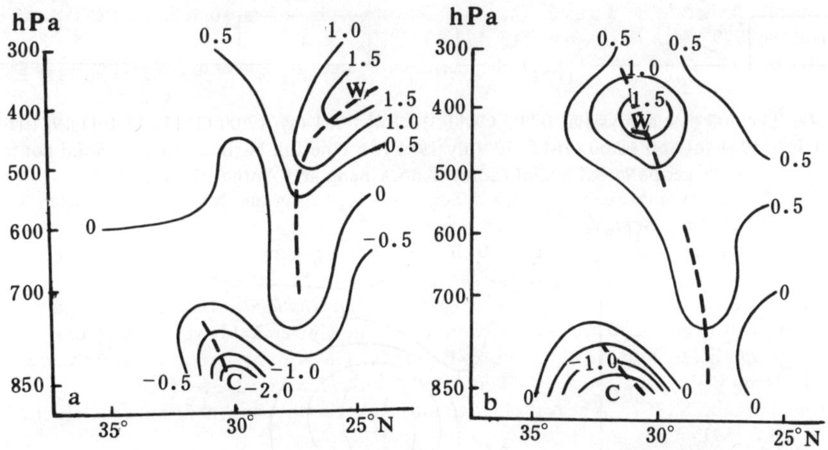

Figure 5.52. North-south cross-sections along 110° E of the temperature departure from the area-averaged values for 00:00 GMT, 5 May 1974 (a) and 00:00 GMT, 14 May 1973 (b). Units: °C.

with the system. The divergence field is characterized by low-level convergence and high-level divergence, with the non-divergenve level found at 6–7 km. Over the vortex region, the upward motion is significant at the mature stage, especially in the southeastern sector where the intense precipitation occurs.

Figure 5.52 is the north-south cross sections of temperature departure for two SW vortices. Below 500 hPa, especially at 850 and 700 hPa, the northern part of the vortices is colder, while the southern part has a deeper layer of warm air, with the axis lines tilting northward or southward. Above 500 hPa, an apparent warm area is situated over the vortices. Thus, the thermal structure of the SW vortex is characterized by warm upper part and cold lower part. The mixing ratio is generally very high in the whole air column of the SW vortex. This is closely related to the intrusion of a warm and moist tongue into the Sichuan Basin from the south. The atmospheric stratification is convectively unstable, with the maximum θ_e at 850 or 700 hPa. This condition is favorable for the development of convection.

References

Anthes, R. A., Haagenson, P., 1983: ‘A comparative numerical simulation of the Sichuan flooding catastrophe (11–15 July 1981)’. Proc. 1st Sino-American Workshop on Mountain Meteorology. The Chinese Academy of Science and U. S. National Academy of Science, Beijing, Amer. Meteorol. Soc., 519–524.

Blackmon, M. L., Wallace, J. M., Lau, N. C., Muellen, S. L., 1977: ‘An observational study of Northern Hemisphere wintertime circulation’. *J. Atmos. Sci.*, **34**, 1040–1053.

Bolin, B., 1950: ‘On the influence of the earth’s orography on the general character of the westerlies’. *Tellus*, **2**, 184–195.

Charney, J. G., Eliassen, A., 1949: ‘A numerical method for predicting the perturbations of the middle latitude westerlies’. *Tellus*, **1**, 38–54.

Chen, Longxun, Reiter, E. R., Feng, Z. Q., 1985: ‘The atmospheric heat source over the Tibetan Plateau: May-August 1979’. *Mon. Wea. Rev.*, **113**, 1771–1790.

Chen Longxun, Li Weiliang, 1985: ‘The atmospheric heat budget in summer over Asia monsoon area’. *Adv. Atmos. Sci.*, **2**, 487–497.

Chen Yuxiang, Ji Liren, Shen Rujin, 1985: ‘The numerical experiments on dynamic forcing by the Tibetan Plateau for various zonal flows’. *Adv. Atmos. Sci.*, **2**, 189–199.

Cressman, G. P., 1981: ‘Circulations of the West Pacific jet stream’. *Mon. Wea. Rev.*, **109**, 2450–2463.

Ding Yihui, Zhao Zhenming, He Shixiu, 1984: ‘Some new aspects in the study on global mean circulation at 200 hPa in summer’. *Kexue Tongbao*, **29**, 1506–1509.

Ding Yihui, Zhao Zhenming, He Shixiu, 1988: ‘A study of long-term mean circulation in the tropics and subtropics at 200 hPa for May-October’. *Sci. Atmos. Sinica*, **12**, 174–181.

Flohn, H., 1968: *Contributions to a meteorology of the Tibetan Highlands*. Atmospheric Science Paper No. 130, Department of Atmospheric Science, Colorado State University, Fort Collins, Colo.

Gao, Y.-X., Tang, M.-C. Luo, S.-W., Shen, Z.-B., Li, C., 1981: ‘Some aspects of recent research on the Qinghai-Xizang Plateau meteorology’. *Bull. Amer. Meteorol. Soc.*, **62**, 31–35.

Hahn, D. G., Manabe, S., 1975: ‘The role of mountains in the South Asian monsoon circulation’. *J. Atmos. Sci.*, **32**, 1515–1541.

He Haiyan, McGinnis, J. W., Song Z. S., Yanai, M., 1987: ‘Onset of the Asian summer monsoon in 1979 and the effect of the Tibetan Plateau’. *Mon. Wea. Rev.*, **115**, 1966–1995.

Hovermale, J. B., 1983: ‘Numerical experiments with the Sichuan flooding catastrophe (11–15 July 1981)’. Proc. 1st Sino-American Workshop on Mountain Meteorology. The Chinese Academy of Science and the U. S. National Academy of Science, Beijing, Amer. Meteorol. Soc., 243–264.

Krishnamurti, T. N., Bhalme, H. N., 1976: ‘Oscillations of a monsoon system. Part I: Observational aspects’. *J. Atmos. Sci.*, **33**, 1937–1954.

Kuo, H. L., Qian, Y. F., 1981: ‘Influence of the Tibetan Plateau on cumulative and diurnal changes of weather and climate in summer’. *Mon. Wea. Rev.*, **109**, 2337–2356.

Kuo, Y.-H., Cheng L. S., Anthes, R. A., 1986: ‘Mesoscale analyses of the Sichuan flood catastrophe, 11–15 July 1981’. *Mon. Wea. Rev.*, **114**, 1984–2003.

Lu Juning, 1984: ‘A composite study of low-level vortex over the Qinghai-Xizang Plateau’. Proc. the

Qinghai-Xizang Plateau Meteorological Experiment (QXPMEX). Vol. 2, Science Press, Beijing. 1984, 195–205.

Luo Huibang, Yanai, M., 1983: 'The large-scale circulation and heat sources over the Tibetan Plateau and surrounding areas during the early summer of 1979. Part I: Precipitation and kinematic analyses'. *Mon. Wea. Rev.*, **111**, 922–944.

Luo Huibang, Yanai, M., 1984: 'The large-scale circulation and heat sources over the Tibetan Plateau and surrounding areas during the early summer of 1979. Part II: Heat and moisture budgets'. *Mon. Wea. Rev.*, **112**, 966–989.

Lyons, S. W., Murakami, T., 1981: 'Orographic influence ofthe Rocky Mountains on the winter circulation over the contiguous United States. Part I: Large-scale aspects'. *J. Meteorol. Soc. Japan*, **59**, 320–336.

Murakami, T., Godbole, R. V., Kelker, R. R., 1970: 'Numerical simulation of the monsoon along 80° E'. In: C. S. Ramage, ed: Proceedings of the Conference on the Summer Monsoon in Southeast Asia, 39–51. Navy Weather Research Facility Report NTIS AD–876–677, Norfolk, Va.

Murakami, T., Unninayar, M. S., 1977: 'Atmospheric circulation during December 1970 through February 1971'. *Mon. Wea. Rev.*, **105**, 1024–1038.

Murakami, T., 1981: 'Orographic influence of the Tibetan Plateau on the Asiatic winter monsoon circulation. Part I: Large-scale aspects'. *J. Meteorol. Soc. Japan.*, **59**, 40–65.

Murakami, T., Ding, Y. H., 1982: 'Wind and temperature changes over Eurasia during the early summer of 1979'. *J. Meteorol. Soc. Japan*, **60**, 183–196.

Nakamura, H., 1976: 'Some problems in reproducing planetary waves by numerical models of the atmosphere'. *J. Meteorol. Soc. Japan*, **54**, 128–146.

Nakamura, H., 1978: 'Dynamic effects of mountains on the general circulation of the atmosphere: II. Effects of mountain barriers on the barotropic jet'. *J. Meteorol. Soc. Japan*, **56**, 341–352.

Nakamura, H., Murakami, T., 1983: 'Orographic effects on cold surges and lee cyclogenesis as revealed by a numerical experiment. Part I: Time mean aspect'. *J. Meteorol. Soc. Japan*, **61**, 524–546.

Newton, C. W., 1956: 'Mechanisms of circulation change during a lee cyclogenesis'. *J. Meteorol.*, **13**, 528–539.

Nitta, Ts., 1983: 'Observational study of heat sources over the eastern Tibetan Plateau during the summer monsoon'. *J. Meteorol. Soc. Japan*, **61**, 606–618.

Qian, Y. F., Yan, H., Wang, Q. Q., Wang, A. Y., 1988: *Numerical study of orographic effect in the planetary atmosphere*. Science Press (in Chinese), Beijing. 217 pp.

Reed, R. J. 1977: 'The structure and properties of African wave disturbances as observed during phase III of GATE'. *Mon. Wea. Rev.*, **105**, 317–313.

Riehl, H., 1967: *Southeast Asia monsoon study*. Colorado State University, Fort Collins, Colorado, 33 pp.

Reiter, E. R., Tang, M., 1984: 'Plateau effects on diurnal circulation pattern'. *Mon. Wea. Rev.*, **112**, 617–637.

Staff Members of the Academia Sinica, 1957: 'On the general circulation over eastern Asia, I'. *Tellus*, **9**, 432–446.

Staff Members of the Academia Sinica, 1958: 'On the general circulation over eastern Asia, II'. *Tellus*, **10**, 58–75.

Tao, S.-Y., Ding, Y. H., 1981: 'Observational evidence of the influence of the Qinghai-Xizang (Tibet) Plateau on the occurrence of heavy rain and severe convective storms in China'. *Bull. Amer. Meteorol. Soc.*, **62**, 23–30.

Wang Bin, Orlanski, I., 1987: 'Study of a heavy rain vortex formed over the eastern flank of the Tibetan Plateau'. *Mon. Wea. Rev.*, **115**, 1370–1393.

Webster, P. J., Zhou, L. C., 1980: 'Seasonal structure of a simple monsoon system'. *J. Atmos. Sci.*, **37**, 354–367.

Weng Duming, 1986: 'An analysis of the characteristic features of the surface heat source and heat balance over the Qinghai-Xizang Plateau from May to August, 1979'. Proc. International Symposium on the Qinghai-Xizang Plateau and Mountain Meteorology. March 20–24, 1984, Science Press, Beijing. 201–218.

Yanai, M., 1987: 'Effects of the Tibetan Plateau on the evolution of the Asian summer monsoon'. The 3rd US-PRC Monsoon Workshop, Kunming, China. 16–23 June, 1987.

Yang, J. C., Tao, S.-Y., Gu, Z. C., Ye, D. Z., 1960: *Meteorology on the Tibetan Plateau.* Science Press, (in Chinese), Beijing. 280 pp.

Yao Lanchang, Luo Siwei, 1984: 'Monthly mean heating fields of the atmosphere and their annual variation over Asia and the Qinghai-Xizang Plateau'. Proc. 1st Sino-American Workshop on Mountain Meteorology. The Chinese Academy of Science and the U. S. National Academy of Science, Beijing, Amer. Meteorol. Soc., 413–423.

Yeh, T.-C., Yan, W. Y., Wu, G. X., 1989: *A diagnostic study of thermal and circulation fields over the Tibetan Plateau in summer.* Ph.D. dissertation. Institute of Atmospheric Physics. 239 pp.

Yeh, T.C., Gao, Y. X., Tang, M. C., Lo, S. W., Shen, C. B., Gao, D. Y., Song, Z. S., Qian, Y. F., Yuan, F. M., Li, G. Q., Ding, Y. H., Chen, Z. T., Zhou. M. Y., Yang, K. J., Wang, Q. Q., 1979: *Meteorology of Qinghai-Xizang (Tibet) Plateau.* Science Press, Beijing. (in Chinese), 278 pp.

Yeh, T.-C., Dao, S.-H., Li, M.-T., 1959: 'The abrupt change of circulation over the Northern Hemisphere during June and October'. In: B. Bolin ed: The Atmosphere and the Sea in Motion, Rossby Memorial, The Rockefeller Institute Press, New York. 249–267.

Yin, M.-T., 1949: 'A synoptic-aerologic study of the onset of the summer monsoon over India and Burma'. *J. Meteorol.*, **6**, 393–400.

Zhang, J. J., Zhu, B. Z., Zhu, F. K., Weng, D. M., Sun, G. W., Lu, J. N., Peng, Y. Q., Wang, Y. Z., 1989 *Advances in the Qinghai-Xizang Plateau Meteorology*. Science Press, Beijing (in Chinese). 268 pp.

Zhou, X.-P., Hu, X.-F., 1982: 'A brief analysis and numerical simulation of the Sichuan extraordinarily heavy rainfall event'. Proc. 1st Sino-American Workshop on Mountain Meteorology. The Chinese Academy of Science and the U. S. National Academy of Science, Beijing. Amer. Meteorol. Soc., 555–565.

Zhu Baozhen, 1957: 'The steady-state perturbations of the westerlies by the large-scale heat sources and sinks and the earth's orography (II)'. *Acta Meteorol. Sinica*, **28**, 198–224.

CHAPTER 6

HEAT, MOISTURE AND ENERGY BUDGETS OVER THE MONSOON REGIONS OF CHINA AND SOME ASPECTS OF MONSOON DYNAMICS

6.1. Moisture and Water Budgets in the Monsoon Regions

Moisture transport and budget are closely related to monsoon activity. As the onset of the summer monsoon occurs, a large amount of moisture is brought into the monsoon region with monsoonal airflows, thus providing a necessary condition for the possible occurrence of monsoon precipitation. The wet episode of the monsoon season usually corresponds to the atmosphere with high moisture content and much precipitation. Therefore, significant attention has been paid to this problem.

Figure 6.1 and 6.2 are the moisture content in a mean air column for May and June 1979–1981 (Chen *et al.*, 1983). It can be seen that in May an area of high moisture content with the maximum of 60 kg m^{-2} is observed over the southern part of the South China Sea and the southeastern part of the Indo-China Peninsula (10° N), while an area of low moisture content is located in the near-equatorial region. In June, the area of high moisture content moves northward up to 22° N, with maximum of 64 kg m^{-2}. In the near-equatorial region, low moisture content conditions still exist. In southern China the moisture content generally augments. Note that another area of high moisture content over the Bay of Bengal, comparable in magnitude to the one over the South China Sea, is created. The above distribution characteristics are closely related to the onset and activity of the summer monsoon in East Asia. The area of high moisture content over the South China Sea, and the Indo-China Peninsula occurs when the summer monsoon sets in over this region. The monsoon over the Bay of Bengal does not occur, at this time, thus the moisture content does not show any significant increase. In June, the summer monsoon over the South China Sea expands northward and westward and meanwhile the summer monsoon over the Bay of Bengal begins to occur, therefore, the moisture content in these two regions shows a significant increase, with the second area of high moisture content produced over the Bay of Bengal. This result also provides further evidence of the existence of the East Asian monsoon system.

The vertical partition of moisture in the air column is nearly the same for May and June, i.e., the moisture content from the surface to 850 hPa accounts for about 40% of the total in the air column, from 850 to 700 hPa about 30%, and above 500 hPa less than 10%. The major portion of the moisture is concentrated in the layer below 700 hPa.

Figure 6.3 and 6.4 show the intergrated moisture transport from the surface to 300 hPa. In May, the maximum moisture transport belt is located in southern China and

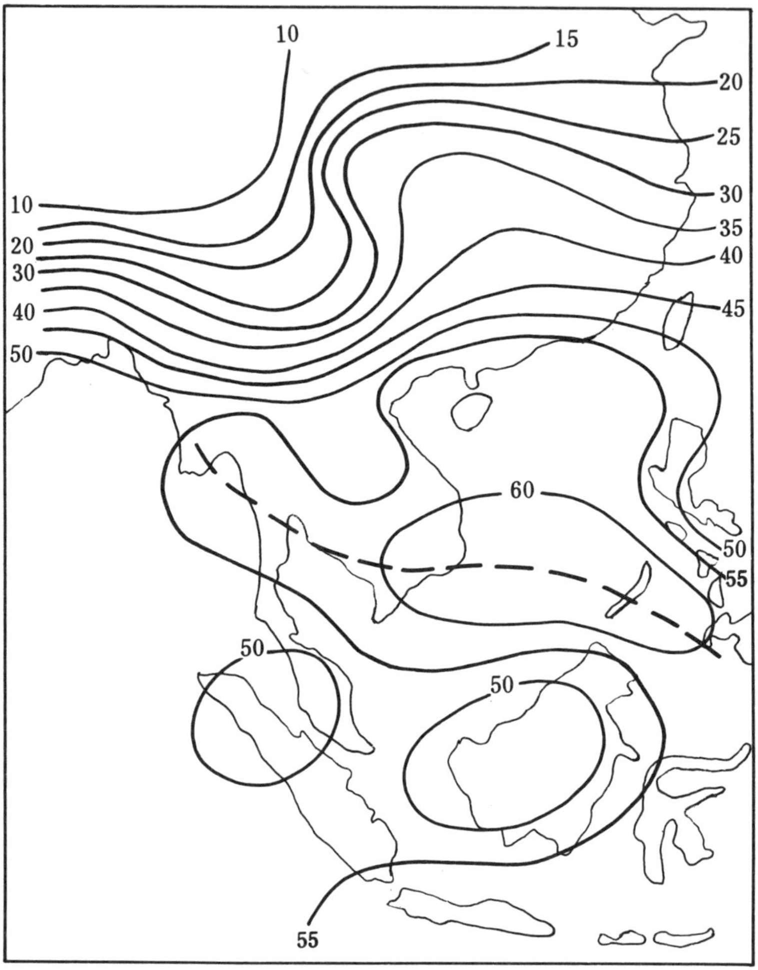

Figure 6.1. The distribution of the moisture content in a mean air column for May 1979–1981. Units: kg m^{-2} (Chen *et al.*, 1983).

rapidly decreases northward. It can be seen that the strong moisture transport belt in the coastal area of southern China does not originate in the cross-equatorial airflow, nor does it come from the airflow associated with the subtropical high over the western Pacific. On the other hand, the moisture could not come from the Bay of Bengal. Thus, the moisture supply in May in southern China mainly has its origin in the northern part of the South China Sea. The moisture transport in June is rather different fron that for May, with two maximum transport areas: the southern part of the Bay of Bengal and the middle and lower Yangtze River Valley . The moisture over the southern part of China is mainly supplied by the transport belt over the middle and northern part of the South China Sea. A part of the moisture supply comes from the Bay of Bengal while the moisture supply received in the middle and lower Yangtze

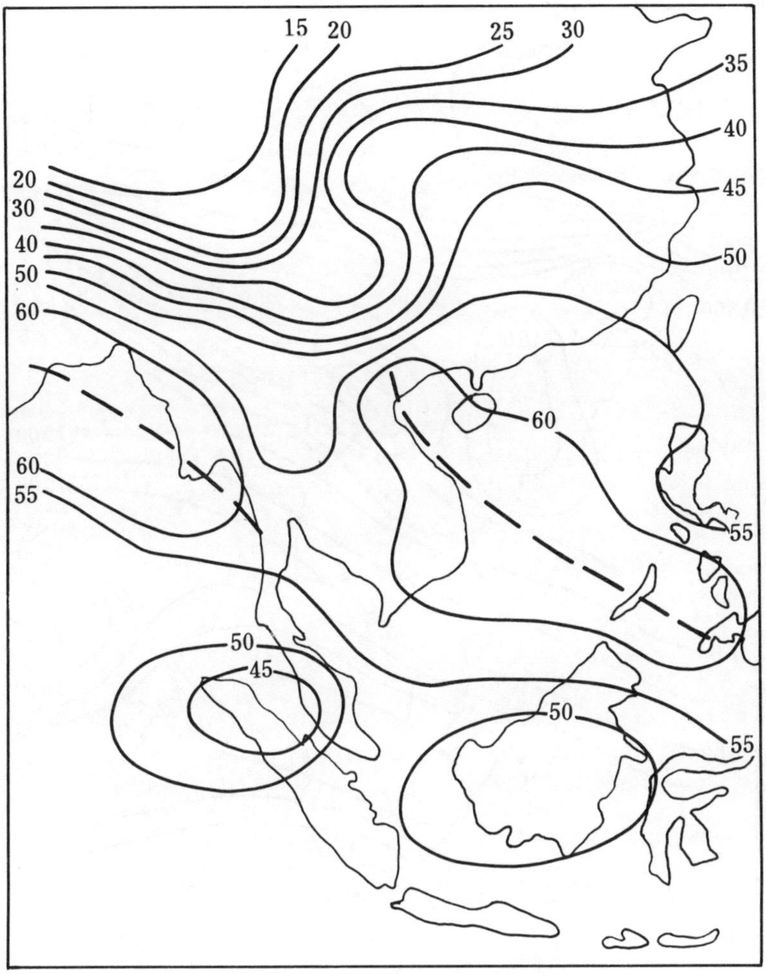

Figure 6.2. The same as Fig. 6.1, but for June (Chen *et al.*, 1983).

River Valleys is partly used for local Meiyu rainfalls and the remaining part further transported to Japan, providing an important moisture source for the rainy reason there.

The moisture transport in July is mainly governed by two branches of airflows: one coming from the Bay of Bengal and the other from the south and southeast. The maximum moisture transport belt stretches from the northern part of the South China Sea to the middle and lower Yangtze River Valleys, with a greater magnitude than that in June (Fig. 6.5) (Shen, 1983). Note that the moisture may be tansported northward into northern China, corresponding to the rainy season there.

Figure 6.6 is the mean July moisture flux divergence for 1974–1978, showing that there is a significant moisture flux convergence area in the Yangtze River and Huaihe

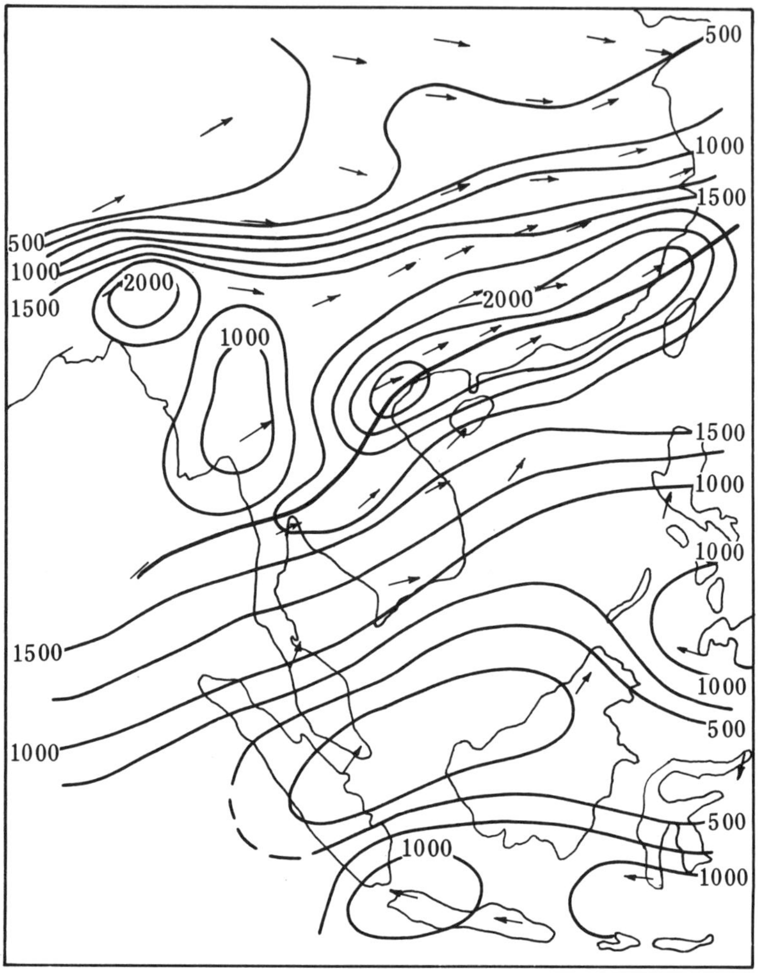

Figure 6.3. The mean moisture transport for May integrated from the surface to 300 hPa for 1979–1981. The arrows represent the mean transport direction for the whole column. Units: g cm^{-2} s^{-1} (Chen *et al.*, 1983).

Valleys, with the moisture flux divergence areas located to the north (the Yellow River and northern China) and to the south (Guangdong and Guangxi Provinces) of this area, respectively. This pattern of moisture flux divergence is favorable for the occurrence and maintenance of precipitation in the Yangtze River and Huaihe Valleys, as a major moisture source.

Some investigators have studied the problem of the origin of moisture transported into China. Previously in the 60's, it was argued that the moisture over the eastern part of the China Main mainly came from the south, but this is only true for individual years.

Based on a study by Chen *et al.* (1983), in May and June the moisture transport into

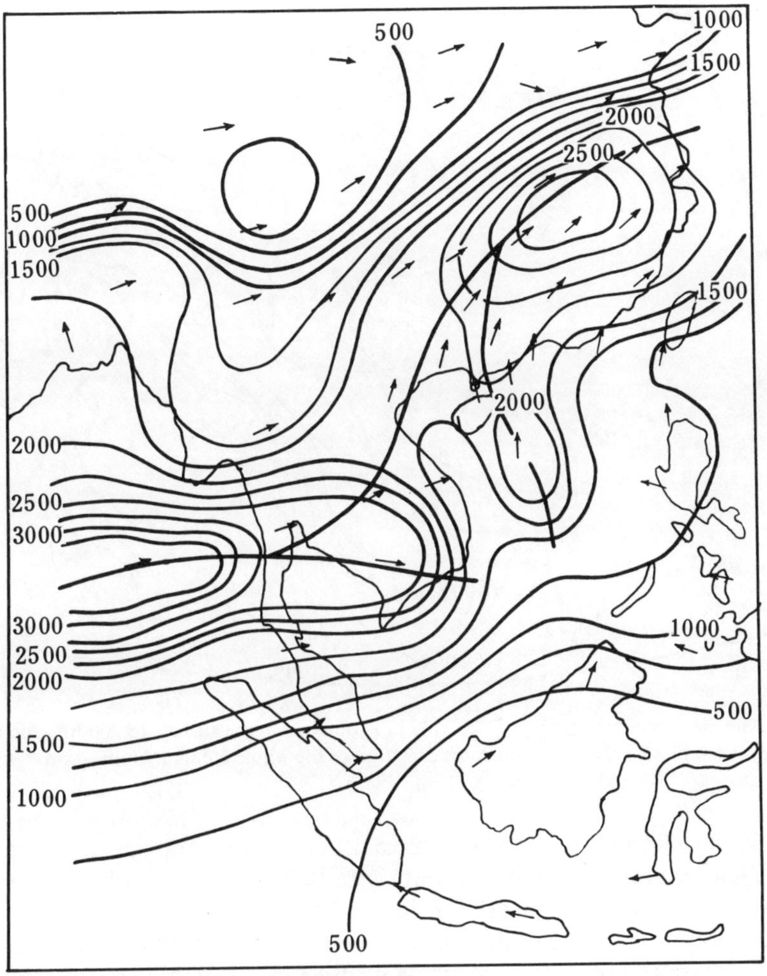

Figure 6.4. The same as Fig. 6.3, but for June (Chen *et al.*, 1983).

southern China and the Yangtze River Valley mainly comes from the southern boundary. In May (Table 6.1), the moisture imput from the south is about one and a half times greater than that from the west and in June this rate increases up to 4–6 times greater. In the southwest region of China, the moisture transport mainly comes

Table 6.1. The ratio of moisture flux from the western boundary and that from the southern boundary (Chen *et al.*, 1983)

Month	Southern China	The Yangtze River Valley	Southwest region
May	1:1.6	1:1.4	24:1
June	1:5.7	1:4.1	2:1

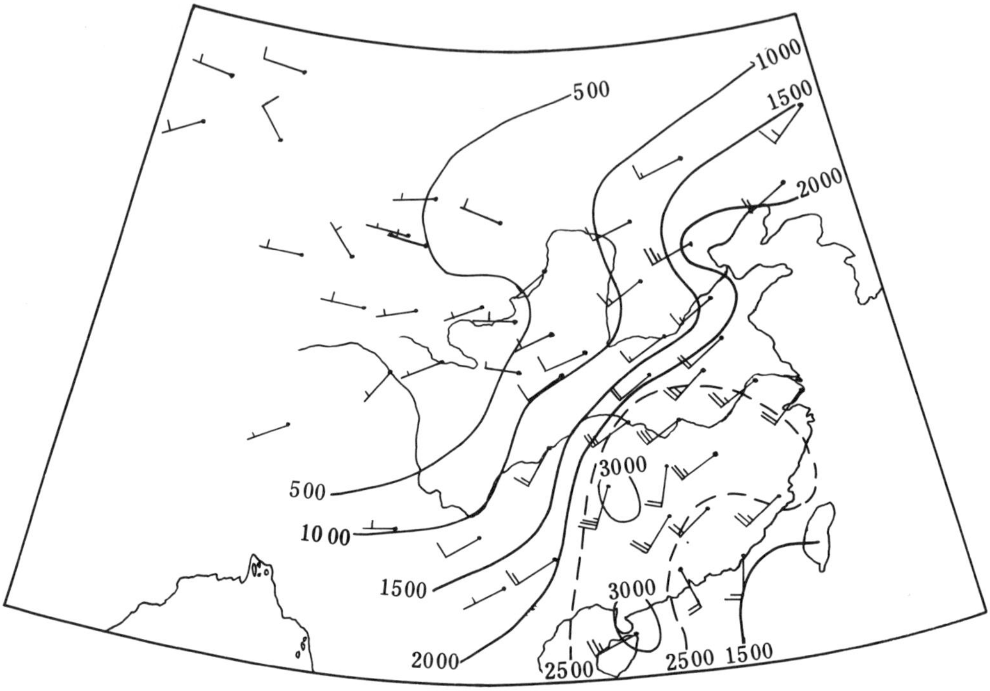

Figure 6.5. The mean moisture transport for July integrated from the surface to 200 hPa for 1974–1978. The wind vectors represent the mean transport direction of the whole column. Units: g cm^{-2} s^{-1} (Shen *et al.*, 1983).

from the western boundary. In May almost all moisture is transported in this region from the western boundary and is about twice that from the southern boundary.

Based on a study by Shen (1983), it has been shown that in recent times the moisture imported during July has also come from the southern, the southwestern and the southeastern boundaries. As in May and June, generally, the moiture coming from the southern boundary is most significant while that coming from the southeastern boundary makes a minimum contribution. An obvious example is the moisture import into the Yangtze River Valley. It can be seen in Table 6.2 that the

Table 6.2. The moisture amount transported into the Yangtze River Valley across the southern and southwestern boundaries in July 1974–1978. Units: 10^7g s^{-1} 200 km^{-1} (Shen *et al.*, 1983)

Year	Southwestern boundary	Southern boundary
1974	2784	4974
1975	3251	4286
1976	2464	3309
1977	2934	5334
1978	1628	1750
Average	2612	3930

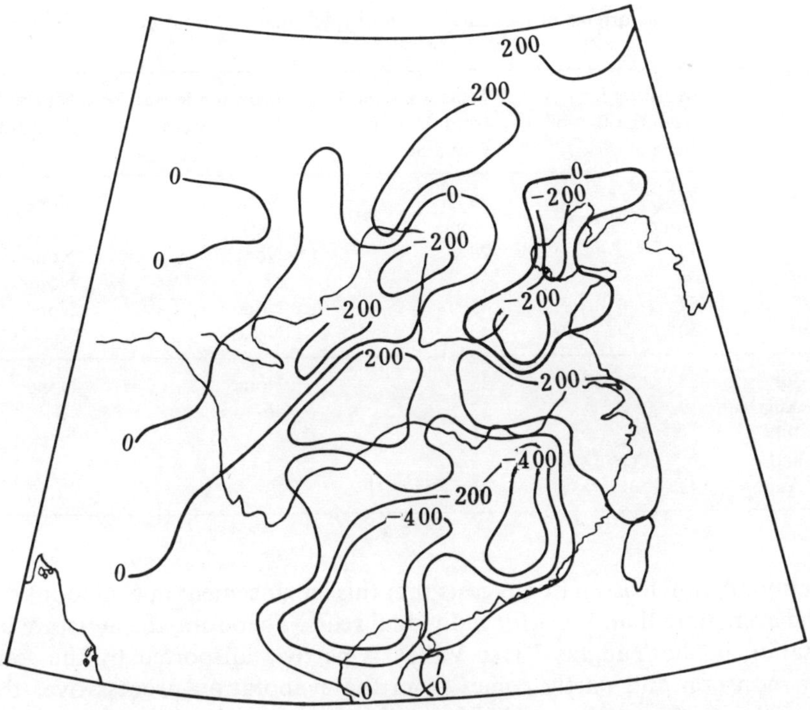

Figure 6.6. The pattern of moisture flux divergence in July averaged for 1974–1978. Units: mm. The "minus" sign denotes divergence (Shen *et al.*, 1983).

maximum moisture transport into this region is produced by importation at the southern boundary, and then the southwestern boundary.

Luo (1983) estimated the percentage of different moisture sources with respect to the total moisture import for July and August (See Table 6.3). For the mean July condition, 67% of the moisture had its origin in the Bay of Bengal and 30% came from the South China Sea. But, for the mean August condition, the moisture import from the western Pacific accounted for 46%, while 32% came from the South China Sea. The percentage of moisture transport from the Bay of Bengal was only 19%. Therefore, the moisture import from the east is very important in August for the rainy season in northern China.

It should be pointed out that the problem of the origin of moisture in fact includes two aspects: the moisture source available for precipitation over the China Main and the contribution of the moisture originating in the China Main to this moisture source.

As discussed above, the moisture is mainly brought in by the southwest monsoonal airflow, being transported into the China Main across the southern and southwestern boundaries. It is very interesting that the moisture amount continuously increases to the south of the Yangtze River as the moisture is transported northward, indicating that, during the northward inflow of moisture, it may be replenished with local

Table 6.3. Statistics of the origins of moisture transport (percentage of total moisture transport) (Luo, 1983)

Year		Moisture import from the Bay of Bengal	Moisture import from the South China Sea	Moisture import from western Pacific	Moisture import from the north
July 1978		71	29	None	None
Aug. 1978		None	None	100	None
July 1979		72	28	None	None
Aug. 1979		6	56	38	None
July 1980		57	43	None	None
Aug. 1980		50	40	None	10
Average	July	67	33	None	None
	Aug.	19	32	46	3
	July and Aug.	43	33	23	2

evaporation. A rough estimate suggests that this enhancement may amount to about 60%, and even more than 100% for individual years. Therefore, the moisture used for precipitation in the Yangtze River Valley is partly transported by the southwest summer monsoon and partly comes from the evaporation process over the land surface in the region to the south of it.

It is rather controversial whether or not the eastern part of China is a source of moisture. Xu's estimate (Xu, 1958) indicated that the eastern part of China is a source of moisture. Based on the estimate for drought years in the Yangtze River Valley, Shen (1983) also obtained the same conclusion as to the source of moisture. A recent estimate by Luo (1983) has suggested that for drought years the Yangtze River Valley is the source of moisture, while it is the sink of moisture for normal or flooding years. This problem remains to be further studied.

The next problem is the regional moisture budget. The moisture budget equation may be written in the following form:

$$R = \Delta F + E + \Delta W, \tag{6.1}$$

where R is the precipitation amount, ΔF is the net moisture flux for the studied region or the moisture flux divergence, E is the regional evaporation, and ΔW the moisture storage in the atmosphere which may be neglected in some cases due to its small magnitude. Chen *et al.* (1983) estimated the moisture budgets for May and June over the South China Sea and the Yangtze River Valley, as shown in Table 6.4. In May and June (especially in May), southern China undergoes a pre-summer rainy season. At this time, it has a maximum net moisture convergence (46,500 tons s^{-1}) and a maximum precipitation amount (235.4 mm $mon.^{-1}$). In June when the Yangtze River Valley has the Meiyu season, it corresponds to the maximum moisture sink or moisture convergence (70,600 tons s^{-1}) and the maximum precipitation amount (256.5 mm $mon.^{-1}$). Local evoporation also makes a large contribution to the precipitation, about one-fourth of the precipitation amount. In this estimate, the moisture storage term is neglected because of the small magnitude in the rainy season. It is to be noted that the moisture budgets for the northern and southern parts of the South China Sea are quite distinctive. In both May and June, the northern part is a moisture source, with moisture transported into the China Main, while the southern part is a strong moisture sink (particularly in June), thus producing 373 mm precipitation, as estimated from Eq. (6.1).

Table 6.4. Mean moisture budgets for various regions in May and June averaged for 1979–1981 (Chen *et al.*, 1983)

Month	Region	ΔF 10^{-4} t. s^{-1}	E mm mon.$^{-1}$	R (estimated) mm mon.$^{-1}$	R' (observed) mm mon.$^{-1}$
	Yangtze River Valley	4.36	59.0	179.2	151.4
	Southern China	4.65	51.3	235.4	234.9
May	North part of South China Sea	−1.14	116.0	91.1	105.6
	South part of South China Sea	6.41	99.3	162.2	not available
	Yangtze River Valley	7.06	68.1	256.5	212.9
	Southern China	3.36	73.2	206.2	198.3
June	North part of South China Sea	−2.61	101.2	44.1	124.3
	South part of South China Sea	28.51	93.9	373.4	not available

Luo (1983) also used Eq. (6.1) to estimate the moisture budget of the southeastern part of China (23–35° N, 100–120° E) in July and August for 1978–1980 (Table 6.5). It can be seen that for 1978 (the drought year), the mean precipitation amount is far smaller than the flooding year (1980). Note the negative ΔF, indicating moisture flux divergence in this region. Thus, the moisture supply or compensation can only be the local evaporation. Table 6.5 shows the relatively large surface evaporation for 1978 which exceeds the precipitation amount in this region. This evidence suggests that the evaporation from the underlying surface is an important moisture source for drought years. As indicated above, this result is in good agreement with that obtained by Xu

(1958). For normal and flooding years (1979 and 1980), in contrast to drought years, the net moisture flux is positive, i.e., moisture flux convergence. In such cases, the region under study is a moisture sink.

Table 6.5. Mean moisture budgets over the southeastern part of China in July and August for 1978–1980 (Luo, 1983)

Year	ΔF 10^4 t.mon^{-1}	E 10^4 t.mon^{-1}	E^* mm cm^{-2}	R 10^4 t.mon^{-1}	R^* mm mon.$^{-1}$	ΔW 10^4 t.mon^{-1}
July 1978	-66.9×10	372.8×10	165	308.2×10	136	2.3×10
Aug. 1978	-77.7×10	411.4×10	182	326.9×10	145	-6.8×10
July 1979	361.6×10	52.8×10	23	398.6×10	176	-15.8×10
Aug. 1979	281.3×10	121.2×10	54	388.9×10	172	-13.6×10
July 1980	171.3×10	278.8×10	123	441.0×10	193	-9.1×10
Aug. 1980	299.9×10	218.3×10	97	502.4×10	222	-15.8×10

E^* and R^*: long-term mean evaporation and rainfall amount, respectively.

Moisture transport and budget in China has a significant inter-annual variability which is connected with great monsoon inter-annual vagary. In particular, for drought and flooding years, the moisture conditions show an obvious contrast. Chen *et al.* (1983) made a contrasting study of moisture conditions between dry months (May 1979 and June 1981) and wet months (June 1979 and May 1981) for southern China and the Yangtze River Valley. For drought years, there was strong eastward moisture transport over the southern part of the Indo-China Peninsula, with the maximum transport axis located along 10° N. Over nearly the whole South China Sea region, there was significant moisture convergence or moisture sink. For flooding years, the opposite case was true. Over the northern part of the South China Sea, the moisture divergence or moisture source was very marked, and the moisture flux coming from the west greatly decreased. It seems to be that the precipitation condition in southern China is opposite to the activity of the summer monsoon over the Bay of Bengal. A preliminary estimate of moisture origin suggests that, for drought years, the moisture import from the Bay of Bengal is larger than that from the South China Sea and the opposite case is true for flooding years. In the southern part of China, drought years were characterized by moisture divergence while flooding years were characterized by convergence. Most of the moisture came from the southern coastal region of southern China. For either drought years or flooding years, the moisture fluxes coming from the cross-equatorial airflow and the western flank of the subtropical high were very weak.

Based on Eq. (6.1), Chen *et al.* (1983) estimated the moisture budget for various subregions (Table 6.6). It can be seen that very significant evaporation occurred over the northern part of the South China Sea, with the areal and monthly mean evaporation amount being 100–200 mm and far greater evaporation for flooding years. This strong evaporation provides another moisture source for transport into the southern China region and creating moisture convergence favorable for precipitation occurrence. The moisture convergence for flooding years in southern China and the Yangtze River region was generally greater than that for drought years, especially for southern China. Over the northern part of the South China Sea, it was moisture source for flooding years while it was a strong moisture sink for drought years. Precipitation amounts are in general agreement with the observed one.

Table 6.6. Monthly mean moisture budgets for drought years and flooding years over various regions* (Chen *et al.*, 1983)

Year	Month	Terms of Eq. 6.1	The Yangtze River Valley	Southern China	Northern part of the South China Sea	Southern part of the South China Sea
Drought years	May	ΔF	3.57	3.37	8.65	22.08
		E	93	49	112	95
		R	191	182	299	311
		R'	148	233	218	not available
	June	ΔF	2.81	325	5.60	37.85
		E	86	78	98	75
		R	161	203	216	434
		R'	236	210	399	not available
Flooding years	May	ΔF	3.21	5.10	−9.45	9.45
		E	104	56	129	116
		R	193	258	0	209
		R'	190	226	82	not available
	June	ΔF	4.19	5.73	−1.20	21.48
		E	100	63	105	105
		R	212	283	80	309
		R'	288	249	69	not available

* R and R' are the estimated and observed rainfall amounts, respectively. Units are the same as for Table 6.5.

Jin (1983) also estimated the moisture transport over the South China Sea for the summer of 1979, arriving at similar conclusions to those of Chen *et al.* This region acts as a moisture source for the China Main, but it is a moisture sink for other oceanic areas because the moisture coming from them first concentrates in this region and is then further tranported into southern China. In June-August 1979, most of the moisture was transported into the South China Sea from the west and the south. At the northern boundary, the moisture flux is always directed northward. However, during the period of active monsoon, the northward moisture transport into southern China increased by 2–3 times. This situation also occurred before and after the onset of the Indian summer. The northward moisture transport post to the onset of the monsoon was as 3.8 times greater than before the onset. Another interesting finding is the periodic behavior of moisture transport over the South China Sea, with the biweekly oscillation (13–14 days) or 8–10 days oscillation dominating. In addition, there exists a shorter-period oscillation (3–4 days) which may be associated with the propagation of tropical waves.

In 1980, the precipitation was very unusual in most of China, with the pre-summer rainy season in southern China ending during the last part of May, with a northward shift of the major rain belt over the Yangtze River Valley and Huaihe River, and even farther north, in the first part of June , and then retreating to the Yangtze River Valley in July and August. The period of the Meiyu season in the Yangtze River Valley is especially prolonged, spanning 83 days, from 9 June to 31 August, and producing a total rainfall amount that was 70–100% greater than normal. This year was the most severe flooding year since 1896, and only second in severity to 1954. An analysis of the moisture flux budget for the summer of 1980 indicates that the unusual and prolonged precipitation event is maintained by the significant moisture flux convergence by monsoonal airflow (Shen and Huang, 1983). Figure 6.7 demonstrates that at the latitude band of 27.5–32.5° N, there was always a considerable moisture flux convergence each month, with the maximum located at about 30° N in July and August, when rain-producing disturbances such as shear-lines are often present, thus causing persistent heavy rainfalls. Based on the estimate of the trajectory of moisture transport, one finds that the moisture transported into the Yangtze River Valley mainly has its origin in the tropical monsoonal airflow over the Bay of Bengal and, secondly, in the western Pacific and the South China Sea. Luo (1983) also obtained a similar result. This condition is rather different from the conditions in May and June as discussed previously.

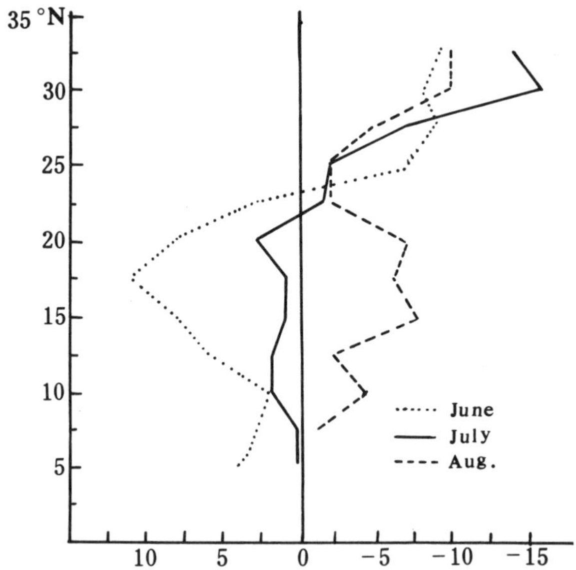

Figure 6.7. Latitudinal profile of moisuture flux divergence averaged for the 125° E longitudinal range for June-August of 1980. Units: g cm^{-2} s^{-1} (Shen and Huang, 1983).

6.2. Heat Budget in the Monsoon Regions

As previously indicated in Chapter 5, numerous authors have investigated the heat sources and sinks in summer over the Asian monsoon region, including Luo and Yanai (1984), and Chen and Li (1985). They pointed out that the strongest heat source in the summer monsoon season is located over the Bay of Bengal and central India rather than over the Tibetan Plateau, because the former has a maximum heating rate of 8°C day^{-1}, while the latter only has a maximum heating rate of about 1°C day^{-1}. Other heat sources are found over the region to the east of the South China Sea and tropical western Pacific. These heat sources play a very important role in driving the regional monsoon circulations.

Firstly, a detailed discussion of the heat budget over the South China Sea and southern China will be given below, mainly based on the study by Ju and Luo (1987). They utilized the dataset for the period April-June 1981 to investigate the relationship between heat sources during the build-up of the summer monsoon in Asia and the tropical circulation. The atmospheric heat source (Q_1 and Q_2) may be estimated with Eqs. (5.9) and (5.10), using large-scale variables.

Figure 6.8 is the temporal variations of the areal mean vertically integrated Q_1 and Q_2, i.e., $[<Q_1>]$ and $[<Q_2>]$. $[<\ >]$ represents the areal average and $<\ \ >$ the vertical intergral. It can be seen that $[<Q_1>]$ and $[<Q_2>]$ have similar trends of variation, indicating that the major contribution of Q_1 was made from the condensational latent heating. Prior to 25 April $[<Q_1>]$ and $[<Q_2>]$ were negligibly small, being generally lower than 20 w m^{-2}. This condition corresponds to the winter situation, i.e., the absence of a heat source over Southeast Asia. After 11 May $[<Q_1>]$ and $[<Q_2>]$ significantly increased as much as one order of magnitude, with $[<Q_1>] > 100$ w m^{-2} and $[<Q_2>] > 75$ w m^{-2}. The analysis of the flow patterns at 850 hPa also show a significant change around 11 May. Therefore, the date of 11 May is defined as the onset day of the summer monsoon of 1981 over Southeast Asia (including the Indo-China Peninsula, the South China Sea, and southern China). As previously indicated, the summer monsoon in this region usually sets in much earlier than its counterpart over the Indian Peninsula.

Figure 6.9 is the mean profiles of $[Q_1]$, $[Q_2]$ and $[\omega]$ for four different stages of the monsoon development. The late spring stage (A) was characterized by the negative $[Q_1]$ in nearly the whole troposphere, indicating the dominance of radiative cooling,

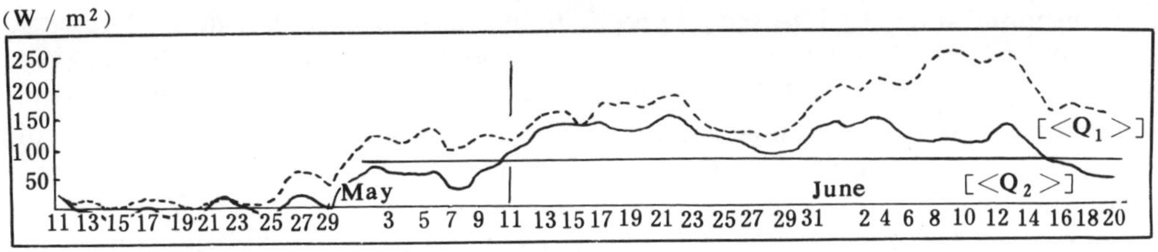

Figure 6.8. The variations of the area-averaged, vertically integrated Q_1 and Q_2 ($[<Q_1>]$ and $[<Q_2>]$) for April-June 1981 with time. Units: w m^{-2} (Ju and Luo, 1987).

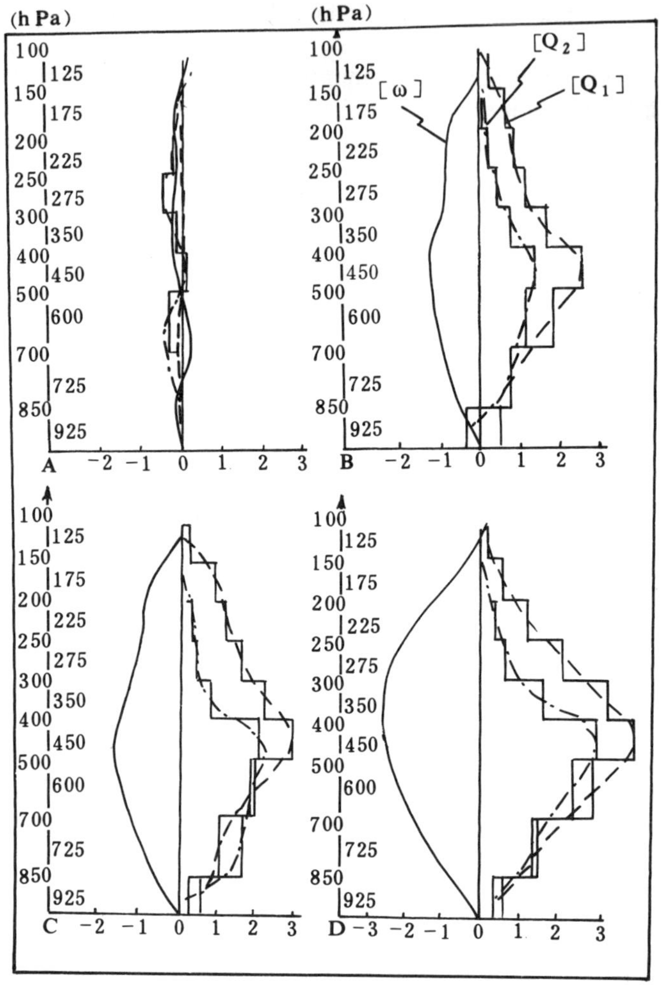

Figure 6.9. Mean profiles of $[Q_1]$, $[Q_2]$ and $[\omega]$ for four stages (A, B, C and D) of summer monsoon development over Southeast Asia (see the text). Units: °C day^{-1} for $[Q_1]$ and $[Q_2]$; hPa h^{-1} for ω (Ju and Luo, 1987).

and the negative $[Q_2]$ below 450 hPa, indicating the presence of evaporation cooling. At the initial stage (B), $[Q_1]$ and $[Q_2]$ began to increase, with $[Q_1]$ having 2.4 °C day^{-1} at 450 hPa and $[Q_2]$ 1.5°C day^{-1}. Due to the insignificant separation of the peaks of $[Q_1]$ and $[Q_2]$, it is inferred that the frontal continuous precipitation could dominate this region. At the onset stage (C), $[Q_1]$ and $[Q_2]$ rapidly enhanced, with the peak of $[Q_1]$ of 2.8°C day^{-1} at 450 hPa and the peak of $[Q_2]$ of 1.6°C day^{-1} at 500 hPa. At the prevailing stage (D) $[Q_1]$ and $[Q_2]$ further increased, with the peaks of $[Q_1]$ and $[Q_2]$ being of 3.8 and 2.9°C day^{-1},respectively. Note the increasing contribution of the condensational heating during the later two stages, but convective activity seems to be somewhat inactive over Southeast Asia and the South China Sea.

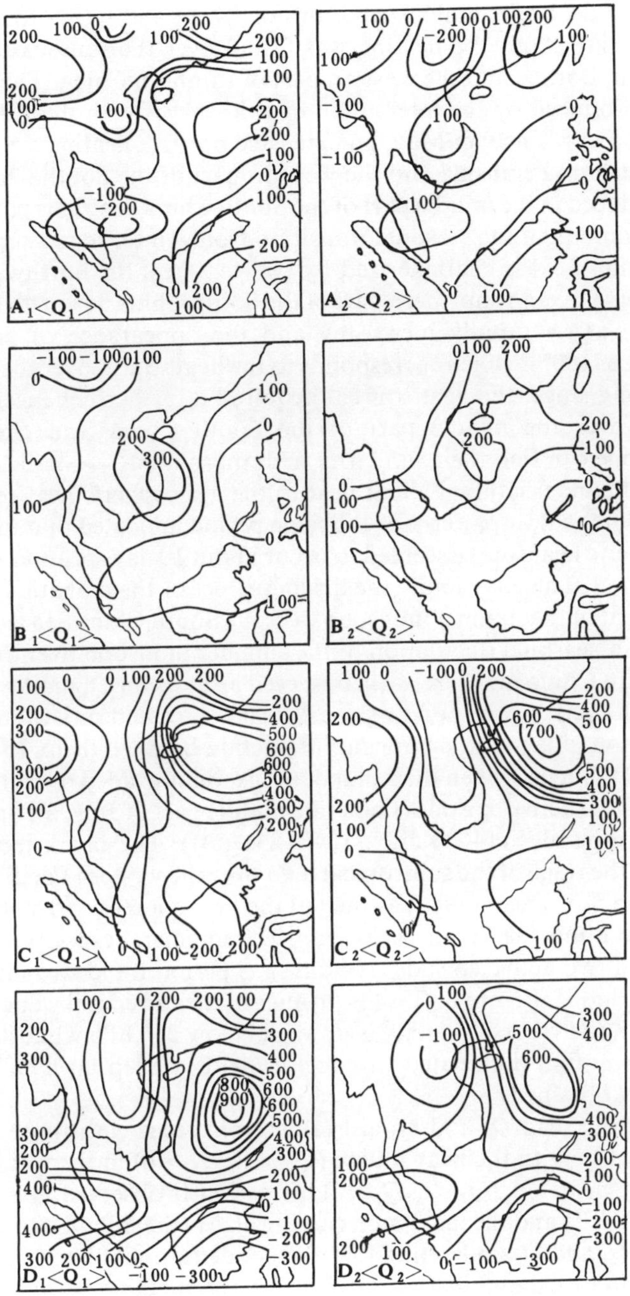

Figure 6.10. The horizontal patterns of Q_1 and Q_2 for the various stages (same as Fig. 6.9) of summer monsoon development of 1981. Units: w m^{-2} (Ju and Luo, 1987).

Figure 6.10 is the horizontal distributions of $<Q_1>$ and $<Q_2>$ for various stages. For the late spring stage, the heating was found over the southeastern part of the Tibetan Plateau, Borneo and the wastern part of southern China. The latter two heat sources were produced by the interaction of cold surges with the ITCZ and frontal activity, respectively. Early in May, an extensive positive heating area was observed over the Indo-China Peninsula and the western part of the South China Sea, with a maximum produced in the middle part of the South China Sea. This pattern of heating reflects the beginning of the seasonal transition from winter to summer. At the onset stage, the heating field is characterized by two features: the northward shift of the heating maximum over the northern part of the South China Sea, with the magnitudes of $<Q_1>$ and $<Q_2>$ rapidly increasing and the appearance of another heating maximum west of 100° E which corresponds to the heat source over the Bay of Bengal. At the prevailing stage, the heat sources acquired a further enhancement when the ITCZ moved over the middle part of the South China Sea. From the $<Q_2>$ distributions, one can find their similarity with those of $<Q_1>$, indicating the major contribution of condensational latent heat to the atmospheric heat sources.

An analysis of the flow pattern for the same period indicated that the development of the atmospheric heat source seemed to occur about 10 days prior to the onset of the summer monsoon. This may imply the driving effect of the diabatic heating field on the establishment of the summer monsoon over the South China Sea and southeastern China. The intra-seasonal fluctuation of the summer monsoon over the South China Sea and southern China may be clearly observed after its onset, mainly showing up in the active-break cycle. Correspondingly, the heat source and sink should also assume a similar variation, due to the significant interaction between them. In the summer of 1979 (31 May-31 August) when the summer monsoon prevailed over the South China Sea, there were two active episodes (June-the middle part of July; 26 July– 14 August) and two break episodes (14–25 July, 15–31 August). The break monsoon usually occurred when the ridge of the subtropical high over the western Pacific extended into the South China Sea. Duan (1983) estimated the heat sources and sinks averaged for the active period and the break period of 1979 summer, respectively, defining the above-described two active episodes (64 days) as period I and two inactive episodes (29 days) as period II. Figure 6.11 is his profiles of mean vertical velocity ($\omega = dp/dt$). During period I, the vertical velocity is upward below 250 hPa while in the period II, the downward motion dominated this region, with a deep layer of mean positive vertical velocity obtained.

The profiles of mean Q_1 and Q_2 are rather consistent with those of the vertical velocity (Fig. 6.12), with the heat source ($Q_1 > 0$, $Q_2 > 0$) in the whole layer during period I and the heat sink ($Q_1 <0$, $Q_2 <0$) for period II observed. Maximum heating was found at 500 hPa and the maximum cooling also at 500 hPa. Note the nearly same magnitudes of Q_1 and Q_2, indicating that the condensational latent heating assumes the major portion of the diabatic heating. For the inactive period of the summer monsoon, the negative Q_1 may be interpreted by the major contribution of longwave radiation cooling to the diabatic heating (Li and Zhong, 1987). They estimated the various components of the diabatic heating field for the northern part of the South China Sea for the same period of the 1979 summer by use of simple physical formula.

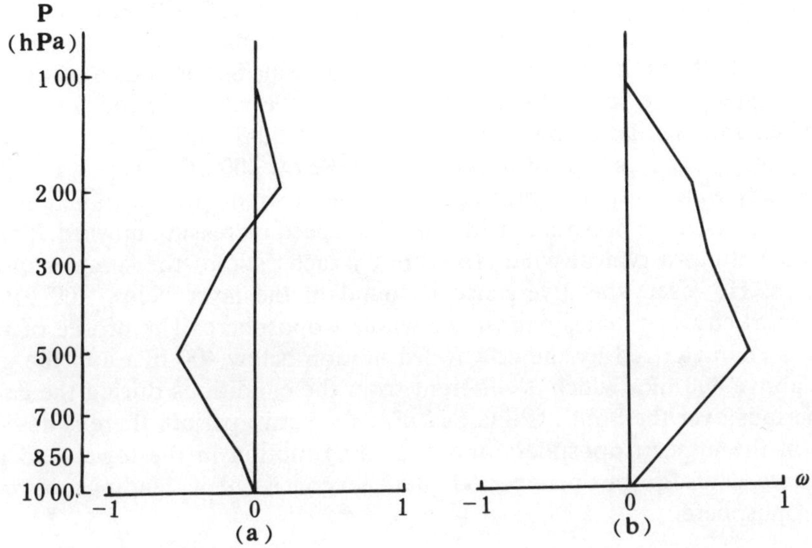

Figure 6.11. The profiles of mean vertical velocity ($\omega = dp/dt$) for period I (left panel) and period II (right panel) of summer of 1979 (see the text). Units: 10^{-3} hPa s^{-1} (Duan, 1983).

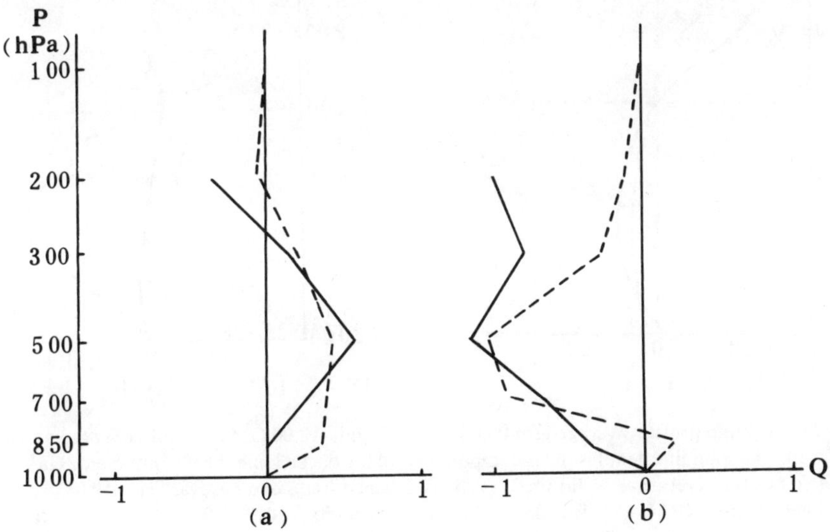

Figure 6.12. The mean profiles of Q_1 (solid line) and Q_2 (dashed line) averaged for period I (left panel) and period II (right panel) of 1979 summer. Units: 10^{-5} cal. g^{-1} s^{-1} (Duan, 1983).

They also pointed out the character of the heat source for the active period and the heat sink for the break period of the summer monsoon. For the former, the condensation heating made the major contribution.

Gao and Zou (1984) selected an active monsoon period, 22–26 July 1982, to study the heat and moisture budgets over southern China, and they obtained a rather different picture from the situation over the South China Sea, as indicated above. The wind structure of the southwest monsoon during the active period over southern China is generally similar to that over India, but the fields of divergence, vorticity, and vertical velocity are nearly opposite (Fig. 6.13). Below 400 hPa, there was a layer of southwesterly wind, with an insignificant change of wind speed with height. Above 400 hPa, the easterlies prevailed, with the wind speed increasing upward. This wind profile constitutes a typical wind structure characteristic of the summer monsoon circulation. However, the divergence is found in the layer below 700 hPa. The negative vorticity dominates almost the whole troposphere. The profile of vertical velocity is characterized by the downward motion below 400 hPa and the upward motion above 400 hPa which is different from the conditions during the active or break periods over the South China Sea. For the active period, there is descending motion in the upper troposphere and ascending motion in the lower and middle troposphere, while for the break period, there is a consistent descending motion in the whole troposphere.

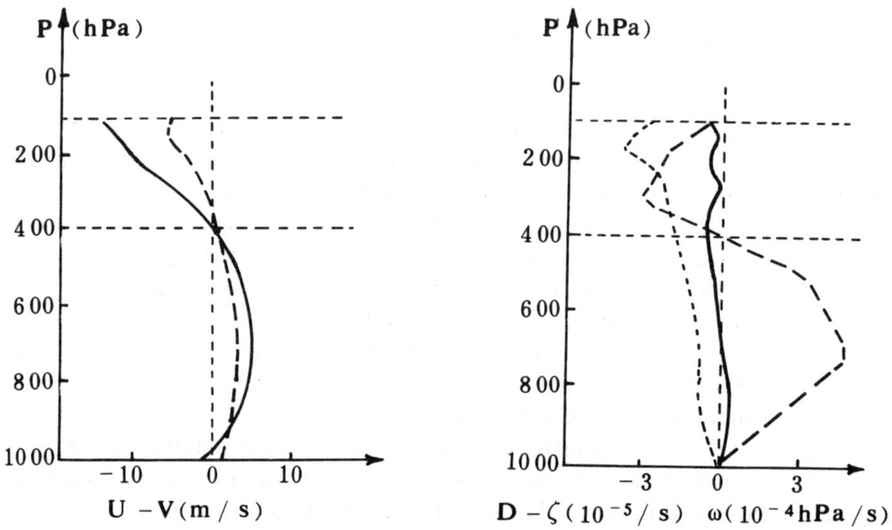

Figure 6.13. (a) Wind profiles averaged for 00:00 GMT, 22 July-12:00 GMT, 26 July 1982 (an active period of monsoon). The solid line denotes the u-component and the dashed line the v-component. Units: m/s; (b) same as (a), but for divergence (solid line), vorticity (dashed line), and vertical velocity ($\omega = dp/dt$, bold dashed line). Units: 10^{-5} s^{-1} for divergence and vorticity; 10^{-4} hPa s^{-1} for vertical velocity (Gao and Zou, 1984).

The profiles of Q_1 and Q_2 (Fig. 6.14) show a significant heat source and a moisture source below about 400 hPa and a heat source and a moisture sink in the easterlies above 400 hPa. During the active period of monsoon over southern China, the sky is generally cloud free, with a very limited amount of precipitation. This fact indicates the negligible contribution of condensation heating to Q_1.

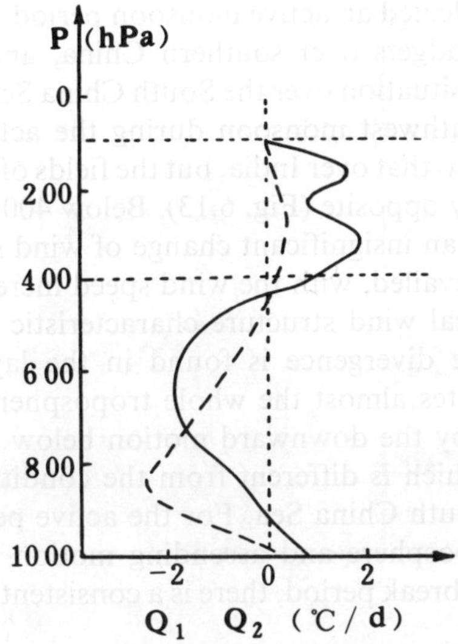

Figure 6.14. The same as Fig. 6.13, but for the mean profiles Q_1 (solid line) and Q_2 (dashed line). Units:°C day^{-1} (Gao and Zou, 1984).

Therefore, it may be inferred that the cooling below 400 hPa is mainly produced by longwave radiative cooling. Gao amd Zou (1984) used a parameterization scheme of radiation, including solar heating, to estimate the contribution of radiative heating (cooling). They obtained that the solar heating is very small, with a maximum found at 850~700 hPa, but the longwave radiative cooling is surprisingly large, with a maximum of −22.5°C day^{-1} in the layer of 850~700 hPa. The total radiative profile is given in Figure 6.15. Apparently, the longwave radiative cooling seems to be overestimated, but its dominating contribution appears to be recongnized. This result obviously supports the view that the active southwest monsoon without much precipitation, is mainly radiatively controlled.

The Meiyu period in the Yangtze River Valley is one of the major rainy seasons in China which is closely related to the activity of the summer monsoon in East Asia. During the process of rainfall of the Meiyu, a large amount of condensational heat may be released, thus making a major contribution to the heat source and sink in this region. Luo and Yanai (1984) estimated Q_1 and Q_2 over the Yangtze River Valley and over eastern China and showed apparent heating and drying (moisture sink) of nearly the same magnitude and nearly the same heights of Q_1 and Q_2 peaks; this, according to Luo and Yanai, is a signature of steady, continuous stratiform rain (Fig. 5.12). This condition is quite different from that in other regions.

Later, Ding and Hu (1988) used a similar method to estimate the heat source and sink over the Yangtze River Valley for the 1984 Meiyu season. They found that, during this year, the precipitation was mainly of a convective nature. This markedly distinctive result reflects a great inter-annual variability of the heat source and sink during the Meiyu season which may result from different large-scale circulation

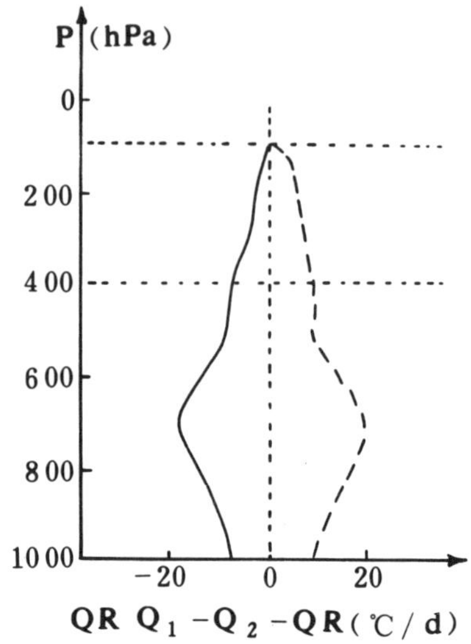

Figure 6.15. The same as Fig. 6.13, but for the profile of total radiative heating or cooling (Q_R) (solid line). Dashed line denotes the curve of Q_1-Q_2-Q_R. Units: °C day^{-1} (Gao and Zou, 1984).

conditions or may be due to different rain-bearing systems. However, for the 1983 Meiyu season, the continuous precipitation seemed to dominate the middle and lower Yangtze River Valley (Ding and Wang, 1988). Figure 6.16 gives the profiles of the vertical p-velocity, Q_1 and Q_2 for the different episodes of the 1983 Meiyu season. For the episode prior to the Meiyu season, the vertical velocity shows the upward motion in the deep layer for 00:00 and 12:00 GMT. The profiles of Q_1 and Q_2 are very similar below 200 hPa, only with the Q_1 peak slightly higher than that of Q_2, indicating the dominance of continuous and stable precipitation. The maximum heating is 6–7°C day^{-1}, at 400–500 hPa. During the Meiyu period, Q_1 and Q_2 have similar shapes, with the peaks of Q_1 and Q_2 found at the same height of 300–400 hPa range. This implies weak convective activity and predominance of continuous precipitation. The precipitation post to the Meiyu period was very weak. Note the downward motion below 600 hPa, which is an unfavorable condition for precipitation. Q_1 and Q_2 are negative below 400 hPa, indicating evaporative cooling. Above 400 hPa, Q_1 and Q_2 are positive, but their magnitudes are very small. During the dry episode, the downward motion dominated nearly the whole troposphere and Q_1 and Q_2 are negative in the deep troposphere. This cooling is created by longwave radiative cooling, because the condensational heating is nearly zero and the sensible heating is confined in the lower layer.

Based on a comparison of Q_1 at 00:00 and 12:00 GMT, one may find their obvious diurnal variations, reflecting similar variations of precipitation (Fig. 6.17). At the preceding episode of the Meiyu season (9–17 June), the diurnal variations of ω, Q_1, and Q_2 were very significant. At 00:00 GMT, the downward motion, cooling and evaporation dominated the deep layer of the troposphere, while at 12:00 GMT, the upward motion, heating and condensation occurred. Therefore, Meiyu rainfalls

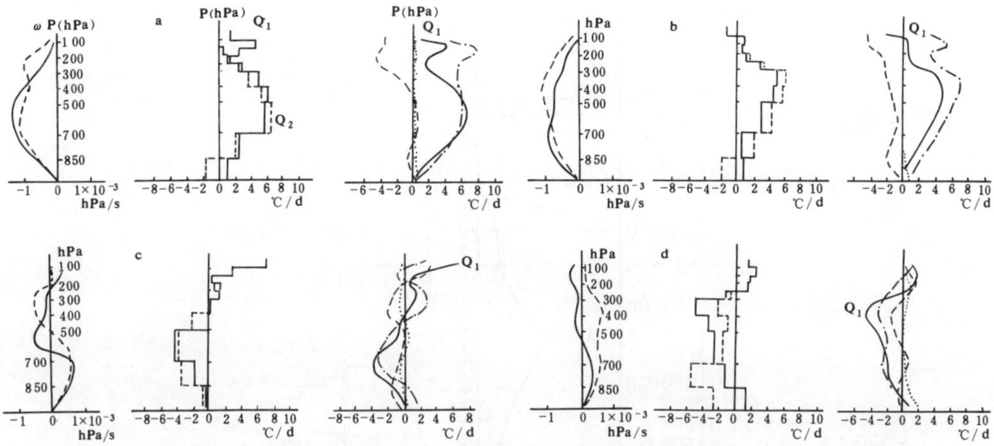

Figure 6.16. The vertical p-velocity (ω) (left), Q_1 and Q_2 (right) averaged for four time spans: (a) the episode prior to Meiyu, (b) during the Meiyu, (c) episode after the Meiyu and (d) the dry spell. The solid lines denote 00:00 GMT values and the dashed lines 12:00 GMT values. The solid lines on the right penal indicate Q_1 and the dashed lines Q_2. The third penals to right represent three components of Q_1. Units: °C day^{-1} for Q_1 and Q_2; hPa s^{-1} for ω (Ding and Wang, 1988).

mainly fell late in afternoon. The second episode (18 June-3 July) of the rainfall also assumes considerable diurnal variations. The diurnal variations for the third episode (4–17 July), the period of the most intense precipitation, were very large, but with a decrease in the magnitudes of ω, Q_1 and Q_2 from 00:00 to 12:00 GMT,.

It is very interesting to note that the heating process during the Meiyu season may be related to the seasonal transition from spring to summer in eastern China (Ding and Hu, 1988). It has been pointed out that the general circulation over East Asia undergoes a distinct seasonal change in the early summer (normally in June). One major event characterizing this change is the almost nearly simultaneous occurrence of the onset of the summer monsoon and the sudden northward shift of the subtropical westerly jet stream in East Asia, and the commencement of the Meiyu in the Yangtze River Valley. This fact reflects the possible thermally-forced effect of the heat source caused by much precipitation on the large-scale circulation in East Asia after the onset of the Meiyu. From Figure 6.18, the time variation of the vertically-integrated Q_1 along 30° N, it can be seen that the variation in the vertically-integrated apparent heat source is in good accordance with that of the precipitation amounts (not shown). The three episodes of the significant heating along 30° N occurred during the Meiyu of 1984.

About two days after the termination of the first episode of significant heating (29 May 9–2 June), the maximum of the westerly winds at 200 hPa to the south of 30° N, suddenly disappeared and rapidly jumped northward to about 40° N. From Figure 6.19, one may see this out-of-phase variation in the westerly winds at 200 hPa at 30° N and 40° N, i.e., with the rapid decrease in the westerly wind speed at 30° N from 3 June corresponding to the concurrent increase in the westerly wind speed at 40° N. The variations are obviously associated with the significant variation in the heat sources at 30° N, as shown in Figure 6.18. Especially, the northward shift of the

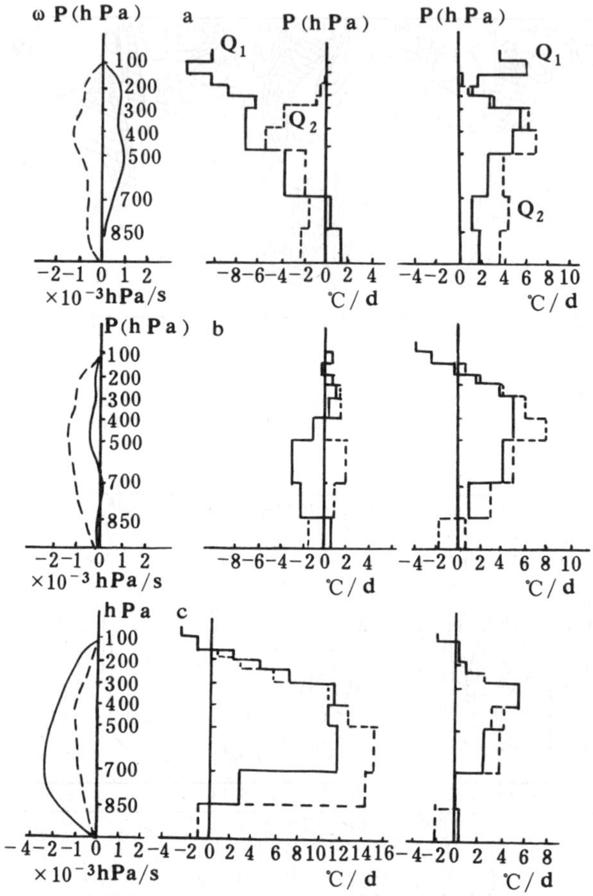

Figure 6.17. The vertical profiles of ω (left), Q_1 and Q_2 at 00:00 GMT (middle) and Q_1 and Q_2 at 12:00 GMT (right) during the 1983 Meiyu period: (a) the rainfall episode from 9 to 17 June, (b) the rainfall episode from 18 June to 3 July, and (c) the rainfall episode from 4 to 17 July. The solid lines in the left panel denote the profile of ω at 00:00 GMT and the dashed lines the profile of ω at 12:00 GMT. Other symbols are the same as Figure 6.16 (Ding and Wang, 1988).

position of the westerly jet stream most significantly occurred during the two major rainfall episodes of Meiyu, i.e., from 4 to 10 June and from 12 to 16 June. Due possibly to the thermal forcing caused by the successive three rainfall processes during the Meiyu, the westerly jet stream at 30° N completely disappeared after 12 June and, meanwhile, the westerly jet stream was steadily established at 40° N.

The above-described variation may provide evidence that the seasonal northward shift of the subtropical westerly jet stream from spring to summer is not only confined at the longitudinal range of the Tibetan Plateau, but also occurs in other places where thermal forcing does exist.

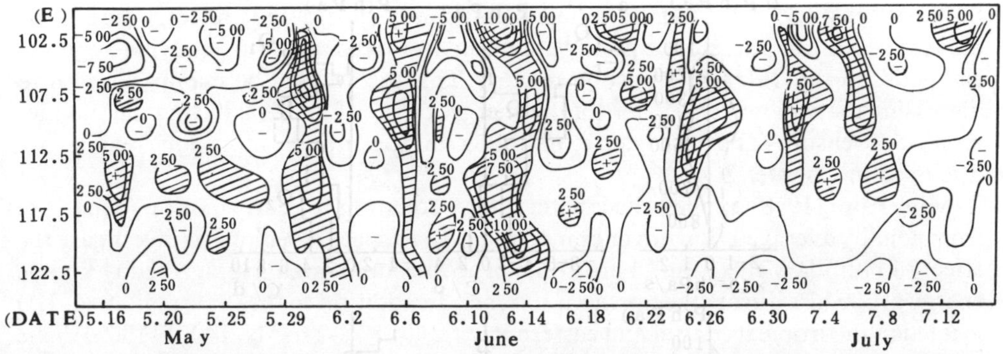

Figure 6.18. The longitude-time cross-section of the vertically integrated Q_1 along 30° N. The areas with values greater than 250 w m^{-2} are shaded. Units: w m^{-2} (Ding and Hu, 1988).

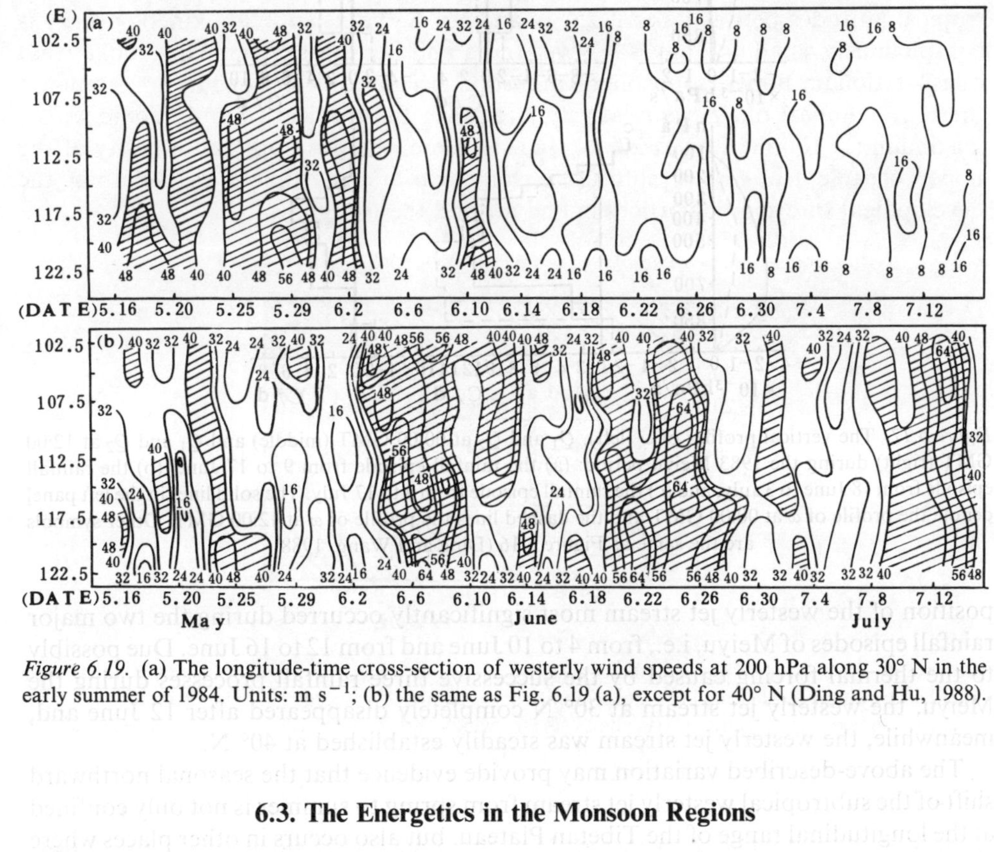

Figure 6.19. (a) The longitude-time cross-section of westerly wind speeds at 200 hPa along 30° N in the early summer of 1984. Units: m s^{-1}; (b) the same as Fig. 6.19 (a), except for 40° N (Ding and Hu, 1988).

6.3. The Energetics in the Monsoon Regions

The study of the energetics over the monsoon regions is an important aspect of monsoon dynamics which may help clarify the development and maintenance of monsoon circulation features and weather systems. It is possible that the nature of the flow patterns and disturbances differs significantly in different regions during the monsoon season, therefore special attention has been paid to studying regional energetics in the different monsoon regions. This section will be mainly devoted to discussing the regional characters of the energetics over East Asia during the summer monsoon season. The energetic conditions in the winter monsoon regime are discussed in Chapter 2.

Murakami (1978) applied the kinetic energy equation for eddy motion (departure from zonally averaged flows over limited regions) to specific regions to investigate the effect of boundary fluxes and barotropic processes upon area-averaged eddy kinetic energy at 200 hPa during the northern summer. In addition to the three regions (I–III) located in the tropical central and eastern Pacific (14.8° S-28.7° N, 140° E-130° W), he selected six other regions (1–6) covering South and Southeast Asia and the Indian Ocean. Figure 6.20 illustrates the regional differences in disturbance activity during summer. The 200 hPa K^* (the transient eddy kinetic energy) is quite large poleward of about 30° N with a maximum near Japan. In contrast, K^* is notably small over the tropical latitudes between about 25° N and 15° S. Transient eddy kinetic energy is conspicuously small ($<20\ m^2\ s^{-2}$) over India and Southeast Asia. This implies that quasi-stationary upper tropospheric monsoon circulation components are dominant there. In comparison, K^* is relatively large over the tropical central Pacific with a maximum ($> 100\ m^2\ s^{-2}$) at 20° N, 165° E. Here, the position and intensity of the upper oceanic trough are subject to some variations, sometimes facilating the development of cyclonic vortices in and around the trough.

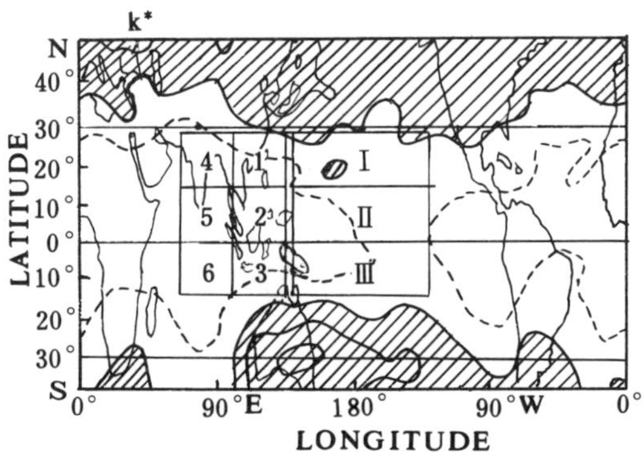

Figure 6.20. Three-year (1970–72) summer mean transient eddy kinetic energy (K^*, $10\ m^2\ s^{-2}$) at 200 hPa. Intervals are 10 units. Shading indicates regions where 200 hPa K^* exceeds 10 units. Dashed lines are for 5 units. Shown are regions I–III and 1–6, to which the kinetic energy equation was applied (Murakami, 1978).

Table 6.7 is the three-year mean (1970–1972) energetics in summer. The main results indicate that area-averaged 200 hPa zonal mean flows are barotropically unstable over the subtropical North Pacific, i.e., I $(K, K') > 0$, implying that the kinetic energy of the eddy motion at 200 hPa tends to increase at the expense of the kinetic energy of zonally averaged motion, but is barotropically stable over the tropical North Pacific and the tropical South Pacific. Over regions 2, 3, and 6 (the South China Sea-Indonesia region and the tropical South Indian Pacific Ocean) there are area-averaged 200 hPa zonal mean flows which are barotropically unstable, while over regions 1, 4 and 5 (southern China and the northern part of the South China Sea, the South Asian continent and the North Indian Ocean) the flows are barotropically stable. Over these region, summer rains are substantial.

Table 6.7. Three-year interaction and kinetic energy fluxes (10 $m^2 s^{-3}$) during the summer (June-August) of 1970–1972, at 200 hPa over regions I–III and 1–6 (Murakami, 1978)

Region	$I(K, K')$*	$F_x(K')$**	$F_y(K')$**
I	0.99	−1.49	0.26
II	−0.33	−0.44	−0.26
III	−0.30	−0.95	0.54
1	−0.19	−0.13	−0.34
2	0.11	−0.27	0.12
3	0.97	−0.59	0.25
4	−0.32	0.33	0.08
5	−0.04	0.35	−0.02
6	0.33	0.32	−0.14

* $I(K, K')$ expresses the interaction between eddies and zonal mean flows over the limited regions.
** $F(K')$ indicates the convergence of eddy kinetic energy fluxes across the boundaries. $F_x(K')$ is the east-west flux convergence and $F_y(K')$ the north-south flux convergence.

A relative large $F_y(K')$ is found over regions 1, 2, and 3. In comparison, regions 4–6 gain substantial amounts of energy via $F_y(K')$ (Fig. 6.21), mainly through large inward transports of 200 hPa K' across the eastern boundaries of regions 4, 5, and 6; namely, upper disturbances over these regions develop through receiving large amounts of energy from regions 1–3 (Philippines, South China Sea, and Indonesia). Of interest in Figure 6.21 is a somewhat large import of 200 hPa K' across the northern boundary of regions 4 and 1. This indicates that lateral coupling is important in maintaining 200 hPa eddy motions over India and its vicinity as well as over southern China and the northern part of the South China Sea. This pronounced import of 200 hPa K' across the northern boundary of region 1 is accomplished by somewhat strong northerlies at 200 hPa over the eastern China-South Japan region. This area may represent a major route through which the summer monsoon circulation interacts with mid-latitude systems.

Inter-annual variability in regional energetics is significant. Over Indonesia and the South Indian Ocean, the upper zonal mean flows were substantially more barotropically unstable in 1972 than in 1970 or 1971. This is contrasted with exceptionally stable 200 hPa zonal mean flows in the summer of 1972 near India where the monsoon rains were abnormally weak. In short, a weak monsoon near India is associated with strong barotropic stability of upper zonal mean flows over the equatorial central Pacific and in the neighborhood of India, while above-average instability occurred over Indonesia and the Indian Ocean.

By utilizing the NMC data over the Asian region during the early summer of 1979,

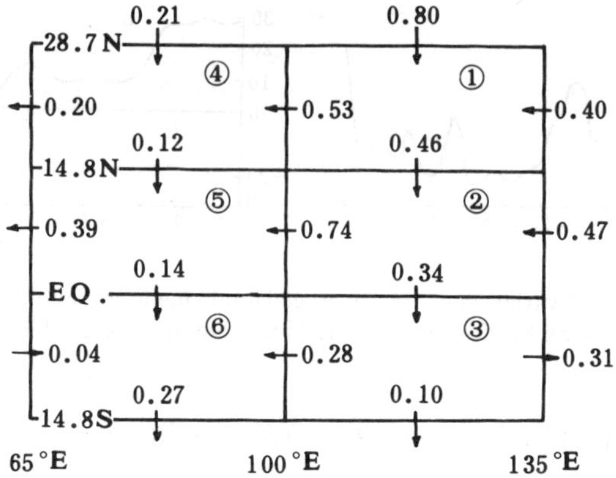

Figure 6.21. Boundary transports of eddy kinetic energy K' at 200 hPa over regions 1–6, expressed in units: 10^{-4} m^2 s^{-2} (Murakami, 1978).

Ding (1984) estimated the budgets of kinetic energy in synoptic-scale disturbances and the general flow over China and Japan. It is found that the time period of maximum generation of eddy kinetic energy corresponded to the respective rainy season for the three studied regions (southern China, central China and Japan). The eddy kinetic energy produced mainly at the upper level (300–100 hPa) and partly at the lower-level was, in a greater part, transported out of all three studied regions to the surroundings. Therefore, they may be regarded as important source regions of eddy kinetic energy during the period of generation. The kinetic energy budget in the general flow is closely related to the activities of the upper jet stream and low-level jet over these regions. In short, the results obtained by Ding's study from the view-point of energetics, show the far-reaching effects of the onset and advance of the summer monsoon on the changes in the general circulation and activities of rain-bearing disturbances in the region of East Asia and Japan.

Yang (1987) further estimated the budgets of the atmospheric energy before and after the onset of the East Asian summer monsoon for 1980 over the South China Sea. Figure 6.22 is the curves of temporal variation of the area-averaged total kinetic energy and total available potential energy at 850 hPa for May and June. It can be seen that, at the stage prior to the onset of the summer monsoon (stage I, 1 April-9 May), the total kinetic energy had a low stable level, while the total available potential energy was positive and large. However, after the onset of the summer monsoon (stage II, 18 May-29 June), the total kinetic energy suddenly increased and then fluctuated around a higher level, while the total available potential energy significantly dropped down to negative values.

Yang used the following set of energy equations to estimate the changes in the energetic features before and after the onset of the summer monsoon:

$$\left[\frac{\partial K}{\partial t}\right] = FH(K) + FV(K) + G(K) + D(K), \tag{6.2}$$

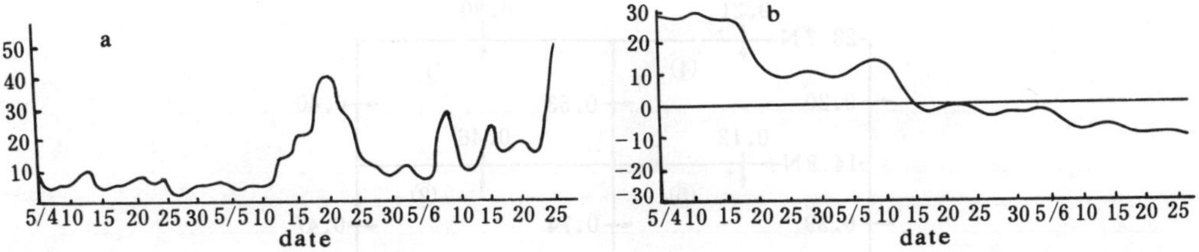

Figure 6.22. (a) The curve of variation of areal averaged total kinetic energy at 850 hPa for May-June 1980. Units: $m^2\,s^{-2}$; (b) same as (a), but for the total available potential energy. Units: $10^6 J\,m^{-2}$ (Yang, 1987).

$$[\frac{\partial A}{\partial t}]=FH(A)+FV(A)+G(A)+C(K,A)+D(A), \tag{6.3}$$

$$[\frac{\partial K_e}{\partial t}]=FH(K_e)+FV(K_e)+C(K_S,K_e)+G(K_e)+D(K_e), \tag{6.4}$$

$$[\frac{\partial A_e}{\partial t}]=FH(A_e)+FV(A_e)+G(A_e)+C(A_s,A_e)+C(K_e,A_e)+D(A_e), \tag{6.5}$$

where [ ] denotes the areal average, "*" the departure from the areal average, and the overbar the time average, $K=1/2(u^2+v^2)$, $K_e=1/2(u^*+v^*2)$,

$$A=\frac{C_p}{g}\int_{P_u}^{P_s} NTdp,\quad A_e=\frac{C_p}{g}\int_{P_u}^{P_s} N^*T^*dp,\quad N=1-\frac{p_r^k}{p^k} \tag{6.6}$$

(N is efficiency factor), P_r reference pressure, K_s and A_s the area-averaged kinetic energy and potential available energy, respectively, P_u and P_s the upper and lower-level pressure of the intergration layer, *FH* and *FV* the horizontal and vertical fluxes of *A* or *K*, *G* generation terms of *A* or *K*, *C* conversion terms of *A* or *K*, and D dissipation terms of *A* or *K*.

For the sake of brevity, only the vertically integrated energy parameters will be given in Tables 6.8–6.11. For the total kinetic energy, their source and sink terms nearly balance out, with other terms being very small, but for stage I, the energy level was much lower than at stage II. After the onset of the summer monsoon, the destruction of the kinetic energy increased by 35%. This condition corresponds to the increase of ageostrophic airflow. The destruction of total kinetic energy is nearly offset by the production of kinetic energy through the interaction between subgrid-scale and grid-scale motion, i.e., the cumulus effect.

Table 6.8. The budget of vertically integrated total kinetic energy. Units: $w\,m^{-2}$ (Yang, 1987)

Layer (hPa)	[∂K/∂t]		FH(K)		FV(K)		G(K)		D(K)	
	I	II	I	II	I	II	I	II	I	II
100–1000	0.05	0.07	0.46	−1.53	−0.04	−0.18	−172.71	−233.7	172.26	234.90

Table 6.9. The budget of the vertically integrated available potential energy. Units: w m^{-2} (Yang, 1987)

Layer (hPa)	$[\partial A/\partial t]$		$FH(A)$		$FV(A)$		$G(K, A)$		$G(A)$		$D(A)$	
	I	II	I	II	I	II	I	II	I	II	I	II
100–1000	−6.1	−8.6	34.0	88.6	8.1	7.3	59.7	−447.8	0.6	1.5	−108.6	341.8

From Table 6.9, it can be seen that the generation of available potential energy by diabatic heating is a small term. At stage I, the total available potential energy mainly came from the horizontal import from outside at high level and the conversion of kinetic energy at the low level. $C(K, A)$ assumed a significant change from stage I to stage II, i.e., from 59.7 to -477.8 w m^{-2}, indicating that a large amount of available potential energy was converted into the kinetic energy. This energy conversion mainly occurred in the layer of 300–700 hPa. Interestingly, $D(A)$ is also large in this layer which is nearly in balance with $C(K, A)$.

The budget of the eddy kinetic energy is quite different from that of the total kinetic energy (Table 6.10). $G(K_e)$ is the main kinetic energy source, in particular at the stage II. This implies the occurrence of a vigorous generation of K_e that is partly (57.10 w m^{-2}) exported outside the studied domain. The maximum generation is found at the high and low levels, respectively. The large generation of K_e at stage III mainly resulted from the energy conversion for the eddy available potential energy. $D(K_e)$ is the main sink of the eddy kinetic energy, with the magnitude at stage III being 7.6 times greater than that at stage I.

Table 6.10. The budget of vertically integrated eddy kinetic energy. Units: w m^{-2} (Yang, 1987)

Layer (hPa)	$[\partial K_e/\partial t]$		$FH(K_e)$		$FV(K_e)$	
	I	II	I	II	I	II
100–1000	−1.31	2.14	3.24	−57.10	−0.59	−2.81

Layer (hPa)	$C(K_s, K_e)$		$G(K_e)$		$D(K_e)$	
	I	II	I	II	I	II
100–1000	−0.01	−6.82	15.19	216.46	−19.14	−149.59

Table 6.11 is the budget of the eddy available potential energy. The major term of A_e resulted from the direct generation by diabatic heating. After the onset of the summer monsoon, G (A_e) increased 16 times, with the maximum generation found in the upper troposphere. As indicated in the discussion of the budget of K_e, the energy conversion from A_e to K_e is a preferred conversion direction. This point is clearly demonstrated in the negative $C(K_e, A_e)$ at stage II. The magnitude of $C(A_s, A_e)$ is the second to $C(K_e, A_e)$, mainly with the conversion from A_e to A_s. $FH(A_e)$ amd D

$FH(A_e)$ are also of importance for the budget of A_e. At stage II, $FH(A_e)$ acted as a A_e source for the surroundings.

For the budget of A_e, at stage I about 60% of A_e was generated by diabatic heating, 16% came from the conversion from A_s and the remaining was contributed by the import from the outside or the energy conversion from A_e. At stage II, about 40% of the generated A_e is converted into K_e, 31% into A_s, and the remainder was exported outside.

Table 6.11. The budget of vertically integrated eddy available potential energy. Units: w m^{-2} (Yang, 1987)

Layer (hPa)	$[\partial A_e/\partial t]$		$FH(A_e)$		$FV(A_e)$		$C(K_e,A_e)$	
	I	II	I	II	I	II	I	II
100–1000	0.07	4.41	1.48	−44.81	−0.18	−3.38	1.43	−74.32

Layer (hPa)	$C(A_s,A_e]$		$G(A_e)$		$D(A_e)$	
	I	II	I	II	I	II
100–1000	2.32	−57.96	9.79	172.09	−14.77	12.79

Duan (1983) estimated the budgets of the eddy kinetic energy and the eddy available potential energy for the 1979 summer. He also found that the energy conversion from the eddy available potential energy to the eddy kinetic energy is the major source of the eddy kinetic energy, with the maximum conversion located in the layer of 500–300 hPa, while the barotropic energy conversion from the mean kinetic energy to the eddy kinetic energy is one order of magnitude smaller than the baroclinic conversion. The vertical transport of the eddy kinetic energy is directed downward. Therefore, the picture of the budget of the eddy kinetic energy over the South China Sea is that the generated kinetic energy in the middle and high troposphere is transported downward to provide the energetic condition for the development of weather systems in the lower troposphere.

A power-spectrum analysis was applied to the time series of the energetic quantities in order to study the dominating oscillation modes and it was revealed that the energy conversion from the eddy available potential energy to the eddy kinetic energy was mainly accomplished with the low frequency mode (30–40 day period) (Fig. 6.23). Similar conditions are also found for the vertical transport of the eddy kinetic energy and the generation of the eddy available potential energy.

The energetics of the monsoon depression in South Asia or tropical depression over East Asia during the summer season is also an important part of monsoon energetics. Surgi (1984) and Ding *et al.* (1987) estimated the energetics of the monsoon depression that occurred over the northern head of the Bay of Bengal during 1–7 July 1979. Surgi averaged the results of the estimated energetics for two separate period;

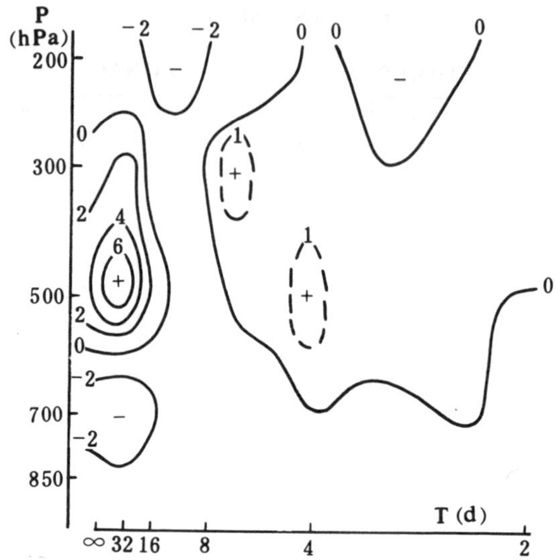

Figure 6.23. The power spectrum of the energy conversion between the eddy available potential energy and the eddy kinetic energy for the summer of 1979 (31 May-31 August) (Duan, 1983).

i.e., for the formative and post-formative periods. Figure 6.24 illustrates the results of Surgi's calculations from 75–105° E, and between the equator and 35° N through the troposphere. These results show that, in the formative stage the eddy kinetic energy K_e is enhanced by the barotropic as well as the baroclinic energy conversions, while the depression stage shows a more dominant role for the baroclinic process. The results of the energetics from multi-level global spectral model forecasts were evaluated by Krishnamurti *et al.* (1983) over a domain extending from 10 to 30° N, 50 to 100° E through the depth of the troposphere. The barotropic conversion $C(K_s, K_e)$ is considerably smaller and is even negative during the formative period compared to the baroclinic conversion. A sharp rise of the latter between 3–7 July 1979 lends support to the observational energetics obtained by Surgi. Furthermore, it was noted by Krishnamurti *et al.* (1983) that the horizontal shear flow initially contributed to a net increase of eddy kinetic energy at the lower levels, especially at the 900 hPa level. Thus, the role of the barotropic mechanism in the very lowest troposphere during the formative stage was intially present over this region.

Ding *et al.* (1987) only estimated the energetics of this depression for the post-formative period (5–7 July 1979). Table 6.12 shows the result averaged for this period. At this stage of development of the depression, the baroclinic conversion is the major mode of eddy energy conversion, which is similar to the results by Surgi (1984) and Krishnamurti *et al.* (1983). The contribution of barotropic energy conversion at this stage is considerably small. Note the large fluxes of energy at the boundaries, indicating the importance of the energy import for the budget of the eddy kinetic energy.

In comparison with the budget of the eddy kinetic energy of the monsoon

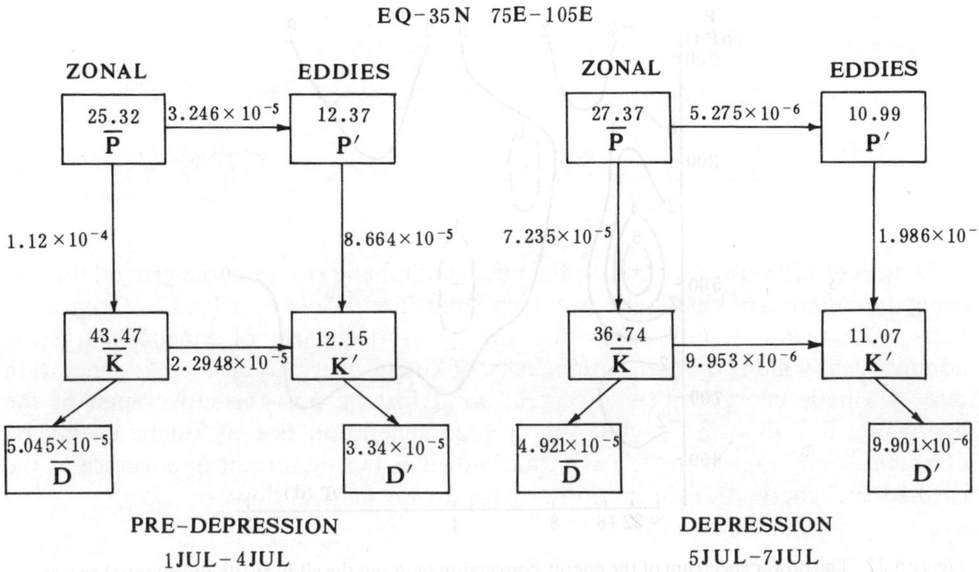

Figure 6.24. Energetics based on pre- and post-onset data sets during two different periods of July 1979. The units of energy quantities are m s^{-1}, while the exchange are in m s^{-1} (Surgi, 1984).

Table 6.12. The budget of the eddy kinetic energy for the post-formative period of the monsoon depression, 5–7 July 1979 throughout the troposphere. Units: 10 w m^{-2} (Ding *et al.*, 1987)

Layer (hPa)	$C(K_z,K_e)$	$C(A_e,K_e)$	$FV(K_e)$	$FV(A_e)$	$FH(K_e)$	$FH(A_e)$	$\partial K_e/\partial t$	$D(K_e)$
Sur.-100	0.24	3.53	0	0	9.18	15.89	−3.63	−34.08

* K_z is the zonally averaged kinetic energy.

depression, the budget of the eddy kinetic energy was estimated for a tropical depression over the South China Sea in the early part of August 1979. The results are given in Table 6.13. It can be seen that the baroclinic conversion of energy is of small negative values, i.e., a weak destruction of the eddy kinetic energy. The barotropic energy conversion contributed much to the energy source which is in remarkable contrast with the case of the monsoon depression over the Bay of Bengal. However, a similar condition is large energy fluxes at boundaries that, in fact, are terms of major energy sources. Therefore, the weak tropical cyclones, either the monsoon depression or the tropical depression, are strongly dependent on their environments. For the mature typhoon, it is documented that the baroclinic conversion and the work term at the boundaries are major energy sources (Ding and Liu, 1985). From the viewpoint of the internal eddy kinetic energy source, the tropical cyclone and the monsoon depression are not fully the same tropical system. For the former, the baroclinic process is very important, while for the latter the barotropic and baroclinic processes are both important. As the intensity of the tropical cyclone increases, the baroclinic process would predominate, especially when it develops into a typhoon-level tropical cyclone.

Table 6.13. The budget of the eddy kinetic energy for the tropical depression over the South China Sea in the early part of August 1979 for the layer of 1000–300 hPa. Units: w m^{-2}

Stage	$C(K_z,K_e)$	$I(A_e,K_e)$	$FH(A_e)$	$FH(K_e)$	$FV(K_e)$	$FV(A_e)$	$\partial K_e/\partial t$	$D(K_e)$
Initial Mature	−1.18	−0.55	2.11	4.05	−0.21	−0.01	0.82	−3.30
	1.40	−0.52	3.24	3.85	0.58	−0.10	−2.31	−10.76

Ding *et al.* (1987) also estimated the budget of kinetic energy in the general flow by using the equation of kinetic energy which is partitioned into divergent and rotational parts. The major results show that the generation term of kinetic energy by nondivergent wind is the main source term of kinetic energy. The average generation rate of kinetic energy by nondivergent wind for the post-formative span of the depression is 7.40 w m^{-2}, while the average generation rate of kinetic energy by divergent wind is 0.6 w m^{-2}. This result indicates the significant importance of the barotropical energy transformation process for the general flow.

6.4. Some Aspects of Monsoon Dynamics and Prediction of the Summer Monsoon Precipitation

The classical concept of the monsoon was put forward as a climatological concept. The prevailing winds over oceans and continents are opposite for summer and winter. It was realized that the monsoon in fact was a type of large-scale thermal circulation induced by thermal contrast between oceans and continents. But, the atmospheric center of action observed near the surface in winter, the Siberian high, is not situated in the middle of the Asian continent; nor is the continental low in summer consistent with the hottest region. This led one to further realize that the monsoon development may be associated with the large-scale topography (mainly the Tibetan Plateau) and the planetary-scale waves aloft. In this sense, the cause of monsoons should be addressed on the planetary-scale basis and in the context of the dynamics of planetary waves.

The distribution of the ocean and continent may create an inhomogeneous pattern of heating; the large-scale topography, as an elevated heat source, may also produce the differential heating field. Then both may exert a thermal effect on the atmospheric motion, thus driving the development of the monsoon circulation. In turn, the atmospheric motion may also influence the contribution to the heating field made by the distribution of oceans and continents. Thus, the thermal effect of this distribution is acomplished through the atmospheric thermal stratification, moisture distribution, etc., that is to say, the thermal effect of the land-sea contrast, the topographical effect, and the atmospheric motion constitute a complex interactive system.

A number of numerical experiments on the effect of diabatic heating on the monsoon circulation have been undertaken. Among them, Kuo and Qian (1981) pointed out the primary importance of the diabatic heating for the formation of monsoon circulation. The diabatic heating also plays an important role in the intraheadlseasonal variation of the monsoon circulation, e.g. in the change from the break to active period (Li *et al.*, 1987). In the diabatic experiment, by use of the initial break monsoon state , Kuo and Qian simulated active monsoon circulation which is characterized by a monsoon trough located over the middle part of the Indian Subcontinent, typical meridional monsoon cell over India and the foothill of the Tibetan Plateau, strong southwest monsoonal airflow, and enhanced monsoon rainfall over middle India. In the adiabatic experiment, however, these conditions cannot be well simulated. Therefore, the diabatic heating created mainly by the thermal effect of the Tibetan Plateau and partly due to the land-sea thermal contrast, is very critical for the development of the summer monsoon over India.

The summer monsoon over the South China Sea and the western Pacific is, to a greater degree, independent of the Indian summer monsoon. This implies that it may be driven by a rather different heating field. Based on the study of the summer monsoon over the South China Sea for June and July 1982 (Xie and Zhu, 1987), the monsoonal airflow over the South China Sea in the early summer (June) had its origin in the oceanic area south of India, but in mid-summer (July), it came from both the cross-equatorial airflow at 105° E and the Indian monsoon. By using the χ–ψ interaction it is revealed that this seasonal difference in the origin of monsoonal airflows mainly results from the different distributions of heat sources and sinks. Early in summer, heating occurs over the South China Sea, while cooling occurs over the Bay of Bengal, thus causing the southwest monsoon to develop eastward from the west. At the same time, the cross-equatorial airflow at 105° E is insignificant. In mid-summer when the South China Sea is heated while the Australian region is cooled, a remarkable cross-equatorial airflow originating in Australia occurs and the southwest monsoon over the South China Sea develops northward from the south. This observational association was verified by numerical experiments for the sensitivity of the South China Sea monsoon activity to differential heating in summer by using a three-level primitive equation regional model. When heating is applied to the South China Sea and cooling to the Bay of Bengal, the monsoonal airflow over the South China Sea mainly comes from the Indian region and the cross-equatorial flow at 105° E is relatively weak. This is basically similar to the flow pattern in the early summer. But, when the heating is applied to the South China Sea and cooling to Autstralia, the monsoon over the South China Sea mainly comes from the Southern Hemisphere, and the cross-equatorial airflow at 105° E is relatively strong. This is basically similar to the flow pattern in mid-summer as described above. Figures 6.25 and 6.26 show the contrasting wind fields for the cases of the zonal and meridional heating gradients, respectively. For the latter, the strong meridional south wind is very evident, while for the former, the zonal westerly wind is very strong, extending from the Indian region to the West-Pacific. This marked sensitivity clearly reflects the primary importance of the thermal forcing in the development of regional monsoon circulation.

The subtropical high over the western Pacific is an important component of the East Asian monsoon system. Its advance and retreat as well as evolution are closely related to the occurrence of precipitation events in China. In Chapter 1, the major aspects of the subtropical high have been discussed. Here, only the possible link

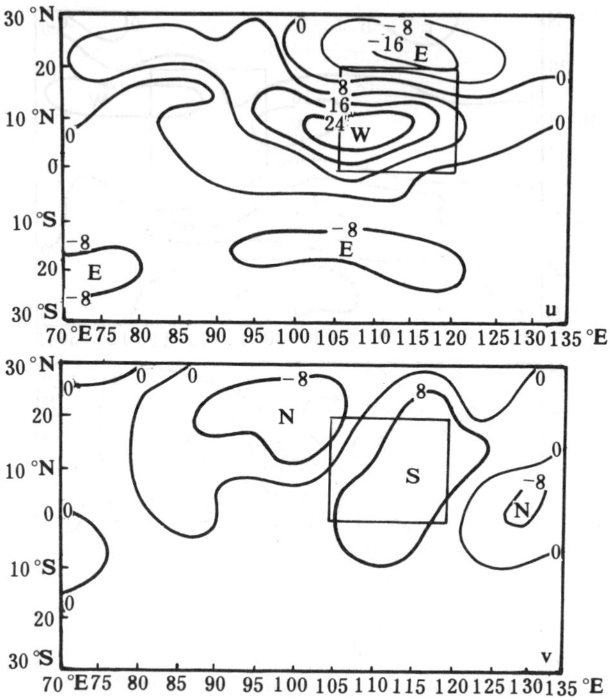

Figure 6.25. The predicted wind fields at 900 hPa for 72 h integration in the case of the zonal heating gradient: (a) zonal wind (top panel); (b) meridional wind (bottom panel). Units: m s^{-1} (Xie and Zhu, 1987).

between its behavior and monsoon activity will be dealt with. Recently, Zhang and Yu (1990) used three-dimensional *E-P* flux to diagnose the medium-range variation of the subtropical high over the western Pacific during the Meiyu season of 1982. Figure 6.27 shows the latitudinal variations of the positions of the subtropical high during the different stages of the Meiyu season of 1982. Overall, the high underwent a significant northward advance. Correspondingly, the regions of positive potential vorticity related to the subtropical high also underwent a similar east-west oscillation (Fig. 6.28). Figure 6.29 presents the cross-sections of $\nabla \cdot F_s$ (divergence of *E-P* flux) along 120° E at these three stages, indicating the meridional transport of the eddy potential vorticity by the relation

$$\nabla \cdot F_s \approx \overline{v'q'}. \tag{6.7}$$

where v' and q' are zonal departures of the meridional wind component and geostrophic potential vorticity. Overbar represents the zonal average. According to Eq. (6.7), the positive area of the wave activity flux at the middle level over subtropics in Figure 6.29 represents the area of northward transport of potential vorticity, and the negative areas at high and low levels represent the areas of the southward transport. The maxima of the northward transport are located at 25° N, 27.5° N, and 30° N for the three different stages, respectively. The northward

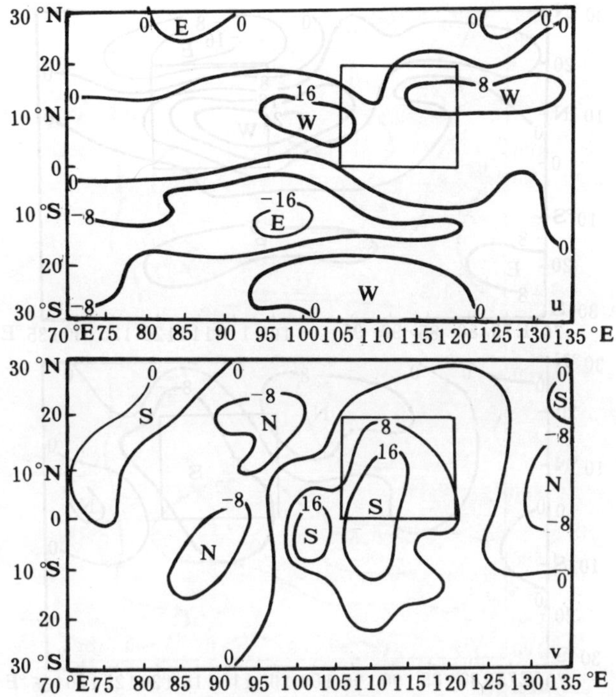

Figure 6.26. The same as Fig. 6.25, but for the case of the meridional heating gradient (Xie and Zhu, 1987).

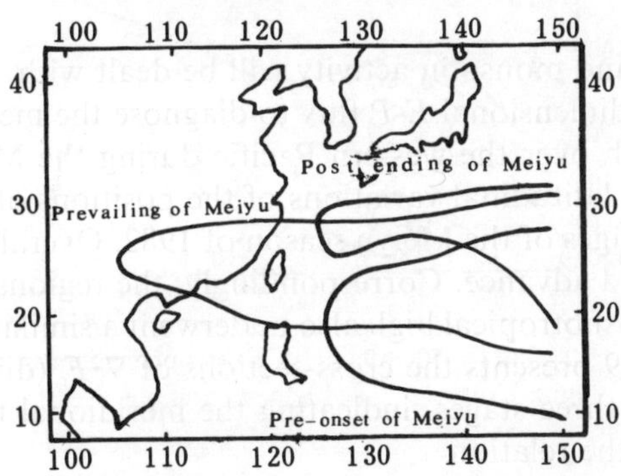

Figure 6.27. The evolution of the 5880 m isopleth of geopotential height for the three stages of the Meiyu season of 1982. The pre-onset period (a): 1-8 July; the Meiyu period (b): 9–25 July and the ending period (c): 26– 31 July, 1982 (Zhang and Yu, 1990).

advance of these maxima indeed corresponds to the twice northward jumps of the subtropical high.

An analysis of the cross-sections of the *E-P* flux has further indicated that over East Asia there exist two paths of northward transport of potential vorticity. They are

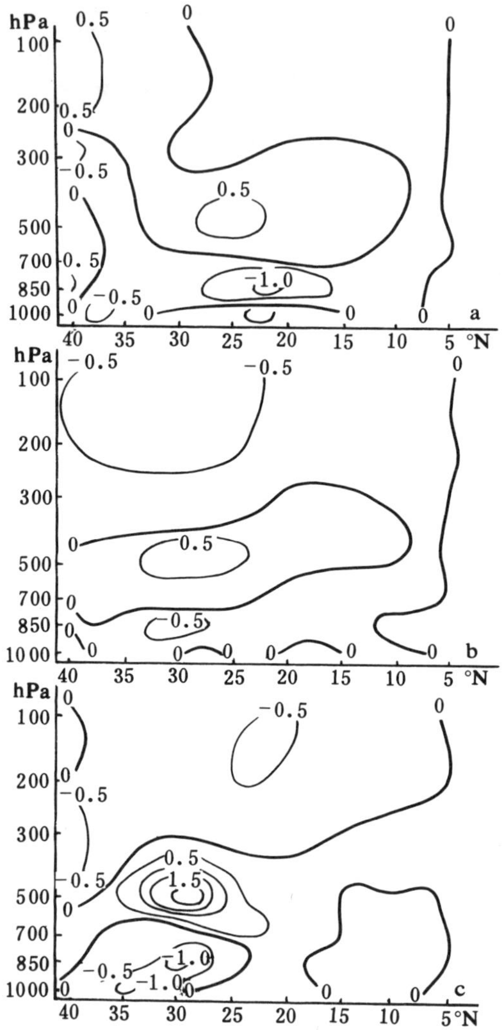

Figure 6.28. The same as Fig. 6.27, but for 500 hPa mean geostrophic potential vorticity fields for the three stages. Units: $10^{-6}\ s^{-1}$ (Zhang and Yu, 1990).

closely connected with the westward extending and northward advance of the subtropical high. An analysis of the wave activity flux also shows that the region of the subtropical high is just the region of convergence of wave energy, and its changes in intensity and range are associated with the intensity and movement of the subtropical high. This wave energy convergence results from a merging of wave energy fluxes from the tropical flow on the south side, the rain belt on the north side, and the heating field on the east side of the Tibetan Plateau.

Another central problem of the monsoon dynamics is the dynamics of the low-frequency oscillations. In Chapter 4, the general characteristics of the main modes of

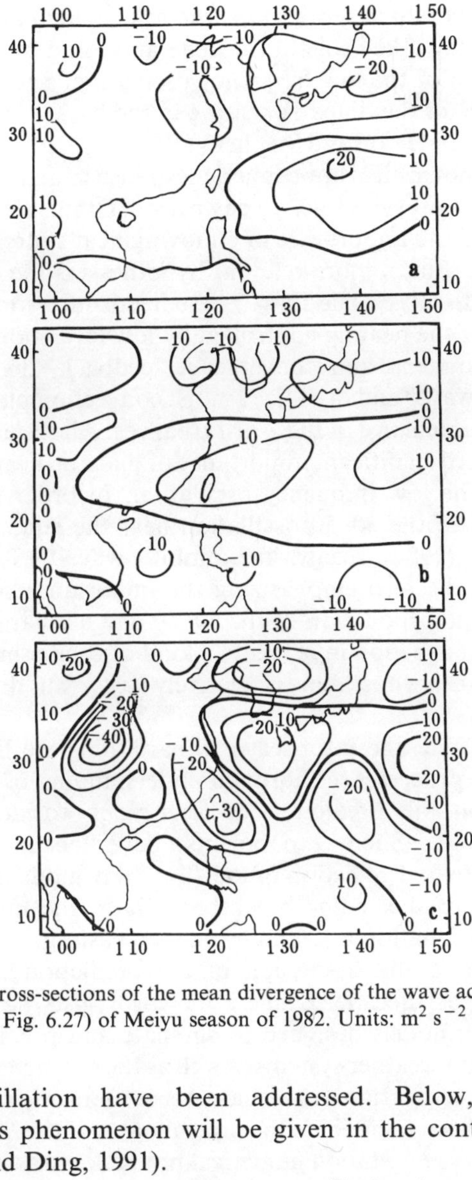

Figure 6.29. Meridional cross-sections of the mean divergence of the wave activity flux at 120° E for the three stages (same as Fig. 6.27) of Meiyu season of 1982. Units: $m^2 s^{-2}$ (Zhang and Yu, 1990).

low frequency oscillation have been addressed. Below, a brief theoretical interpretation of this phenomenon will be given in the context of the 40–50 day oscillation (Wang and Ding, 1991).

Many essential features of the 40–50 day oscillation have been simulated using various versions of GCMs. These numerical simulations demonstrate that the model 40–50 day low frequency oscillation is caused by the eastward movements of planetary scale wavenumbers one or two circulation modes which are primarily confined to 15° latitude from the equator and vertically penetrate the entire troposphere with a baroclinic structure. The condensational heating interacting with the dynamic process is responsible for generation and maintenance of these long-lasting planetary scale modes. In addition, the surface latent and sensible heat fluxes into the boundary layer which destabilizes the atmosphere ahead of the ascending zone were shown to play a key role in the generation and maintenance of the slow northward propagation of the low frequency oscillation. Several investigators have further studied the role of the diabatic heating in the basic dynamic processes of the 40–50 day oscillation. Anderson and Stephens (1987) regard the 40–50 day oscillation as a manifestation of the tropical atmospheric response to an imposed stationary heat source pulsating in intensity with a 40–50 day period. Others (Hayashi and Miyahara, 1987) have emphasized the importance of a moving heat source with either constant or oscillating strength. But, a more relevent hypothesis is the consideration of heat sources interacting with large scale flow. These heat sources may be represented by evaporation heating or the heating controlled by low level moisture convergence. In the context of this condensation convergence feedback, the interaction between equatorial large-scale waves and organized moist convection plays an essential role in the generation and maintenance of these disturbances. All the above described works have remarkedly stressed significant importance of diabatic heating for the formation and development of the low frequency oscillation. In order to illustrate the slow eastward propagation of the 30–50 oscillation near the equatorial region ($\sim$10 m s^{-1}), Lau and Peng (1987) regard the mobile wave-CISK as the generation mechanism of this mode, also emphasizing the importance of the condensational heating feedback for the production of the 30–50 day oscillation. Li (1983) pointed out that the cumulus convection may create a kind of oscillating-type unstable wave through CISK mechanism whose scale is relatively large, thus driving several periodic oscillations.

In view of the fact that the intra-seasonal variations are linked to a variety of fundermental physical processes, a thorough understanding of the phenomenon has a potential impact on the prediction of long-range variations of the tropical atmospheric circulation. Such experiments have been undertaken recently (Krishnamurti *et al.*, 1990). Prediction of the 40–50 day mode is a rather challenging task and will be one of the major breakthroughs in the numerical modeling of atmospheric general circulation and long-range forecasting.

Next, we will turn to the discussion of the prediction of summer monsoon precipitation. In the past decade, Chinese meteorologists have developed several limited-area numerical models designed to simulate and predict the evolution and structure of precipitating weather systems as well as the characters of the precipitation area and amount of precipitation in the East Asian monsoon region. In this regard, numerous works have been focused on the precipitation of the Meiyu season. Starting from 1984, operational precipitation numerical prediction was issued at national and regional levels by meteorological services using the limited-area primitive equation models on short-term basis, respectively, with a special emphasis on the prediction of the Meiyu precipitation or heavy rainfalls.

During the period 11–15 July 1981, an excessively heavy rainfall occurred over the Sichuan Basin in China. Observational studies showed that this heavy rainfall was largely related to the formation of a medium-scale vortex (SW vortex) on the eastern slope of the Tibetan Plateau. Several investigators have simulated this heavy rainfall event (Zhou and Hu, 1983; Chen and Dell'Osso, 1985; Kuo *et al.*, 1988) by using different numerical models. Zhou and Hu (1983) pointed out the importance of initial field and reasonable treatment of terrain features for adequate simulation of this event. They found that the main source of water vapor is found over the South China Sea rather than the Bay of Bengal. The simulation by Kuo *et al.* (1988) indicated that the SW vortex completely formed within a broad southwesterly monsoon flow, possibly induced by the strong variations in the terrain elevation. As the southwesterly monsoon current impinged upon the meso-scale Yungui Plateau, which extends from the southeastern corner of the main Tibetan Plateau, the low-level airflow was blocked. The air aloft descended into the Sichuan Basin to the lee of the Plateau, creating cyclonic relative vorticity through vertical stretching. Differential surface friction and diabatic effects were not responsible for the initiation of the vortex. However, both latent heating and surface fluxes of sensible and latent heat were essential to the full development of the SW vortex (Anthes, 1990). The model predicted a 48 h rainfall maximum of 213 mm over the Sichuan Basin, which is in good agreement with observations.

The meso-scale or intermediate-scale vortex are frequently observed to develop and propagate along the Meiyu front, and some of them may bring about heavy rainfall, as indicated in Section 5.8. The numerical experiments by Zhou *et al.* (1984), using a fine-mesh model, suggested that at the genesis stage of the vortex or low, dynamic conditions such as a change in wind field, is important while, at the development stage, the condensational heating, especially convective heating, has a significant effect. Li *et al.* (1987) also simulated the Meiyu weather systems and associated heavy rain. During 19–20 June 1982 heavy rainfall fell over the Yangtze River Valley, with more than 300 mm falling near the city of Wuhan. Li *et al.* successfully simulated sub-synoptic scale convergence lines at low and middle levels, strong cyclonic circulation and the divergent circulation aloft and predicted the position of the major rainbelt for this case. They particularly stressed the critical role of the interactions of the latent heat release and the low and upper-level jets for the maintenance of the slowly-moving convective systems. Without latent heat, the model does not maintain the concentrated region of high relative vorticity , the low-level jet, and the intense rainfall. The important effect of latent heat release has been documented by the numerical simulations of Zhang (1990). He stressed that both large-scale and convective heating are very important for the correct simulation of the Meiyu rainfall and low-level jets. In addition, even for 24 h precipitation prediction, the radiative forcing cannot be neglected, especially for nocturnal rainfall episodes.

Recently, Qian *et al.* (1991) used a global model at the UK Meteorological Office to numerically simulate the Meiyu onset and its rain belt. They found that the onset of the Meiyu and its rain belt are sensitive to the Tibetan Plateau and the water vapor conditions over the Bay of Bengal. For the case without the Plateau or suppressed moisture supply from the Bay of Bengal, the position of the rain belt was not simulated in a realistic manner and the predicted rainfall amount is far lower than observed.

In summary, in large-scale perspective, the above-described numerical simulations have identified at least four factors which are critical for the occurrence and maintenance of summer monsoon precipitation over China: the monsoonal airflow (mainly southwesterly airflow) and the low-level jet which is generally embedded in the large-scale monsoonal airflow, an abundant moisture supply mainly coming from the Bay of Bengal and the South China Sea, the large-scale tterrain feature , in particular the Tibetan Plateau, and the latent heat release including large-scale and convective heating.

At present, there are two models of numerical precipitation prediction for operational use in China (Zou, 1989). One is the limited-area precipitation prediction model operated at the National Meteorological Center, Beijing, mainly for the prediction of precipitation distribution and occurrence timing over a rather large area, and its products are disseminated all over China. The other is the operational model of limited-area numerical precipitation prediction for the Yangtze and Huaihe River Valleys operated by the Wuhan Weather Observatory, which has been designed and developed mainly for forecasting rainstorms in the Meiyu season. Table 6.14 is a comparison of the basic characters of these two models.

Table 6.14. Comparison between two operational limited area models (Zou, 1989)

	Wuhan	Beijing
Validity	0–24 h 12–36 h	0–24 h 12–36 h
Initial time	00:00(GMT)	00:00(GMT)
Time of issue	07:00	08:00
Area	Limited E-W 2900 km N-S 23,000 km	Limited 5000 km 3400 km
Projection	Lambert	Polar-conical
Equation	Primitive	Primitive
Grid space	100 km	190.5 km
Layers	5	5
Initial wind field	Observed	Derived
Topography	Included	Not included
Boundary	Fixed	Nested

A preliminary test and verification have been made on the forecasting capabilities and performance of these two models. The results show that both are capable of forecasting precipitation in central China, and both are successful in "yes or no" forecasting, as well as predicting rainfall areas. Their common problem, however, is that the predicted rainfall areas are unrealistically large, while the rainfall intensity is too low. Relatively speaking, the rainfall areas predicted by the Wuhan model is smaller, and the location of rainfall, rainfall maxima, and intensity are all better than the forecasts made by the Beijing limited-area model for the same areas, especially for the 12–36 h forecasts of the maximum rainfall amount. It is therefore obvious that an increase of horizontal resolution can improve precipitation prediction to a certain

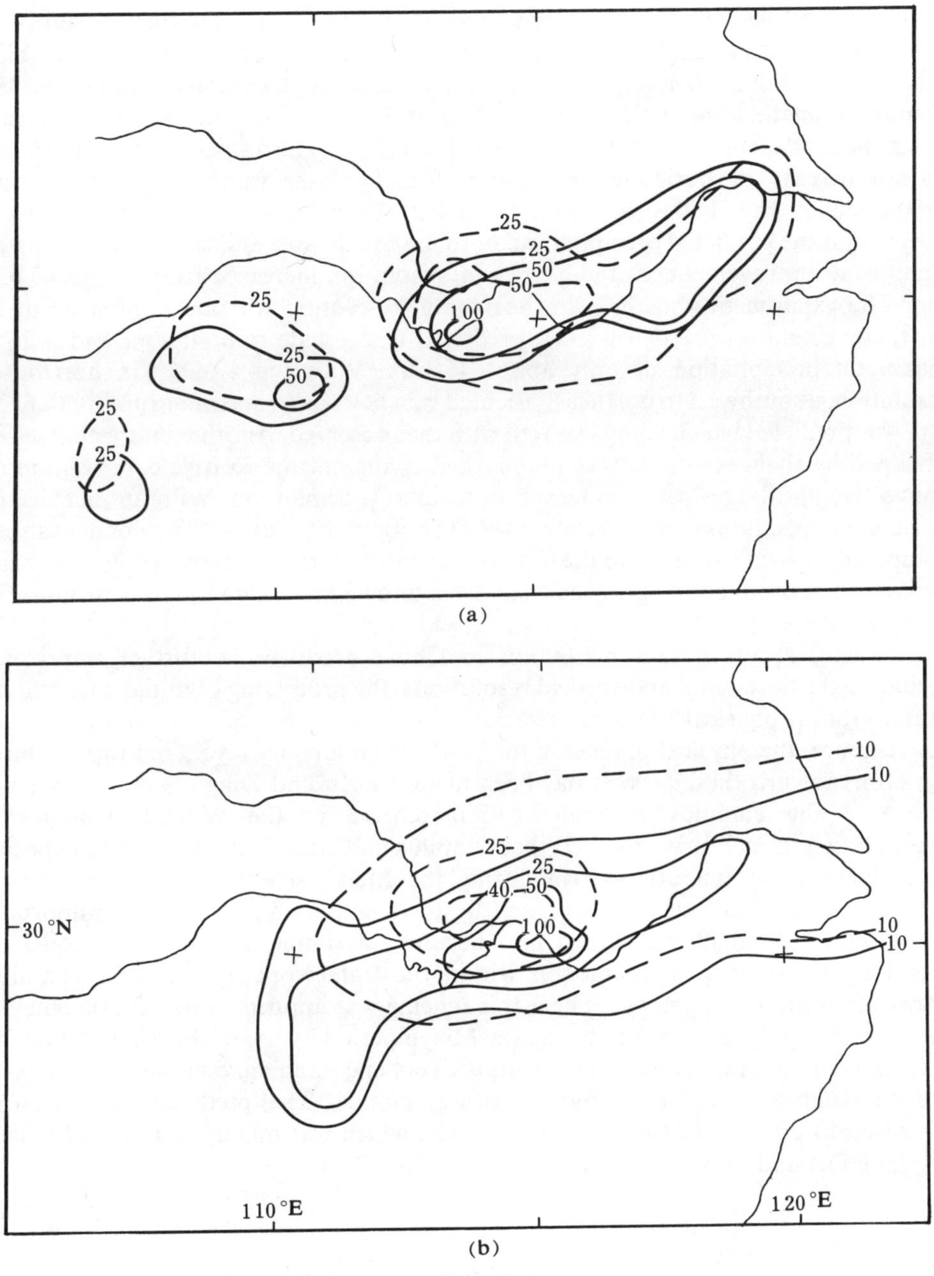

Figure 6.30. Predicted (dashed lines) and observed (solid lines) 24 h precipitation patterns associated with a low-level shear line, (a) for 00:00 GMT, 2 July-00:00 GMT, 3 July 1987, and (b) for 12:00 GMT, 2 July-12:00 GMT, 3 July 1987. Units: mm (Zou, 1989).

degree, as well as have a predictive capability of heavy rainfalls. Figure 6.30 is an example, indicating that the predicted area of heavy rainfall and its maximum, which is associated with a low-level shearline, are quite consistent with observations, but the amount is smaller, only half (56 mm day^{-1}) of the observed. The next 24 h forecast is also basically comparable. Recently, Guo *et al.* (1990) also stressed the importance of an increase in horizontal resolution for the improvement of precipitation prediction by the Beijing limited-area model. Their sensitivity experiments have shown that the predicted precipitation, both in the rain area and the amount, showed significant improvements as the horizontal resolution increased from 381 to 47.625 km. The experiment with 381 km horizontal resolution for one rainfall event in northern China on 9 August 1984, predicted a too wide rainfall area and a 24 h maximum precipitation of only about 46 mm. With the 47.625 km horizontal resolution, a southwest to northeast oriented rain belt with a maximum amount of 234 mm was predicted which compares well with that observed. Another interesting result obtained by their experiments is the marked change of the relative contribution of convective and large-scale condensation to total precipitation. With an increase of horizontal resolution, the relative contribution of convective condensation, computed by Kuo's scheme, to the total precipitation decreases remarkably, while the precipitation produced by grid-scale condensation will take a dominating proportion of the total precipitation .

Numerical precipitation prediction in China needs to be further improved, including the horizontal and vertical resolutions, the processing of initial data and its initialization, physical parameterization, the prediction of the moisture flux convergence, the physical process in the planetary boundary layer, and topography. An effort toward these aspects has been made by Xu and Ding (1990). They have improved the cumulus parameterization scheme in the Wuhan limited-area precipitation model. It is observed that cumulus and stratiform clouds often coexist in a rain-bearing synoptic system during the Meiyu season, and the associated precipitation amount may be amplified by their interaction. With some important processes of cloud physics taken into account, a simple seeder-feeder model to describe this process of interaction within the framework of Kuo-type cumulus parameterization is suggested. Then, this scheme was applied to six experiments on the prediction of precipitation during the Meiyu period by use of the Wuhan limited-area precipitation model. And the results are very encouraging, especially for cases of the coexistence of cumulus and stratiform clouds. The predicted rainfall was increased by 30–40% of total rainfall amount, which was mainly caused by the ice-crystal effect and, to a lesser extent, the collision effect.

References

Anderson, J. R., Stephens, R. D., 1987: 'The response of the tropical atmosphere to low frequency thermal forcing'. *J. Atmos. Sci.*, **44**, 676–686.

Anthes, R. A., 1990: 'Recent applications of the Penn State/NCAR mesoscale model to synoptic, mesoscale, and climate studies'. *Bull. Amer. Meteorol. Soc.*, **71**, 1610–1629.

Chen Longxun, Li Weiliang, 1985: 'The atmospheric heat budget in summer over Asia monsoon area'. *Adv. Atmos. Sci.*, **2**, 487–498.

Chen Shixun, Gao Shaofeng, Yang Song, 1983: 'The source of water vapour in the southern part of China in May and June and its abnormality. Proc. of the Sympos. Summer Monsoon in South and East Asia', 10–15 October 1982, Kunming, China. Yunnan People's Press.

Chen Shixun, Gao Shaofeng, Yang Song, Guo Yingqing, 1983: 'The transport and balance of water vapour in the area south of the Yangtze River in May and June'. Proc. of Sympos. Summer Monsoon in South and East Asia, 10–15 October 1982, Kunming, China. Ynunan People's Press.

Chen Shoujun, Dell'Osso, L., 1984: 'Numerical prediction of the heavy rainfall vortex over Eastern Asia monsoon region'. *J. Meteorol. Soc. Japan*, **62**, 730–747.

Ding Yihui, 1984: 'Large-scale balance of kinetic energy over eastern China and Japan during the pre- and post-onset phases of summer monsoon of 1979'. *Sci. Atmos. Sinica*, **8**, 272–282.

Ding Yihui, Liu Yeuzhen, 1985: 'A study of the budget of kinetic energy in typhoons. Part I: The budget of total kinetic energy and eddy kinetic energy'. *Sci. Sinica (Series B)*, **28**, 956–966.

Ding Yihui, Zhang Baoyan, Fu Xiuqin, 1987: 'A study on the budget of kinetic energy of a monsoon depression over the Bay of Bengal'. *J. Tropical Meteorol.*, **3**, 20–30.

Ding Yihui, Hu Jian, 1988: 'The variation of the heat sources in East Asia in the early summer of 1984 and their effects on the large-scale circulation in East Asia'. *Adv. Atmos. Sci.*, **5**, 171–180.

Ding Yihui, Wang Xiaofang, 1988: 'An analysis of distribution of apparent heat sources and sinks over the middle reaches of the Yangtze River during the Meiyu season in 1983'. *J. Tropical Meteorol.*, **4**, 134–145.

Duan Tinyang, 1983: 'An analysis of the atmospheric heat source and the energy budget over the South China Sea in summer 1979'. Proc. of Sympos. Summer Monsoon in South and East Asia, 15–21 August 1980. Hangzhou, China, Yunnan People's Press, 102–115.

Gao Shaofeng, Zou Meien, 1984: 'Heat and momentum budgets in southern China during active SW monsoon period'. *J. Tropical Meteorol.*, **1**, 14–23.

Guo Xiaorong, Yan Zhihui, Zhong Guoan, 1990: BMC limited area model: operational application and research. *East Asia and Western Pacific Meteorology and Climate*, World Scientific, Singapore. 572 pp.

Hayashi, Y, Miyahara, S. 1987: 'A three dimension linear response of the tropical intraseasonal oscillation'. *J. Meteorol. Soc. Japan*, **65**, 843–852.

Jin Zuhui, 1983:' The water vapour budget over the southern China in summer 1979'. Proc. of Sympos. Summer Monsoon in South and East Asia, 15–21, August, 1980. Hangzhou, China, Yunnan People's Press, 1983.

Ju Jianhua, Luo Huibang, 1987: 'The structures of the atmospheric heat suorces and the relationship between the heat sources and the tropical circulation in the formation of the summer monsoon in East Asia'. Collected Papers on Meteorological Science and Technology, No. 10 (Monsoon Meteorology), China Meteorol. Press, Beijing. 235 pp.

Krishnamurti, T. N., Rasch, R. J., Pan, H. L., Chu, S. H., Ingles, K., 1983: 'Details of low latitude medium-range numerical weather prediction using a global spectral model'. I: Formation of a monsoon depression. *J. Meteorol. Soc. Japan*, **61**, *188–207*.

Krishnamurti, T. N., Sabramanyum, M. Ousterhof, D. K., Daughenbaugh, G., 1990: 'Predictability of low frequency modes'. Meteorol. Atmos. Phys., **44**, 63–83.

Kuo, H. L., Qian, Y. F., 1981: 'Influence of the Tibetan Plateau on cumulative and diurnal change of the weather and climate in summer'. *Mon. Wea. Rev.*, **109**, 2237–2356.

Kuo, Y.-H., Cheng, L., Bao, J.-W., 1988: 'Numerical simulation of the 1981 Sichuan flood catastrophe. Part I: Evolution of a mesoscale southwest vortex', *Mon. Wea. Rev.*, **116**, 2481–2504.

Lau, K. M., Peng, L., 1987: 'Origin of low frequency (intraseasonal) oscillations in the tropical atmosphere. Part I: Basic theory'. *J. Atmos. Sci.*, **44**, 950–972.

Li Chongyin, 1983: 'CISK instability oscillating type convection'. *Sci. Sinica (Series B)*, **26**, 857–865.

Li Jun, Qiu Zhang, Wang Anyu, Gao Youxi, 1987: The numerical experiments for effects of diabatic heating on the change of break summer monsoon to active one in India. Collected Papers of Meteorological Science and Technology, No. 10 (Monsoon Meteorology), China Meteorol. Press, Beijing, pp. 173–182.

Li Weiliang, Zhong Minsen, 1983: 'The futures of medium-range oscillation about atmospheric heat budget over the north part of the South China Sea in summer 1979'. Collected Papers on Meteorological

Science and Technology, No. 10 (Monsoon Meteorology). China Meteorol. Press, Beijing. 235 pp.
Li Y.-F., A. Huang, Gao, K., 1987: 'The role of convective heating in heavy rain system during Mei-rains season'. *Sci. Sinica (Series B)*, **30**, 182–197.
Luo H. B., Yanai, M., 1984: 'The large-scale circulation and heat sources over the Tibetan Plateau and surrounding areas during the early summer of 1979. Part II: Heat and moisture budgets'. *Mon. Wea. Rev.*, **112**, 966–986.
Luo Shaohua, 1983: 'The influence of summer monsoon on the precipitation in the middle and lower reaches of the Yangtze River in 1980'. Proc. of Sympos. Summer Monsoon in South and East Asia, 10–15 October 1982, Kumming, China, Yunnan People's Press, 1983.
Luo Shaohua, 1983: 'The transfer of water vapour and the water balance over southeastern China in summer'. Proc. of Sympos. Summer Monsoon in South and East Asia, 15–21 August 1980, Hangzhou, China, Yunnan People's Press, 1983.
Murakami, T., 1 978: 'Regional energetics of the summer circulations'. *Mon. Wea. Rev.*, **106**, 614–628.
Qian Zhengan, Bromley, R. A., Agles, M. A., 1991: 'Some analyses and numerical simulations of Meiyu in East Asia in 1983'. *Adv. Atmos. Sci.*, **8**, 67–78.
Shen Jianzhu, 1983: 'Atmospheric water vapour budget in summer monsoon over China'. Proc. of Sympos. Summer Monsoon in South and East Asia, 10–15 October 1982, Kunming, China, Yunnan People's Press, 1983.
Shen Ruigui, Huang Genshen, 1983: 'The relationship between the monsoon circulation and the water vapour transfer in the rainfall period in southern China during summer 1980'. Proc. of Sympos.Summer Monsoon in South and East Asia, 15–21 August 1980, Hangzhou, China, Yunnan People's Press, 1983.
Surgi, N., 1984: 'The structure and dynamics of a monsoon depression'. Proc. 15th Conf. Hurricanes and Tropical Meteorol., Miami, Amer. Meteor. Soc., pp. 354–358.
Wang Bin, Ding Yihui, 1992: 'The synoptic and dynamic aspects of the Madden-Julian oscillation: A review'. *Adv. Atmos. Sci.*, **9**, 93–111.
Xie Lian, Zhu Qiangen, 1987: 'The numerical experiments for sensitivity of the South China Sea monsoon activity to differential heating in summer'. Collected Papers of Meteorological Science and Technology, No. 10 (Monsoon Meteorology). China Meteorol. Press, Beijing. 183–194.
Xu Suying, 1958: 'The vapour transport and water balance in China'. *Acta Meteorol. Sinica*, **29**, 33–43.
Xu Xihua, Ding Yihui, 1990: A cumulus parameterization scheme with consideration of interaction of stratiform clouds and cumulus and its application to precipitation numerical prediction. *Sci. Sinica (Series B)*, 998–1008.
Yang Guixun, 1987: 'The energy budgets before and after the establishment of the summer monsoon over the South China Sea'. *J. Topical Meteorol.*, **3**, 31–41.
Zhan Qihe, Yu Shihua, 1990: 'Diagnosis of the medium-range variation of the subtropical high over the western Pacific during a Meiyu process by three dimensional E-P flux'. *Adv. Atmos. Sci.*, **7**, 463–474.
Zhang Binchen, 1990: *A Study of the Meiyu rainstorms over the Yangtze Valley*. China Meteorol. Press, Beijing. 269 pp.
Zhou Xiaoping, Hu Xingfang, 1983: 'A brief analysis and numerical simulation of the Sichuan extraordinarily heavy rainfall event'. Proc. 1st Sino-American Workshop on Mountain Meteorol., Science Press, Beijing, 697 pp.
Zhou Xiaoping, Zhao Sixiong, Zhang Baoyan, 1984: 'A numerical simulation of the meso-low formation on Meiyu front'. *Sci. Atmos. Sinica.*, **8**, 353–361.
Zou Jingmeng, 1989: The monitoring and prediction of rainstorms and severe convective weather systems in China. *Mesoscale Forecasting and Its Application*. WMO No. 712, 1989, pp. 49–86.

INDEX

ATMOSPHERIC SCIENCES LIBRARY

1. F.T.M. Nieuwstadt and H. van Dop (eds.): *Atmospheric turbulence and aire pollution modelling.* 1982; rev. ed. 1984 ISBN Pb (1984) 90-277-1807-5
2. L.T. Mateev: *Cloud dynamics.* Translated from Russian. 1984 ISBN 90-277-1737-0
3. H. Flohn and R. Fantechi (eds.): *The climate of Europe: Past, present and future.* Natural and man-induced climate changes: A European perspective. 1984 ISBN 90-277-1745-1
4. V.E. Zuev, A.A. Zemlyanov, Yu.D. Kopytin and A.V. Kuzikovskii: *High-power laser radiation in atmospheric aerosols.* Nonlinear optics of aerodispersed media. Translated from Russian. 1985 ISBN 90-277-1736-2
5. G. Brasseur and S. Solomon: *Aeronomy of the middle atmosphere.* Chemistry and physics of the stratosphere and mesosphere. 1984; rev. ed. 1986 ISBN (1986) 90-277-2343-5; Pb 90-277-2344-3
6. E.M. Feigelson (ed.): *Radiation in a cloudy atmosphere.* Translated from Russian. 1984 ISBN 90-277-1803-2
7. A.S. Monin: *An introduction to the theory of climate.* Translated from Russian. 1986 ISBN 90-277-1935-7
8. S. Hastenrath: *Climate and circulation of the tropics.* 1985; rev. ed. 1988 ISBN 90-277-2026-6
9. M.I. Budyko: *The evolution of the biosphere.* Translated from Russian. 1986 ISBN 90-277-2140-8
10. R.S. Bortkovskii: *Air-sea exchange of heat and moisture during storms.* Translated from Russian, rev. ed. 1987 ISBN 90-277-2346-X
11. V.E. Zuev and V.S. Komarov: *Statistical models of the temperature and gaseous components of the atmosphere.* Translated from Russian. 1987 ISBN 90-277-2466-0
12. H. Volland: *Atmospheric tidal and planetary waves.* 1988 ISBN 90-277-2630-2
13. R.B. Stull: *An introduction to boundary layer meteorology.* 1988 ISBN 90-277-2768-6; Pb 90-277-2769-4
14. M.E. Berlyand: *Prediction and regulation of air pollution.* Translated from Russian, rev. ed. 1991 ISBN 0-7923-1000-4
15. F. Baer, N.L. Canfield and J.M. Mitchell (eds.): *Climate in human perspective.* A tribute to Helmut E. Landsberg (1906-1985). 1991 ISBN 0-7923-1072-1
16. Ding Yihui: *Monsoons over China.* 1994 ISBN 0-7923-1757-2

KLUWER ACADEMIC PUBLISHERS – DORDRECHT / BOSTON / LONDON